T0269395

Bocconi & Springer Series

Volume 5

Bocconi & Springer Series

THE ONLINE VERSION OF THE BOOKS PUBLISHED IN THE SERIES IS AVAILABLE ON SpringerLink

1. G. Peccati, M.S. Taqqu
   Wiener Chaos: Moments, Cumulants and Diagrams. A survey with computer implementation
   2011, XIV + 274 pp, ISBN 978-88-470-1678-1
2. A. Pascucci
   PDE and Martingale Methods in Option Pricing
   2011, XVIII + 720 pp, ISBN 978-88-470-1780-1
3. F. Hirsch, C. Profeta, B. Roynette, M. Yor
   Peacocks and Associated Martingales, with Explicit Constructions
   2011, XXXII + 384 pp, ISBN 978-88-470-1907-2
4. I. Nourdin
   Selected Aspects of Fractional Brownian Motion
   2012, X + 122 pp, ISBN 978-88-470-2822-7
5. J. Baldeaux, E. Platen
   Functionals of Multidimensional Diffusions with Applications to Finance
   2013, XXIV + 426 pp, ISBN 978-3-319-00746-5

For further information:
www.springer.com/series/8762

Jan Baldeaux • Eckhard Platen

Functionals of Multidimensional Diffusions with Applications to Finance

BOCCONI
UNIVERSITY
PRESS

Jan Baldeaux
University of Technology Sydney
Haymarket, Australia

Eckhard Platen
University of Technology Sydney
Haymarket, Australia

ISSN 2039-1471
ISSN 2039-148X (electronic)
Bocconi & Springer Series
ISBN 978-3-319-03334-1
ISBN 978-3-319-00747-2 (eBook)
DOI 10.1007/978-3-319-00747-2
Springer Cham Heidelberg New York Dordrecht London

Mathematics Subject Classification: 60J60, 62P05, 60H10, 65C05, 43A80

Printed on acid-free paper

Springer is part of Springer Science+Business Media (www.springer.com)

To Ilse and Dieter
To Cherie

Preface

Diffusion processes can be employed to model many phenomena arising in the natural and social sciences, such as biological and financial quantities. When modeling these phenomena, it is often natural to employ a number of driving Wiener or Bessel processes, placing us immediately in a multi- and often high-dimensional setting. The key questions that then typically arise concern a range of functionals for such models. The focus of this research monograph is, therefore, on tractable multidimensional models with functionals that have explicit solutions. After transformations of Brownian motion, as applied for the Black-Scholes model, it will be natural to concentrate in this book on models that are in some sense transformations of squares of Brownian motions, such as Bessel processes, square root processes and affine processes. Additionally, tractable diffusion processes will be studied which have been recently discovered via Lie symmetry methods. Numerical methods will be presented that allow to evaluate efficiently and accurately a wide range of functionals of multidimensional diffusions. The importance of these functionals and methods will be demonstrated in applications to finance. However, the same functionals and numerical methods can be of relevance in many other areas of application.

Given the ubiquitousness of multidimensional Wiener and Bessel processes, it is obvious that particular functionals of Wiener and Bessel processes are of importance in many different areas. However, the multidimensional nature of the processes, especially if non-trivial dependence structures are modeled between their drivers, often means that these functionals are difficult to compute. Especially closed-form solutions are rarely available. Consequently, numerical methods have to be usually employed to compute these important functionals.

The contribution of this monograph is fourfold: Firstly, it collects in a systematic way existing results on functionals of tractable processes from the literature. These results are mostly of closed form and so far often only more widely known for one-dimensional processes or very special multidimensional processes exhibiting a trivial dependence structure. Secondly, the book provides approaches which empower the reader to obtain systematically closed form solutions for various problems of interest. Thirdly, it recalls powerful numerical methods from the literature, and discusses how to apply these to the stochastic processes and functionals studied in this text. Finally, it suggests how to exploit the availability of closed form solutions in finance for particular models when numerically solving more general models even in high-dimensional situations.

Our systematic approach to developing closed form solutions proceeds around the following ideas: In the one-dimensional setting, we recall mathematical methods developed by Craddock and collaborators, see Craddock (2009), Craddock and Lennox (2007), Craddock and Lennox (2009), and Craddock and Platen (2004). In particular, we employ Lie symmetry group methods to compute transition densities of stochastic processes of interest. Furthermore, we study solvable affine models, in the sense introduced by Grasselli and Tebaldi, see Grasselli and Tebaldi (2008), which include a wide range of functionals of affine models for which explicit solutions can be obtained.

As often as we have access to explicit transition densities, we employ Monte Carlo and quadrature methods, which allow us to solve integration problems associated with functionals of interest. To quantify functionals rather generally, we remark that explicit solutions derived in this text serve as a useful check for these methods and allow us to tailor these methods.

Besides considering functionals of multidimensional Wiener and Bessel processes, which can be applied in very different areas, we focus on functionals of multidimensional Wiener and Bessel processes occurring in finance. We show how methods developed in this text can be used to solve typical problems under the benchmark approach. Within this book, we discuss several classes of tractable diffusions, which do not satisfy the classical Lipschitz and linear growth conditions, often assumed to be in force when studying diffusion models. The ability to compute important functionals of these tractable diffusions allows us to access a rich modeling world. In fact, advanced realistic long-term financial modeling under classical Lipschitz and linear growth conditions may potentially turn out to be not realistic enough for typical risk management tasks.

The purpose of Chap. 1 is to demonstrate that the book can be used to solve practical problems arising in finance under the benchmark approach. Chapters 2 and 3 summarize the current literature in the area. Chapter 4 introduces Lie Symmetry Group methods, an important tool that can be used to compute functionals of one- and multidimensional diffusions. Chapter 5 builds on Chap. 4 and we show how to compute explicitly important functionals of diffusions. In Sect. 5.5, we give a first application of these results to problems in finance. Chapter 6 applies the results from Chap. 5 to stochastic volatility models, where we present a simulation method based on Lie Symmetry methods. We then continue our study of multidimensional diffusions and turn to affine processes. We summarize the existing literature on affine processes in Chaps. 7 and 8. Chapter 9 presents the novel approach to affine processes due to Grasselli & Tebaldi. This approach analyzes when functionals of multidimensional affine processes can be computed analytically, and hence is of upmost importance to the topic of this book. As this approach is recent, we illustrate it using several examples in Sects. 9.4, 9.5, 9.6, and 9.7. Finally, we discuss a flexible class of multidimensional affine processes, the Wishart processes. Unlike the classical affine processes discussed in Chaps. 7, 8, and 9, Wishart processes do not assume values in the Euclidean space, but are matrix-valued. To fully appreciate the flexibility of Wishart processes, we provide an introduction to matrix-valued processes in Chap. 10. It seems that introductions to Lie symmetry methods for diffusions, to matrix-valued processes, and to Wishart processes have not been discussed in such a book form before. We hope that this monograph will be a valuable reference for readers interested in these topics. Finally, Chap. 14 integrates the material covered in the preceding chapters and demonstrates how it can be used to solve problems in finance entailing credit risk.

The remaining chapters are supporting chapters. We survey numerical methods, including Monte Carlo and quadrature methods and computational tools, which complement the methods described in Chaps. 4 to 11. Chapters 15, 16 and 17 are self-contained and aimed to summarize key results on stochastic processes, time-homogeneous scalar diffusions, and the distinction between martingales and strict local martingales. The material on stochastic processes is included to make this book more self-contained and easily readable without forcing some readers to rely on other sources. The material on time-homogeneous scalar diffusions is used in a few places in this book. In those cases, the reader is simply referred to the corresponding results in Chap. 16. Finally, Chap. 17 discusses when local martingales are true martingales or strict local martingales, which is an important theme of this book and relevant for the benchmark approach to finance.

The formulas in the book are numbered according to the chapter and section where they appear. Assumptions, theorems, lemmas, definitions and corollaries are numbered sequentially in each section. The most common notations are listed at the beginning and an *Index of Keywords* is given at the end of the book. Some readers may find the *Author Index* at the end of the book useful. *Suggestions for the Reader* are given after the content to give a guidance to the use of the book for different groups of readers. The *basic notation* used in the book is summarized after these suggestions.

We conclude with the remark that the practical application and theoretical understanding of functionals of multidimensional diffusions are an area of ongoing research. This book shall stimulate interest and further work on such methods and their application.

We would like to thank several colleagues and friends for their collaboration in related research and valuable suggestions on the manuscript, including Leung Lung Chan, Mark Craddock, Ke Du, Kevin Fergusson, Martino Grasselli, Hardy Hully, Katja Ignatieva, Monique Jeanblanc, Constantinos Kardaras, Renata Rendek, Wolfgang Runggaldier, and Stefan Tappe. Particular thanks go to the late Katrin Platen for her help and suggestions regarding the initial outlay of this book.

It is greatly appreciated if readers could forward any suggestions, errors, misprints or suggested improvements to: JanBaldeaux@gmail.com. The interested reader is likely to find in future updated information about functionals of multidimensional diffusions via a link on Jan Baldeaux's homepage.

Sydney
February 2013

Jan Baldeaux
Eckhard Platen

Suggestions for the Reader

The material of this book has been arranged in a way that should make it accessible to as wide a readership as possible. Prospective readers will have different backgrounds and objectives. The following four groups of suggestions are aimed to help using the book efficiently.

(i) Let us begin with those readers who aim for a sufficient understanding to apply multidimensional diffusions in their field of application, which may not necessarily be finance. Deeper mathematical issues are avoided in the following suggested sequence of introductory reading, which provides a guide to the book for those without a strong mathematical background.

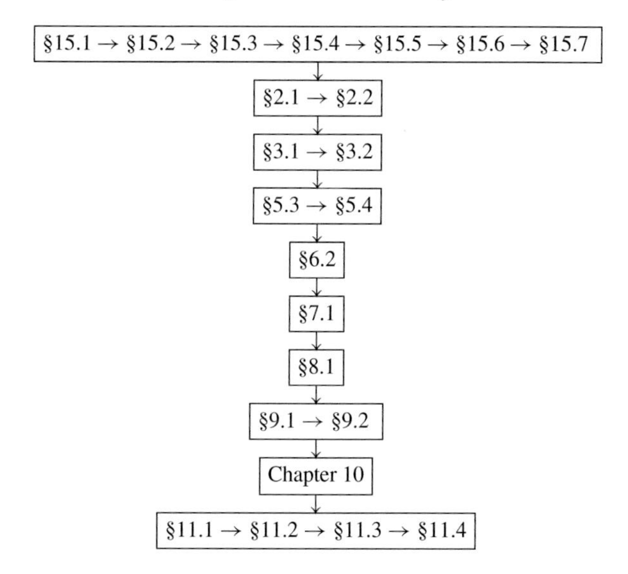

(ii) Engineers, quantitative analysts and others with a more technical background in mathematical and quantitative methods who are interested in applying multidimensional diffusions, and in implementing functionals thereof or studying novel diffusion processes could use the book according to the following suggested flowchart. Without too much emphasis on proofs the selected material provides the underlying mathematics and core results.

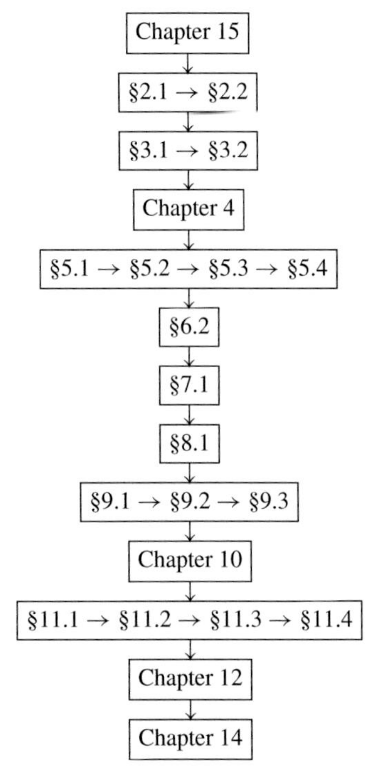

(iii) Readers with strong mathematical background and mathematicians most likely omit the introductory Chap. 15. The following flowchart focuses on the theoretical aspects of the computation of functionals of multidimensional diffusions while avoiding well-known and more applied topics.

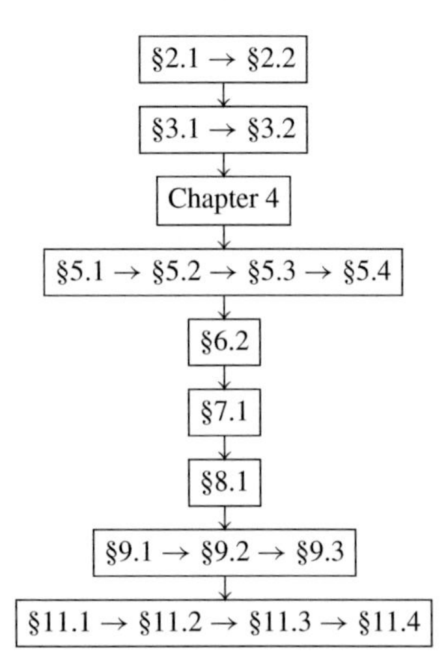

(iv) Financial engineers, quantitative analysts, risk managers, fund managers, insurance professionals and others who have no strong mathematical background and are interested in finance, insurance and other areas of risk management will find the following flowchart helpful. It suggests the reading for an introduction into quantitative methods in finance and related areas with focus on in many ways explicitly tractable models.

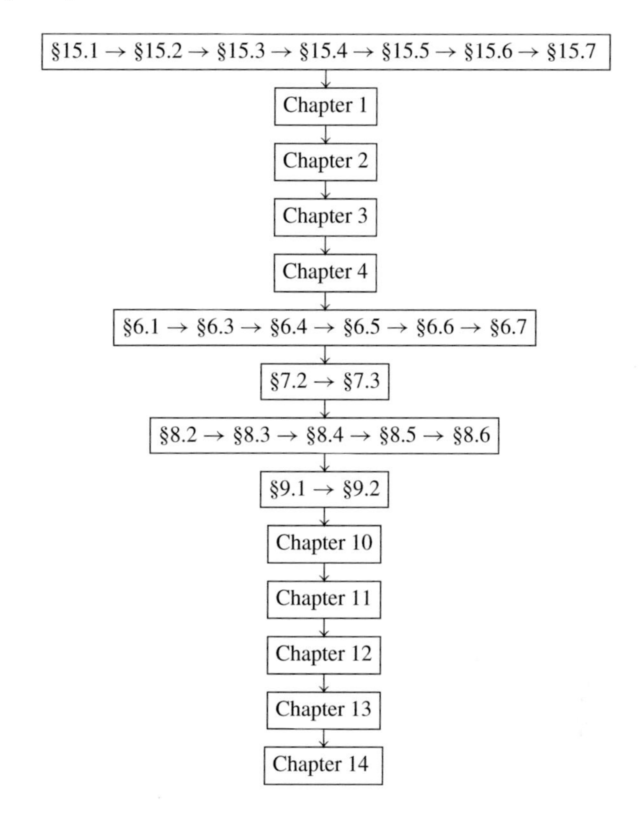

Contents

Basic Notation

μ_X	mean of X;
σ_X^2, $\mathrm{Var}(X)$	variance of X;
$\mathrm{Cov}(X, Y)$	covariance of X and Y;
$\inf\{\cdot\}$	greatest lower bound;
$\sup\{\cdot\}$	smallest upper bound;
$\max(a, b) = a \vee b$	maximum of a and b;
$\min(a, b) = a \wedge b$	minimum of a and b;
$(a)^+ = \max(a, 0)$	maximum of a and 0;
$\boldsymbol{x}^\top$	transpose of a vector or matrix $\boldsymbol{x}$;
$\boldsymbol{x} = (x^1, x^2, \ldots, x^d)^\top$	column vector $\boldsymbol{x} \in \Re^d$ with ith component x^i;
$\lvert\boldsymbol{x}\rvert$	absolute value of $\boldsymbol{x}$ or Euclidean norm;
$\boldsymbol{A} = [a^{i,j}]_{i,j=1}^{k,d}$	$(k \times d)$-matrix $\boldsymbol{A}$ with ijth component $a^{i,j}$;
$\det(\boldsymbol{A})$	determinant of a matrix $\boldsymbol{A}$;
$\boldsymbol{A}^{-1}$	inverse of a matrix $\boldsymbol{A}$;
$(\boldsymbol{x}, \boldsymbol{y})$	inner product of vectors $\boldsymbol{x}$ and $\boldsymbol{y}$;
$\mathcal{N} = \{1, 2, \ldots\}$	set of natural numbers;
∞	infinity;
(a, b)	open interval $a < x < b$ in $\Re$;
$[a, b]$	closed interval $a \le x \le b$ in $\Re$;
$\Re = (-\infty, \infty)$	set of real numbers;
$\Re^+ = [0, \infty)$	set of nonnegative real numbers;
$\Re^d$	d-dimensional Euclidean space;
Ω	sample space;
$\emptyset$	empty set;
$A \cup B$	the union of sets A and B;
$A \cap B$	the intersection of sets A and B;
$A \backslash B$	the set A without the elements of B;
$\mathcal{E} = \Re \backslash \{0\}$	$\Re$ without origin;
$[X, Y]_t$	covariation of processes X and Y at time t;
$[X]_t$	quadratic variation of process X at time t;
$n! = 1 \cdot 2 \cdot \ldots \cdot n$	factorial of n;

$[a]$	largest integer not exceeding $a \in \Re$;
i.i.d.	independent identically distributed;
a.s.	almost surely;
f'	first derivative of $f : \Re \to \Re$;
f''	second derivative of $f : \Re \to \Re$;
$f : Q_1 \to Q_2$	function f from Q_1 into Q_2;
$\frac{\partial u}{\partial x^i}$	ith partial derivative of $u : \Re^d \to \Re$;
$\left(\frac{\partial}{\partial x^i}\right)^k u$	kth order partial derivative of u with respect to x^i;
$\exists$	there exists;
$F_X(\cdot)$	distribution function of X;
$f_X(\cdot)$	probability density function of X;
$\phi_X(\cdot)$	characteristic function of X;
$\mathbf{1}_A$	indicator function for event A to be true;
$N(\cdot)$	Gaussian distribution function;
$\Gamma(\cdot)$	gamma function;
$\Gamma(\cdot;\cdot)$	incomplete gamma function;
$(\text{mod}\, c)$	modulo c;
$\mathcal{A}$	collection of events, sigma-algebra;
$\underline{\mathcal{A}}$	filtration;
$E(X)$	expectation of X;
$E(X \mid \mathcal{A})$	conditional expectation of X under $\mathcal{A}$;
$P(A)$	probability of A;
$P(A \mid B)$	probability of A conditioned on B;
$\in$	element of;
$\notin$	not element of;
$\neq$	not equal;
$\approx$	approximately equal;
$a \ll b$	a is significantly smaller than b;
$\lim_{N\to\infty}$	limit as N tends to infinity;
$\liminf_{N\to\infty}$	lower limit as N tends to infinity;
$\limsup_{N\to\infty}$	upper limit as N tends to infinity;
$\imath$	square root of -1, imaginary unit;
$\delta(\cdot)$	Dirac delta function at zero;
$\boldsymbol{I}$	unit matrix;
$\text{sgn}(x)$	sign of $x \in \Re$;
$\mathcal{L}_T^2$	space of square integrable, progressively measurable functions on $[0, T] \times \Omega$;
$\mathcal{B}(U)$	smallest sigma-algebra on U;
$\ln(a)$	natural logarithm of a;
MM	Merton model;
MMM	minimal market model;
GIG	generalized inverse Gaussian;
GH	generalized hyperbolic;
VG	variance gamma;

GOP	growth optimal portfolio;
NP	numéraire portfolio;
EWI	equi-value weighted index;
ODE	ordinary differential equation;
SDE	stochastic differential equation;
PDE	partial differential equation;
PIDE	partial integro differential equation;
$I_\nu(\cdot)$	modified Bessel function of the first kind with index ν;
$K_\lambda(\cdot)$	modified Bessel function of the third kind with index λ;
Δ	time step size of a time discretization;
$\binom{i}{l} = \frac{i!}{l!(i-l)!}$	combinatorial coefficient;
$\mathcal{C}^k(\mathcal{R}^d, \mathcal{R})$	set of k times continuously differentiable functions;
$\mathcal{C}^k_P(\mathcal{R}^d, \mathcal{R})$	set of k times continuously differentiable functions which, together with their partial derivatives of order up to k, have at most polynomial growth;
$\mathcal{L}^1(\mathcal{A}, P)$	set of $\mathcal{A}$-measurable random variables with first moments under P;
$\mathcal{M}_{m,n}(\Re)$	set of all $m \times n$ matrices with entries in $\Re$;
$\mathcal{M}_n(\Re)$	set of all $n \times n$ matrices with entries in $\Re$;
$GL(p)$	group of all invertible matrices of $\mathcal{M}_p(\Re)$;
$\mathcal{S}_p$	linear subspace of all symmetric matrices of $\mathcal{M}_p(\Re)$;
$\mathcal{S}_p^+$	set of all symmetric positive definite matrices of $\mathcal{M}_p(\Re)$;
$\mathcal{S}_p^-$	set of all symmetric negative definite matrices of $\mathcal{M}_p(\Re)$;
$\overline{\mathcal{S}_p^+}$	closure of $\mathcal{S}_p^+$ in $\mathcal{M}_p(\Re)$, that is the set of all symmetric positive semidefinite matrices of $\mathcal{M}_p(\Re)$;
$K, K_1, \ldots, \tilde{K}, C, C_1, \ldots, \tilde{C}, \ldots$	letters such as these represent finite positive real constants that can vary from line to line. All these constants are assumed to be independent of the time step size Δ.

Chapter 1
A Benchmark Approach to Risk Management

To provide for this book a relevant field of application for functionals of multidimensional diffusions, problems of pricing and hedging will be discussed in a general financial modeling framework. This chapter introduces a unified continuous time framework for financial and insurance modeling. It can be applied to portfolio optimization, derivative pricing, financial modeling, actuarial pricing and risk measurement. It is based on the benchmark approach presented in Platen and Heath (2010) and the concept of benchmarked risk minimization, see Du and Platen (2012a). The best performing, strictly positive portfolio is chosen as natural benchmark for asset allocation and also as natural numéraire for pricing. This portfolio is the growth optimal portfolio (GOP), which maximizes expected growth or log-utility. Furthermore, it is also the numéraire portfolio (NP) such that any nonnegative portfolio, denominated in units of the NP turns out to be a supermartingale. This fundamental property leads to a natural pricing rule under the real world probability measure, which identifies the minimal replicating price. We alert the reader to an important property of the benchmark approach, namely that an equivalent risk neutral probability measure need not exist. This provides the modeler with significant freedom compared to the classical risk neutral approach. Not only models will be accommodated that are covered by most of the classical no-arbitrage pricing literature in financial mathematics, including, for instance, Karatzas and Shreve (1998) and Björk (1998), but also a much wider range of models will be allowed, which go beyond the classical risk neutral paradigm, see e.g. Loewenstein and Willard (2000), Fernholz and Karatzas (2005), Platen (2002), Karatzas and Kardaras (2007), and Galesso and Runggaldier (2010). The focus of this book will be on tractable models, which may go beyond the classical no-arbitrage world, and the computation of their functionals.

1.1 A Continuous Market Model

Many applications we will discuss involve financial instruments in a continuous market, where prices of traded securities do not exhibit jumps and their

J. Baldeaux, E. Platen, *Functionals of Multidimensional Diffusions with Applications to Finance*, Bocconi & Springer Series 5, DOI 10.1007/978-3-319-00747-2_1,

denomination is in units of domestic currency. We first consider such a market containing $d \in \mathcal{N}$ sources of continuous *traded uncertainty*. Later we will also consider markets with jumps in price processes expressed in several currencies. Continuous traded uncertainty is represented by d independent standard Wiener processes $W^k = \{W_t^k, t \geq 0\}$, $k \in \{1, 2, \ldots, d\}$. These are defined on a filtered probability space $(\Omega, \mathcal{A}, \underline{\mathcal{A}}, P)$. The filtration $\underline{\mathcal{A}} = (\mathcal{A}_t)_{t \geq 0}$ is assumed to satisfy the usual conditions and $\mathcal{A}_0$ is the trivial initial σ-algebra.

1.1.1 Primary Security Accounts

A *primary security account* is an investment account, consisting of only one kind of security. It is used to model the evolution of wealth due to the ownership of primary securities, with all dividends and income reinvested. We denote the time t value of the jth risky primary security account by S_t^j, for $j \in \{1, 2, \ldots, d\}$ and $t \geq 0$. The 0th primary security account $S^0 = \{S_t^0, t \geq 0\}$ is the domestic locally riskless savings account, which continuously accrues at the adapted short term interest rate process $r = \{r_t, t \geq 0\}$.

To specify the dynamics of primary securities in the given financial market, we assume without loss of generality that the jth primary security account value S_t^j satisfies the SDE

$$dS_t^j = S_t^j \left(a_t^j \, dt + \sum_{k=1}^{d} b_t^{j,k} \, dW_t^k \right) \tag{1.1.1}$$

for $t \geq 0$ with initial value $S_0^j > 0$ and $j \in \{1, 2, \ldots, d\}$. We assume that the *appreciation rate process* a^j and the *generalized volatility processes* $b^{j,k}$ take almost surely finite values and are predictable, $k, j \in \{1, 2, \ldots, d\}$. We assume the models to be such that a unique strong solution of the system of SDEs (1.1.1) exists, see Chap. 15.7.

1.1.2 Market Price of Risk

To securitize the different sources of *traded uncertainty* properly, i.e. to avoid redundant primary security accounts, we introduce the *generalized volatility matrix* $\boldsymbol{b}_t = [b_t^{j,k}]_{j,k=1}^d$ for all $t \geq 0$ and make the following assumption:

Assumption 1.1.1 *The volatility matrix $\boldsymbol{b}_t$ is* invertible *for Lebesgue-almost-every $t \geq 0$.*

Assumption 1.1.1 allows us to introduce the *market price of risk* vector

$$\boldsymbol{\theta}_t = \left(\theta_t^1, \ldots, \theta_t^d\right)^\top = \boldsymbol{b}_t^{-1}[\boldsymbol{a}_t - r_t \, \mathbf{1}] \tag{1.1.2}$$

for $t \geq 0$. Here $\boldsymbol{a}_t = (a_t^1, \ldots, a_t^d)^\top$ denotes the *appreciation rate vector* and $\mathbf{1} = (1, \ldots, 1)^\top$ the *unit vector*. Using (1.1.2), we can rewrite the SDE for the jth primary security account in the form

$$dS_t^j = S_t^j \left(r_t\, dt + \sum_{k=1}^{d} b_t^{j,k} \left(\theta_t^k\, dt + dW_t^k \right) \right) \tag{1.1.3}$$

for $t \geq 0$ and $j \in \{0, 1, \ldots, d\}$. We observe that θ_t^k denotes the market price of risk with respect to the kth Wiener process W^k. For $j = 0$ in (1.1.3) we denote by S_t^0 the savings account, which is locally riskless with $b_t^{0,k} = 0$ for all $k \in \{1, 2, \ldots, d\}$ and $t \geq 0$. The market price of risk plays a central role and determines the risk premium that risky securities command from the perspective of the domestic currency.

1.1.3 Black-Scholes Model

The standard market model, which has served for several decades practitioners and theoreticians, can be obtained by simply assuming the short rate process, the volatility processes and the market price of risk processes to represent deterministic functions of time. The resulting Black-Scholes model, see Black and Scholes (1973), is a highly tractable model, in particular, when the just mentioned deterministic functions are constant. Unfortunately, the Black-Scholes model has many deficiencies. This book will discuss other highly tractable models that reflect more realistically the observed dynamics of financial markets.

1.1.4 Portfolios

The vector process $\boldsymbol{S} = \{\boldsymbol{S}_t = (S_t^0, \ldots, S_t^d)^\top,\ t \geq 0\}$ characterizes the evolution of all primary security accounts. We call a predictable stochastic process $\boldsymbol{\delta} = \{\boldsymbol{\delta}_t = (\delta_t^0, \ldots, \delta_t^d)^\top,\ t \geq 0\}$ a strategy if the Itô integral $I_{\delta,S}(t) = \sum_{j=0}^{d} \int_0^t \delta_s^j\, dS_s^j$ of the corresponding gains from trade exists, see Chap. 15.4. Here δ_t^j denotes the number of units of the jth primary security account held at time $t \geq 0$ in the portfolio S_t^δ, $j \in \{0, 1, \ldots, d\}$, and we denote by

$$S_t^\delta = \sum_{j=0}^{d} \delta_t^j\, S_t^j \tag{1.1.4}$$

the time t value of the *portfolio process* $S^\delta = \{S_t^\delta,\ t \geq 0\}$. A strategy $\boldsymbol{\delta}$ and the corresponding portfolio process S^δ are called *self-financing* if

$$dS_t^\delta = \sum_{j=0}^{d} \delta_t^j\, dS_t^j \tag{1.1.5}$$

for all $t \geq 0$. This means that all changes in the portfolio value are due to gains or losses from trading in primary security accounts. We remind the reader that $\boldsymbol{\delta}$ is assumed to be a predictable process. For simplicity, we consider only self-financing portfolios when discussing the above continuous market model. When we will later model price processes with jumps and perform benchmarked risk minimization, we will also allow portfolios that may no longer be self-financing.

1.2 Best Performing Portfolio as Benchmark

For a given strategy $\boldsymbol{\delta}$ with strictly positive self-financing portfolio process S^δ we use $\pi^j_{\delta,t}$ to denote the fraction of wealth invested in the jth primary security account at time t, that is

$$\pi^j_{\delta,t} = \delta^j_t \frac{S^j_t}{S^\delta_t} \tag{1.2.6}$$

for $t \geq 0$ and $j \in \{0, 1, \ldots, d\}$. We note that the fractions can be negative and always sum to one, that is

$$\sum_{j=0}^{d} \pi^j_{\delta,t} = 1. \tag{1.2.7}$$

In terms of the vector of fractions $\boldsymbol{\pi}_{\delta,t} = (\pi^1_{\delta,t}, \ldots, \pi^d_{\delta,t})^\top$ we obtain from (1.1.5), (1.1.3) and (1.2.6) the SDE

$$dS^\delta_t = S^\delta_t \big\{ r_t\, dt + \boldsymbol{\pi}^\top_{\delta,t} \boldsymbol{b}_t (\boldsymbol{\theta}_t\, dt + d\boldsymbol{W}_t) \big\} \tag{1.2.8}$$

for a strictly positive portfolio process S^δ, where $t \geq 0$ and

$$d\boldsymbol{W}_t = \big(dW^1_t, \ldots, dW^d_t \big)^\top.$$

We now use the Itô formula to obtain the SDE for the logarithm of the portfolio,

$$d\ln\big(S^\delta_t\big) = g^\delta_t\, dt + \sum_{k=1}^{d} \sum_{j=1}^{d} \pi^j_{\delta,t} b^{j,k}_t\, dW^k_t \tag{1.2.9}$$

for $t \geq 0$. The *growth rate* at time t for S^δ_t is then given by

$$g^\delta_t = r_t + \sum_{k=1}^{d} \left[\sum_{j=1}^{d} \pi^j_{\delta,t}\, b^{j,k}_t\, \theta^k_t - \frac{1}{2} \left(\sum_{j=1}^{d} \pi^j_{\delta,t}\, b^{j,k}_t \right)^2 \right] \tag{1.2.10}$$

for $t \geq 0$.

We now define for $t \geq 0$ the fractions

$$\boldsymbol{\pi}_{\delta_*,t} = \big(\pi^1_{\delta_*,t}, \ldots, \pi^d_{\delta_*,t} \big)^\top = \big(\boldsymbol{\theta}^\top_t \boldsymbol{b}^{-1}_t \big)^\top \tag{1.2.11}$$

of a particular strictly positive portfolio S^{δ_*}, which we will identify below as the *growth optimal portfolio* (GOP). By (1.2.8) and (1.2.11) it follows that $S_t^{\delta_*}$ satisfies the SDE

$$\begin{aligned} dS_t^{\delta_*} &= S_t^{\delta_*}\big(r_t\,dt + \boldsymbol{\theta}_t^\top(\boldsymbol{\theta}_t\,dt + d\mathbf{W}_t)\big) \\ &= S_t^{\delta_*}\left(r_t\,dt + \sum_{k=1}^d \theta_t^k\big(\theta_t^k\,dt + dW_t^k\big)\right) \end{aligned} \tag{1.2.12}$$

for $t \geq 0$, with $S_0^{\delta_*} > 0$. We now define the GOP in the given continuous financial market.

Definition 1.2.1 A strictly positive portfolio process $S^{\underline{\delta}}$ that maximizes the growth rate g_t^δ, see (1.2.10), of all strictly positive portfolio processes S^δ such that

$$g_t^\delta \leq g_t^{\underline{\delta}} \tag{1.2.13}$$

almost surely for all $t \geq 0$ is called a *GOP*.

The proof of the following result is given in Platen and Heath (2010).

Corollary 1.2.2 *Under the Assumption* 1.1.1 *the portfolio process* $S^{\delta_*} = \{S_t^{\delta_*}, t \geq 0\}$ *satisfying* (1.2.13) *is a GOP. Its value process is uniquely determined.*

The GOP is in many ways the best performing portfolio of the given investment universe. Let us briefly mention one of its most striking properties, which is given by the fact that the path of the GOP outperforms in the long run the path of any other strictly positive portfolio almost surely.

Theorem 1.2.3 *The GOP S^{δ_*} has almost surely the largest long term growth rate in comparison with that of any other strictly positive portfolio S^δ in the sense that*

$$\limsup_{T\to\infty} \frac{1}{T}\ln\left(\frac{S_T^{\delta_*}}{S_0^{\delta_*}}\right) \geq \limsup_{T\to\infty} \frac{1}{T}\ln\left(\frac{S_T^{\delta}}{S_0^{\delta}}\right) \tag{1.2.14}$$

almost surely.

The proof of this result and further results on outstanding performance properties of the GOP are given in Platen and Heath (2010). The GOP is also called Kelly portfolio, see Kelly (1956) and MacLean et al. (2011).

Section 1.5 will describe a Diversification Theorem, which states that under general assumptions, a sequence of well-diversified portfolios approximates the GOP. Therefore, from a practical perspective one can say that a global well diversified portfolio appears to be a good proxy for the GOP. For instance, the MSCI world accumulation index is such a tradable portfolio that one can use as a reasonable proxy for the GOP for various purposes. An even better performing proxy of the GOP is described in Platen and Rendek (2012), see Sect. 1.5.

1.3 Supermartingale Property and Pricing

1.3.1 Benchmarked Portfolios

In the following, the "best" performing portfolio of our continuous financial market, the GOP, will be employed as *benchmark* for various risk management tasks. For portfolio investing, it can serve as a benchmark in the classical sense. Furthermore, it can be used as numéraire for pricing in conjunction with the real world probability measure. Finally, in the area of risk measurement the GOP is ideally suited to play the role of the required broad based index or well diversified portfolio that should be used when measuring general market risk in the regulatory sense, see Platen and Stahl (2003). We call prices expressed in units of S^{δ_*} benchmarked prices. It follows from the Itô formula and relations (1.2.8) and (1.2.12) that a benchmarked portfolio process $\hat{S}^\delta = \{\hat{S}^\delta_t,\ t \geq 0\}$, with

$$\hat{S}^\delta_t = \frac{S^\delta_t}{S^{\delta_*}_t} \tag{1.3.15}$$

for $t \geq 0$, satisfies the SDE

$$\begin{aligned} d\hat{S}^\delta_t &= \sum_{k=1}^d \left(\sum_{j=1}^d \delta^j_t\, \hat{S}^j_t\, b^{j,k}_t - \hat{S}^\delta_t \theta^k_t \right) dW^k_t \\ &= \hat{S}^\delta_t \sum_{k=1}^d \left(\sum_{j=1}^d \pi^j_{\delta,t} b^{j,k}_t - \theta^k_t \right) dW^k_t \end{aligned} \tag{1.3.16}$$

for $t \geq 0$.

The SDE (1.3.16) describes the dynamics of a benchmarked portfolio. As an example, the benchmarked savings account $\hat{S}^0_t$ satisfies the SDE

$$d\hat{S}^0_t = -\hat{S}^0_t \sum_{k=1}^d \theta^k_t\, dW^k_t \tag{1.3.17}$$

for $t \geq 0$, since $b^{0,k}_t = 0$, for all $k \in \{1, 2, \ldots, d\}$.

1.3.2 Supermartingale Property

We are now in a position to state the mathematically most important property of a financial market model, namely its *supermartingale property*.

Theorem 1.3.1 *Any nonnegative benchmarked portfolio process* $\hat{S}^\delta$ *is an* $(\underline{\mathcal{A}}, P)$-*supermartingale, that is,*

$$\hat{S}^\delta_t \geq E\big(\hat{S}^\delta_\tau \,\big|\, \mathcal{A}_t\big) \tag{1.3.18}$$

for all bounded stopping times $\tau \in [0, \infty)$ *and* $t \in [0, \tau]$.

A proof of this theorem can be found in Platen and Heath (2010). It has the important economical interpretation that the GOP is the "best" performing portfolio, and any other portfolio denominated in units of the GOP can trend only downwards or has at most no trend. The benchmark that satisfies the supermartingale property (1.3.18) is also called the *numéraire portfolio* (NP), see Long (1990) and Becherer (2001). We emphasize the fundamental fact that nonnegative benchmarked portfolios are supermartingales, even in general semimartingale markets, as long as a finite numéraire portfolio exists, see Platen (2004) and Karatzas and Kardaras (2007). We will provide in Sect. 1.4 such a financial market modeling framework.

1.3.3 Strong Arbitrage

For several decades, pricing rules have been based on excluding some classical form of arbitrage, which generally has been linked to the Fundamental Theorem of Asset Pricing, see e.g. Ross (1976), Harrison and Kreps (1979), and Delbaen and Schachermayer (2006). In the following, we will consider a notion of strong arbitrage and show that pricing based on excluding strong arbitrage makes sense in our more general setting.

Definition 1.3.2 A nonnegative portfolio that starts at zero and reaches at some later time a strictly positive value with strictly positive probability is called a *strong arbitrage*.

From Theorem 1.3.1 it is known that all nonnegative benchmarked portfolios are supermartingales. Since a nonnegative supermartingale, which starts at zero always remains at zero, we obtain the following conclusion:

Corollary 1.3.3 *There is no nonnegative portfolio that is a strong arbitrage.*

Corollary 1.3.3 shows that pricing on the basis of excluding strong arbitrage does not work in our general market since it is in any case excluded. By assuming the existence of an NP we obtain a far richer modeling world than the classical theory can provide without excluding anything what the classical approach can provide. Consequently, free lunches with vanishing risk in the sense of Delbaen and Schachermayer (1998), see also Delbaen and Schachermayer (2006), can occur in our general market. Also some other weak forms of classical arbitrage may arise, see Loewenstein and Willard (2000). This does not create any problems from a practical point of view. On the contrary, it enables us to better reflect real market behavior, especially in the long run, by significantly extending the universe of potential models.

1.3.4 Law of the Minimal Price

The previous subsection suggested that classical forms of arbitrage can be present in our market, hence we cannot rely on the classical no-arbitrage pricing methodology, mainly expressed via the risk neutral pricing formula, see Harrison and Kreps (1979), Karatzas and Shreve (1998) or Björk (1998).

To present a more general pricing concept applicable in our general setting, we firstly define a *fair* price process.

Definition 1.3.4 A security price process $V = \{V_t,\ t \in [0,\infty)\}$ is called *fair* if its benchmarked value $\hat{V}_t = \frac{V_t}{S_t^{\delta_*}}$ forms an $(\underline{\mathcal{A}}, P)$-martingale.

In a family of supermartingales sharing the same nonnegative value at some future bounded stopping time, it is the martingale that attains the minimal possible price, see Revuz and Yor (1999) or Platen and Heath (2010). This basic fact yields naturally the *Law of the Minimal Price*:

Corollary 1.3.5 *Consider a bounded stopping time $\tau \in (0,\infty)$ and a given future $\mathcal{A}_\tau$-measurable payoff H to be paid at time τ, where $E(\frac{H}{S_\tau^{\delta_*}} | \mathcal{A}_0) < \infty$. If there exists a fair nonnegative portfolio S^δ with $S_\tau^\delta = H$ almost surely, then this is the minimal possible nonnegative portfolio that replicates the payoff.*

Clearly, a fair portfolio provides the least expensive possibility for an investor to reach some future payoff H to be delivered at time τ. We alert the reader to the fact that in our general setting there exist self-financing portfolios that are not fair. This means that the classical Law of One Price needs to be abandoned and can be substituted by the above Law of the Minimal Price.

1.3.5 Real World Pricing

Consider the payoff H payable at a bounded stopping time $\tau \in [0,\infty)$, which is assumed to be $\mathcal{A}_\tau$-measurable and to satisfy $E(\frac{H}{S_\tau^{\delta_*}}) < \infty$. We apply the Law of the Minimal Price to the payoff H and conclude that its fair price, denoted by $U_H(t)$ at time $t \in [0,\tau]$, is then the minimal possible price and given by the *real world pricing formula*

$$U_H(t) = S_t^{\delta_*} E\left(\frac{H}{S_\tau^{\delta_*}} \,\middle|\, \mathcal{A}_t \right). \tag{1.3.19}$$

We point out that this formula represents an absolute pricing rule. In our approach, pricing is an investment decision. The payoff is valued relative to the "best" performing portfolio under the real world expectation. We remind the reader that its numéraire is the GOP, which is the numéraire portfolio, and in this sense the best performing portfolio. We price under the real world probability measure and not under an assumed risk neutral probability measure, as is often the case in the literature. Due to the supermartingale property, any other self-financing replicating portfolio process can only be more expensive than the above fair price process. We conclude that in a competitive market the real world pricing formula provides the economically correct price since it is the minimal possible price when replicating a payoff.

1.3.6 Risk Neutral Pricing

It is well known, see Platen and Heath (2010), that real world pricing and risk neutral pricing are equivalent as long as the candidate Radon-Nikodym derivative

$$\Lambda_\theta(t) = \frac{S_t^0}{S_t^{\delta_*}} \frac{S_0^{\delta_*}}{S_0^0} \tag{1.3.20}$$

for the putative risk neutral probability measure forms an $(\underline{\mathcal{A}}, P)$-martingale. An application of the Bayes rule to the real world pricing formula (1.3.19) for maturity date $\tau = T$ yields in this case the *risk neutral pricing formula*

$$U_H(t) = E\left(\frac{\Lambda_\theta(T)}{\Lambda_\theta(t)} \frac{S_t^0}{S_T^0} H \,\middle|\, \mathcal{A}_t \right) = S_t^0 E_\theta\left(\frac{H}{S_T^0} \,\middle|\, \mathcal{A}_t \right). \tag{1.3.21}$$

Here E_θ denotes the expectation under the here assumed risk neutral probability measure P_θ, with $\frac{dP_\theta}{dP}|_{\mathcal{A}_T} = \Lambda_\theta(T)$. This shows that classical risk neutral pricing is under appropriate assumptions a special case of real world pricing.

We remark that when the benchmarked savings account is a strict supermartingale, $\Lambda_\theta(t)$ does not form a martingale and risk neutral prices can be significantly more expensive than fair prices, in particular, for long-dated securities. We refer to Platen and Heath (2010) and Chap. 3 in Platen and Bruti-Liberati (2010) for examples and more details on this issue.

We conclude by noting that the *actuarial pricing formula*, see Bühlmann and Platen (2003), arises from the real world pricing formula (1.3.21) as another special case when for fixed maturity $\tau = T$ the random payoff H is independent of $S_T^{\delta_*}$, yielding

$$U_H(t) = P(t, T) E(H \mid \mathcal{A}_t). \tag{1.3.22}$$

Here $P(t, T) = S_t^{\delta_*} E((S_T^{\delta_*})^{-1} \mid \mathcal{A}_t)$ is the fair zero coupon bond, which represents the discount factor. Thus, real world pricing unifies actuarial and classical risk neutral pricing. More precisely, it makes pricing an investment decision. We will see in the next section that the real world pricing formula is the natural pricing rule also in the case of not fully replicable contingent claims. Throughout the book we will price a wide range of payoffs under a number of tractable models by using the real world pricing formula.

1.4 Benchmarked Risk Minimization

In this section, we discuss the problem of hedging not perfectly replicable contingent claims by using a benchmark, namely the numéraire portfolio, as numéraire. *Benchmarked risk minimizing* (BRM) strategies, as introduced in Du and Platen (2012a), are employed to solve this problem. The concept of benchmarked risk minimization generalizes classical risk minimization, as pioneered in Föllmer and Sondermann (1986), Föllmer and Schweizer (1991), and Schweizer (1995). Benchmarked risk minimization employs the real world probability measure as pricing measure and the numéraire portfolio as numéraire and benchmark to identify the minimal possible price for a contingent claim. Furthermore, the resulting profit and loss is only driven by uncertainty that is orthogonal to traded uncertainty, and forms a local martingale that starts at zero. Consequently, benchmarked profit and losses, when pooled and sufficiently different, can become in total asymptotically negligible through diversification. This property is highly desirable from a risk management point of view. It makes benchmarked risk minimization the least expensive method for pricing and hedging a diversified pool of not fully replicable benchmarked contingent claims.

We present benchmarked risk minimization in a diffusion setting to illustrate the approach. The general semimartingale case is covered in Du and Platen (2012a). We consider a filtered probability space $(\Omega, \mathcal{A}, \underline{\mathcal{A}}, P)$ which carries an m'-dimensional Brownian motion

$$W = \{W_t = (W_t^1, W_t^2, \ldots, W_t^m, W_t^{m+1}, \ldots, W_t^{m'}), t \geq 0\},$$

$m \in \{0, 1, \ldots, m' - 1\}$. We assume the existence of d primary securities, which satisfy Eq. (1.1.3),

$$dS_t^j = S_t^j \left(r_t + \sum_{k=1}^{m} b_t^{j,k} (\theta_t^k \, dt + dW_t^k) \right), \tag{1.4.23}$$

for $t \geq 0$ and $j \in \{0, 1, \ldots, d\}$. For $j = 0$, we again make the assumption that $b^{0,k} = 0$ for $k \in \{1, \ldots, m\}$, so S^0 denotes the locally riskless savings account. Furthermore, the dynamics of the GOP are given by Eq. (1.2.12),

$$dS_t^{\delta_*} = S_t^{\delta_*} \left(r_t \, dt + \sum_{k=1}^{m} \theta_t^k (\theta_t^k \, dt + dW_t^k) \right), \tag{1.4.24}$$

for $t \geq 0$ with $S_0^{\delta_*} > 0$. Hence the benchmarked security $\hat{S}^j = \{\hat{S}_t^j = \frac{S_t^j}{S_t^{\delta_*}}, t \geq 0\}$ satisfies the SDE

$$d\hat{S}_t^j = \hat{S}_t^j \sum_{k=1}^{m} \sigma_t^{j,k} \, dW_t^k,$$

where $\sigma_t^{j,k} = b_t^{j,k} - \theta_t^k$, for $t \geq 0$. Note that the Wiener processes $W^{m+1}, \ldots, W^{m'}$ do not drive traded benchmarked wealth. We recall the notion of a self-financing portfolio from Eq. (1.1.5). Now, let us introduce a class of strategies that can form non-self-financing portfolios.

Definition 1.4.1 A dynamic trading strategy $\mathbf{v}$, initiated at time $t = 0$, is an $\Re^{d+1}$-valued stochastic process $\mathbf{v} = \{\mathbf{v}_t = (\eta_t, \vartheta_t^1, \ldots, \vartheta_t^d),\ t \in [0, \infty)\}$, where its subvector process $\boldsymbol{\vartheta} = \{\boldsymbol{\vartheta}_t = (\vartheta_t^1, \ldots, \vartheta_t^d),\ t \in [0, \infty)\}$ describes the number of units invested in the respective primary security accounts $\hat{S}_t^1, \ldots, \hat{S}_t^d$ to form the benchmarked self-financing part $\sum_{k=1}^d \vartheta_t^k \hat{S}_t^k$ of a corresponding benchmarked price process $\hat{V}^{\mathbf{v}} = \{\hat{V}_t^{\mathbf{v}},\ t \in [0, \infty)\}$ with value

$$\hat{V}_t^{\mathbf{v}} = \sum_{k=1}^d \vartheta_t^k \hat{S}_t^k + \eta_t \tag{1.4.25}$$

of units at time $t \in [0, \infty)$. Here $\mathbf{v}$ is assumed to be an $\Re^d$-valued, predictable process such that

$$\int_0^t \sum_{k=1}^d \sum_{j=1}^d \vartheta_s^k \vartheta_s^j \, d\big[\hat{S}^k, \hat{S}^j\big]_s < \infty \quad P\text{-a.s.} \tag{1.4.26}$$

for all $t \geq 0$. The $\underline{\mathcal{A}}$-adapted, real valued process $\eta = \{\eta_t,\ t \in [0, \infty)\}$, starting with $\eta_0 = 0$, monitors the benchmarked non-self-financing part of the continuous benchmarked price process $\hat{V}^{\mathbf{v}}$, which then satisfies the equation

$$\hat{V}_t^{\mathbf{v}} = \hat{V}_0^{\mathbf{v}} + \int_0^t \sum_{k=1}^d \vartheta_s^k \, d\hat{S}_s^k + \eta_t \tag{1.4.27}$$

for $t \in [0, \infty)$, where $\hat{V}^{\mathbf{v}}$ is assumed to form an $(\underline{\mathcal{A}}, P)$-supermartingale and the Itô integral exists as vector Itô integral.

We note, regarding Eq. (1.4.25), that we have

$$d\big[\hat{S}^k, \hat{S}^j\big]_t = \hat{S}_t^j \hat{S}_t^k \sum_{l=1}^m \sigma_t^{k,l} \sigma_t^{j,l} \, dt.$$

The benchmarked gains from trade during the time interval $[0, t]$, $t \in [0, \infty)$, from holding ϑ_s^j units of the jth primary security account for $j \in \{1, \ldots, d\}$ at time s, $s \in [0, t]$, are given by the Itô integral

$$\int_0^t \vartheta_s^k \, d\hat{S}_s^k. \tag{1.4.28}$$

We emphasize that a dynamic trading strategy generates via its self-financing part (1.4.28) benchmarked gains from trade in a manner that does not require outside funds and also does not generate extra funds. However, in general, capital has to be added or removed from a respective portfolio to match with its benchmarked value over time the evolution of a given benchmarked price process $\hat{V}^{\mathbf{v}}$. We will see that for risk management tasks it is enough to monitor the units of the NP that are added or removed from the portfolio to match a desired price process without requiring to hold physically these units. In particular, via the process η the investor monitors the adapted, cumulative "virtual" capital inflow and outflow from the respective portfolio as it theoretically results from targeting a desired price process.

Note, when there is no inflow or outflow of capital in a dynamic trading strategy, then one deals with a self-financing portfolio, as described in Eq. (1.1.5). More generally, when allowing extra capital inflows and outflows, one obtains from Definition 1.4.1 the following result.

Corollary 1.4.2 *For a dynamic trading strategy*

$$\mathbf{v} = \left\{\mathbf{v}_t = \left(\eta_t, \vartheta_t^1, \vartheta_t^2, \dots, \vartheta_t^d\right), \, t \in [0, \infty)\right\},$$

as introduced in Definition 1.4.1 *with benchmarked price process* $\hat{V}^{\mathbf{v}}$, *the corresponding benchmarked portfolio at time t is the sum*

$$\hat{V}_t^{\mathbf{v}} = \hat{S}_t^{\delta} = \sum_{k=1}^{d} \delta_t^k \hat{S}_t^k \tag{1.4.29}$$

with

$$\delta_t^k = \vartheta_t^k + \eta_t \delta_{*,t}^k. \tag{1.4.30}$$

Here $\delta_{*,t}^k$, $k \in \{1, \dots, d\}$, *denotes at time t the number of units of the kth primary security account needed to form the NP, that is, one has for the benchmarked NP*

$$1 = \hat{S}_t^{\delta_*} = \sum_{k=1}^{d} \delta_{*,t}^k \hat{S}_t^k,$$

for $t \in [0, \infty)$.

We now address claims which are not fully replicable in the sense that there may not exist a self-financing strategy that delivers the claim P-a.s. In particular, we aim to identify the least expensive way of delivering targeted contingent claims through hedging while removing asymptotically the total hedge error in a large trading book.

Definition 1.4.3 For a dynamic trading strategy

$$\mathbf{v} = \left\{\mathbf{v}_t = \left(\eta_t, \vartheta_t^1, \dots, \vartheta_t^d\right), \, t \in [0, \infty)\right\},$$

as introduced in Definition 1.4.1, with corresponding benchmarked price $\hat{V}_t^{\mathbf{v}} = \hat{S}_t^{\delta} = \sum_{k=1}^d \delta_t^k \hat{S}_t^k$ at time $t \in [0, \infty)$ according to Eq. (1.4.29), the corresponding *benchmarked profit and loss* (P&L) process

$$\hat{C}^{\delta} = \left\{\hat{C}_t^{\delta_*}, \, t \in [0, \infty)\right\}$$

is defined as the benchmarked price minus the benchmarked gains from trading the self-financing part of the respective portfolio, see Eq. (1.4.28), minus the initial benchmarked price, that is,

$$\hat{C}_t^{\delta} = \hat{S}_t^{\delta} - \sum_{k=1}^{d} \int_0^t \vartheta_u^k \, d\hat{S}_u^k - \hat{S}_0^{\delta}$$

for $t \in [0, \infty)$.

From Definitions 1.4.1 and 1.4.3 one obtains directly the following statement.

Corollary 1.4.4 *For a dynamic trading strategy*

$$\mathbf{v} = \left\{\mathbf{v}_t = \left(\eta_t, \vartheta_t^1, \ldots, \vartheta_t^d\right), \, t \in [0, \infty)\right\}$$

the corresponding benchmarked P&L process $\hat{C}^\delta = \{\hat{C}_t^\delta, \, t \in [0, \infty)\}$ *coincides with the adapted process* $\eta = \{\eta_t, \, t \in [0, \infty)\}$ *that monitors the cumulative inflow and outflow of extra capital.*

For simplicity, in the current chapter the hedging and, thus, the benchmarked P&L process $\hat{C}^\delta$ for a given benchmarked portfolio $\hat{S}^\delta$ is assumed to start at the initial time $t = 0$. Therefore, the benchmarked P&L has initial value $\hat{C}_0^\delta = \eta_0 = 0$ and monitors at time t with $\hat{C}_t^\delta = \eta_t$ the adapted accumulated benchmarked capital that flew in or out of the benchmarked portfolio process $\hat{V}^\mathbf{v} = \hat{S}^\delta$ until this time. In other words, $\hat{C}_t^\delta$ represents the benchmarked external costs incurred by the portfolio $\hat{S}^\delta$ over the time period $[0, t]$ after the hedge was set up at the initial time zero. Intuitively, the adapted process η can be interpreted as benchmarked hedge error.

Now assume for the moment that a financial institution holds a large number of independent benchmarked P&Ls, which are independent, square-integrable martingales started at zero. By increasing the number of benchmarked P&Ls in such a trading book, it follows by the Law of Large Numbers, which refers to the real world probability measure, that the resulting total benchmarked P&L process will become asymptotically negligible. In this manner, the benchmarked total P&L of the trading book can be asymptotically removed via diversification. The insight that such removal is possible will be crucial for benchmarked risk minimization. Capturing this observation is the aim of the following remark:

Remark 1.4.5 Benchmarked P&Ls should preferably be driftless, and, thus, local martingales, starting at initiation with value zero.

According to Remark 1.4.5, a benchmarked P&L should be locally in the mean self-financing. Mean-self-financing turns out to be an extremely useful notion, which was introduced in Schweizer (1991).

Definition 1.4.6 A dynamic trading strategy

$$\mathbf{v} = \left\{\mathbf{v}_t = \left(\eta_t, \vartheta_t^1, \ldots, \vartheta_t^d\right), \, t \in [0, \infty)\right\}$$

is called *locally real world mean-self-financing* if its adapted process η forms an $(\underline{\mathcal{A}}, P)$-local martingale starting at zero.

Having introduced the concepts of P&L and mean-self-financing, we are now in a position to discuss pricing and hedging. Recall from Definition 1.4.1 that dynamic trading strategies form benchmarked nonnegative price processes that are consistent with the fact that the NP is the "best" performing portfolio in the sense that benchmarked price processes form supermartingales. Furthermore, note at this stage that for a given benchmarked price process a corresponding dynamic trading strategy remains potentially exposed to some ambiguity concerning what forms its self-financing part and what constitutes its non-self-financing part, see Eq. (1.4.30). This ambiguity will be removed by focusing below on benchmarked P&Ls with fluctuations that are "orthogonal" to those of all self-financing benchmarked portfolios, and thus, intuitively have no chance to be removed via hedging. To formalize this idea we introduce the following notion.

Definition 1.4.7 A dynamic trading strategy

$$\mathbf{v} = \left\{\mathbf{v}_t = \left(\eta_t, \vartheta_t^1, \ldots, \vartheta_t^d\right), \, t \in [0, \infty)\right\}$$

has an *orthogonal* benchmarked P&L $\eta = \{\eta_t, \, t \in [0, \infty)\}$ if η is orthogonal to benchmarked traded wealth in the sense that $\eta_t \int_0^t \sum_{k=1}^d \bar{\vartheta}_s^k \, d\hat{S}_s^k$ forms an $(\underline{\mathcal{A}}, P)$-local martingale for every predictable self-financing strategy $\bar{\boldsymbol{\vartheta}} = \{\bar{\boldsymbol{\vartheta}}_t = (\bar{\vartheta}_t^1, \ldots, \bar{\vartheta}_t^d), t \in [0, \infty)\}$ satisfying Eq. (1.4.26).

In some sense, all hedgeable uncertainty is removed from an orthogonal benchmarked P&L. To fix the so far identified desirable properties of dynamic trading strategies, let us define the following set.

Definition 1.4.8 Fix a maturity date $T \in (0, \infty)$. For a given benchmarked contingent claim $\hat{H}_T \in \mathcal{L}^1(\mathcal{A}_T, P)$, the set of $\mathcal{A}_T$-measurable random variables with finite first moments, define the set $\hat{\mathcal{V}}_{\hat{H}_T}$ of locally real world mean-self-financing dynamic trading strategies $\mathbf{v}$, which deliver $\hat{H}_T$ with orthogonal benchmarked P&L and corresponding benchmarked price $\hat{V}_t^{\mathbf{v}} = \hat{S}_t^\delta$ for all $t \in [0, T]$, satisfying Eqs. (1.4.27), (1.4.28), (1.4.29), and (1.4.30).

There may exist several nonnegative benchmarked hedge portfolios that could deliver a given benchmarked contingent claim. The following concept of *benchmarked risk minimizing* (BRM) strategies selects the most economical benchmarked price process, which is the least expensive possible price process with the above identified desirable properties.

Definition 1.4.9 For a given benchmarked contingent claim $\hat{H}_T \in \mathcal{L}^1(\mathcal{A}_T, P)$ a dynamic trading strategy $\tilde{\mathbf{v}} = \{\tilde{\mathbf{v}}_t = (\tilde{\eta}_t, \tilde{\vartheta}_t^1, \ldots, \tilde{\vartheta}_t^d), \, t \in [0, T]\} \in \hat{\mathcal{V}}_{\hat{H}_T}$ with corresponding benchmarked price process $\hat{V}^{\tilde{\mathbf{v}}} = \hat{S}^{\tilde{\delta}}$ is called *benchmarked risk minimizing* (BRM) if for all dynamic trading strategies $\mathbf{v} \in \hat{\mathcal{V}}_{\hat{H}_T}$, with $\hat{S}_t^\delta$ satisfying Eq. (1.4.29), the price $\hat{S}_t^{\tilde{\delta}}$ is minimal in the sense that

$$\hat{S}_t^{\tilde{\delta}} \leq \hat{S}_t^\delta \tag{1.4.31}$$

P-a.s. for all $t \in [0, T]$.

As required by Eq. (1.4.31) and similarly as in Corollary 1.3.5, we can exploit the fact that the martingale among the nonnegative supermartingales contained in $\hat{\mathcal{V}}_{\hat{H}_T}$ yields the minimal possible benchmarked price process, see Revuz and Yor (1999). Therefore, we have directly the following result:

Corollary 1.4.10 *For a given benchmarked contingent claim $\hat{H}_T \in \mathcal{L}^1(\mathcal{A}_T, P)$ a BRM dynamic trading strategy $\mathbf{v} = \{\mathbf{v}_t = (\eta_t, \vartheta_t^1, \dots, \vartheta_t^d),\ t \in [0, T]\}$ forms with the corresponding benchmarked price process $\hat{V}^{\mathbf{v}}$ an $(\underline{\mathcal{A}}, P)$-martingale, that is,*

$$\hat{V}_t^{\mathbf{v}} = \hat{S}_t^{\delta} = E(\hat{H}_T \mid \mathcal{A}_t)$$

P-a.s. for $t \in [0, T]$.

We now discuss the implementation of BRM strategies. In particular, it will be extremely useful to have access to martingale representations. For the diffusion based models discussed in this book, these are available for many standard claims. Since a martingale representation of a benchmarked contingent claim, which separates the hedgeable and the orthogonal nonhedgeable part, is crucial for practical hedging, we introduce the following notion:

Definition 1.4.11 A benchmarked contingent claim $\hat{H}_T \in \mathcal{L}^1(\mathcal{A}_T, P)$ is called *regular* if it has for all $t \in [0, T]$ a martingale representation of the form

$$\hat{H}_T = E(\hat{H}_T \mid \mathcal{A}_t) + \sum_{k=1}^{d} \int_t^T \vartheta_{\hat{H}_T}^j(s)\, d\hat{S}_s^j + \eta_{\hat{H}_T}(T) - \eta_{\hat{H}_T}(t) \tag{1.4.32}$$

P-a.s. with some predictable vector process

$$\boldsymbol{\vartheta}_{\hat{H}_T} = \left\{\boldsymbol{\vartheta}_{\hat{H}_T}(t) = \left(\vartheta_{\hat{H}_T}^1(t), \dots, \vartheta_{\hat{H}_T}^d(t)\right),\ t \in [0, T]\right\}$$

satisfying Eq. (1.4.26), and some local martingale

$$\eta_{\hat{H}_T} = \left\{\eta_{\hat{H}_T}(t),\ t \in [0, T]\right\}$$

with

$$\eta_{\hat{H}_T}(0) = 0.$$

Furthermore, for any predictable process $\boldsymbol{\vartheta} = \{\boldsymbol{\vartheta}_t = (\vartheta_t^1, \dots, \vartheta_t^d),\ t \in [0, T]\}$, satisfying Eq. (1.4.26), the product process $Z^{\mathbf{v},\hat{H}_T} = \{Z_t^{\mathbf{v},\hat{H}_T},\ t \in [0, T]\}$ with

$$Z_t^{\mathbf{v},\hat{H}_T} = \eta_{\hat{H}_T}(t) \sum_{k=1}^{d} \int_0^t \vartheta_s^k\, d\hat{S}_s^k,$$

$t \in [0, T]$, forms an $(\underline{\mathcal{A}}, P)$-local martingale.

Combining Definition 1.4.9, Corollary 1.4.10, and Definition 1.4.11, the concept of benchmarked risk minimization allows us to obtain in a straightforward manner the following statement.

Corollary 1.4.12 *For a regular benchmarked contingent claim* $\hat{H}_T \in \mathcal{L}^1(\mathcal{A}_T, P)$ *there exists a BRM strategy*

$$\mathbf{v} = \left\{\mathbf{v}_t = \left(\eta_{\hat{H}_T}(t), \vartheta^1_{\hat{H}_T}(t), \ldots, \vartheta^{d+1}_{\hat{H}_T}(t)\right), \, t \in [0,T]\right\} \in \hat{\mathcal{V}}_{\hat{H}_T}$$

with corresponding benchmarked portfolio process

$$\hat{S}_T^{\delta_{\hat{H}_T}} = \hat{V}_T^{\mathbf{v}} = \hat{H}_T \quad P\text{-a.s.}$$

The benchmarked price at time $t \in [0,T]$ *is determined by the real world pricing formula*

$$\hat{S}_t^{\delta_{\hat{H}_T}} = \hat{V}_t^{\mathbf{v}} = E(\hat{H}_T \mid \mathcal{A}_t), \tag{1.4.33}$$

yielding within the set of strategies $\hat{\mathcal{V}}_{\hat{H}_T}$ *the minimal possible price process. The resulting benchmarked P&L at time* $t \in [0,T]$ *is given by*

$$\hat{C}_t^{\delta_{\hat{H}_T}} = \eta_{\hat{H}_T}(t).$$

This process is orthogonal to benchmarked traded wealth in the sense that the product $\hat{C}_t^{\delta_{\hat{H}_T}} \int_0^t \sum_{k=1}^d \vartheta_s^k \, d\hat{S}_s^k$ *forms an* $(\underline{\mathcal{A}}, P)$*-local martingale for every predictable self-financing strategy* $\boldsymbol{\vartheta} = \{\boldsymbol{\vartheta}_t = (\vartheta_t^1, \ldots, \vartheta_t^d)\, t \in [0,T]\}$, *satisfying Eq.* (1.4.26). *In terms of the martingale representation* (1.4.32), *the components of the strategy* $\mathbf{v}$ *are obtained by the number* $\eta_{\hat{H}_T}(t)$ *of units of the NP to be monitored in the nonhedgeable part of* $\hat{H}_T$, *and the number of units* $\vartheta^j_{\hat{H}_T}(t)$ *of the primary security account,* $j \in \{1, 2, \ldots, d\}$, $t \in [0,T]$, *to be held at time* t *in the self-financing hedgeable part of* $\hat{H}_T$.

The self-financing hedgeable part of the benchmarked price process $\hat{S}_t^{\delta_{\hat{H}_T}}$ forms a local martingale and has at time $t \in [0,T]$ the benchmarked value

$$E(\hat{H}_T \mid \mathcal{A}_t) - \eta_{\hat{H}_T}(t) = \sum_{k=1}^d \int_0^t \vartheta^k_{\hat{H}_T}(s)\, d\hat{S}_s^k.$$

The vector of units $\mathbf{v}_{\hat{H}_T}(t) = (\vartheta^1_{\hat{H}_T}(t), \ldots, \vartheta^d_{\hat{H}_T}(t))$ to be held in the primary security accounts follows by making the benchmarked P&L orthogonal to benchmarked traded wealth. Due to the possible presence of redundant primary security accounts, the self-financing strategy $\boldsymbol{\vartheta}_{\hat{H}_T}$ may not be unique. We emphasize that BRM strategies yield for not fully replicable contingent claims the real world pricing formula (1.4.33), which for fully replicable claims was given in (1.3.19).

The above results demonstrate that based on the existence of a martingale representation for a regular benchmarked contingent claim $\hat{H}_T$, one obtains via benchmarked risk minimization a unique minimal price process together with a hedging strategy that makes its benchmarked price process an $(\underline{\mathcal{A}}, P)$-martingale. The benchmarked P&L of a regular benchmarked contingent claim is a local martingale and orthogonal to any benchmarked self-financing portfolio, in the sense that their product becomes a local martingale.

We conclude this section on BRM with an illustrative example. We consider the NP given by Eq. (1.4.24), where we assume for simplicity that $r_t = 0$, $t \geq 0$. Additionally, we consider a contingent claim H_T with fixed maturity $T \in [0, \infty)$. We assume H_T depends on $W_T^{m+1}, \ldots, W_T^{m'}$, so that its conditional expectation $H_t = E(H_T | \mathcal{A}_t)$, $t \in [0, T]$, is independent from S^δ. In its first step, a BRM strategy requires to apply the real world pricing formula (1.4.33) to obtain the benchmarked price process $\hat{S}_t^{\delta_{\hat{H}_T}}$ of the benchmarked claim $\hat{H}_T = H_T \hat{S}_T^0 = H_T (S_T^{\delta_*})^{-1}$ at time $t \in [0, T]$, that is, we have by the assumed independence of H_T and $\hat{S}_T^0$ that

$$\hat{S}_t^{\delta_{\hat{H}_T}} = E(\hat{H}_T \mid \mathcal{A}_t) = E(H_T \mid \mathcal{A}_t) E(\hat{S}_T^0 \mid \mathcal{A}_t) = H_t \hat{P}_T(t). \tag{1.4.34}$$

We point out that $\hat{P}_T(t) = E(\hat{S}_T^0 \mid \mathcal{A}_t)$ is the real world price at time t of a zero coupon bond with maturity T, which satisfies

$$d\hat{P}_T(t) = \frac{\partial \hat{P}_T(t)}{\partial \hat{S}^0} d\hat{S}_t^0.$$

When denoting by $\vartheta^0_{\hat{H}_T}(t)$ the number of units of the savings account that the BRM strategy holds at time $t \in [0, T]$, then the benchmarked P&L satisfies by Eq. (1.4.3) and the product rule for $H_t \hat{P}_T(t)$ the SDE

$$d\hat{C}_t^{\delta_{\hat{H}_T}} = d\hat{S}_t^{\delta_{\hat{H}_T}} - \vartheta^1_{\hat{H}_T}(t)\, d\hat{S}_t^0 = \left(H_t \frac{\partial \hat{P}_T(t)}{\partial \hat{S}^0} - \vartheta^0_{\hat{H}_T}(t) \right) d\hat{S}_t^0 + \hat{P}_T(t)\, dH_t.$$

Recall that the second step of benchmarked risk minimization requires that the benchmarked P&L has to be orthogonal to the benchmarked traded wealth. A benchmarked self-financing portfolio $\hat{S}_t^\vartheta$, which invests at time t the number ϑ_t^0 of units of the savings account and the remainder of its wealth in the NP, satisfies the SDE

$$d\hat{S}_t^\vartheta = \vartheta_t^0\, d\hat{S}_t^0.$$

Since the processes $\hat{S}^\vartheta$, $\hat{S}^0$, and $\hat{C}^{\delta_{\hat{H}_T}}$ are $(\underline{\mathcal{A}}, P)$-local martingales, the product $\hat{C}^{\delta_{\hat{H}_T}} \hat{S}^\vartheta$ satisfies by the product rule in the given continuous market an SDE with zero drift if the covariation of $\hat{C}^{\delta_{\hat{H}_T}}$ and $\hat{S}^\vartheta$ vanishes for all $t \in [0, \infty)$, that is

$$\left[\hat{S}^\vartheta, \hat{C}^{\delta_{\hat{H}_T}}\right]_t = \int_0^t \left(H_s \frac{\partial \hat{P}_T(s)}{\partial \hat{S}^0} - \vartheta^0_{\hat{H}_T}(s) \right) \vartheta^0_{\hat{H}_T}(s)\, d\left[\hat{S}^0, \hat{S}^0\right]_s.$$

Therefore, the benchmarked P&L is orthogonal to traded wealth in the sense of Definition 1.4.7 if

$$\vartheta^0_{\hat{H}_T}(t) = H_t \frac{\partial \hat{P}_T(t)}{\partial \hat{S}^0}. \tag{1.4.35}$$

Finally, we remark that the benchmarked P&L

$$\eta_{\hat{H}_T}(t) = \hat{C}_t^{\delta_{\hat{H}_T}} = \int_0^t \hat{P}_T(s)\, dH_s \tag{1.4.36}$$

is a local martingale that starts at zero, as required in Definition 1.4.11. Hence in Eqs. (1.4.34), (1.4.35), and (1.4.36), we have identified the key quantities for the characterization of the martingale representation of a regular benchmarked contingent claim $\hat{H}_T$, as required in Definition 1.4.11. Finally, we remark that ϑ_t^0 units of the savings account are held at time t, and

$$\vartheta^1_{\hat{H}_T}(t) = \hat{S}_t^{\delta_{\hat{H}_T}} - \eta_{\hat{H}_T}(t) - \vartheta^0_{\hat{H}_T}(t)\hat{S}_t^0$$

units of the NP are held, see Corollary 1.4.12. Note that the hedging strategy with $\vartheta^0_{\hat{H}_T}(t)$ given in (1.4.35) depends via H_t on the evolving information about the nonhedgeable part of the claim. Classical risk minimization ignores such evolving information, see Du and Platen (2012a).

1.5 Diversification

This chapter has presented a consistent approach to asset pricing and risk management. The approach is based on the existence of the NP or GOP. The purpose of this section is to discuss the construction of a proxy of the NP, where we follow Platen and Rendek (2012), see also Platen (2005) and Sect. 3.4 in Platen and Bruti-Liberati (2010). In particular, we present the *naive diversification theorem* (NDT) by Platen and Rendek (2012). Essentially, the theorem states that an equi-weighted index (EWI) approximates the NP of a given set of stocks when the number of constituents is large and the given investment universe is well securitized. This can be interpreted as meaning that the risk factors driving the underlying risky securities are sufficiently different. An important upshot of the approximation of the GOP using an EWI is the following: when funds approximate the EWI, they stabilize the market. Important liquidity is provided in case the market crashes. On the other hand, popular assets would be sold when asset bubbles emerge. In both extreme cases, an EWI can serve as a stabilizing factor in the financial market architecture since it provides important liquidity.

We continue to rely on a filtered probability space $(\Omega, \mathcal{A}, \underline{\mathcal{A}}, P)$ and we represent traded uncertainty using independent standard Wiener processes $W^k = \{W_t^k, t \in [0, \infty)\}$, where $k \in \mathcal{N}$. In what follows, we consider a sequence of markets indexed by the number d of risky primary securities. To be precise, we write for the jth primary security account in the dth market $S^j_{(d)}(t)$, which satisfies an SDE as given in (1.1.1). The primary securities include a savings account $S^0_{(d)}(t) = \exp\{\int_0^t r_s\, ds\}$ for $t \geq 0$. Here $r = \{r_t, t \geq 0\}$ denotes an adapted short rate process, which we assume for simplicity to be the same in each market. We include d nonnegative, risky primary security account processes $S^j_{(d)} = \{S^j_{(d)}(t), t \geq 0\}$, $j \in \{1, 2, \ldots, d\}$, in the dth market, each of which is driven by the Wiener processes $W^1, W^2, \ldots, W^d$. In the dth market a given strategy δ and the volatilities of market prices of risk depend typically on d. For simplicity, we shall suppress these dependencies in our notation and only mention it when required.

For the dth market, we assume that there exists a unique GOP $S^{\delta_*}_{(d)} = \{S^{\delta_*}_{(d)}(t), t \geq 0\}$, satisfying the SDE (1.2.12), where we fix $S^{\delta_*}_{(d)}(0) = 1$. We obtain the following dynamics for the jth benchmarked primary security in the dth market at time t,

$$\frac{d\hat{S}^j_{(d)}(t)}{\hat{S}^j_{(d)}(t)} = \sum_{k=1}^{d} \sigma^{j,k}_{(d)}(t)\, dW^k_t, \tag{1.5.37}$$

where

$$\sigma^{j,k}_{(d)}(t) = b^{j,k}_{(d)}(t) - \theta^k_{(d)}(t),$$

for $t \geq 0$, $d \in \mathcal{N}$ and $j, k \in \{1, 2, \dots, d\}$. The benchmarked self-financing portfolio process $\hat{S}^{\delta}_{(d)} = \{\hat{S}^{\delta_*}_{(d)}(t),\, t \geq 0\}$ with strategy

$$\boldsymbol{\delta} = \{\boldsymbol{\delta}_t = (\delta^1_t, \delta^2_t, \dots, \delta^d_t),\, t \geq 0\}$$

is driven by the SDE

$$d\hat{S}^{\delta}_{(d)}(t) = \sum_{j=1}^{d} \delta^j_t\, d\hat{S}^j_{(d)}(t),$$

which is driftless. If we introduce the fractions

$$\pi^j_{\delta,(d)}(t) = \frac{\delta^j_t\, \hat{S}^j_{(d)}(t)}{\hat{S}^{\delta_*}_{(d)}(t)},$$

where $\sum_{j=1}^{d} \pi^j_{\delta,(d)}(t) = 1$, then

$$\begin{aligned}
\frac{d\hat{S}^{\delta}_{(d)}(t)}{\hat{S}^{\delta_*}_{(d)}(t)} &= \sum_{j=1}^{d} \delta^j_t \frac{d\hat{S}^j_{(d)}(t)}{\hat{S}^{\delta}_{(d)}(t)} \\
&= \sum_{j=1}^{d} \pi^j_{\delta,(d)}(t) \sum_{k=1}^{d} \sigma^{j,k}_{(d)}(t)\, dW^k_t.
\end{aligned}$$

The dth equi-weighted index (EWId) invests the fractions

$$\pi^j_{\delta_{EWId},t} = \begin{cases} \frac{1}{d} & \text{for } j \in \{1, 2, \dots, d\} \\ 0 & \text{otherwise.} \end{cases} \tag{1.5.38}$$

Since the benchmarked NP is a constant, we have

$$\frac{d\hat{S}^{\delta_*}_{(d)}(t)}{\hat{S}^{\delta_*}_{(d)}(t)} = \sum_{k=1}^{d} \sum_{j=1}^{d} \pi^j_{\delta_*,t} \sigma^{j,k}_{(d)}(t)\, dW^k_t = 0. \tag{1.5.39}$$

It is now the aim to construct sequences of portfolios that approximate the NP in a mathematically precise and practically useful sense. The limits of the return processes of such sequences of benchmarked portfolios should then be constant with value zero, due to (1.5.39). More precisely, the *return process* $\hat{Q}^{\delta}_{(d)} = \{\hat{Q}^{\delta}_{(d)}(t), t \geq 0\}$ of a benchmarked portfolio $\hat{S}^{\delta}_{(d)}$, given by the SDE

$$d\hat{Q}^{\delta}_{(d)}(t) = \frac{d\hat{S}^{\delta}_{(d)}(t)}{\hat{S}^{\delta}_{(d)}(t)}, \tag{1.5.40}$$

for $t \geq 0$ with $\hat{Q}^{\delta}_{(d)}(0) = 0$, has to have small fluctuations to be a good proxy for the NP.

Definition 1.5.1 A sequence $(\hat{S}^{\delta}_{(d)})_{d\in\{1,2,\ldots\}}$ of strictly positive benchmarked portfolios is called a sequence of *benchmarked approximate numéraire portfolios* if for each $\varepsilon > 0$ and $t \geq 0$ one has

$$\lim_{d\to\infty} P\left(\frac{d}{dt}\langle\hat{Q}^{\delta}_{(d)}\rangle_t > \varepsilon\right) = 0.$$

The intuition here is that if one can construct a sequence of benchmarked portfolios, where the quadratic variation of the return process vanishes asymptotically, then the limit can only be represented by the constant one, that is, the benchmarked NP.

It seems reasonable to say that the returns of a benchmarked primary security account express its specific or idiosyncratic traded uncertainty against the market as a whole. Due to the natural structure of the market with different types of economic activity in different sectors of the economy, it is reasonable to assume that a particular specific uncertainty drives only the returns of a restricted number of benchmarked primary security accounts. If this is the case, then one could say that the securitization of the market is sufficiently developed and a diversification effect can be expected. To capture this property of a market in a mathematically precise manner, one can introduce the following notion:

Definition 1.5.2 A financial market is well-securitized if there exists a real number $q > 0$ and a stochastic process $\underline{\sigma}^2 = \{\underline{\sigma}^2_t, t \geq 0\}$ with finite mean such that for all $d, k \in \{1, 2, \ldots\}$, and $t \geq 0$, one has

$$\frac{1}{d}\left|\sum_{j=1}^{d}\sigma^{j,k}_{(d)}\right|^2 \leq \frac{1}{d^q}\underline{\sigma}^2_t. \tag{1.5.41}$$

We point out that for the following naive diversification theorem (NDT) to hold, we require a weaker assumption than Eq. (1.5.41), namely we require

$$\lim_{d\to\infty} P\left(\frac{1}{d^2}\sum_{j=1}^{d}\sum_{k=1}^{d}\left(\sigma^{j,k}_{(d)}(t)\right)^2 > \varepsilon\right) = 0, \tag{1.5.42}$$

for all $\varepsilon > 0$ and $t \geq 0$. In fact, condition (1.5.42) is necessary and sufficient for the sequences of EWIs with fractions given by Eq. (1.5.38) to be a sequence of approximate numéraire portfolios.

Theorem 1.5.3 (Naive Diversification Theorem) *The sequence of EWIs, with fractions given by Eq.* (1.5.38), *is a sequence of approximate numéraire portfolios if and only if condition* (1.5.42) *holds. Furthermore, condition* (1.5.41) *is sufficient.*

Proof To begin with, note that the return process of the dth benchmarked EWI has at time t the value

$$\hat{Q}^{\delta_{EWId}}_{(d)}(t) = \sum_{j=1}^{d} \frac{1}{d} \sum_{k=1}^{d} \int_0^t \sigma^{j,k}_{(d)}(s)\, dW^k_s.$$

The quadratic variation of this return process is then of the form

$$\langle \hat{Q}^{\delta_{EWId}}_{(d)} \rangle_t = \frac{1}{d^2} \int_0^t \sum_{k=1}^{d} \sum_{j=1}^{d} \big(\sigma^{j,k}_{(d)}(s)\big)^2\, ds.$$

Hence we have

$$\lim_{d\to\infty} P\left(\frac{d}{dt}\langle \hat{Q}^{\delta_{EWId}} \rangle_t > \varepsilon\right) = \lim_{d\to\infty} P\left(\frac{1}{d^2} \sum_{j=1}^{d} \sum_{k=1}^{d} \big(\sigma^{j,k}_{(d)}(t)\big)^2 > \varepsilon\right),$$

for all $\varepsilon > 0$, $t \geq 0$, which completes the first part of the proof. Regarding the second part, note that from Jensen's inequality, we obtain

$$\frac{1}{d^2} \sum_{k=1}^{d} \sum_{j=1}^{d} \big(\sigma^{j,k}_{(d)}(t)\big)^2 \leq \frac{1}{d} \sum_{k=1}^{d} \left| \frac{1}{\sqrt{d}} \sum_{j=1}^{d} \sigma^{j,k}_{(d)}(t) \right|^2.$$

From condition (1.5.41) and the Markov inequality, we get for all $\varepsilon > 0$ and $t \geq 0$,

$$\begin{aligned}\lim_{d\to\infty} P\left(\frac{1}{d} \sum_{k=1}^{d} \left| \frac{1}{\sqrt{d}} \sum_{l=1}^{d} \sigma^{j,k}_{(d)}(t) \right|^2 > \varepsilon\right) &\leq \lim_{d\to\infty} P\left(\frac{1}{d^q} \underline{\sigma}^2_t > \varepsilon\right) \\ &\leq \lim_{d\to\infty} \frac{1}{d^q} \frac{1}{\varepsilon} E\big(\underline{\sigma}^2_t\big) = 0.\end{aligned}$$

This completes the proof. □

One should emphasize that the statement of the NDT is quite robust. Under condition (1.5.42), it covers a wide range of models. To some extend, the NDT is model independent since no particular assumptions about the underlying market model have been made. By imposing different assumptions, eventually similar to those in Platen (2005), one can prove separately the convergence of sequences of more general diversified portfolios towards the NP.

Chapter 2
Functionals of Wiener Processes

In this chapter, we discuss scalar- and multidimensional processes, which are based on the Wiener process, and consequently apply them in the context of the benchmark approach.

2.1 One-Dimensional Functionals of Wiener Processes

We summarize well-known SDEs and transition densities for models and processes closely related to the Wiener process or Brownian motion, including:

- the Bachelier model;
- the Black-Scholes model;
- the Ornstein-Uhlenbeck-process;
- the geometric Ornstein-Uhlenbeck-process.

Also we collect results from the literature on *functionals* of *Wiener processes* and add new results and presentations. We remark that parts of this section are based on Borodin and Salminen (2002), Jeanblanc et al. (2009), Chap. 3, and Platen and Heath (2010), Chap. 4.

2.1.1 Wiener Process

The Wiener process is a continuous Markov process and has the following transition density:

$$p(s, x; t, y) = \frac{1}{\sqrt{2\pi(t-s)}} \exp\left\{-\frac{(y-x)^2}{2(t-s)}\right\}, \tag{2.1.1}$$

for $t \in [0, \infty)$, $s \in [0, t]$ and $x, y \in \Re$. For the purpose of illustration, we display some transition densities in Fig. 2.1.1 as functions of time t and final value y, where we set the initial time to $s = 0$ and the initial value to $x = 0$.

J. Baldeaux, E. Platen, *Functionals of Multidimensional Diffusions with Applications to Finance*, Bocconi & Springer Series 5, DOI 10.1007/978-3-319-00747-2_2,

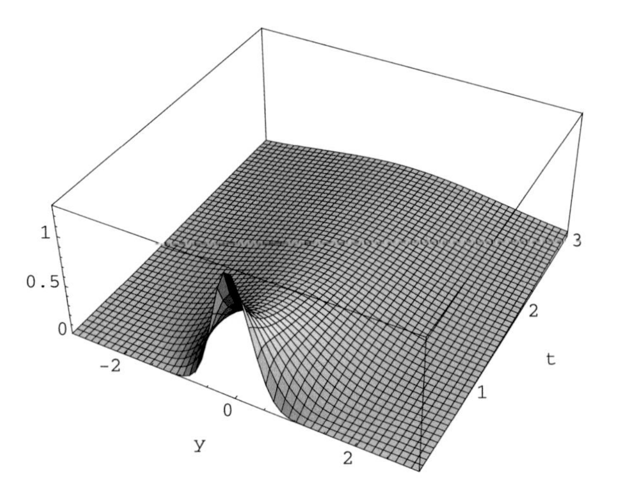

Fig. 2.1.1 Probability densities for the standard Wiener process

The Wiener process enjoys the *strong Markov property*, which allows us to formulate the following lemma:

Lemma 2.1.1 *For a finite stopping time τ, the process $\tilde{W} = \{\tilde{W}_t,\ t \geq 0\}$, where*

$$\tilde{W}_t = W_{\tau+t} - W_\tau, \tag{2.1.2}$$

is a Wiener process with respect to its natural filtration.

We now introduce the following notation

$$\begin{aligned} T_a &= \inf\{t \geq 0\colon\ W_t = a\} \\ M_t &= \sup_{0\leq s\leq t} W_s \\ m_t &= \inf_{0\leq s\leq t} W_s. \end{aligned}$$

The following proposition, commonly referred to as reflection principle, employs Lemma 2.1.1 and the symmetry of the Wiener process, see Lemma 15.1.3.

Proposition 2.1.2 *Let $y \geq 0$, $x \leq y$, then one has*

$$P(W_t \leq x,\ M_t \geq y) = P(W_t \geq 2y - x). \tag{2.1.3}$$

For a proof, see e.g. Jeanblanc et al. (2009), Proposition 3.1.1.1. Next, we discuss the joint distribution of (M_t, W_t), see Theorem 3.1.1.2 in Jeanblanc et al. (2009).

Proposition 2.1.3 *For a Brownian motion W_t and its running maximum M_t, the following formulas hold*:

$$P(W_t \leq x, M_t \leq y) = N\left(\frac{x}{\sqrt{t}}\right) - N\left(\frac{x-2y}{\sqrt{t}}\right), \quad y \geq 0,\ x \leq y,$$

$$P(W_t \le x, M_t \le y) = P(M_t \le y) = N\left(\frac{y}{\sqrt{t}}\right) - N\left(\frac{-y}{\sqrt{t}}\right), \quad y \ge 0,\ x \ge y,$$
$$P(W_t \le x, M_t \le y) = 0, \quad y \le 0.$$

The distribution of (W_t, M_t) *is given by*

$$P(W_t \in dx,\ M_t \in dy) = \mathbf{1}_{y \ge 0} \mathbf{1}_{x \le y} \frac{2(2y-x)}{\sqrt{2\pi t^3}} \exp\left\{-\frac{(2y-x)^2}{2t}\right\} dx\, dy.$$

The law of the maximum satisfies the following equality, see Proposition 3.1.3.1 in Jeanblanc et al. (2009),

$$P(M_t \le y) = N\left(\frac{y}{\sqrt{t}}\right) - N\left(\frac{-y}{\sqrt{t}}\right), \quad y \ge 0.$$

We remark that the law of the maximum of a process finds important applications in derivative pricing, see Sect. 2.3.

Proposition 2.1.4 *For a Brownian motion* W_t *and its running minimum* m_t, *the following formulas hold*:

$$P(W_t \ge x,\ m_t \ge y) = N\left(\frac{-x}{\sqrt{t}}\right) - N\left(\frac{2y-x}{\sqrt{t}}\right), \quad y \le 0,\ x \ge y$$
$$P(W_t \ge x,\ m_t \ge y) = N\left(\frac{-y}{\sqrt{t}}\right) - N\left(\frac{y}{\sqrt{t}}\right), \quad y \le 0,\ x \le y$$
$$P(W_t \ge x,\ m_t \ge y) = 0, \quad y \ge 0.$$

The law of the minimum satisfies, for $y \le 0$,

$$P(m_t \ge y) = N\left(\frac{-y}{\sqrt{t}}\right) - N\left(\frac{y}{\sqrt{t}}\right).$$

Finally, we turn to hitting times, which are also used in derivative pricing, for example when studying rebates, see Sect. 2.3.

Proposition 2.1.5 *Let* T_y *be the first hitting time of* $y \in \Re$ *for a standard Brownian motion. Then for* $\lambda > 0$,

$$E\left(\exp\left\{-\frac{\lambda^2}{2} T_y\right\}\right) = \exp\{-|y|\lambda\}.$$

We can also compute the density

$$P(T_y \in dt) = \frac{x}{\sqrt{2\pi t^3}} \exp\left\{-\frac{x^2}{2t}\right\} \mathbf{1}_{t \ge 0}\, dt.$$

This section concludes with results on integrals of Brownian motion, taken from Borodin and Salminen (2002). Such formulas are useful when studying Asian options and related contracts, such as Australian options, see Sect. 2.3:

$$P\left(\int_0^t W_s\,ds \in dy\right) = \frac{\sqrt{3}}{\sqrt{2\pi t^3}} \exp\left\{-\frac{3y^2}{2t^3}\right\} dy$$

$$P\left(\int_0^t W_s\,ds \in dy,\ W_t \in dz\right) = \frac{\sqrt{3}}{\pi t^2} \exp\left\{-\frac{z^2}{2t} - \frac{3(2y-zt)^2}{2t^3}\right\} dy\,dz.$$

2.1.2 Bachelier Model

The results in the previous section can be extended to the case

$$X_t = \nu t + W_t, \quad t \geq 0,$$

a Brownian motion with drift. This process corresponds to the *Bachelier model*, which models the stock price S_t via

$$S_t = S_0 + \mu t + \sigma W_t, \quad t \geq 0,$$

see Bachelier (1900). Again, we employ the notation

$$T_a = \inf\{t \geq 0\colon\ X_t = a\}$$
$$M_t = \sup_{0\leq s\leq t} X_s$$
$$m_t = \inf_{0\leq s\leq t} X_s.$$

We start our discussion with the transition density of the process X,

$$p(s,x;t,y) = \frac{1}{\sqrt{2\pi(t-s)}} \exp\left\{-\frac{(y-x-\nu(t-s))^2}{2(t-s)}\right\}, \tag{2.1.4}$$

for $t \in [0,\infty)$, $s \in [0,t]$ and $x, y \in \Re$. The following result corresponds to Proposition 2.1.3 and uses Proposition 3.2.1.1 and Corollary 3.2.1.2 from Jeanblanc et al. (2009).

Proposition 2.1.6 *For a Brownian motion with drift X_t and its running maximum M_t, the following formulas hold*:

$$P(X_t \leq x, M_t \leq y) = N\left(\frac{x-\nu t}{\sqrt{t}}\right) - \exp\{2\nu y\} N\left(\frac{x-2y-\nu t}{\sqrt{t}}\right), \quad y \geq 0,\ x \leq y.$$

The density of (W_t, M_t) is given by

$$P(X_t \in dx,\ M_t \in dy)$$
$$= \mathbf{1}_{x<y}\mathbf{1}_{0<y} \frac{2(2y-x)}{\sqrt{2\pi t^3}} \exp\left\{\nu x - \frac{1}{2}\nu^2 t - \frac{(2y-x)^2}{2t}\right\} dx\,dy.$$

Furthermore, the law of the maximum satisfies

$$P(M_t \leq y) = N\left(\frac{y-\nu t}{\sqrt{t}}\right) - \exp\{2\nu y\} N\left(\frac{-y-\nu t}{\sqrt{t}}\right), \quad y \geq 0.$$

Next, we present results corresponding to Proposition 2.1.4.

Proposition 2.1.7 *For a Brownian motion with drift X_t and its running minimum m_t, the following formulas hold*:

$$P(X_t \geq x,\ m_t \geq y) = N\left(\frac{-x+\nu t}{\sqrt{t}}\right) - \exp\{2\nu y\} N\left(\frac{-x+2y+\nu t}{\sqrt{t}}\right).$$

Furthermore, the law of the minimum is given by

$$P(m_t \geq y) = N\left(\frac{-y+\nu t}{\sqrt{t}}\right) - \exp\{2\nu y\} N\left(\frac{y+\nu t}{\sqrt{t}}\right), \quad y \leq 0.$$

We now turn to hitting times, see Eq. (3.2.3) in Jeanblanc et al. (2009).

Proposition 2.1.8 *Let T_y be the first hitting time of the level y for a Brownian motion with drift. Then*

$$P(T_y \in dt) = \frac{|y|}{\sqrt{2\pi t^3}} \exp\left\{-\frac{1}{2t}(y-\nu t)^2\right\} \mathbf{1}_{t\geq 0}\, dt.$$

Furthermore,

$$E\left(\exp\left\{-\frac{\lambda^2}{2} T_y\right\}\right) = \exp\{\nu y\} \exp\left\{-|y|\sqrt{\nu^2+\lambda^2}\right\}.$$

Results on integrals of Brownian motion with drift can be found in Borodin and Salminen (2002), see Eqs. (1.8.4) and (1.8.8) in their Appendix 1,

$$P\left(\int_0^t X_s\, ds \in dy\right) = \frac{\sqrt{3}}{\sqrt{2\pi t^3}} \exp\left\{-\frac{3(y-\nu t^2/2)^2}{2t^3}\right\} dy$$

$$P\left(\int_0^t X_s\, ds \in dy, X_t \in dz\right) = \frac{\sqrt{3}}{\pi t^2} \exp\left\{-\frac{(z-\nu t)^2}{2t} - \frac{3(2y-zt)^2}{2t^3}\right\} dy\, dz.$$

We now derive the transition density of a Brownian motion with drift killed at $z \in \Re$. To do so, we firstly recall Lemma 2.1 from Hulley and Platen (2008), which requires us to introduce the following notation: let $Y = \{Y_t,\ t \geq 0\}$ be a regular one-dimensional time-homogeneous diffusion process, whose state space is an interval $I \subseteq \Re$, which is typically $\Re$, $[0, \infty)$ or $(0, \infty)$ and which starts at $x \in I$. We shall denote the transition density of Y with respect to its speed measure by $q(.,.,.)$, where we omit the dependence on the initial time $s = 0$, so that

$$P(Y_t \in A) = \int_A q(t, x, y) m(y)\, dy,$$

for all $t \geq 0$ and $x \in I$ and for every Borel set $A \in \mathcal{B}(I)$. Furthermore, for any $z \in I$, let

$$T_z^Y := \inf\{t > 0\colon\ Y_t = z\}$$

be the first-passage time of Y to z. We shall denote its density with respect to the Lebesgue measure by $p_z(.,.)$, so that

$$P\left(T_z^Y \leq t\right) = \int_0^t p_z(x, s)\, ds.$$

Furthermore, let $\tilde{q}_z(.,.,.)$ denote its transition density, with respect to the speed measure of Y killed at z, so that

$$P\big(Y_t \in A,\, T_z^Y > t\big) = \int_A \tilde{q}_z(t,x,y)m(y)\,dy,$$

for all $A \in \mathcal{B}(I)$. We are now in a position to state Lemma 2.1 from Hulley and Platen (2008):

Lemma 2.1.9 *Let $x, y, z \in I$ and suppose that $t > 0$. Then*

$$q(t,x,y) = \tilde{q}_z(t,x,y) + \int_0^t p_z(x,s)q(t-s,z,y)\,ds. \tag{2.1.5}$$

Intuitively speaking, the first term in (2.1.5) corresponds to those trajectories which travel from x to y without visiting z, whereas the second includes those trajectories which do visit z between 0 and t. We now use Lemma 2.1.9 to derive the density of a Brownian motion with drift started at x killed at z. We remark that this density will be employed in the pricing of Barrier options under the Black-Scholes model in Sect. 2.3. From Borodin and Salminen (2002), we obtain for a Brownian motion with drift

$$X_t = \nu t + W_t$$

started at x and

$$T_a = \inf\{t > 0\colon\, X_t = a\}$$

that

$$q(t,x,y) = \frac{1}{2\sqrt{2\pi t}}\exp\left\{-\mu(y+x) - \frac{\mu^2 t}{2} - \frac{(x-y)^2}{2t}\right\} \tag{2.1.6}$$

and

$$p_z(x,t) = \frac{|z-x|}{\sqrt{2\pi}t^{3/2}}\exp\left\{-\frac{(z-x-\mu t)^2}{2t}\right\} \tag{2.1.7}$$

and hence the following corollary:

Corollary 2.1.10 *For a Brownian motion with drift $X = \{X_t, t \geq 0\}$ started at x, we have*

$$\tilde{q}_z(t,x,y) = \begin{cases} \frac{1}{2\sqrt{2\pi t}}\exp\{-\mu(x+y) - \frac{\mu^2 t}{2}\} & \\ \quad \times \big(\exp\{-\frac{(x-y)^2}{2t}\} - \exp\{-\frac{(x+y-2z)^2}{2t}\}\big) & y, x > z \\ 0 & y < z \leq x \end{cases} \tag{2.1.8}$$

and

$$\tilde{q}_z(t,x,y) = \begin{cases} 0 & y > z > x \\ \frac{1}{2\sqrt{2\pi t}} \exp\{-\mu(x+y) - \frac{\mu^2 t}{2}\} & \\ \quad \times \left(\exp\{-\frac{(x-y)^2}{2t}\} - \exp\{-\frac{(x+y-2z)^2}{2t}\}\right) & x, y < z. \end{cases} \tag{2.1.9}$$

Proof Assume that $x, y > z$ or $x, y < z$, then from Lemma 2.1.9, we need to compute

$$\begin{aligned}
&\int_0^t p_z(x,s) q(t-s,z,y)\, ds \\
&\quad = \int_0^t \frac{|z-x|}{\sqrt{2\pi} t^{3/2}} \exp\left\{-\frac{(z-x-\mu s)^2}{2s}\right\} \frac{1}{2\sqrt{2\pi(t-s)}} \\
&\qquad \times \exp\left\{-\mu(y+x) - \frac{\mu^2(t-s)}{2} - \frac{(z-y)^2}{2(t-s)}\right\} ds \\
&\quad = \frac{|z-x|}{2(2\pi)} \exp\left\{-\mu(x+y) - \frac{\mu^2 t}{2}\right\} \int_0^t \frac{\exp\{-\frac{(z-x)^2}{2s} - \frac{(z-y)^2}{2(t-s)}\}}{s^{3/2}\sqrt{t-s}}\, ds.
\end{aligned}$$

Noting that for $y, x > z$ and $x, y < z$ we have $\frac{z-x}{z-y} > 0$, we employ the following change of variables

$$\sqrt{t/s - 1}\sqrt{\frac{z-x}{z-y}} = \xi,$$

to obtain

$$\begin{aligned}
&\int_0^t p_z(x,s) q(t-s,z,y)\, ds \\
&\quad = \frac{|z-x|}{2\pi} \exp\left\{-\mu(x+y) - \frac{\mu^2 t}{2} - \frac{((z-y)^2 + (z-x)^2)^2}{2t}\right\} \frac{1}{t}\sqrt{\frac{z-y}{z-x}} \\
&\qquad \times \int_0^\infty \exp -\frac{1}{2}\left(\frac{1}{\xi^2} + \xi^2\right) \frac{(z-x)(z-y)}{t}\, d\xi \\
&\quad = \frac{|z-x|}{2\pi} \exp\left\{-\mu(x+y) - \frac{((z-y)^2 + (z-x)^2)}{2t}\right\} \frac{1}{t}\sqrt{\frac{z-y}{z-x}} \\
&\qquad \times \exp\left\{-\frac{(z-x)(z-y)}{t}\right\} \sqrt{\frac{\pi}{2}} \sqrt{\frac{t}{(z-x)(z-y)}} \\
&\quad = \frac{1}{2\sqrt{2\pi t}} \exp\left\{-\mu(x+y) - \frac{\mu^2 t}{2} - \frac{((z-y) + (z-x))^2}{2t}\right\},
\end{aligned}$$

where we used *MATHEMATICA* to arrive at the second last equation. Consequently,

$$\tilde{q}_z(t,x,y) = \frac{1}{2\sqrt{2\pi t}} \exp\left\{-\mu(x+y) - \frac{\mu^2 t}{2}\right\}$$

$$\times\left(\exp\left\{-\frac{(x-y)^2}{2t}\right\}-\exp\left\{-\frac{(2z-x-y)^2}{2t}\right\}\right)$$

for $x, y > z$ and $x, y < z$. □

Now, we focus on occupation times, firstly deriving the result for standard Brownian motion and subsequently for Brownian motion with drift. Occupation times measure the amount of time a stochastic process spends above or below a particular level. They have important applications in finance, as there are products whose pay-offs depend on the amount of time the asset price spends above or below a particular barrier. We are particularly interested in obtaining the distribution of occupation times explicitly. The approach to obtain such distributions we present here is based on Jeanblanc et al. (2009) and is motivated by the following result, see Theorem 2.5.1.1 in Jeanblanc et al. (2009): for convenience, we use E_x to denote the expectation with respect to the probability distribution of a Brownian motion started at x.

Theorem 2.1.11 *Let $\alpha \in \Re^+$ and let $k : \Re \to \Re^+$ and $g : \Re \to \Re$ be continuous functions and let g be bounded. Then the function*

$$f(x)=E_x\left(\int_0^\infty g(W_t)\exp\left\{-\alpha t-\int_0^t k(W_s)\,ds\right\}dt\right) \tag{2.1.10}$$

is piecewise twice differentiable and satisfies the differential equation

$$(\alpha+k)f=\frac{1}{2}f''+g. \tag{2.1.11}$$

We firstly consider $A_t^+ := \int_0^t \mathbf{1}_{[0,\infty)}(W_s)\,ds$, which measures the amount of time the standard Brownian motion $W = \{W_t, t \in [0,\infty)\}$ spends above 0 during the time interval $[0,t]$. Consider an exponentially distributed random variable τ, $\tau \sim \text{Exp}(\lambda)$, which is independent of W. Clearly,

$$E_x\left(\exp\{-\beta A_\tau^+\}\right)=\lambda f(x),$$

where

$$f(x):=E_x\left(\int_0^\infty \exp\left\{-\alpha t-\beta\int_0^t \mathbf{1}_{[0,\infty)}(W_s)\,ds\right\}dt\right).$$

However, $f(x)$ can be interpreted as a double Laplace transform of the density of A_τ^+, with respect to occupation time and the upper limit of the time interval. Inversion of the double Laplace transform will provide us with the desired density. Theorem 2.1.11 provides us with a useful expression for f, which can be inverted, if necessary numerically. To illustrate the technique used to obtain the distribution of occupation times, we present below the proof of the next result, see also Proposition 2.5.2.1 in Jeanblanc et al. (2009).

Proposition 2.1.12 *The law of $A_t^+ := \int_0^t \mathbf{1}_{[0,\infty)}(W_s)\,ds$ is given by*

$$P\left(A_t^+\in ds\right)=\frac{ds}{\pi\sqrt{s(t-s)}}\mathbf{1}_{0\le s<t}.$$

Proof We set $k(x) = \beta \mathbf{1}_{x\geq 0}$ and $g(x) = 1$ in Theorem 2.1.11. Then we obtain

$$f(x) = E_x\biggl(\int_0^\infty \exp\biggl\{-\alpha t - \beta \int_0^t \mathbf{1}_{[0,\infty)}(W_s)\,ds\biggr\}dt\biggr),$$

which solves

$$\begin{cases} \alpha f(x) = \frac{1}{2} f''(x) - \beta f(x) + 1, & x \geq 0 \\ \alpha f(x) = \frac{1}{2} f''(x) + 1, & x \geq 0. \end{cases}$$

In Jeanblanc et al. (2009), an explicit solution for $f(x)$ is obtained. We are particularly interested in the special case

$$f(0) = \int_0^\infty \exp\{-\alpha t\} E_0\bigl(e^{-\beta A_t^+}\bigr)\,dt = \frac{1}{\sqrt{\alpha(\alpha+\beta)}}. \tag{2.1.12}$$

However, we recall

$$\int_0^\infty e^{-\alpha t}\biggl(\int_0^\infty du\, \mathbf{1}_{s<t} \frac{\exp\{-\beta u\}}{\pi\sqrt{u(t-u)}}\biggr)dt = \frac{1}{\sqrt{\alpha(\alpha+\beta)}},$$

so we can explicitly invert the double Laplace transform (2.1.12) to complete the proof. □

We remark that the same technique can be used to compute the corresponding result for the occupation time of a Brownian motion with drift. Let $X_t = \nu t + W_t$, and consider the occupation time of this Brownian motion above the level $L > 0$

$$A_t^{+,L,\nu} = \int_0^t \mathbf{1}_{X_s > L}\,ds,$$

and we define $A_t^{-,L,\nu}$ analogously. Using the same idea as before, together with the relevant Feynman-Kac result, we get

$$\begin{aligned} P\bigl(A_t^{-,0,\nu} \in du\bigr) &= \biggl(\sqrt{\frac{2}{\pi u}} \exp\biggl\{-\frac{\nu^2}{2}u\biggr\} - 2\nu\Theta(\nu\sqrt{u})\biggr) \\ &\quad \times \biggl(\nu + \frac{1}{\sqrt{2\pi(t-u)}} \exp\biggl\{-\frac{\nu^2}{2}(t-u)\biggr\} - \nu\Theta(\nu\sqrt{t-u})\biggr), \end{aligned} \tag{2.1.13}$$

where $\Theta(x) = \frac{1}{\sqrt{2\pi}} \int_x^\infty \exp\{-\frac{y^2}{2}\}\,dy$. Finally,

$$P\bigl(A_t^{-,L,\nu} \leq u\bigr) = \int_0^u \varphi(s, L; \nu) P\bigl(A_{t-s}^{-,0,\nu} < u - s\bigr)\,ds,$$

where $\varphi(s, L; \nu)$ is the density $P(T_L \in ds)/ds$, where T_L denotes the first time the Brownian motion with drift hits the level L, $T_L = \inf\{t\colon\ X_t = L\}$, and

$$\varphi(s, L; \nu) = \frac{L}{\sqrt{2\pi s^3}} \exp\biggl\{-\frac{1}{2s}(y - \nu s)^2\biggr\}\mathbf{1}_{s\geq 0}.$$

Finally, the law of $A_t^{+,L,\nu}$ follows from $A_t^{+,L,\nu} + A_t^{-,L,\nu} = t$.

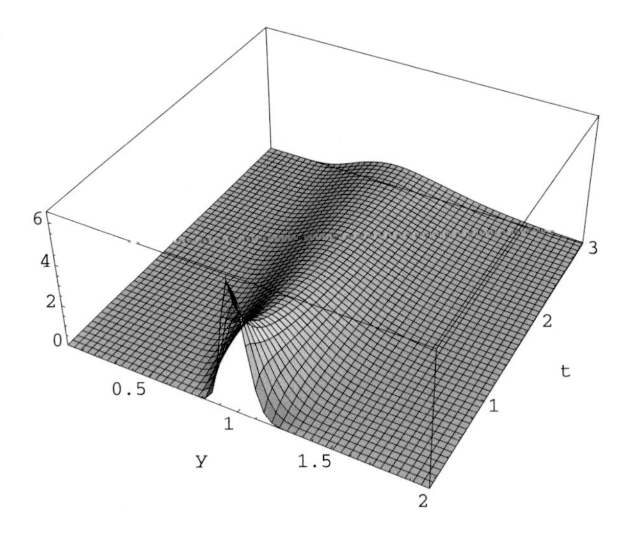

Fig. 2.1.2 Transition density for geometric Brownian motion

2.1.3 Geometric Brownian Motion

Geometric Brownian motion is a process of significant importance in finance, as the Black-Scholes model (BSM), Black and Scholes (1973), is based on it, see also Sect. 2.3. We can describe geometric Brownian motion via the SDE

$$dX_t = X_t\left(\left(g + \frac{1}{2}b^2\right)dt + b\,dW_t\right), \tag{2.1.14}$$

subject to $X_0 > 0$. Equation (2.1.14) can be explicitly solved to yield

$$X_t = X_0 \exp(gt + bW_t).$$

Its transition density function satisfies

$$p(s, x; t, y) = \frac{1}{\sqrt{2\pi(t-s)}by}\exp\left\{-\frac{(\ln(y) - \ln(x) - g(t-s))^2}{2b^2(t-s)}\right\}, \tag{2.1.15}$$

for $t \in [0, \infty)$, $s \in [0, t]$ and $x, y \in (0, \infty)$. Figure 2.1.2 shows the transition density for a geometric Brownian motion with growth rate $g = 0.05$, volatility $b = 0.2$ and initial value $x = 1$ at time $s = 0$ for the period from 0.1 to 3 years.

The corresponding laws of first hitting times, maximum, and minimum follow easily from the corresponding results for a Brownian motion with drift. Regarding the integrals, we have the following result, see Yor (2001) and Pintoux and Privault (2011):

$$\begin{aligned}
&P\left(\int_0^t \exp\{\sigma W_s - p\sigma^2 s/2\}\,ds \in du,\, W_t \in dy\right) \\
&\quad = \frac{\sigma}{2}\exp\{-p\sigma y/2 - p^2\sigma^2 t/8\}\exp\left\{-2\frac{1 + \exp\{\sigma y\}}{\sigma^2 u}\right\} \\
&\qquad \times \theta\left(\frac{4\exp\{\sigma y/2\}}{\sigma^2 u}, \frac{\sigma^2 t}{4}\right)\frac{du}{u}\,dy,
\end{aligned}$$

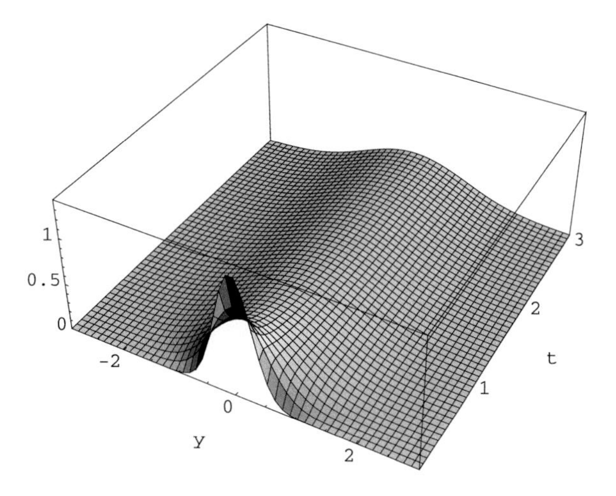

Fig. 2.1.3 Transition density of standard OU process starting at $(s, x) = (0, 0)$

where $p = -\frac{2g}{b^2}$, $u > 0$, $y \in \Re$ and

$$\theta(v,t) = \frac{v \exp\{\frac{\pi^2}{2t}\}}{\sqrt{2\pi^3 t}} \int_0^\infty \exp\left\{-\frac{\xi^2}{2t} - v\cosh(\xi)\right\} \sinh(\xi) \sin\left(\frac{\pi\xi}{t}\right) d\xi, \quad v > 0.$$

2.1.4 Ornstein-Uhlenbeck Process

The *Ornstein-Uhlenbeck process* is also a process of importance in finance and forms the basis of the *Vasiček model*, see Vasiček (1977). We consider the standard Ornstein-Uhlenbeck process,

$$dX_t = -X_t\,dt + \sqrt{2}\,dW_t,$$

where $X_0 = x \in \Re$. Its transition density is Gaussian,

$$p(s,x;t,y) = \frac{1}{\sqrt{2\pi(1 - e^{-2(t-s)})}} \exp\left\{-\frac{(y - xe^{-(t-s)})^2}{2(1 - e^{-2(t-s)})}\right\}, \quad (2.1.16)$$

for $t \in [0,\infty)$, $s \in [0,t]$ and $x, y \in \Re$, with mean $x\,e^{-(t-s)}$ and variance $1 - e^{-2(t-s)}$.

To illustrate the stochastic dynamics of this process we show in Fig. 2.1.3 the transition density of a standard OU process for the period from 0.1 to 3 years with initial value $x = 0$ at time $s = 0$. As can be seen from Fig. 2.1.3 the transition densities for the standard OU process seem to stabilize after a period of about one year. In fact, as can be seen from (2.1.16) these transition densities asymptotically approach, as $t \to \infty$, a standard Gaussian density. This is in contrast, for example, to transition densities for the Wiener process, which do not converge to a *stationary density*, see (2.1.1) and Fig. 2.1.1. For illustration, we plot in Fig. 2.1.4 the transition

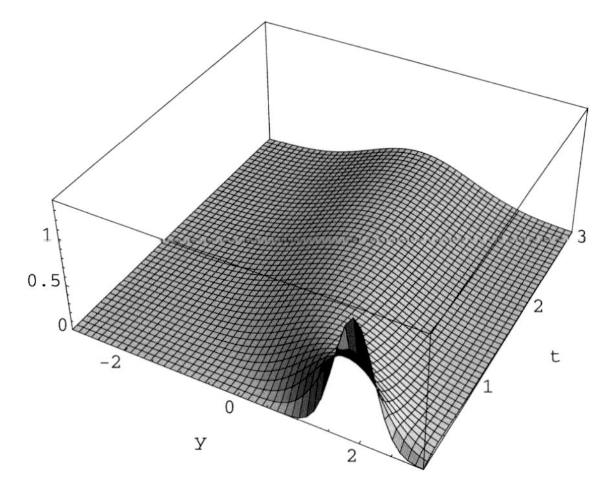

Fig. 2.1.4 Transition density of standard OU process starting at $(s, x) = (0, 2)$

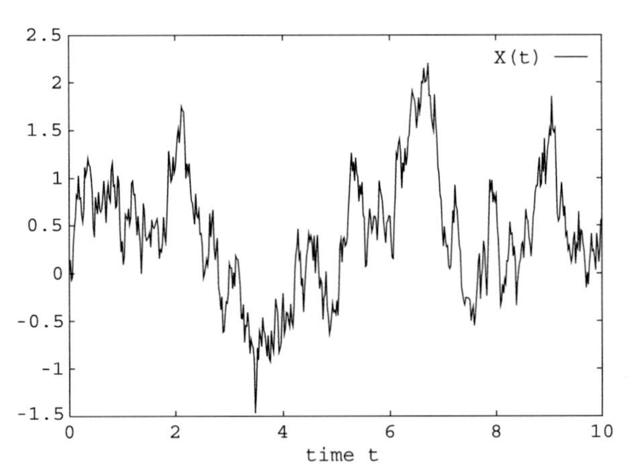

Fig. 2.1.5 Path of a standard Ornstein-Uhlenbeck process

density for a standard OU process that starts at the initial value $x = 2$ at time $t = 0$. Note how the transition density evolves towards a median that is close to 0.

In Fig. 2.1.5 a path of a standard OU process is shown. It can be observed that this trajectory fluctuates around some reference level. Indeed, as already indicated, the standard OU process has a stationary density. This can be seen from (2.1.16) when $t \to \infty$. Note also that the Gaussian property of the standard OU process means that the process itself and even a scaled and shifted OU process may become negative. We now recall Proposition 3.4.1.1 from Jeanblanc et al. (2009), which characterizes the first hitting time of the level 0,

$$T_0 = \inf\{t \geq 0\colon\ X_t = 0\}.$$

Proposition 2.1.13 *The density function of T_0 is given by*

$$f(t) = \frac{x}{2\sqrt{\pi}} \exp\left\{\frac{x^2}{4}\right\} \exp\left\{\frac{1}{2}\left(t - \frac{x^2}{2}\coth(t)\right)\right\} \left(\frac{1}{\sinh(t)}\right)^{\frac{3}{2}}.$$

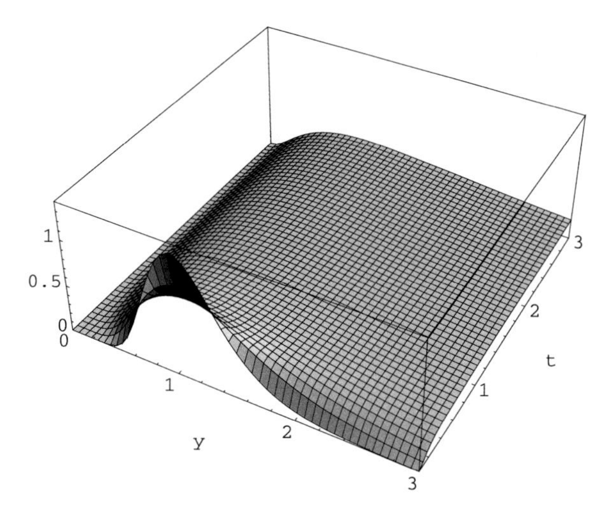

Fig. 2.1.6 Transition density of a geometric Ornstein-Uhlenbeck process

Furthermore, integrals of the Ornstein-Uhlenbeck process are of importance in finance, as they impact bond prices for example. Defining

$$n(t,T) = \left(1 - \exp\left\{-(T-t)\right\}\right),$$

we have that

$$\int_0^T X_s\,ds \sim N\left(n(0,T)X_0, 2\int_0^T n^2(u,T)\,du\right).$$

2.1.5 Geometric Ornstein-Uhlenbeck Process

Exponentiating an Ornstein-Uhlenbeck process, as discussed in the previous subsection, we obtain a geometric Ornstein-Uhlenbeck process. Its transition density is lognormal satisfying

$$p(s,x;t,y) = \frac{1}{y\sqrt{2\pi(1-e^{-2(t-s)})}} \exp\left\{-\frac{(\ln(y) - \ln(x)e^{-(t-s)})^2}{2(1-e^{-2(t-s)})}\right\}, \qquad (2.1.17)$$

for $t \in [0,\infty)$, $s \in [0,t]$ and $x, y \in (0,\infty)$. In Fig. 2.1.6 we display the corresponding probability densities for the time period from 0.1 to 3 years with initial value $x = 1$ at the initial time $s = 0$. In this case the transition density converges over time to a limiting lognormal density as stationary density, as can be seen from (2.1.17). Figure 2.1.7 shows a trajectory for the geometric OU process. We note that it stays positive and shows large fluctuations for large values. Since it is the exponential of an Ornstein-Uhlenbeck process, one can use the result on the hitting time from the previous subsection.

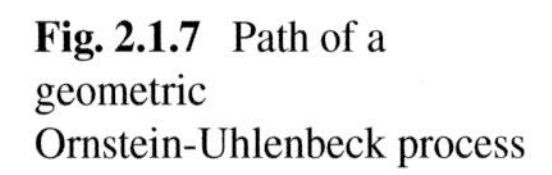

Fig. 2.1.7 Path of a geometric Ornstein-Uhlenbeck process

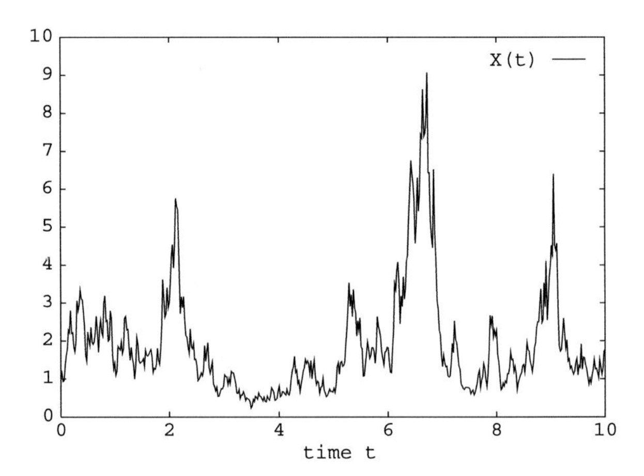

2.2 Functionals of Multidimensional Wiener Processes

In this section, we discuss functionals of multidimensional Wiener processes or Brownian motions, in particular their SDEs and transition densities. When modeling complex systems, such as a financial market, it is often necessary to employ a multidimensional stochastic process to model the uncertainty. It is crucial to understand the dependence structure between the individual stochastic processes, hence we briefly discuss copulas before discussing stochastic processes.

2.2.1 Copulas

Each multivariate distribution function has its, so called *copula*, which characterizes the dependence structure between the components. Roughly speaking, the copula is the joint density of the components when they are each transformed into uniformly $U(0,1)$ distributed random variables. Essentially, every multivariate distribution has a corresponding copula. Conversely, each copula can be used together with some given marginal distributions to obtain a corresponding multivariate distribution function. This is a consequence of Sklar's theorem, see for instance McNeil et al. (2005).

If, for instance, $\boldsymbol{Y} \sim N_d(\boldsymbol{\mu}, \boldsymbol{\Omega})$ is a Gaussian random vector, then the copula of $\boldsymbol{Y}$ is the same as the copula of $\boldsymbol{X} \sim N_d(\mathbf{0}, \boldsymbol{\Omega})$, where $\mathbf{0}$ is the zero vector and $\boldsymbol{\Omega}$ is the correlation matrix of $\boldsymbol{Y}$. By the definition of the d-dimensional Gaussian copula we obtain

$$\begin{aligned} C_{\boldsymbol{\Omega}}^{Ga} &= P\big(N(X_1) \le u_1, \ldots, N(X_d) \le u_d\big) \\ &= N_{\Omega}\big(N^{-1}(u_1), \ldots, N^{-1}(u_d)\big), \end{aligned} \tag{2.2.18}$$

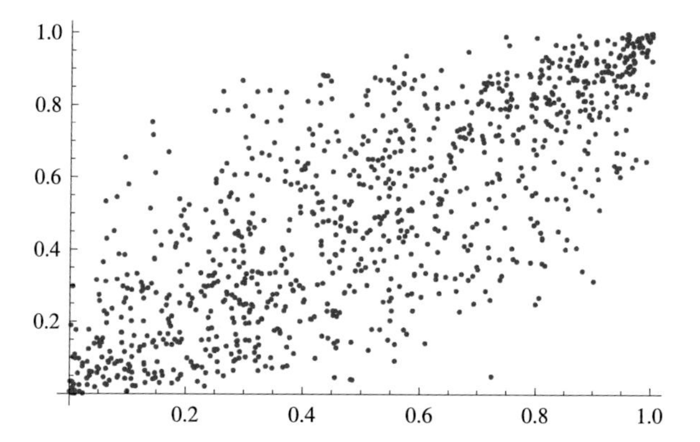

Fig. 2.2.8 Gaussian copula with parameter $\varrho = 0.5$

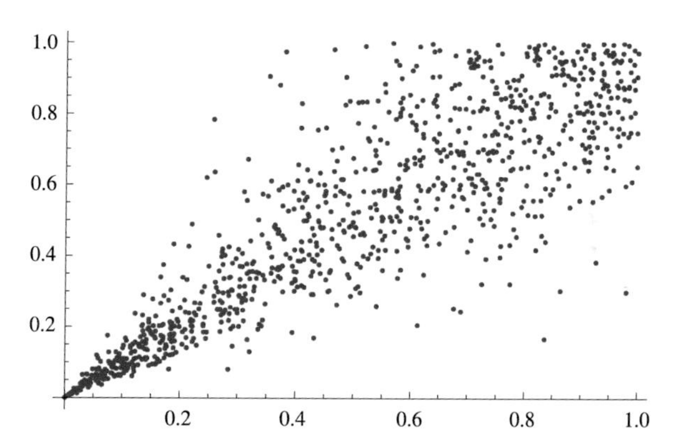

Fig. 2.2.9 Clayton copula with parameter $\theta = 0.5$

where N denotes the standard univariate normal distribution function and N_{Ω} denotes the joint distribution function of $\boldsymbol{X}$. Hence, in two dimensions we obtain

$$C_{\boldsymbol{\Omega}}^{Ga}(u_1, u_2) = \int_{-\infty}^{N^{-1}(u_1)} \int_{-\infty}^{N^{-1}(u_2)} \frac{1}{2\pi(1-\varrho^2)^{1/2}} \times \exp\left\{ \frac{-(s_1^2 - 2\varrho s_1 s_2 + s_2^2)}{2(1-\varrho^2)} \right\} ds_1\, ds_2, \qquad (2.2.19)$$

where $\varrho \in [-1, 1]$ is the correlation parameter in $\boldsymbol{\Omega}$. In Fig. 2.2.8, we simulate from a Gaussian copula with parameter $\varrho = 0.5$.

Another example of a copula is the Clayton copula. This copula can be expressed in the d-dimensional case as

$$C_{\theta}^{Cl} = \left(u_1^{-\theta} + \cdots + u_d^{-\theta} - d + 1\right)^{-1/\theta}, \quad \theta \geq 0, \qquad (2.2.20)$$

where the limiting case $\theta = 0$ is the d-dimensional independence copula. For purposes of comparison, in Fig. 2.2.9, we simulate from a Clayton copula with $\theta = 0.5$. It is evident from Figs. 2.2.8 and 2.2.9, that the Gaussian copula does not allow for tail dependence, whereas the Clayton copula does.

Moreover, d-dimensional Archimedian copulas can be expressed in terms of Laplace-Stieltjes transforms of distribution functions on $\Re^+$. If F is a distribution

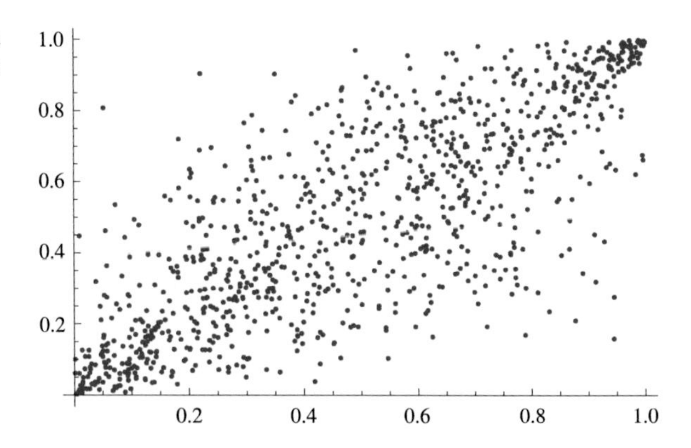

Fig. 2.2.10 Student t copula with four degrees of freedom and $\rho = 0.8$

function on $\Re^+$ satisfying $F(0) = 0$, then the Laplace-Stieltjes transform can be expressed by

$$\hat{F}(t) = \int_0^\infty e^{-tx}\, dF(x), \quad t \geq 0. \tag{2.2.21}$$

Using the Laplace-Stieltjes transform the d-dimensional Archimedian copula has the form

$$C^{Ar}(u_1, \ldots, u_d) = E\left(\exp\left\{-V \sum_{i=1}^{d} \hat{F}^{-1}(u_i)\right\}\right) \tag{2.2.22}$$

for strictly positive random variables V with Laplace-Stieltjes transform $\hat{F}$. We show in Fig. 2.2.10 the Student t copula for four degrees of freedom, which has been shown in Ignatieva et al. (2011) to reflect extremely well the dependence of log-returns of well-diversified indices in different currencies. Compared to Fig. 2.2.8, we notice a marked difference in the tails of the distribution, the Student t copula allows for higher dependence in the extreme values.

A simulation method follows directly from this representation, see Marshall and Olkin (1988). More examples of multidimensional copulas can be found in McNeil et al. (2005).

Note that each of the following transition densities relate to their own copulas. We will list the transition densities for:

- Multidimensional Wiener processes;
- Multidimensional geometric Brownian motions;
- Multidimensional OU-processes;
- Multidimensional geometric OU-processes.

It is well-known that more analytical results are available for one-dimensional than for multidimensional processes. Hence it is important to have access to the transition densities, so that important functionals can be computed numerically, using e.g. the techniques to be presented in Chap. 12.

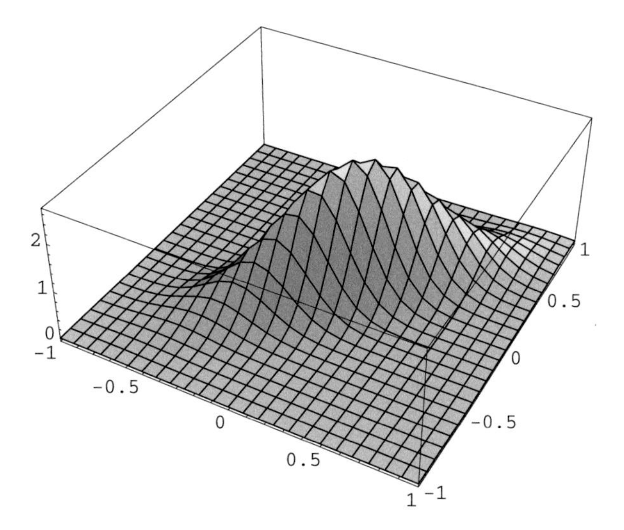

Fig. 2.2.11 Bivariate transition density of the two-dimensional Wiener process for fixed time step $\Delta = 0.1$, $x_1 = x_2 = 0.1$ and $\varrho = 0.8$

2.2.2 *Multidimensional Wiener Process*

As a first example of a continuous multidimensional stochastic process, whose transition density can be expressed explicitly, we focus on the d-dimensional Wiener process. This fundamental stochastic process has a multivariate Gaussian transition density of the form

$$p(s, \boldsymbol{x}; t, \boldsymbol{y}) = \frac{1}{(2\pi(t-s))^{d/2}\sqrt{\det \boldsymbol{\Sigma}}} \exp\left\{\frac{(\boldsymbol{y}-\boldsymbol{x})^\top \boldsymbol{\Sigma}^{-1}(\boldsymbol{y}-\boldsymbol{x})}{2(t-s)}\right\}, \quad (2.2.23)$$

for $t \in [0, \infty)$, $s \in [0, t]$ and $\boldsymbol{x}, \boldsymbol{y} \in \Re^d$. Here $\boldsymbol{\Sigma}$ is a normalized covariance matrix. Its copula is the Gaussian copula (2.2.18), which is simply derived from the corresponding multivariate Gaussian distribution function. In the bivariate case with correlated Wiener processes this transition probability density simplifies to

$$\begin{aligned} p(s, x_1, x_2; t, y_1, y_2) \\ = \frac{1}{2\pi(t-s)\sqrt{1-\varrho^2}} \\ \times \exp\left\{-\frac{(y_1-x_1)^2 - 2(y_1-x_1)(y_2-x_2)\varrho + (y_2-x_2)^2}{2(t-s)(1-\varrho^2)}\right\}, \end{aligned} \quad (2.2.24)$$

for $t \in [0, \infty)$, $s \in [0, t]$ and $x_1, x_2, y_1, y_2 \in \Re$. Here the correlation parameter ϱ varies in the interval $[-1, 1]$. In the case of correlated Wiener processes one can first simulate independent Wiener processes and then form from these, by linear transforms, correlated ones.

In Fig. 2.2.11 we illustrate the bivariate transition density of the two-dimensional Wiener process for the time increment $\Delta = t - s = 0.1$, initial values $x_1 = x_2 = 0.1$ and correlation $\varrho = 0.8$. One can also generate dependent Wiener processes that have a joint distribution with a given copula.

2.2.3 *Transition Density of a Multidimensional Geometric Brownian Motion*

The multidimensional geometric Brownian motion is a componentwise exponential of linearly transformed Wiener processes. Given a vector of correlated Wiener processes $\boldsymbol{W}$ with the transition density (2.2.23) we consider the following transformation

$$\boldsymbol{S}_t = \boldsymbol{S}_0 \exp\{\boldsymbol{a}t + \boldsymbol{B}\boldsymbol{W}_t\}, \tag{2.2.25}$$

for $t \in [0, \infty)$, where the exponential is taken componentwise. Here $\boldsymbol{a}$ is a vector of length d, while the elements of the matrix $\boldsymbol{B}$ are as follows

$$B^{i,j} = \begin{cases} b^j & \text{for } i = j \\ 0 & \text{otherwise,} \end{cases} \tag{2.2.26}$$

where $i, j \in \{1, 2, \ldots, d\}$. Then the transition density of the above defined geometric Brownian motion has the following form

$$\begin{aligned} &p(s, \boldsymbol{x}; t, \boldsymbol{y}) \\ &= \frac{1}{(2\pi(t-s))^{d/2}\sqrt{\det \boldsymbol{\Sigma}}\prod_{i=1}^{d} b^i y_i} \\ &\quad \times \exp\left\{-\frac{(\ln(\boldsymbol{y}) - \ln(\boldsymbol{x}) - \boldsymbol{a}(t-s))^\top \boldsymbol{B}^{-1}\boldsymbol{\Sigma}^{-1}\boldsymbol{B}^{-1}}{2} \right. \\ &\quad \times \left. \frac{(\ln(\boldsymbol{y}) - \ln(\boldsymbol{x}) - \boldsymbol{a}(t-s))}{t-s}\right\} \end{aligned} \tag{2.2.27}$$

for $t \in [0, \infty)$, $s \in [0, t]$ and $\boldsymbol{x}, \boldsymbol{y} \in \Re^d_+$. Here the logarithm is understood componentwise. In the bivariate case this transition density takes the particular form

$$\begin{aligned} &p(s, x_1, x_2; t, y_1, y_2) \\ &= \frac{1}{2\pi(t-s)\sqrt{1-\varrho^2}b^1 b^2 y_1 y_2} \\ &\quad \times \exp\left\{-\frac{(\ln(y_1) - \ln(x_1) - a^1(t-s))^2}{2(b^1)^2(t-s)(1-\varrho^2)}\right\} \\ &\quad \times \exp\left\{-\frac{(\ln(y_2) - \ln(x_2) - a^2(t-s))^2}{2(b^2)^2(t-s)(1-\varrho^2)}\right\} \\ &\quad \times \exp\left\{\frac{(\ln(y_1) - \ln(x_1) - a^1(t-s))(\ln(y_2) - \ln(x_2) - a^2(t-s))\varrho}{b^1 b^2(t-s)(1-\varrho^2)}\right\}, \end{aligned}$$

for $t \in [0, \infty)$, $s \in [0, t]$ and $x_1, x_2, y_1, y_2 \in \Re^+$, where $\varrho \in [-1, 1]$.

In Fig. 2.2.12 we illustrate the bivariate transition density of the two-dimensional geometric Brownian motion for the time increment $\Delta = t - s = 0.1$, initial values $x_1 = x_2 = 0.1$, correlation $\varrho = 0.8$, volatilities $b^1 = b^2 = 2$ and growth parameters $a^1 = a^2 = 0.1$.

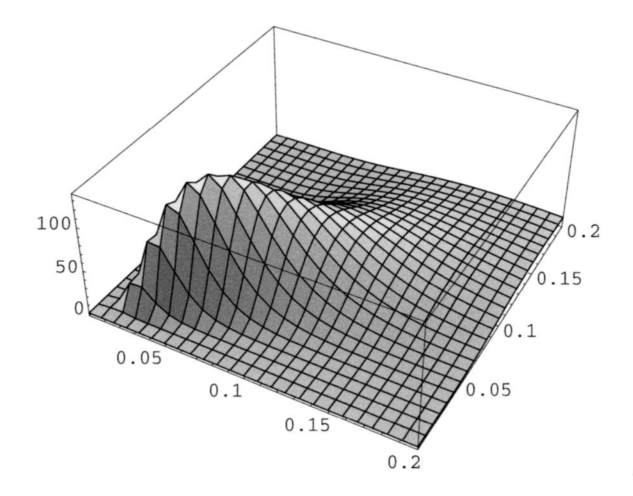

Fig. 2.2.12 Bivariate transition density of the two-dimensional geometric Brownian motion for $\Delta = 0.1$, $x_1 = x_2 = 0.1$, $\varrho = 0.8$, $b^1 = b^2 = 2$ and $a_1 = a_2 = 0.1$

2.2.4 Transition Density of a Multidimensional OU-Process

Another example is the standard d-dimensional *Ornstein-Uhlenbeck* (OU)-*process*. This process has a Gaussian transition density of the form

$$p(s, \boldsymbol{x}; t, \boldsymbol{y}) = \frac{1}{(2\pi(1 - e^{-2(t-s)}))^{d/2}\sqrt{\det \boldsymbol{\Sigma}}} \times \exp\left\{-\frac{(\boldsymbol{y} - \boldsymbol{x}e^{-(t-s)})^\top \boldsymbol{\Sigma}^{-1}(\boldsymbol{y} - \boldsymbol{x}e^{-(t-s)})}{2(1 - e^{-2(t-s)})}\right\}, \quad (2.2.28)$$

for $t \in [0, \infty)$, $s \in [0, t]$ and $\boldsymbol{x}, \boldsymbol{y} \in \Re^d$, with mean $\boldsymbol{x}e^{-(t-s)}$ and covariance matrix $\boldsymbol{\Sigma}(1 - e^{-2(t-s)})$, $d \in \{1, 2, \ldots\}$, see e.g. Platen and Bruti-Liberati (2010). In the bivariate case the transition density of the standard OU-process is expressed by

$$p(s, x_1, x_2; t, y_1, y_2) = \frac{1}{2\pi(1 - e^{-2(t-s)})\sqrt{1 - \varrho^2}} \times \exp\left\{-\frac{(y_1 - x_1 e^{-(t-s)})^2 + (y_2 - x_2 e^{-(t-s)})^2}{2(1 - e^{-2(t-s)})(1 - \varrho^2)}\right\} \times \exp\left\{\frac{(y_1 - x_1 e^{-(t-s)})(y_2 - x_2 e^{-(t-s)})\varrho}{(1 - e^{-2(t-s)})(1 - \varrho^2)}\right\}, \quad (2.2.29)$$

for $t \in [0, \infty)$, $s \in [0, t]$ and $x_1, x_2, y_1, y_2 \in \Re$, where $\varrho \in [-1, 1]$.

2.2.5 Transition Density of a Multidimensional Geometric OU-Process

The transition density of a d-dimensional *geometric OU-process* can be obtained from the transition density of the multidimensional OU-process by applying the exponential transformation. Therefore, it can be expressed as

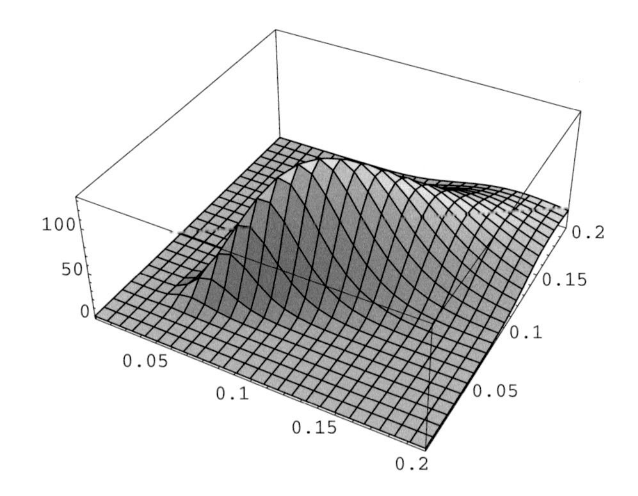

Fig. 2.2.13 Bivariate transition density of the two-dimensional geometric OU-process for $\Delta = 0.1$, $x_1 = x_2 = 0.1$ and $\varrho = 0.8$

$$p(s, \boldsymbol{x}; t, \boldsymbol{y}) = \frac{1}{(2\pi(1 - e^{-2(t-s)}))^{d/2} \sqrt{\det \boldsymbol{\Sigma}} \prod_{i=1}^{d} y_i} \times \exp\left\{ -\frac{(\ln(\boldsymbol{y}) - \ln(\boldsymbol{x}) e^{-(t-s)})^\top \boldsymbol{\Sigma}^{-1} (\ln(\boldsymbol{y}) - \ln(\boldsymbol{x}) e^{-(t-s)})}{2(1 - e^{-2(t-s)})} \right\}, \tag{2.2.30}$$

for $t \in [0, \infty)$, $s \in [0, t]$ and $\boldsymbol{x}, \boldsymbol{y} \in \Re_+^d$, $d \in \{1, 2, \ldots\}$. In the bivariate case the transition density of the multidimensional geometric OU-process is of the form

$$\begin{aligned} p(s, x_1, x_2; t, y_1, y_2) &= \frac{1}{2\pi(1 - e^{-2(t-s)}) \sqrt{1 - \varrho^2} y_1 y_2} \\ &\quad \times \exp\left\{ -\frac{(\ln(y_1) - \ln(x_1) e^{-(t-s)})^2 + (\ln(y_2) - \ln(x_2) e^{-(t-s)})^2}{2(1 - e^{-2(t-s)})(1 - \varrho^2)} \right\} \\ &\quad \times \exp\left\{ \frac{(\ln(y_1) - \ln(x_1) e^{-(t-s)})(\ln(y_2) - \ln(x_2) e^{-(t-s)})\varrho}{(1 - e^{-2(t-s)})(1 - \varrho^2)} \right\}, \end{aligned} \tag{2.2.31}$$

for $t \in [0, \infty)$, $s \in [0, t]$ and $x_1, x_2, y_1, y_2 \in \Re_+$, where $\varrho \in [-1, 1]$.

In Fig. 2.2.13 we illustrate the bivariate transition density of the two-dimensional geometric OU-process for the time increment $\Delta = t - s = 0.1$, initial values $x_1 = x_2 = 0.1$ and correlation $\varrho = 0.8$. It is now obvious how to obtain the transition density of the componentwise exponential of other Gaussian vector processes.

2.3 Real World Pricing Under the Black-Scholes Model

In this section, we continue to discuss a continuous financial market as introduced in Chap. 1. We illustrate real world pricing under the benchmark approach using the Black-Scholes model (BSM), see Black and Scholes (1973) and Sect. 1.1. The resulting explicit formulas are of importance not only for the BSM but also for more general models when used in variance reduction techniques, see Platen and Bruti-Liberati (2010). In addition, we illustrate that real world pricing does, in fact, recover the well-known risk neutral pricing as special case and is hence consistent with the classical approach. Finally, we remark that we could, of course, in the case of the BSM perform the relevant change of measure to directly obtain the risk neutral prices. However, this section aims to illustrate real world pricing, and hence we proceed by computing the expected value in (1.3.19) directly in the case of the BSM.

To alleviate notation we define the *benchmarked volatility* $\sigma_t^{j,k}$ by setting

$$\sigma_t^{0,k} = \theta_t^k \tag{2.3.32}$$

for $j = 0$ and $k \in \{1, 2, \dots, d\}$, and

$$\sigma_t^{j,k} = \theta_t^k - b_t^{j,k} \tag{2.3.33}$$

for $k \in \{1, 2, \dots, d\}$ and $j \in \{1, 2, \dots, d\}$, $t \geq 0$. Consequently, it follows from (1.2.12) that the SDE governing the dynamics of the GOP becomes

$$dS_t^{\delta_*} = S_t^{\delta_*}\left(r_t\,dt + \sum_{k=1}^{d} \sigma_t^{0,k}\left(\sigma_t^{0,k}\,dt + dW_t^k\right)\right), \tag{2.3.34}$$

which can be solved explicitly to yield

$$S_t^{\delta_*} = S_0^{\delta_*} \exp\left\{\int_0^t \left(r_s + \frac{1}{2}\sum_{k=1}^{d}(\sigma_s^{0,k})^2\right) ds + \sum_{k=1}^{d}\int_0^t \sigma_s^{0,k}\,dW_s^k\right\} \tag{2.3.35}$$

for all $t \geq 0$. Furthermore, the jth benchmarked primary security account $\hat{S}_t^j = \frac{S_t^j}{S_t^{\delta_*}}$ can be shown to satisfy

$$d\hat{S}_t^j = -\hat{S}_t^j \sum_{k=1}^{d} \sigma_t^{j,k}\,dW_t^k, \tag{2.3.36}$$

for all $j \in \{0, 1, \dots, d\}$ and $t \geq 0$, with $\hat{S}_0^j = S_0^j$, which follows from (1.3.16) by setting $\pi_{\delta,t}^i = 1$ for $i = j$ and $\pi_{\delta,t}^i = 0$ otherwise. Consequently, we obtain the following explicit expression for the jth benchmarked primary security account

$$\hat{S}_t^j = \hat{S}_0^j \exp\left\{-\frac{1}{2}\int_0^t \sum_{k=1}^{d}(\sigma_s^{j,k})^2\,ds - \sum_{k=1}^{d}\int_0^t \sigma_s^{j,k}\,dW_s^k\right\} \tag{2.3.37}$$

for $j \in \{0, 1, \dots, d\}$ and $t \geq 0$.

We now illustrate that under the benchmark approach, the benchmarked primary security accounts $\hat{S}_t^j$, $j \in \{0, 1, \dots, d\}$ are the pivotal objects of study: in particular, specifying the savings account S_t^0 and the benchmarked primary security accounts suffices to determine the entire investment universe. The ratio $S_t^{\delta_*} = \frac{S_t^0}{\hat{S}_t^0}$, for all $t \geq 0$, see (1.3.15), expresses the GOP and NP in terms of the savings account and the benchmarked savings account. The product $S_t^j = \hat{S}_t^j S_t^{\delta_*}$ recovers each primary security account from the corresponding benchmarked primary security account and the GOP for each $j \in \{1, 2, \ldots, d\}$ and $t \geq 0$.

Next, we introduce the processes $|\sigma^j| = \{|\sigma_t^j|, t \geq 0\}$ for $j \in \{0, 1, \ldots, d\}$, by setting

$$|\sigma_t^j| = \sqrt{\sum_{k=1}^{d} (\sigma_t^{j,k})^2}. \tag{2.3.38}$$

These processes enable us to introduce the *aggregate continuous noise processes* $\hat{W}^j = \{\hat{W}_t^j, t \in [0, \infty)\}$ for $j \in \{0, 1, \ldots, d\}$, defined by

$$\hat{W}_t^j = \sum_{k=1}^{d} \int_0^t \frac{\sigma_s^{j,k}}{|\sigma_s^j|} dW_s^k. \tag{2.3.39}$$

An application of Lévy's Theorem for the characterization of the Wiener process, see Chap. 15, Theorem 15.3.3, allows us to conclude that $\hat{W}^j$ is a Wiener process for each $j \in \{0, 1, \ldots, d\}$. We point out that the Wiener processes $\hat{W}^0, \hat{W}^1, \ldots, \hat{W}^d$ can be correlated. Furthermore, we enforce Assumption 1.1.1, so that the volatility matrix $\boldsymbol{b}_t = [b_t^{j,k}]_{j,k=1}^d$ becomes invertible for all $t \geq 0$. Note that so far in this section the short rate and volatility processes are not specified and remain still general.

2.3.1 The Black-Scholes Model

The stylized *Black-Scholes model* (BSM) arises if we assume that all parameter processes, that is, the short rate and the volatilities, are constant, i.e. if we set $r_t = r$ and $\sigma_t^{j,k} = \sigma^{j,k}$ for each $j \in \{0, 1, \ldots, d\}$, $k \in \{1, 2, \ldots, d\}$ and $t \geq 0$. Consequently, (2.3.34) and (2.3.36) become in this case

$$S_t^{\delta_*} = S_0^{\delta_*} \exp\left\{r\, t + \frac{t}{2}|\sigma^0|^2 + |\sigma^0|\, \hat{W}_t^0\right\} \tag{2.3.40}$$

and

$$\hat{S}_t^j = S_0^j \exp\left\{-\frac{t}{2}|\sigma^j|^2 - |\sigma^j|\hat{W}_t^j\right\} \tag{2.3.41}$$

for each $j \in \{0, 1, \ldots, d\}$ and all $t \geq 0$. From (2.3.41) it is clear that the benchmarked primary security accounts $\hat{S}_t^j$, $j \in \{0, 1, \ldots, d\}$, are continuous martingales, as they are driftless geometric Brownian motions. As this holds, in particular, for the benchmarked savings account, the Radon-Nikodym derivative process $\Lambda_\theta(t) = \frac{\hat{S}_t^0}{\hat{S}_0^0}$ in (1.3.20) is an $(\underline{\mathcal{A}}, P)$-martingale. We conclude that the standard risk neutral pricing approach could, therefore, be used for derivative pricing under the BSM making use of the risk neutral pricing formula (1.3.21). Finally, we emphasize that we do not advocate the BSM as a reasonably realistic description of observed market dynamics. However, given its familiarity, it is useful for illustrating real world pricing under the benchmark approach for classical models, which produces the same answers as risk neutral pricing. Furthermore, the fact that explicit formulas can be obtained for many derivatives is extremely useful in practice. We will derive below explicit formulas and descriptions of a range of derivative prices under the BSM by using real world pricing.

2.3.2 Zero Coupon Bonds

We firstly demonstrate how to price a standard default-free *zero coupon bond* that pays one unit of the domestic currency at its maturity date $T \in [0, \infty)$. It follows from the real world pricing formula (1.3.19) that the value of the zero coupon bond at time t is given by

$$P_T(t) = S_t^{\delta_*} E\left(\frac{1}{S_T^{\delta_*}} \,\Big|\, \mathcal{A}_t\right) = \frac{1}{\hat{S}_t^0} E\left(\exp\left\{-\int_t^T r_s\,ds\right\} \hat{S}_T^0 \,\Big|\, \mathcal{A}_t\right) \tag{2.3.42}$$

for all $t \in [0, T]$. Since $\hat{S}_t^0 = \frac{S_t^0}{S_t^{\delta_*}}$ is an $(\underline{\mathcal{A}}, P)$-martingale and $r_t = r$ is constant we obtain

$$P_T(t) = \exp\{-r(T-t)\} \frac{1}{\hat{S}_t^0} E\big(\hat{S}_T^0 \mid \mathcal{A}_t\big) = \exp\{-r(T-t)\} \tag{2.3.43}$$

for all $t \in [0, T]$. As expected, this is the usual bond pricing formula that is determined by the deterministic short rate r, which one can also obtain via risk neutral pricing, see (1.3.20) and Harrison and Kreps (1979). As long as the benchmarked savings account, and with this the Radon-Nikodym derivative of the risk neutral measure, is a martingale one obtains this classical zero coupon bond price.

2.3.3 Forward Contracts

We now aim to price a *forward contract* with the delivery of one unit of the jth primary security account at the maturity date T, which is written or initiated at time $t \in [0, T]$ for $j \in \{0, 1, \ldots, d\}$. The value of the forward contract written at initiation time t is zero by definition. The real world pricing formula (1.3.19) yields then the following relation, which determines the forward price $F_T^j(t)$ at time $t \in [0, T]$ via the relation

$$S_t^{\delta_*} E\left(\frac{F_T^j(t) - S_T^j}{S_T^{\delta_*}} \,\Big|\, \mathcal{A}_t\right) = 0. \tag{2.3.44}$$

By (2.3.42) and $\hat{S}_T^j = \frac{S_T^j}{S_T^{\delta_*}}$, we obtain

$$F_T^j(t) = \frac{S_t^{\delta_*} E(\hat{S}_T^j \mid \mathcal{A}_t)}{S_t^{\delta_*} E(\frac{1}{S_T^{\delta_*}} \mid \mathcal{A}_t)} = \frac{S_t^j}{P_T(t)} \frac{1}{\hat{S}_t^j} E\big(\hat{S}_T^j \mid \mathcal{A}_t\big) \tag{2.3.45}$$

for a given $t \in [0, T]$. Again, as the benchmarked primary security accounts are $(\underline{\mathcal{A}}, P)$-martingales under classical models as the BSM, it follows using (2.3.43) that

$$F_T^j(t) = S_t^j \exp\{r(T-t)\} \tag{2.3.46}$$

for all $t \in [0, T]$. This is then also the standard risk neutral formula for the forward price, see for instance Musiela and Rutkowski (2005).

2.3.4 Asset-or-Nothing Binaries

Binary options can be considered to be building blocks for several more complex derivatives. This is useful to know when it comes to the valuation and hedging of various exotic options, see e.g. Ingersoll (2000), Buchen (2004), and Baldeaux and Rutkowski (2010).

The derivative contract we consider in this subsection is an *asset-or-nothing binary* on a market index, which we interpret here as the GOP. At its maturity date T, this derivative pays its holder one unit of the market index if its value is greater than the strike K, and nothing otherwise. Using the real world pricing formula (1.3.19) and (2.3.40), we obtain under the BSM

$$\begin{aligned} A_{T,K}(t) &= S_t^{\delta_*} E\left(\mathbf{1}_{\{S_T^{\delta_*} \geq K\}} \frac{S_T^{\delta_*}}{S_T^{\delta_*}} \,\middle|\, \mathcal{A}_t\right) \\ &= S_t^{\delta_*} P\big(S_T^{\delta_*} \geq K \mid \mathcal{A}_t\big) \\ &= S_t^{\delta_*} P\left(S_t^{\delta_*} \exp\left\{\left(r + \frac{1}{2}|\sigma^0|^2\right)(T-t) + |\sigma^0|\big(\hat{W}_T^0 - \hat{W}_t^0\big)\right\} \geq K \,\middle|\, \mathcal{A}_t\right) \\ &= S_t^{\delta_*} N(d_1) \end{aligned} \tag{2.3.47}$$

for all $t \in [0, T]$, where

$$d_1 = \frac{\ln(\frac{S_t^{\delta_*}}{K}) + (r + \frac{1}{2}|\sigma^0|^2)(T-t)}{|\sigma^0|\sqrt{T-t}} \tag{2.3.48}$$

and $N(\cdot)$ denotes the Gaussian distribution function.

2.3.5 Bond-or-Nothing Binaries

In this subsection, we consider pricing a *bond-or-nothing binary*, which pays the strike $K \in \Re^+$ at maturity T in the event that the market index at time T is not less than K. As before, the market index is interpreted as the GOP.

$$\begin{aligned}
B_{T,K}(t) &= S_t^{\delta_*} E\left(\mathbf{1}_{\{S_T^{\delta_*} \geq K\}} \frac{K}{S_T^{\delta_*}} \,\Big|\, \mathcal{A}_t\right) \\
&= S_t^{\delta_*} E\left(\mathbf{1}_{\{S_T^{\delta_*} \geq K\}} K \frac{\hat{S}_T^0}{\hat{S}_0^0} \frac{\hat{S}_0^0}{S_T^0} \,\Big|\, \mathcal{A}_t\right) \\
&= S_t^{\delta_*} \frac{\hat{S}_0^0}{S_T^0} E\big(\mathbf{1}_{\{S_T^{\delta_*} \geq K\}} K \Lambda_{|\sigma^0|}(T) \,\big|\, \mathcal{A}_t\big).
\end{aligned} \tag{2.3.49}$$

Under the BSM, making use of Girsanov's theorem and the Bayes rule facilitates pricing, in particular, we recall that the benchmarked savings account is an $(\underline{\mathcal{A}}, P)$-martingale and one has the Radon-Nikodym derivative process $\Lambda_{|\sigma^0|} = \{\Lambda_{|\sigma^0|}(t),\ t \in [0,T]\}$, where

$$\Lambda_{|\sigma^0|}(t) = \frac{\hat{S}_t^0}{\hat{S}_0^0} = \exp\left\{-\frac{t}{2}|\sigma^0|^2 - |\sigma^0|\hat{W}_t^0\right\}. \tag{2.3.50}$$

This process is used to define a measure $P_{|\sigma^0|}$ via

$$\frac{dP_{|\sigma^0|}}{dP} = \Lambda_{|\sigma^0|}(T), \tag{2.3.51}$$

by setting

$$P_{|\sigma^0|}(A) = E\big(\Lambda_{|\sigma^0|}(T)\mathbf{1}_A\big) = E_{|\sigma^0|}(\mathbf{1}_A) \tag{2.3.52}$$

for $A \in \mathcal{A}_T$. We use $E_{|\sigma^0|}$ to denote the expectation with respect to $P_{|\sigma^0|}$. By Girsanov's theorem, $W^{|\sigma^0|} = \{W_t^{|\sigma^0|},\ t \in [0,T]\}$, where

$$W_t^{|\sigma^0|} = \hat{W}_t^0 + |\sigma^0| t \tag{2.3.53}$$

is a standard Brownian motion on the filtered probability space $(\Omega, \mathcal{A}, \underline{\mathcal{A}}, P_{|\sigma^0|})$. This yields for (2.3.49) the relations

$$\begin{aligned}
B_{T,K}(t) &= S_t^{\delta_*} \frac{\hat{S}_0^0}{S_T^0} K P_{|\sigma^0|}\big(S_T^{\delta_*} \geq K \,\big|\, \mathcal{A}_t\big) E\big(\Lambda_{|\sigma^0|}(T) \,\big|\, \mathcal{A}_t\big) \\
&= S_t^{\delta_*} \frac{\hat{S}_0^0}{S_T^0} K P_{|\sigma^0|}\big(S_T^{\delta_*} \geq K \,\big|\, \mathcal{A}_t\big) \frac{\hat{S}_t^0}{\hat{S}_0^0} \\
&= K \exp\{-r(T-t)\} \\
&\quad \times P_{|\sigma^0|}\left(W_{T-t}^{|\sigma^0|} \geq \frac{\ln(\frac{K}{S_t^{\delta_*}}) + (r - \frac{1}{2}|\sigma^0|^2)(T-t)}{|\sigma^0|} \,\Big|\, \mathcal{A}_t\right) \\
&= K \exp\{-r(T-t)\} N(d_2)
\end{aligned}$$

for $t \in [0, T]$, where

$$d_2 = \frac{\ln(\frac{S_t^{\delta_*}}{K}) + (r - \frac{1}{2}|\sigma^0|^2)(T-t)}{|\sigma^0|\sqrt{T-t}} \tag{2.3.54}$$

and $N(\cdot)$ is again the Gaussian distribution function.

2.3.6 European Options

We now focus on pricing a European call option with maturity $T \in [0, \infty)$ and strike $K \in \Re^+$ on a market index, which is again interpreted as the GOP. Invoking the real world pricing formula (1.3.19), and recalling the previously obtained binaries, we obtain the price of the European call option

$$\begin{aligned} c_{T,K}(t) &= S_t^{\delta_*} E\left(\frac{(S_T^{\delta_*} - K)^+}{S_T^{\delta_*}} \,\bigg|\, \mathcal{A}_t\right) \\ &= S_t^{\delta_*} E\left(\mathbf{1}_{\{S_T^{\delta_*} \geq K\}} \frac{S_T^{\delta_*} - K}{S_T^{\delta_*}} \,\bigg|\, \mathcal{A}_t\right) \\ &= A_{T,K}(t) - B_{T,K}(t) \end{aligned} \tag{2.3.55}$$

for all $t \in [0, T]$. Combining (2.3.47) and (2.3.54) gives

$$c_{T,K}(t) = S_t^{\delta_*} N(d_1) - K \exp\{-r(T-t)\} N(d_2) \tag{2.3.56}$$

for all $t \in [0, T]$, where d_1 and d_2 are given by (2.3.48) and (2.3.54), respectively.

The above explicit formula corresponds to the original pricing formula for a European call on a stock under the BSM, as given in Black and Scholes (1973). Similarly, the price of a European put option is given by

$$p_{T,K}(t) = K \exp\{-r(T-t)\} N(-d_2) - S_t^{\delta_*} N(-d_1),$$

for all $t \in [0, T]$.

2.3.7 Rebates

In this subsection, we consider the valuation of a *rebate* written on a market index, which is again interpreted as the GOP. This claim pays one unit of the domestic currency as soon as the index hits a certain level, assuming this occurs before a contracted expiry date $T > 0$. Following Hulley and Platen (2008), the trigger level for the rebate is a deterministic barrier $Z_t := z \exp\{rt\}$, for some $z > 0$. We mention the fact that the deterministic barrier grows at the risk-free rate, which is economically attractive, as it makes the price of the rebate dependent on the performance of the index relative to that of the savings account. We make use of the following stopping times

$$\sigma_{z,t} := \inf\{u > 0\colon\ S^{\delta_*}_{t+u} = Z_{t+u}\} \tag{2.3.57}$$

and

$$\tau_z := \inf\{t > 0\colon\ Y_t = z\}, \tag{2.3.58}$$

where $Y = \{Y_t, t \geq 0\}$ satisfies

$$Y_t = x \exp\left\{\frac{1}{2}\left|\sigma^0\right|^2 t + \left|\sigma^0\right| \hat{W}^0_t\right\} \tag{2.3.59}$$

and $x := \exp\{-rt\} S^{\delta_*}_t$. Furthermore, we introduce the auxiliary process $X = \{X_t,\ t \geq 0\}$, where

$$X_t = \nu t + \hat{W}^0_t, \tag{2.3.60}$$

and $\nu = \frac{1}{2}|\sigma^0|$. This means X is a Brownian motion with drift. Additionally, we define

$$T_a := \inf\{t > 0\colon\ X_t = a\}. \tag{2.3.61}$$

It is easily seen that we have the following equality in distribution

$$\sigma_{z,t} \stackrel{d}{=} \tau_z \stackrel{d}{=} T_{\tilde{z}} \tag{2.3.62}$$

under P, where $\tilde{z} := \ln(\frac{z}{x}) \frac{1}{|\sigma^0|}$.

First, we consider the valuation of a *perpetual rebate*, for which $T = \infty$. It follows by applying real world pricing that

$$\begin{aligned} R_{\infty,z}(t) &= S^{\delta_*}_t E\left(\frac{1}{S^{\delta_*}_{t+\sigma_{z,t}}} \,\middle|\, \mathcal{A}_t\right) \\ &= \frac{S^{\delta_*}_t}{Z_t} E\big(\exp\{-r\sigma_{z,t}\} \,\big|\, \mathcal{A}_t\big) \\ &= \frac{S^{\delta_*}_t}{Z_t} E\big(\exp\{-r\tau_z\} \,\big|\, \mathcal{A}_t\big). \end{aligned}$$

Making use of the known moment generating function of $T_{\tilde{z}}$, see Proposition 2.1.8, we get

$$E\big(\exp\{-rT_{\tilde{z}}\} \,\big|\, \mathcal{A}_t\big) = \left(\frac{z}{x}\right)^{1/2} \exp\left\{-\left|\ln\left(\frac{z}{x}\right)\right| \frac{\sqrt{2r + (\frac{|\sigma^0|}{2})^2}}{|\sigma^0|}\right\} \tag{2.3.63}$$

and hence

$$R_{\infty,z}(t) = \left(\frac{S^{\delta_*}_t}{Z_t}\right)^{\frac{1}{2}} \exp\left\{-\left|\ln(Z_t) - \ln\big(S^{\delta_*}_t\big)\right| \frac{\sqrt{2r + (\frac{|\sigma^0|}{2})^2}}{|\sigma^0|}\right\}. \tag{2.3.64}$$

Now we turn our attention to the rebate with finite maturity $T < \infty$. Using the real world pricing formula (1.3.19) we obtain

$$
\begin{aligned}
R_{T,z}(t) &= S_t^{\delta_*} E\left(\frac{\mathbf{1}_{t+\sigma_{z,t} \le T}}{S_{t+\sigma_{z,t}}^{\delta_*}} \,\Big|\, \mathcal{A}_t \right) \\
&= \frac{S_t^{\delta_*}}{Z_t} E\big(\mathbf{1}_{\sigma_{z,t} \le T-t} \exp\{-r\sigma_{z,t}\} \,\big|\, \mathcal{A}_t \big) \\
&= \frac{S_t^{\delta_*}}{Z_t} E\big(\mathbf{1}_{T_{\tilde{z}} \le T-t} \exp\{-rT_{\tilde{z}}\} \,\big|\, \mathcal{A}_t \big) \\
&= \frac{S_t^{\delta_*}}{Z_t} \int_0^{T-t} \exp\{-ru\} \frac{|\tilde{z}|}{\sqrt{2\pi} u^{3/2}} \exp\left\{ -\frac{(\tilde{z} - \nu u)^2}{2u} \right\} du,
\end{aligned}
$$

where the last equality employs the distribution of $T_{\tilde{z}}$. Using the change of variables $l := u^{-1/2}$, we obtain

$$
\begin{aligned}
&\int_0^{T-t} \frac{\exp\{-ru - \frac{(\tilde{z}-\nu u)^2}{2u}\}}{u^{3/2}} du \\
&= 2 \int_{(T-t)^{-1/2}}^{\infty} \exp\left\{ -\frac{(\tilde{z})^2}{2} l^2 - \left(r + \frac{\nu^2}{2} \right) l^{-2} \right\} dl \\
&= 2 \frac{\exp\{-2\sqrt{bc}\} \sqrt{\pi} (\mathit{erfc}(a\sqrt{b} - \frac{\sqrt{c}}{a}) + \exp\{4\sqrt{bc}\} \mathit{erfc}(a\sqrt{b} + \frac{\sqrt{c}}{a}))}{4\sqrt{b}},
\end{aligned}
\tag{2.3.65}
$$

where $a := (T-t)^{-1/2}$, $b := \frac{(\tilde{z})^2}{2}$ and $c := (r + \frac{\nu^2}{2})$ and c is assumed to be positive. Furthermore, *erfc* denotes the complement of the error function *erf*, i.e. $\mathit{erfc}(z) = 1 - \mathit{erf}(z)$, where $\mathit{erf}(z) = \frac{2}{\pi} \int_0^z \exp\{-t^2\} dt$. Finally, we remark that (2.3.65) can also be easily confirmed using *Mathematica*.

2.3.8 Barrier Options

In this subsection we consider a *barrier option* on a market index, the GOP or NP. As in the previous subsection, the payoff of this contingent claim is determined by whether or not the index hits a certain level prior to its maturity $T > 0$. In particular, we consider a European call with strike price $K > 0$, that is knocked out if the index breaches the same deterministic barrier Z as in the previous subsection, sometime before expiry.

Using the real world pricing formula and the notation introduced in the previous subsection, we obtain the following price for this claim

$$
\begin{aligned}
C_{T,K,z}^{uo}(t) &= S_t^{\delta_*} E\left(\mathbf{1}_{t+\sigma_{z,t} > T} \frac{(S_T^{\delta_*} - K)^+}{S_T^{\delta_*}} \,\Big|\, \mathcal{A}_t \right) \\
&= S_t^{\delta_*} E\left(\mathbf{1}_{\sigma_{z,t} > T-t} \left(1 - \frac{K}{S_T^{\delta_*}} \right)^+ \,\Big|\, \mathcal{A}_t \right)
\end{aligned}
$$

$$= S_t^{\delta_*} E\left(\mathbf{1}_{\tau_z > T-t}\left(1 - \frac{K\exp\{-rT\}}{Y_{T-t}}\right)^+ \Bigg| \mathcal{A}_t\right)$$

$$= S_t^{\delta_*} E\left(\mathbf{1}_{T_{\tilde{z}} > T-t}\left(1 - \frac{k}{x\exp\{\sigma X_{T-t}\}}\right)^+ \Bigg| \mathcal{A}_t\right) \qquad (2.3.66)$$

where, as in the previous subsection, $\nu := \frac{1}{2}|\sigma^0|$, $x := S_t^{\delta_*}\exp\{-rt\}$, $k := K \times \exp\{-rT\}$, $\tilde{z} := \ln(\frac{z}{x})\frac{1}{|\sigma^0|}$, $\sigma := |\sigma^0|$ and $X = \{X_t,\, t \geq 0\}$, where X_t is given by (2.3.60) and T_a denotes the first time the process X hits the level a. As X is a Brownian motion with drift and T_a denotes the associated first hitting time of the level a, we can apply Corollary 2.1.10 to obtain the price of the above Barrier option. We remark that an alternative derivation of this formula, based on Girsanov's theorem and the Bayes' rule, is presented in Musiela and Rutkowski (2005).

Following Hulley and Platen (2008), we find it convenient to distinguish the following two cases: firstly $S_t^{\delta_*} \leq Z_t \Leftrightarrow x \leq z$, in which case we deal with an *up-and-out call* and $S_t^{\delta_*} \geq Z_t \Leftrightarrow x \geq z$, in which case we deal with a *down-and-out call*. Regarding the up-and-out call, we remark that the Brownian motion with drift X started at 0 killed at $\tilde{z}$ lives on the domain $(-\infty, \tilde{z})$. Finally, setting $a := \ln(\frac{k}{x})\frac{1}{|\sigma^0|}$, we obtain from (1.3.19) the following pricing formula for an up-and-out call option:

$$C^{uo}_{T,K,z}(t) = S_t^{\delta_*}\int_a^{\tilde{z}}\left(1 - \frac{k}{x\exp\{\sigma y\}}\right)\tilde{q}_{\tilde{z}}(T-t, 0, y)m(y)\,dy$$

$$= S_t^{\delta_*}\int_a^{\tilde{z}}\left(1 - \frac{k}{x\exp\{\sigma y\}}\right)\frac{1}{\sqrt{2\pi(T-t)}}\exp\left\{\nu y - \frac{\nu^2(T-t)}{2}\right\}$$

$$\times\left(\exp\left\{-\frac{y^2}{2(T-t)}\right\} - \exp\left\{-\frac{(y-2\tilde{z})^2}{2(T-t)}\right\}\right)dy \qquad (2.3.67)$$

using the fact that the speed measure of a Brownian motion with drift is given by $m(y) = 2\exp\{2\nu y\}$, see Borodin and Salminen (2002). Consequently, to compute the price of a barrier option, we need to compute four integrals: firstly, we calculate

$$I_1 = S_t^{\delta_*}\int_a^{\tilde{z}}\frac{1}{\sqrt{2\pi(T-t)}}\exp\left\{\nu y - \frac{\nu^2(T-t)}{2} - \frac{y^2}{2(T-t)}\right\}dy$$

$$= S_t^{\delta_*}\left(N\left(\frac{\tilde{z} - \nu(T-t)}{\sqrt{T-t}}\right) - N\left(\frac{a - \nu(T-t)}{\sqrt{T-t}}\right)\right). \qquad (2.3.68)$$

The second integral is given by

$$I_2 = -S_t^{\delta_*}\int_a^{\tilde{z}}\frac{1}{\sqrt{2\pi(T-t)}}\exp\left\{\nu y - \frac{\nu^2(T-t)}{2} - \frac{(y-2\tilde{z})^2}{2(T-t)}\right\}dy$$

$$= -S_t^{\delta_*}\exp\{2\tilde{z}\nu\}\left(N\left(\frac{-\tilde{z} - \nu(T-t)}{\sqrt{T-t}}\right) - N\left(\frac{a - 2\tilde{z} - \nu(T-t)}{\sqrt{T-t}}\right)\right)$$

$$= -Z_t\left(N\left(\frac{-\tilde{z} - \nu(T-t)}{\sqrt{T-t}}\right) - N\left(\frac{a - 2\tilde{z} - \nu(T-t)}{\sqrt{T-t}}\right)\right), \qquad (2.3.69)$$

which is obtained by completing the square. Next the third integral can be computed as follows:

$$
\begin{aligned}
I_3 &= -S_t^{\delta_*}\frac{k}{x}\int_a^{\tilde{z}} \frac{\exp\{y(\nu-\sigma)-\frac{\nu^2(T-t)}{2}-\frac{y^2}{2(T-t)}\}}{\sqrt{2\pi(T-t)}}\,dy \\
&= -S_t^{\delta_*}\frac{k}{x}\exp\left\{\frac{\sigma^2(T-t)}{2}-\nu\sigma(T-t)\right\}\left(N\left(\frac{\tilde{z}-(\nu-\sigma)(T-t)}{\sqrt{T-t}}\right)\right. \\
&\quad \left.-N\left(\frac{a-(\nu-\sigma)(T-t)}{\sqrt{T-t}}\right)\right) \\
&= -K\exp\{-r(T-t)\}\left(N\left(\frac{\tilde{z}-(\nu-\sigma)(T-t)}{\sqrt{T-t}}\right)\right. && (2.3.70)\\
&\quad \left.-N\left(\frac{a-(\nu-\sigma)(T-t)}{\sqrt{T-t}}\right)\right), && (2.3.71)
\end{aligned}
$$

where we also completed the square. Finally, the fourth integral is given by

$$
\begin{aligned}
I_4 &= S_t^{\delta_*}\frac{k}{x}\int_a^{\tilde{z}} \frac{1}{\sqrt{2\pi(T-t)}}\exp\left(y(\nu-\sigma)-\frac{\nu^2(T-t)}{2}-\frac{(y-2\tilde{z})^2}{2(T-t)}\right)dy \\
&= S_t^{\delta_*}\frac{k}{x}\exp\left(2\tilde{z}(\nu-\sigma)-\nu\sigma(T-t)+\frac{\sigma^2(T-t)}{2}\right) \\
&\quad \times\left(N\left(\frac{-\tilde{z}-(\nu-\sigma)(T-t)}{\sqrt{T-t}}\right)-N\left(\frac{a-2\tilde{z}-(\nu-\sigma)(T-t)}{\sqrt{T-t}}\right)\right) \\
&= S_t^{\delta_*}\frac{k}{z}\left(N\left(\frac{-\tilde{z}-(\nu-\sigma)(T-t)}{\sqrt{T-t}}\right)-N\left(\frac{a-2\tilde{z}-(\nu-\sigma)(T-t)}{\sqrt{T-t}}\right)\right).
\end{aligned}
\tag{2.3.72}
$$

Obviously, the price of the up-and-out call is given by the sum of the four terms in (2.3.68), (2.3.69), (2.3.71), and (2.3.72). In summary, this yields the explicit formula

$$C^{uo}_{T,K,z}\left(t,S_t^{\delta_*}\right)=I_1+I_2+I_3+I_4. \tag{2.3.73}$$

We now turn our attention to the down-and-out call, i.e. the case $S_t \geq Z_t \Leftrightarrow x \geq z$. As for the up-and-out call, we remark that the Brownian motion with drift started at 0 killed at $\tilde{z}$ lives on the domain $(\tilde{z},\infty)$. Recalling that $a=\ln(\frac{k}{x})\frac{1}{|\sigma^0|}$, we obtain the following pricing formula for a down-and-out call option from (1.3.19),

$$
\begin{aligned}
C^{do}_{T,K,z}(t) &= S_t^{\delta_*}\int_{\tilde{z}\vee a}^{\infty}\left(1-\frac{k}{x\exp\{\sigma y\}}\right)\tilde{q}_{\tilde{z}}(T-t,0,y)m(y)\,dy \\
&= S_t^{\delta_*}\int_{\tilde{z}\vee a}^{\infty}\left(1-\frac{k}{x\exp\{\sigma y\}}\right)\frac{1}{\sqrt{2\pi(T-t)}}\exp\left\{\nu y-\frac{\nu^2(T-t)}{2}\right\} \\
&\quad \times\left(\exp\left\{-\frac{y^2}{2(T-t)}\right\}-\exp\left\{-\frac{(y-2\tilde{z})^2}{2(T-t)}\right\}\right)dy.
\end{aligned}
\tag{2.3.74}
$$

As for the up-and-out call, the pricing of the down-and-out call entails the computation of the following four integrals

$$\bar{I}_1 = S_t^{\delta_*} \int_{\tilde{z} \vee a}^{\infty} \frac{\exp\{\nu y - \frac{\nu^2(T-t)}{2} - \frac{y^2}{2(T-t)}\}}{\sqrt{2\pi(T-t)}} dy$$
$$= S_t^{\delta_*} N\left(-\frac{(\tilde{z} \vee a) + \nu(T-t)}{\sqrt{T-t}}\right). \tag{2.3.75}$$

The second integral is given by

$$\bar{I}_2 = -S_t^{\delta_*} \int_{\tilde{z} \vee a}^{\infty} \exp\left\{\nu y - \frac{\nu^2(T-t)}{2} - \frac{(y-2\tilde{z})^2}{2(T-t)}\right\} dy$$
$$= -S_t^{\delta_*} \exp\{2\tilde{z}\nu\} N\left(-\frac{(\tilde{z} \vee a) + 2\tilde{z} + \nu(T-t)}{\sqrt{T-t}}\right)$$
$$= -Z_t N\left(-\frac{(\tilde{z} \vee a) + 2\tilde{z} + \nu(T-t)}{\sqrt{T-t}}\right). \tag{2.3.76}$$

The third integral is given by

$$\bar{I}_3 = -S_t^{\delta_*} \frac{k}{x} \int_{\tilde{z} \vee a}^{\infty} \exp\left\{\nu y - \sigma y - \frac{\nu^2(T-t)}{2} - \frac{y^2}{2(T-t)}\right\} dy$$
$$= -S_t^{\delta_*} \frac{k}{x} \exp\left\{\frac{\sigma^2(T-t)}{2} - \nu\sigma(T-t)\right\} N\left(-\frac{(\tilde{z} \vee a) + (\nu-\sigma)(T-t)}{\sqrt{T-t}}\right)$$
$$= -K \exp\{-r(T-t)\} N\left(\frac{-(\tilde{z} \vee a) + (\nu-\sigma)(T-t)}{\sqrt{T-t}}\right), \tag{2.3.77}$$

and the last integral is given by

$$\bar{I}_4 = S_t^{\delta_*} \frac{k}{x} \exp\left\{2\tilde{z}(\nu-\sigma) - \nu\sigma(T-t) + \frac{\sigma^2(T-t)}{2}\right\}$$
$$\times \int_{\tilde{z} \vee a}^{\infty} \frac{\exp\{-\frac{(y-(2\tilde{z}+(\nu-\sigma)(T-t)))^2}{2(T-t)}\}}{\sqrt{2\pi(T-t)}} dy$$
$$= S_t^{\delta_*} \frac{k}{x} \exp\left\{2\tilde{z}(\nu-\sigma) - \nu\sigma(T-t) + \frac{\sigma^2(T-t)}{2}\right\}$$
$$\times N\left(\frac{-(\tilde{z} \vee a) + 2\tilde{z} + (\nu-\sigma)(T-t)}{\sqrt{T-t}}\right)$$
$$= S_t^{\delta_*} \frac{k}{z} N\left(\frac{-(\tilde{z} \vee a) + 2\tilde{z} + (\nu-\sigma)(T-t)}{\sqrt{T-t}}\right). \tag{2.3.78}$$

It follows for the down-and-out call option the formula

$$c_{T,K,z}^{do}(t, S_t^{\delta_*}) = \bar{I}_1 + \bar{I}_2 + \bar{I}_3 + \bar{I}_4.$$

By the same methodology one obtains also other barrier options.

2.3.9 Lookback Options

In this subsection, we consider the valuation of a *lookback option* written on a market index, which is again interpreted as the GOP. A standard lookback call option pays

$$\left(S_T^{\delta_*} - m_T^{S^{\delta_*}}\right)^+ = S_T^{\delta_*} - m_T^{S^{\delta_*}},$$

where $m_T^{S^{\delta_*}} = \min_{t\in[0,T]} S_t^{\delta_*}$. We remark that lookback options are always exercised. In Musiela and Rutkowski (2005), the price of a lookback option is derived via a measure change. In this subsection, we proceed by directly integrating the relevant probability density function derived in Sect. 2.1. For ease of presentation, we consider the pricing of a call option at time $t = 0$, but consequently present the formulas for the general case. The real world pricing formula (1.3.19) gives the following price $LC(0)$ for a lookback call option

$$LC(0) = S_0^{\delta_*} E\left(\frac{(S_T^{\delta_*} - m_T^{S^{\delta_*}})^+}{S_T^{\delta_*}}\right) = S_0^{\delta_*} - S_0^{\delta_*} E\left(\frac{m_T^{S^{\delta_*}}}{S_T^{\delta_*}}\right).$$

From (2.3.40),

$$m_T^{S^{\delta_*}} = S_0^{\delta_*} \min_{t\in[0,T]} \exp\left\{\left(r + \frac{\sigma^2}{2}\right)t + \sigma \hat{W}_t^0\right\},$$

where $\sigma := |\sigma^0|$ and hence

$$\frac{m_T^{S^{\delta_*}}}{S_T^{\delta_*}} = \exp\left\{\min_{t\in[0,T]}\left(\left(r + \frac{\sigma^2}{2}\right)(t - T) + \sigma\left(\hat{W}_t^0 - \hat{W}_T^0\right)\right)\right\}.$$

From the time reversibility of Brownian motion, see Chap. 15, we have the following equality in distribution,

$$\min_{t\in[0,T]}\left(\left(r + \frac{\sigma^2}{2}\right)(t - T) + \sigma\left(\hat{W}_t^0 - \hat{W}_T^0\right)\right) \overset{d}{=} \min_{\tau\in[0,T]}\left(-\left(r + \frac{\sigma^2}{2}\right)\tau + \sigma \hat{W}_\tau^0\right).$$

We use the notation

$$X_t = \nu t + \hat{W}_t^0,$$

where $\nu = -\frac{(r+\frac{\sigma^2}{2})}{\sigma}$, and recall from Sect. 2.1, that the probability density of $\min_{t\in[0,T]} X_t$ satisfies

$$P\left(m_T^X \in dy\right) = \left(\phi\left(\frac{-y + \nu T}{\sqrt{T}}\right)\frac{1}{\sqrt{T}} + 2\nu \exp\{2\nu y\} N\left(\frac{y + \nu T}{\sqrt{T}}\right) + \exp\{2\nu y\}\frac{1}{\sqrt{T}}\phi\left(\frac{y + \nu T}{\sqrt{T}}\right)\right) dy.$$

Hence

$$
\begin{aligned}
E\left(\frac{m_T^{S^{\delta_*}}}{S_T^{\delta_*}}\right) &= E\left(\exp\{\sigma m_T^X\}\right) \\
&= \int_{-\infty}^{0} \exp\{\sigma y\}\phi\left(\frac{-y+\nu T}{\sqrt{T}}\right)\frac{1}{\sqrt{T}}\,dy \\
&\quad + \int_{-\infty}^{0} \exp\{\sigma y\}2\nu\exp\{2\nu y\}N\left(\frac{y+\nu T}{\sqrt{T}}\right)dy \\
&\quad + \int_{-\infty}^{0} \exp\{\sigma y\}\exp\{2\nu y\}\frac{1}{\sqrt{T}}\phi\left(\frac{y+\nu T}{\sqrt{T}}\right)dy.
\end{aligned}
$$

We now compute these three integrals

$$
\begin{aligned}
I_1 &= \int_{-\infty}^{0} \exp\{\sigma y\}\phi\left(\frac{-y+\nu T}{\sqrt{T}}\right)\frac{1}{\sqrt{T}}\,dy \\
&= \int_{-\infty}^{0} \exp\{\sigma y\}\frac{\exp\{-\frac{1}{2}(\frac{y-\nu T}{\sqrt{T}})^2\}}{\sqrt{2\pi T}}\,dy \\
&= \exp\left\{\frac{T}{2}\left(\sigma^2+2\sigma\nu\right)\right\}\int_{-\infty}^{-(\nu+\sigma)\sqrt{T}} \frac{\exp\{-\frac{z^2}{2}\}}{\sqrt{2\pi}}\,dz \\
&= \exp\{-rT\}N\left(-(\nu+\sigma)\sqrt{T}\right) \\
&= \exp\{-rT\}N(d-\sigma\sqrt{T}),
\end{aligned}
$$

where

$$
d = \frac{(r+\frac{1}{2}\sigma^2)\sqrt{T}}{\sigma}. \tag{2.3.79}
$$

Regarding the second integral, we introduce

$$
I_2 = \int_{-\infty}^{0} \exp\{\sigma y\}2\nu\exp\{2\nu y\}N\left(\frac{y+\nu T}{\sqrt{T}}\right)dy.
$$

Using integration by parts, we obtain

$$
\begin{aligned}
&\int_{-\infty}^{0} \exp\{\sigma y\}2\nu\exp\{2\nu y\}N\left(\frac{y+\nu T}{\sqrt{T}}\right)dy \\
&= 2\nu\left(\frac{N(\nu\sqrt{T})}{2\nu+\sigma} - \int_{-\infty}^{0} \frac{\exp\{y(2\nu+\sigma)\}}{(2\nu+\sigma)}\frac{\exp\{-\frac{1}{2}\frac{(y+\nu T)^2}{T}\}}{\sqrt{2\pi T}}\,dy\right) \\
&= \frac{2\nu}{2\nu+\sigma}\left(N(\nu\sqrt{T}) - \exp\{-rT\}N\left(-(\nu+\sigma)\sqrt{T}\right)\right) \\
&= N(-d) + \frac{\sigma^2}{2r}N(-d) - \exp\{-rT\}N(d-\sigma\sqrt{T}) \\
&\quad - \frac{\exp\{-rT\}\sigma^2}{2r}N(d-\sigma\sqrt{T}),
\end{aligned}
$$

where the quantity d is defined in (2.3.79). Finally, regarding the third integral, one has

$$
\begin{aligned}
I_3 &= \int_{-\infty}^{0} \exp\{\sigma y\} \exp\{2\nu y\} \frac{1}{\sqrt{T}} \phi\left(\frac{y+\nu T}{\sqrt{T}}\right) dy \\
&= \int_{-\infty}^{0} \exp\{y(2\nu+\sigma)\} \frac{\exp\{-\frac{1}{2T}(y+\nu T)^2\}}{\sqrt{2\pi T}} dy \\
&= \int_{-\infty}^{0} \frac{\exp\{-\frac{(y-(\nu+\sigma)T)^2}{2T}\}}{\sqrt{2\pi T}} \exp\left\{\frac{T}{2}(2\nu\sigma+\sigma^2)\right\} dy \\
&= \exp\{-rT\} N(d-\sigma\sqrt{T}).
\end{aligned}
$$

Hence, we obtain

$$
\begin{aligned}
&E\big(\exp\{\sigma m_T^X\}\big) \\
&= \exp\{-rT\} N(d-\sigma\sqrt{T}) + N(-d) + \frac{\sigma^2}{2r} N(-d) - \exp\{-rT\} N(d-\sigma\sqrt{T}) \\
&\quad - \exp\{-rT\} \frac{\sigma^2}{2r} N(d-\sigma\sqrt{T}) + \exp\{-rT\} N(d-\sigma\sqrt{T}) \\
&= N(-d) + \frac{\sigma^2}{2r} N(-d) - \exp\{-rT\} \frac{\sigma^2}{2r} N(d-\sigma\sqrt{T}) \\
&\quad + \exp\{-rT\} N(d-\sigma\sqrt{T}) \\
&= 1 - N(d) + \frac{\sigma^2}{2r} N(-d) - \exp\{-rT\} \frac{\sigma^2}{2r} N(d-\sigma\sqrt{T}) \\
&\quad + \exp\{-rT\} N(d-\sigma\sqrt{T}).
\end{aligned}
$$

The time 0 price of a lookback call option is then given by

$$
\begin{aligned}
LC(0) = S_0^{\delta_*} \bigg(& N(d) - \frac{\sigma^2}{2r} N(-d) - \exp\{-rT\} N(d-\sigma\sqrt{T}) \\
& + \exp\{-rT\} \frac{\sigma^2}{2r} N(d-\sigma\sqrt{T}) \bigg).
\end{aligned}
$$

We now recall for a general $t < T$ the following result from Musiela and Rutkowski (2005), see their Proposition 6.7.1.

Proposition 2.3.1 *Assume that $r > 0$. Then the price at time $t < T$ of a European lookback call option equals*

$$
\begin{aligned}
LC(t) = S_t^{\delta_*} & N\left(\frac{\ln(S_t^{\delta_*}/m_t^{S^{\delta_*}}) + r_1(T-t)}{\sigma\sqrt{T-t}}\right) \\
& - m_t^{S^{\delta_*}} N\left(\frac{\ln\left(\frac{S_t^{\delta_*}}{m_t^{S^{\delta_*}}}\right) + r_2(T-t)}{\sigma\sqrt{T-t}}\right)
\end{aligned}
$$

$$
\begin{aligned}
&- \frac{S_t^{\delta_*}\sigma^2}{2r} N\left(\frac{\ln(\frac{m_t^{S^{\delta_*}}}{S_t^{\delta_*}}) - r_1(T-t)}{\sigma\sqrt{T-t}}\right) \\
&+ \exp\{-r(T-t)\} \frac{S_t^{\delta_*}\sigma^2}{2r} \left(\frac{m_t^{S^{\delta_*}}}{S_t^{\delta_*}}\right)^{2r\sigma^{-2}} N\left(\frac{\ln(\frac{m_t^{S^{\delta_*}}}{S_t^{\delta_*}}) + r_2(T-t)}{\sigma\sqrt{T}}\right),
\end{aligned}
$$

where $r_{1,2} = r \pm \frac{1}{2}\sigma^2$.

The payoff of a lookback put option is given by

$$
(M_T^{S^{\delta_*}} - S_T^{\delta_*})^+ = M_T^{S^{\delta_*}} - S_T^{\delta_*},
$$

where $M_T^{\delta_*} = \max_{t\in[0,T]} S_t^{\delta_*}$.

Proposition 2.3.2 *Assume that* $r > 0$. *The price of a European lookback put option at time* $t < T$ *equals*

$$
\begin{aligned}
LP(t) = &- S_t^{\delta_*} N\left(-\frac{\ln(\frac{S_t^{\delta_*}}{M_t^{S^{\delta_*}}}) + r_1(T-t)}{\sigma\sqrt{T-t}}\right) \\
&+ M_t^{S^{\delta_*}} \exp\{-r(T-t)\} N\left(-\frac{\ln(\frac{S_t^{\delta_*}}{M_t^{S^{\delta_*}}}) + r_2(T-t)}{\sigma\sqrt{T-t}}\right) \\
&+ \frac{S_t^{\delta_*}\sigma^2}{2r} N\left(\frac{\ln(\frac{S_t^{\delta_*}}{M_t^{S^{\delta_*}}}) + r_1(T-t)}{\sigma\sqrt{T-t}}\right) \\
&- \exp\{-r(T-t)\} \frac{S_t^{\delta_*}\sigma^2}{2r} \left(\frac{M_t^{S^{\delta_*}}}{S_t^{\delta_*}}\right)^{2r\sigma^{-2}} N\left(\frac{\ln(\frac{S_t^{\delta_*}}{M_t^{S^{\delta_*}}}) - r_2(T-t)}{\sigma\sqrt{T-t}}\right),
\end{aligned}
$$

where again $r_{1,2} = r \pm \frac{1}{2}\sigma^2$.

2.3.10 Asian Options

In this subsection, we consider *Asian options* on a market index, the GOP. Unlike the derivatives presented in the previous subsections, the pay-off of Asian options is based on average values of the market index. In particular, the pay-off of an Asian call option is given by

$$
\left(\frac{1}{T}\int_0^T S_u^{\delta_*}\,du - K\right)^+
$$

and

$$
\left(K - \frac{1}{T}\int_0^T S_u^{\delta_*}\,du\right)^+.
$$

We point out that closed-form solutions, as presented in the preceding subsections, are not available for Asian options. However, using the explicitly derived joint density of $(\int_0^T S_u^{\delta_*}\,du, S_T^{\delta_*})$ from Sect. 2.1, we can obtain an integral representation for the price. In particular, using the notation

$$\begin{aligned}
&P\left(\int_0^t \exp\{\sigma \hat{W}_s^0 - p\sigma^2 s/2\}\,ds \in du,\ \hat{W}_t^0 \in dy\right)\\
&\quad = \frac{\sigma}{2}\exp\{-p\sigma y/2 - p^2\sigma^2 t/8\}\exp\left\{-2\frac{1+\exp\{\sigma y\}}{\sigma^2 u}\right\}\\
&\qquad \times \theta\left(\frac{4\exp\{\sigma y/2\}}{\sigma^2 u}, \frac{\sigma^2 t}{4}\right)\frac{du}{u}\,dy\\
&\quad = f(y,u)\,dy\,du,
\end{aligned}$$

where $p = -(1 + \frac{2r}{|\sigma^0|^2})$ and $\sigma := |\sigma^0|$, we obtain from the real-world pricing formula (1.3.19) the following representation for the price of a call option at time 0, struck at K with maturity T,

$$\begin{aligned}
C_{T,K}^A(0) &= S_0^{\delta_*} E\left(\frac{\left(\frac{\int_0^T S_u^{\delta_*}\,du}{T} - K\right)^+}{S_T^{\delta_*}}\right)\\
&= \frac{S_0^{\delta_*}}{T}\int_0^\infty\int_0^\infty \frac{(u - \frac{TK}{S_0^{\delta_*}})^+}{\exp\{-p\sigma^2 T/2 + \sigma y\}} f(y,u)\,dy\,du. \qquad (2.3.80)
\end{aligned}$$

The above expression needs to be computed numerically, using e.g. the techniques to be presented in Chap. 12. Finally, we alert the reader to a quasi-analytical result shown in Geman and Yor (1993). They computed the Laplace transform with respect to time to maturity. We point out that the proof uses a connection between geometric Brownian motion and time-changed Bessel processes, also referred to as Lampert's Theorem, see Theorem 6.2.4.1 in Jeanblanc et al. (2009). The following result appeared as Proposition 6.8.1 in Musiela and Rutkowski (2005) and is based on Eq. (3.10) in Geman and Yor (1993).

Proposition 2.3.3 *The price of an Asian call option admits the representation*

$$C_{T,K}^A(t) = \frac{4\exp\{-r(T-t)\}S_t^{\delta_*}}{\sigma^2 T} C^w(h,q)$$

where

$$w = \frac{2r}{\sigma^2} - 1, \qquad h = \frac{\sigma^2}{4}(T-t), \qquad q = \frac{\sigma^2}{4S_t^{\delta_*}}\left(KT - \int_0^t S_u^{\delta_*}\,du\right).$$

Moreover, the Laplace transform of $C^w(h,q)$ with respect to h is given by the formula

$$\int_0^\infty \exp\{-\lambda h\}C^w(h,q)\,dh = \int_0^{\frac{1}{2q}} \left(d\exp\{-x\}x^{\gamma-2}(1-2qx)^{\gamma+1}\right)dx,$$

where $\mu = \sqrt{2\lambda + w^2}$, $\gamma = \frac{1}{2}(\mu - w)$, and $d = (\lambda(\lambda - 2 - 2\nu)\Gamma(\gamma - 1))^{-1}$.

We remark that the techniques to be presented in Sect. 13.5 can be used to invert the above Laplace transform.

2.3.11 Australian Options

Australian options are closely related to Asian options. In this case the pay-off depends on the quotient of the average of the market index over a specific time interval and the market index at maturity, i.e. the quotient $\frac{\int_0^T S_u^{\delta_*}\,du}{S_T^{\delta_*}}$, see Handley (2000), Handley (2003), Moreno and Navas (2008), and Ewald et al. (2011). In the BSM framework, a connection between Australian and Asian options is known to exist, see Ewald et al. (2011). The real-world pricing formula (1.3.19) yields the following expression for an Australian call option on the market index:

$$C_{T,K}^{AU}(t) = S_t^{\delta_*} E\left(\left(\frac{\int_0^T S_u^{\delta_*}\,du}{T S_T^{\delta_*}} - K\right)^+ \frac{1}{S_T^{\delta_*}} \,\Big|\, \mathcal{A}_t\right).$$

We now follow Ewald et al. (2011),

$$C_{T,K}^{AU}(t) = S_t^{\delta_*} E\left(\frac{\left(\frac{\int_0^T S_u^{\delta_*}\,du}{T} - K S_T^{\delta_*}\right)^+}{(S_T^{\delta_*})^2} \,\Big|\, \mathcal{A}_t\right).$$

Next we introduce the same auxiliary measure as for the bond-or-nothing binaries, i.e. we recall the Radon-Nikodym derivative process from (2.3.50),

$$\Lambda_{|\sigma^0|}(t) = \frac{\hat{S}_t^0}{\hat{S}_0^0} = \exp\left\{-\frac{t}{2}|\sigma^0|^2 - |\sigma^0|\hat{W}_t^0\right\},$$

and define the measure $P_{|\sigma^0|}$ via

$$\frac{dP_{|\sigma^0|}}{dP} = \Lambda_{|\sigma^0|}(T),$$

by setting

$$P_{|\sigma^0|}(A) = E\big(\Lambda_{|\sigma^0|}(T)\mathbf{1}_A\big) = E_{|\sigma^0|}(\mathbf{1}_A)$$

for $A \in \mathcal{A}_T$. We use $E_{|\sigma^0|}$ to denote the expectation with respect to $P_{|\sigma^0|}$. By Girsanov's theorem, $W^{|\sigma^0|} = \{W_t^{|\sigma^0|},\ t \in [0,T]\}$, where

$$W_t^{|\sigma^0|} = \hat{W}_t^0 + |\sigma^0|t$$

is a standard Brownian motion on the filtered probability space $(\Omega, \mathcal{A}, \underline{\mathcal{A}}, P_{|\sigma^0|})$. Hence

$$C_{T,K}^{AU}(t) = \exp\{-r(T-t)\}E_{|\sigma^0|}\left(\frac{\left(\frac{\int_0^T S_u^{\delta_*}\,du}{T} - K S_T^{\delta_*}\right)^+}{S_T^{\delta_*}} \,\Big|\, \mathcal{A}_t\right). \qquad (2.3.81)$$

We remark that under $P_{|\sigma^0|}$, the dynamics of S^{δ_*} are given by

$$S_t^{\delta_*} = S_0^{\delta_*} \exp\left\{ \left(r - |\sigma^0|^2\right)t + \frac{t}{2}|\sigma^0|^2 + |\sigma^0| W_t^{|\sigma^0|} \right\}.$$

Hence, comparing (2.3.81) to (2.3.80), we point out that computing (2.3.81) amounts to pricing an Asian option with variable strike, but at a different interest rate, namely $r - |\sigma^0|^2$. Clearly, this relation required the candidate measure $P_{|\sigma^0|}$ to be equivalent to P. This does not hold for all models considered in this book, see e.g. Chap. 3. However, assuming suitable integrability conditions are satisfied, we can express the price of an Australian option as an integral over the relevant probability density function and use the techniques from Chap. 12.

2.3.12 Exchange Options

In this subsection, we price *exchange options* on the market index, i.e. the option to exchange the market index denominated in one currency for the market index denominated in another currency. This is our first example of a derivative whose payoff is a functional of two assets. We point out that in the classical literature, see e.g. Margrabe (1978), such contracts are often priced by computing prices under an appropriately chosen probability measure. This is not the case under the benchmark approach, where we only need to compute prices under one measure, the real world probability measure. In particular, we define the time t exchange price as

$$X_t^{i,j} = \frac{S_t^{\delta_*,i}}{S_t^{\delta_*,j}},$$

where $S_t^{\delta_*,i}$ denotes the GOP denominated in currency i, and $S_t^{\delta_*,j}$ denotes the GOP denominated in currency j. We assume that the dynamics of the GOP in currency k are given by

$$dS_t^{\delta_*,k} = S_t^{\delta_*,k}\left(\left(r^k + |\sigma^k|^2\right)dt + |\sigma^k|\, d\hat{W}_t^k\right), \tag{2.3.82}$$

where $k \in \{i, j\}$ and $d[\hat{W}^i, \hat{W}^j]_t = \rho\, dt$. The joint transition density of $S^{\delta_*,i}$ and $S^{\delta_*,j}$ was derived in Sect. 2.2. For deriving the following result, we employ a change of variables. This reduces the computation to one which involves the standard Gaussian bivariate density. As with European call options on the GOP, we find it convenient to firstly price asset binary options on an exchange price, and subsequently bond binary options on an exchange price. We use the notation $A^i_{T,K}(t)$ for an asset binary option on an exchange price in the ith currency, which, based on the real world pricing formula, satisfies

$$A^i_{T,K}(t) = S_t^{\delta_*,i} E\left(\frac{S_T^{\delta_*,i}}{S_T^{\delta_*,j}} \frac{1}{S_T^{\delta_*,i}} \mathbf{1}_{\frac{S_T^{\delta_*,i}}{S_T^{\delta_*,j}} \geq K} \,\Big|\, \mathcal{A}_t \right).$$

Using $q(t, z_t^j, z_t^i, T, z_T^i, z_T^j)$ to denote the joint density of $(S_t^{\delta_*,j}, S_t^{\delta_*,i})$ to $(S_T^{\delta_*,j}, S_T^{\delta_*,i})$, we have

$$\begin{aligned}
A^i_{T,K}(t) &= E\left(\frac{S_t^{\delta_*,i}}{S_T^{\delta_*,i}} X_T^{i,j} \mathbf{1}_{X_T^{i,j} \geq K} \,\middle|\, \mathcal{A}_t\right) \\
&= X_t^{i,j} E\left(\frac{S_t^{\delta_*,j}}{S_T^{\delta_*,j}} \mathbf{1}_{S_T^{\delta_*,j} \leq S_T^{\delta_*,i}/K} \,\middle|\, \mathcal{A}_t\right) \\
&= X_t^{i,j} \int_0^\infty \int_0^{\frac{z_T^i}{K}} \frac{z_t^i}{z_T^i} q\big(t, z_t^j, z_t^i, T, z_T^j, z_T^i\big)\, dz_T^j \, dz_T^i.
\end{aligned}$$

We now use a change of variables to perform computations in terms of the bivariate Gaussian density

$$u_T^k = \frac{\ln(\frac{z_T^k}{z_t^k}) - (r_k + \frac{1}{2}|\sigma^k|^2)(T-t)}{|\sigma^k|\sqrt{T-t}}, \quad k \in \{i, j\}.$$

Hence we obtain

$$A^i_{T,K}(t) = X_t^{i,j} \int_{-\infty}^\infty \int_{-\infty}^{\bar{d}_1(X_t^{i,j})} \frac{z_t^j}{z_T^j} p\big(u_T^j, u_T^i, \rho\big)\, du_T^j \, du_T^i,$$

where

$$\bar{d}_1(x) = \frac{\ln(\frac{x}{K}) - (r_j - r_i + \frac{1}{2}(|\sigma^j|^2 - |\sigma^i|^2))(T-t)}{|\sigma^j|\sqrt{T-t}} + u_T^i \frac{|\sigma^i|}{|\sigma^j|},$$

and

$$p(z_1, z_2, \rho) = \frac{1}{2\pi\sqrt{1-\rho^2}} \exp\left\{-\frac{(z_1^2 - 2\rho z_1 z_2 + z_2^2)}{2(1-\rho^2)}\right\}$$

denotes the density of two correlated standard Gaussian random variables. We now set

$$\begin{aligned}
\bar{u}_T^j &= u_T^j + |\sigma^j|\sqrt{T-t}, \\
\bar{u}_T^i &= u_T^i + \rho|\sigma^j|\sqrt{T-t},
\end{aligned}$$

which allows us to write

$$A^i_{T,K}(t) = X_t^{i,j} \exp\{-r_j(T-t)\} \int_{-\infty}^\infty \int_{-\infty}^{\tilde{d}_1(X_t^{i,j})} p\big(\bar{u}_T^j, \bar{u}_T^i, \rho\big)\, d\bar{u}_T^j \, d\bar{u}_T^i, \quad (2.3.83)$$

where

$$\begin{aligned}
\tilde{d}_1(x) &= \frac{\ln(\frac{x}{K}) + (r_i - r_j + \frac{1}{2}(|\sigma^j|^2 - 2\rho|\sigma^j||\sigma^i| + |\sigma^i|^2))(T-t)}{|\sigma^j|\sqrt{T-t}} + \bar{u}_T^i \frac{|\sigma^i|}{|\sigma^j|} \\
&= \hat{d}\big(X_t^{i,j}\big) + \bar{u}_T^i \frac{|\sigma^i|}{|\sigma^j|}.
\end{aligned}$$

The expression in equation (2.3.83) can be interpreted as the probability that a standard normal random variable Z_1 is less than a constant $\hat{d}_1(X_t^{i,j})$ plus another standard normal random variable Z_2 multiplied by $\frac{|\sigma^i|}{|\sigma^j|}$, i.e.

$$P\left(Z^1 < \hat{d}_1(X_t^{i,j}) + \frac{|\sigma^i|}{|\sigma^j|} Z^2\right).$$

But since $Z^1 - \frac{|\sigma^i|}{|\sigma^j|} Z^2$ is normal with mean zero and variance $\frac{\sigma_{i,j}^2}{|\sigma^j|^2}$, we obtain the following result:

$$A_{T,K}^i(t) = X_t^{i,j} \exp\{-r_j(T-t)\} N\big(d_1(X_t^{i,j})\big),$$

where

$$d_1(X_t^{i,j}) = \frac{\ln(\frac{X_t^{i,j}}{K}) + (r_i - r_j + \frac{1}{2}\sigma_{i,j}^2)(T-t)}{\sigma_{i,j}\sqrt{T-t}},$$

$$\sigma_{i,j}^2 = |\sigma^i|^2 - 2\rho|\sigma^i||\sigma^j| + |\sigma^j|^2.$$

Using similar calculations, we obtain the following result for a binary bond option on the exchange price,

$$B_{T,K}^i(t) = E\left(\frac{S_t^{\delta_*,i}}{S_T^{\delta_*,i}} \mathbf{1}_{X_T^{i,j} > K} \,\Big|\, \mathcal{A}_t\right) = \exp\{-r_i(T-t)\} N\big(d_2(X_t^{i,j})\big),$$

where

$$d_2(X_t^{i,j}) = \frac{\ln(\frac{X_t^{i,j}}{K}) + (r_i - r_j - \frac{1}{2}\sigma_{i,j}^2)(T-t)}{\sigma_{i,j}\sqrt{T-t}}.$$

Finally, we arrive at prices for call and put options in the ith currency on an exchange price at time t with expiry T and strike price K,

$$\begin{aligned} c_{T,K}^i(t) &= X_t^{i,j} \exp\{-r_j(T-t)\} N\big(d_1(X_t^{i,j})\big) \\ &\quad - K \exp\{-r_i(T-t)\} N\big(d_2(X_t^{i,j})\big), \\ p_{T,K}^i(t) &= -X_t^{i,j} \exp\{-r_j(T-t)\} N\big(-d_1(X_t^{i,j})\big) \\ &\quad + K \exp\{-r_i(T-t)\} N\big(-d_2(X_t^{i,j})\big). \end{aligned}$$

2.3.13 American Options

The derivative contracts discussed until now were all *European style options*, i.e. could only be exercised at maturity. We now briefly discuss *American style options*, which allow the holder to exercise the option at any time before maturity. This additional feature makes the pricing of American options more difficult than the pricing of European options. For more information on the mathematics of American option pricing, we refer the reader to McKean (1965), van Moerbeke (1976), Bensoussan (1984), Karatzas (1988), Karatzas (1989), and to Myneni (1992) for a survey. We point out that closed-form solutions similar to the ones derived in the preceding subsections are not available for American options, except for the perpetual case, see for example the discussion in Musiela and Rutkowski (2005). However, we also refer the reader to Zhu (2006).

Consequently, for results that provide almost closed-form solutions numerical methods have to be employed to price American options. A popular method involves restricting the dates at which the option can be exercised to a finite set, i.e. turning the American option into a *Bermudan option*. Using dynamic programming, one can compute prices via backward induction. This in turn can be done via Monte Carlo simulation, see e.g. Broadie and Glasserman (1997). In this context, the transition densities collected in this book are of importance, as they are used to perform the simulation step. Furthermore, the Monte Carlo technique is of course general, one only needs to have access to the relevant transition densities.

There exist more explicit formulas for derivatives under the BSM. It is mainly its explicitly known transition density and the well researched area of functionals of Brownian motions that give access to such a rich set of pricing formulas for the standard market model. It is unfortunate that the BSM provides only a poor reflection of the real market dynamics, in particular, over longer periods of time and for extreme market movements. Therefore, it is essential to find more realistic tractable market models with a similar set of explicit formulas.

Chapter 3
Functionals of Squared Bessel Processes

In this chapter, we discuss important classes of stochastic processes, namely squared Bessel processes and their relatives. These processes turn out to have important applications in financial modeling, as we will demonstrate in Sect. 3.3. Fortunately, these processes are susceptible to Lie symmetry methods, as will be demonstrated in Chaps. 5 and 8, and are also solvable in the sense of Chap. 11. This is important, as it allows us to formulate realistic financial models based on squared Bessel processes and at the same time produce closed-form solutions for important financial derivatives in this class of models.

In Sect. 3.1, we firstly recall results on the squared Bessel process, and subsequently for related processes, in particular:

- the Bessel process;
- the square-root process;
- the $3/2$ process;
- the CEV process.

Most results presented in Sect. 3.1 have appeared in the literature before. However, the aim of this chapter is to present the squared Bessel process as a central object for modeling, which is tractable, and from which one can consequently derive other tractable processes. Furthermore, we point out that several more complex functionals, such as integrals of the processes discussed in this chapter, can be derived via Lie symmetry methods, which will be discussed in Chap. 5.

3.1 One-Dimensional Functionals of Squared Bessel Processes

Let $\delta \in \mathcal{N}$, $x \geq 0$, and introduce δ independent Brownian motions $W^1, W^2, \ldots, W^\delta$, started at $w^1 \in \Re$, $w^2 \in \Re, \ldots, w^\delta \in \Re$, and set

$$x = \sum_{k=1}^{\delta} (w^k)^2.$$

J. Baldeaux, E. Platen, *Functionals of Multidimensional Diffusions with Applications to Finance*, Bocconi & Springer Series 5, DOI 10.1007/978-3-319-00747-2_3,

Now we set

$$X_t = \sum_{k=1}^{\delta} (W_t^k + w^k)^2, \quad t \geq 0. \tag{3.1.1}$$

It follows from the Itô formula, that

$$dX_t = \delta dt + 2 \sum_{k=1}^{\delta} (W_t^k + w^k)\, dW_t^k,$$

where $X_0 = x$. We now set

$$dW_t = \frac{1}{\sqrt{X_t}} \sum_{k=1}^{\delta} (w^k + W_t^K)\, dW_t^k,$$

and find that

$$[W]_t = t,$$

hence W is a Brownian motion, by Lévy's characterization theorem, see Theorem 15.3.3. Consequently, X satisfies the SDE

$$dX_t = \delta dt + 2\sqrt{X_t}\, dW_t,$$

where $X_0 = x$, and we refer to X as a *squared Bessel process* of dimension δ. The reason for the nomenclature is the fact that many quantities pertaining to the squared Bessel process can be expressed in terms of Bessel functions. The above definition can be extended to the case $\delta \geq 0$, and we recall Definition 6.1.2.1 from Jeanblanc et al. (2009):

Definition 3.1.1 For every $\delta \geq 0$ and $x \geq 0$, the unique strong solution to the equation

$$X_t = x + \delta t + 2 \int_0^t \sqrt{X_s}\, dW_s, \quad X_t \geq 0,$$

is called a squared Bessel process with dimension δ started at x.

The dimension δ of the squared Bessel process plays a pivotal role when studying its path properties, in particular its boundary behavior around 0. Table 3.1.1 summarizes the behavior at zero for a squared Bessel process. We refer the reader to Sect. 6.1 in Jeanblanc et al. (2009) for a more detailed discussion, in particular to Proposition 6.1.3.1. In this book, we adopt the convention from Revuz and Yor (1999) and specify the boundary condition at 0 to be reflection. Of course, other specifications such as absorption or killing are also possible, see e.g. Borodin and Salminen (2002). Furthermore, we point out that for $\delta > 2$, X is transient, i.e. X_t goes to infinity as t goes to infinity. For $\delta = 2$, the process never reaches zero, but hits every ball centered around 0 after some time. The squared Bessel process enjoys the following scaling property, see also Sect. 8.7 in Platen and Heath (2010), Proposition 6.1.4.1 in Jeanblanc et al. (2009):

Table 3.1.1 Path properties of the squared Bessel process

Dimension	Property of the squared Bessel process X
$\delta \geq 2$	X does not reach 0
$0 < \delta < 2$	X reaches 0 and reflects
$\delta = 0$	X reaches 0 and is absorbed

Proposition 3.1.2 *If $\{X_t,\ t \geq 0\}$ is a squared Bessel process of dimension δ started at x, then $\{\frac{1}{c} X_{ct},\ t \geq 0\}$ is a squared Bessel process of dimension δ started at $\frac{x}{c}$.*

Also, the sum of independent squared Bessel processes is a squared Bessel process:

Proposition 3.1.3 *If $X = \{X_t,\ t \geq 0\}$ is a squared Bessel process of dimension δ_1 started at x_1 and $Y = \{Y_t,\ t \geq 0\}$ is a squared Bessel process of dimension δ_2 started at x_2, where X and Y are independent, then $Z = \{Z_t,\ t \geq 0\}$, where $Z_t = X_t + Y_t$, is a squared Bessel process of dimension $\delta = \delta_1 + \delta_2$ started at $x = x_1 + x_2$.*

We now show the following relationship in law between squared Bessel processes of different dimensions. We firstly have the following absolute continuity relationship, see Proposition 6.1.5.1 in Jeanblanc et al. (2009). We use P_x^δ to denote the law of a squared Bessel process of dimension δ started at x. In the following, we also use the *index of the squared Bessel process*, given by $\nu = \frac{\delta}{2} - 1$,

$$P_x^\delta \mid_{\mathcal{A}_t} = \left(\frac{X_t}{x}\right)^{\frac{\nu}{2}} \exp\left\{-\frac{\nu^2}{2}\int_0^t \frac{ds}{X_s}\right\} P_x^2 \Big|_{\mathcal{A}_t},$$

for $\delta \geq 2$. Similar relations also exist for processes which can hit zero. In particular, we recall from Borodin and Salminen (2002) or Platen and Heath (2010) for $\delta > 2$ the relation, where $\tau = \inf\{t \geq 0\colon\ X_t = 0\}$,

$$P_x^{4-\delta} \mid_{\mathcal{A}_t \cap \{t < \tau\}} = \left(\frac{x}{X_t}\right)^\nu P_x^\delta \Big|_{\mathcal{A}_t}, \tag{3.1.2}$$

for all $t \in (0, \infty)$. In principle, on the left hand side of the above relationship we consider squared Bessel processes with absorption at zero, and on the right hand side squared Bessel processes that never reach zero. The same relationship (3.1.2) yields for $\delta < 2$ the equation

$$P_x^\delta \mid_{\mathcal{A}_t \cap \{t < \tau\}} = \left(\frac{x}{X_t}\right)^{-\nu} P_x^{4-\delta} \Big|_{\mathcal{A}_t}. \tag{3.1.3}$$

We now turn to the transition density of the squared Bessel process. In particular, we find it useful to recall the following equality in law, for $\nu > -1$ or equivalently $\delta > 0$,

$$\frac{X_t}{t} \stackrel{d}{=} \chi_\delta^2\left(\frac{x}{t}\right),$$

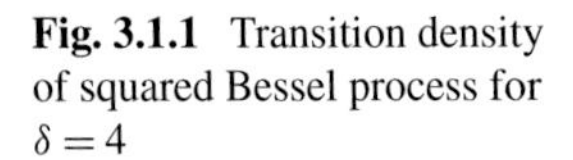

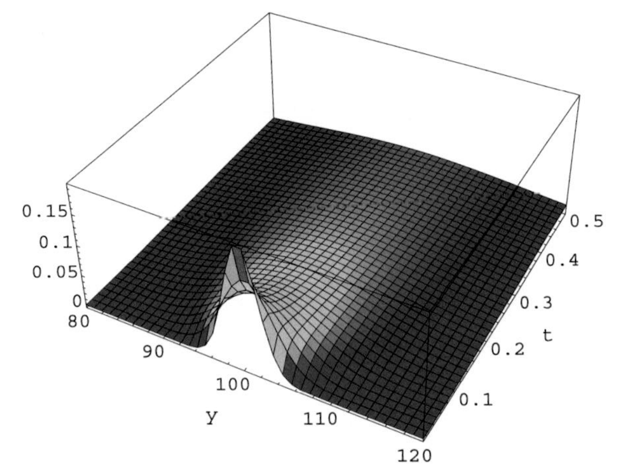

Fig. 3.1.1 Transition density of squared Bessel process for $\delta = 4$

where $\chi^2_\delta(\lambda)$ denotes a non-central χ^2 random variable with δ degrees of freedom and non-centrality parameter λ, see e.g. Platen and Bruti-Liberati (2010). From this equality, we obtain the transition density for a squared Bessel process X starting at the time $s \in [0, \infty)$ in x being at time $t \in (s, \infty)$ in y as

$$p_\delta(s, x; t, y) = \frac{1}{2(t-s)} \left(\frac{y}{x}\right)^{\frac{\nu}{2}} \exp\left\{-\frac{x+y}{2(t-s)}\right\} I_\nu\left(\frac{\sqrt{xy}}{t-s}\right), \tag{3.1.4}$$

see Revuz and Yor (1999), where I_a is the modified Bessel function of the first kind with index a, and we recall that $\nu = \frac{\delta}{2} - 1$. In Fig. 3.1.1 we show the transition density of a squared Bessel process of dimension four, $\delta = 4$, which means index $\nu = 1$, starting at $x = 100$.

For $\delta = 0$, in which case the squared Bessel process is absorbed at 0, one obtains for $x > 0$ and $t \in [0, \infty)$

$$p_0(0, x; t, y) = \frac{1}{2t} \left(\frac{y}{x}\right)^{\frac{\nu}{2}} \exp\left\{-\frac{x+y}{2t}\right\} I_1\left(\frac{\sqrt{xy}}{t}\right). \tag{3.1.5}$$

So far, we have focused on the case $\delta \geq 0$. However, we alert the reader to the fact that squared Bessel processes can also be defined for $\delta < 0$, see Sect. 6.2.6 in Jeanblanc et al. (2009).

We conclude this section with a result on hitting times of squared Bessel processes, where we assume that the process is started at a positive x, see Proposition 6.2.3.1 in Jeanblanc et al. (2009).

Proposition 3.1.4 *Let $\lambda > 0$, $b > 0$, and $T_b := \inf\{t \geq 0\colon X_t = b\}$. Then*

$$E\big(\exp\{-\lambda T_b\}\big) = \left(\frac{b}{x}\right)^{\frac{\nu}{2}} \frac{K_\nu(\sqrt{2\lambda x}\,)}{K_\nu(\sqrt{b2\lambda b}\,)}, \quad b \leq x$$

$$E\big(\exp\{-\lambda T_b\}\big) = \left(\frac{b}{x}\right)^{\frac{\nu}{2}} \frac{I_\nu(\sqrt{2x\lambda}\,)}{I_\nu(\sqrt{b2\lambda}\,)}, \quad x \leq b,$$

where K_ν and I_ν are the modified Bessel functions of the second and first kind, respectively.

3.1.1 Bessel Processes

We obtain the *Bessel process* $R = \{R_t,\ t \geq 0\}$ by taking the square root of the squared Bessel process $X = \{X_t,\ t \geq 0\}$. Properties such as scaling, relations between Bessel processes of different dimensions, and first hitting times, of course, carry over in an obvious way from the squared Bessel process. For completeness, we present the transition density of a Bessel process, which follows easily from (3.1.4), (3.1.5) and the functional relationship between a squared Bessel process and a Bessel process. We point out that the concept of a dimension δ and an index ν carries over in the obvious fashion from a squared Bessel process to a Bessel process.

For $\nu > -1$ or equivalently $\delta > 0$, the transition density of a Bessel process is given by

$$p_\delta(s, x; t, y) = \frac{y}{(t-s)}\left(\frac{y}{x}\right)^\nu \exp\left\{-\frac{x^2+y^2}{2(t-s)}\right\} I_\nu\left(\frac{x\,y}{t-s}\right),$$

and for $\nu = -1$ or equivalently $\delta = 0$, we have

$$p_0(0, x; t, y) = \frac{x}{t}\exp\left\{-\frac{x^2+y^2}{2t}\right\} I_1\left(\frac{x\,y}{t}\right).$$

3.1.2 Square-Root Processes

In this subsection, we study the *square-root (SR) process*, given by the SDE

$$dr_t = \kappa(\theta - r_t)\,dt + \sigma\sqrt{r_t}\,dW_t, \tag{3.1.6}$$

where $r_0 = x$, and we assume that $\kappa\theta > 0$, which results in $r_t \geq 0$ for $x \geq 0$. The process (3.1.6) is employed in the area of interest rate modeling. For example, it is used to model the short rate in the *CIR model*, see Cox et al. (1985). But also in the area of equity modeling, it is used e.g. to model stochastic volatility in the *Heston model*, see Heston (1993). The square-root process has been linked to the squared Bessel process in two ways: firstly by transformation of space-time, and secondly by a change of law, see Pitman and Yor (1982) for both approaches. The following result is Proposition 6.3.1.1 from Jeanblanc et al. (2009) and illustrates the transformation of space-time:

Proposition 3.1.5 *The square-root process* (3.1.6) *is a squared Bessel process transformed by the following space-time changes,*

$$r_t = \exp\{-\kappa t\} X_{c(t)},$$

where $c(t) = \frac{\sigma^2}{4\kappa}(\exp\{\kappa t\} - 1)$, *and* $X = \{X_t,\ t \geq 0\}$ *is a squared Bessel process of dimension* $\delta = \frac{4\kappa\theta}{\sigma^2}$.

Now, we discuss the change of law, for which we introduce the notation ${}^{\kappa}P_x^{\kappa\theta}$ to denote the law of the square-root process that solves (3.1.6) for $\sigma = 2$. We point out that by a simple change of time $A(t) = \frac{4t}{\sigma^2}$ we can reduce (3.1.6) to a square-root process for which $\sigma = 2$. Formally, setting $Y_t = r_{A(t)}$, we obtain

$$dY_t = \kappa'(\theta - Y_t)\,dt + 2\sqrt{Y_t}\,dB_t,$$

where $\kappa' = \frac{4\kappa}{\sigma^2}$ and $B = \{B_t,\ t \geq 0\}$ is a Wiener process.

Proposition 3.1.6 *Let* ${}^{\kappa}P_x^{\kappa\theta}$ *denote the law of the square-root process that solves* (3.1.6) *for* $\sigma = 2$, *then*

$${}^{\kappa}P_x^{\kappa\theta}\,\Big|_{\mathcal{A}_t} = \exp\left\{\frac{\kappa}{4}(x + \kappa\theta t - X_t) - \frac{\kappa^2}{8}\int_0^t X_s\,ds\right\} P_x^{\kappa\theta}\,\Big|_{\mathcal{A}_t}.$$

As with the squared Bessel process, the distribution of the square-root process is linked to the non-central χ^2-distribution: let

$$c(t) = \frac{\sigma^2}{4\kappa}\big(\exp\{\kappa t\} - 1\big)$$

and $\alpha = \frac{x}{c(t)}$, then $\frac{r_t \exp\{\kappa t\}}{c(t)} \sim \chi_\delta^2(\alpha)$, where $\delta = \frac{4\kappa\theta}{\sigma^2}$. We now present the transition density of the square-root process,

$$\begin{aligned} p_\delta(s, x; t, y) &= \frac{\exp\{\kappa\tau\}}{2c(\tau)}\left(\frac{y\exp\{\kappa\tau\}}{x}\right)^{\frac{\nu}{2}} \exp\left\{-\frac{x + y\exp\{\kappa\tau\}}{2c(\tau)}\right\} \\ &\quad \times I_\nu\left(\frac{1}{c(\tau)}\sqrt{xy\exp\{\kappa t\}}\right)\mathbf{1}_{y\geq 0}, \end{aligned}$$

where $c(t) = \frac{\sigma^2}{4\kappa}(\exp\{\kappa t\} - 1)$, $\nu = \frac{\delta}{2} - 1$, $\delta = \frac{4\kappa\theta}{\sigma^2}$ and $\tau = t - s$.

Figure 3.1.2 shows the transition density of an SR process for the period from 0.1 to 3.0 years, with initial value $r_0 = 1.0$, reference level $\theta = 1.0$ and parameters $\kappa = 2$ and $\sigma = \sqrt{2}$. This means that we consider an SR process of dimension $\delta = \frac{4\times 2\times 1}{2} = 4$. Figure 3.1.3 displays a sample path for the SR process.

We conclude this subsection by remarking that the speed of mean reversion κ does not need to be a constant, but can be a function of time. This results in the *inhomogeneous square-root process*, see Sect. 6.3.5 in Jeanblanc et al. (2009). In particular, if $\tilde{r} = \{\tilde{r}_t,\ t \geq 0\}$ solves the SDE

$$d\tilde{r}_t = \big(a - \lambda(t)\tilde{r}_t\big)\,dt + \sigma\sqrt{\tilde{r}_t}\,dW_t,$$

and $\tilde{r}_0 = x$ for a continuous function $\lambda(t)$ and $a > 0$, then

$$\{\tilde{r}_t,\ t \geq 0\} \stackrel{d}{=} \left\{\frac{1}{l(t)}X_{\tilde{c}(t)},\ t \geq 0\right\},$$

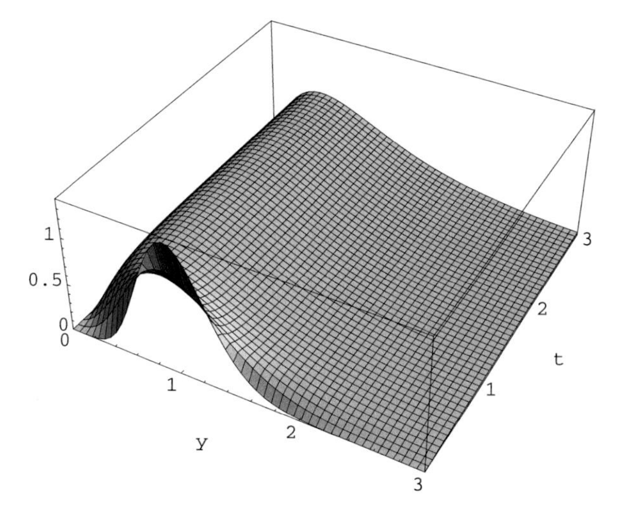

Fig. 3.1.2 Transition density of a square root process

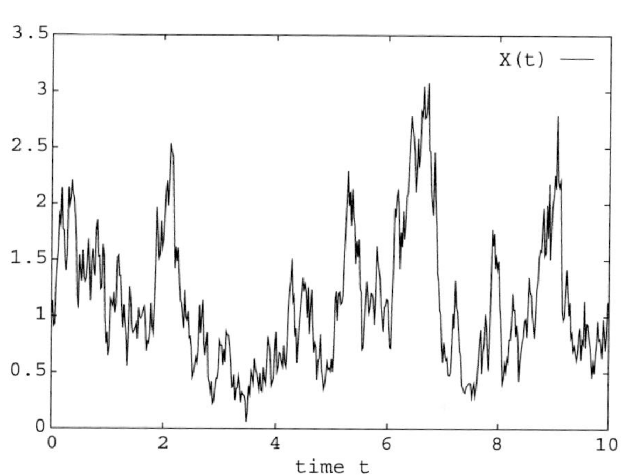

Fig. 3.1.3 Sample path of a square root process of dimension four

where $l(t) = \exp\{\int_0^t \lambda(s)\,ds\}$, $\tilde{c}(t) = \frac{\sigma^2}{4}\int_0^t l(s)\,ds$ and $X = \{X_t,\ t \ge 0\}$ is a squared Bessel process of dimension $\frac{4a}{\sigma^2}$.

3.1.3 3/2 Processes

We now turn to the $3/2$ *process*, which is another tractable process. Its SDE is given by

$$dv_t = \kappa v_t(\theta - v_t)\,dt + \sigma v_t^{3/2}\,dW_t,$$

where $v_0 = v > 0$. The $3/2$ process has been used to model short-rates, see e.g. Ahn and Gao (1999), Platen (1999), but also to model equities, in particular it is used to model stochastic volatility in the $3/2$ *model*, see e.g. Carr and Sun (2007), Itkin and Carr (2010), Lewis (2000). The 3/2 model turns out to be a functional of the square-root process. Defining $r_t = \frac{1}{v_t}$, we obtain

$$dr_t = \left(\kappa + \sigma^2 - \kappa\theta r_t\right) dt - \sigma \sqrt{r_t}\, dW_t,$$

where $r_0 = \frac{1}{v}$. This relationship is useful and allows us to transfer results obtained for the square-root process to the 3/2 model. In particular, we obtain its transition density

$$p_\delta(s, x; t, y) = y^2 \frac{\exp\{\tilde{\kappa}\tau\}}{2\tilde{c}(\tau)} \left(\frac{x \exp\{\tilde{\kappa}\tau\}}{y}\right)^{\frac{\nu}{2}} \exp\left\{-\frac{x^{-1} + y^{-1}\exp\{\tilde{\kappa}\tau\}}{2\tilde{c}(\tau)}\right\}$$
$$\times I_\nu\left(\frac{1}{\tilde{c}(\tau)}\sqrt{(xy)^{-1}\exp\{\tilde{\kappa}t\}}\right)\mathbf{1}_{y>0},$$

where $\tilde{\kappa} = \kappa\theta$, $\tilde{\theta} = \frac{\kappa+\sigma^2}{\kappa\theta}$, $\tilde{c}(t) = \frac{\sigma^2}{4\tilde{\kappa}}(\exp\{\tilde{\kappa}t\} - 1)$ and $\tau = t - s$. Furthermore, it is clear that the dimension of the square-root process $r = \{r_t,\, t \geq 0\}$ is given by $\delta = 4 + \frac{4\kappa}{\sigma^2}$. We point out that if the process r can reach zero, then the process v explodes, and that this is determined by the dimension of the process r.

We alert the reader to the fact that the link between a square-root process and the 3/2 process can also be used for more advanced functionals: for example, the *Broadie-Kaya algorithm*, which is used to simulate integrals of the square-root process, see Chap. 6, can be modified to handle the 3/2 model, which entails simulating integrals of the 3/2 process.

3.1.4 Constant Elasticity of Variance Processes

In this subsection, we study the *constant elasticity of variance (CEV) process*, where we again follow the presentation in Jeanblanc et al. (2009), and write

$$dZ_t = Z_t\left(\mu\, dt + \sigma Z_t^\beta\, dW_t\right),$$

$Z_0 = z > 0$. This process was studied e.g. in Cox (1996) for $\beta < 0$, in Engel and MacBeth (1982) for $\beta > 0$ and in Delbaen and Shirakawa (2002) for $-1 < \beta < 0$. We point out that for $\beta = -\frac{1}{2}$ we recover the square-root process and for $\beta = 0$ the geometric Brownian motion. As we have already studied these processes in detail, we focus on the following three cases:

- $\beta < -\frac{1}{2}$;
- $-\frac{1}{2} < \beta < 0$;
- $\beta > 0$.

The reason for this distinction is given in Table 3.1.2, which summarizes Lemma 6.4.4.1 in Jeanblanc et al. (2009). The result is obtained by relating the CEV process to the squared Bessel process, which for negative dimensions is stopped at 0. We now recall Lemma 6.4.3.1 from Jeanblanc et al. (2009), which links the CEV process to the squared Bessel process.

Table 3.1.2 Path properties of the CEV process

Value of β	Property of the CEV process Z
$\beta > 0$	Z does not reach 0
$\beta < 0$	Z reaches 0 almost surely
$-\frac{1}{2} < \beta < 0$	Z reaches 0 and is absorbed
$\beta < -\frac{1}{2}$	Z reaches 0 and reflects

Proposition 3.1.7 *For $\beta > 0$, or $\beta < -\frac{1}{2}$, a CEV process is a deterministic time-change of a power of a squared Bessel process,*

$$S_t = \exp\{\mu t\}(X_{c(t)})^{-\frac{1}{2\beta}},$$

where $X = \{X_t,\ t \geq 0\}$ is a squared Bessel process of dimension $\delta = 2 + \frac{1}{\beta}$ and $c(t) = \frac{\beta\sigma^2}{2\mu}(\exp\{2\mu\beta t\} - 1)$. If $-\frac{1}{2} < \beta < 0$, then

$$S_t = \exp\{\mu t\}(X_{c(t)})^{-\frac{1}{2\beta}}, \quad t \leq T_0,$$

where T_0 denotes the first time that the squared Bessel process hits 0.

We now give the transition density of the CEV process, for any β and $y > 0$:

$$\begin{aligned} p(s,x;t,y) &= \frac{|\beta|}{c(\tau)} \exp\left\{\mu\tau\left(2\beta + \frac{1}{2}\right)\right\} y^{-\frac{3}{2}-2\beta} x^{\frac{1}{2}} \\ &\quad \times \exp\left\{-\frac{1}{2c(\tau)}\left(x^{-2\beta} + y^{-2\beta}\exp\{2\mu\beta\tau\}\right)\right\} \\ &\quad \times x^{\frac{1}{2}} y^{-2\beta-\frac{3}{2}} I_{|\nu|}\left(\frac{y^{-\beta}\exp\{\beta\mu\tau\}x^{-\beta}}{c(\tau)}\right), \end{aligned}$$

where $\tau = t - s$ and $c(t)$ is defined as in Proposition 3.1.7.

We point out that in Sect. 3.3, we will introduce the *transformed constant elasticity of variance process*, and apply it to finance. Also for the transformed constant elasticity of variance process, the link with the squared Bessel process can be established and turns out to be crucial in deriving explicit pricing formulas.

3.2 Functionals of Multidimensional Squared Bessel Processes

The squared Bessel process sits at the heart of the developments in Sect. 3.1, in fact, all other processes in this section can be related to it. Consequently, when considering multidimensional processes, we need to find a suitable multidimensional generalization of this process.

The *Wishart process* introduced in Bru (1991), turns out to be a suitable generalization: recall that we introduced the squared Bessel process as the sum of squared Wiener processes, see (3.1.1). In this section, we introduce the Wishart process as a matrix product of Wiener processes, which generalizes the idea of summing up squared Wiener processes to obtain a squared Bessel process.

In particular, consider for two integers $n, p \geq 1$, the $n \times p$ matrix $\boldsymbol{W}_t$, whose elements are independent scalar valued Brownian motions. Also, assume that $\boldsymbol{W}_0 = \boldsymbol{C}$ is the initial state matrix. Now, define a Wishart process $\boldsymbol{S} = \{\boldsymbol{S}_t,\ t \geq 0\}$ of dimension p, index n and initial state $\boldsymbol{S}_0$, which we denote by $WIS_p(\boldsymbol{C}^\top \boldsymbol{C}, n, \boldsymbol{0}, \boldsymbol{I})$, see Sect. 11.2, where

$$\boldsymbol{S}_t = \boldsymbol{W}_t^\top \boldsymbol{W}_t, \qquad \boldsymbol{S}_0 = \boldsymbol{C}^\top \boldsymbol{C}.$$

We point out that $\boldsymbol{S}_t$ satisfies the SDE

$$d\boldsymbol{S}_t = n\boldsymbol{I}\,dt + d\boldsymbol{W}_t^\top \sqrt{\boldsymbol{S}_t} + \sqrt{\boldsymbol{S}_t}\,d\boldsymbol{W}_t. \tag{3.2.7}$$

Hence, for a given t, $\boldsymbol{S}_t$ follows the non-central Wishart distribution, see Definition 10.3.12, $\mathcal{W}_p(n, t\boldsymbol{I}_p, t^{-1}\boldsymbol{S}_0)$, which can be easily proven using Theorem 10.3.15. For a definition of this distribution we refer the reader to Chap. 10, Definition 10.3.12, in particular since the definition involves matrix valued special functions, which are defined in Chap. 10.

As the non-central Wishart distribution is the natural generalization of the non-central χ^2 distribution, the Wishart process appears as the natural candidate to generalize the squared Bessel process. The above discussion can be extended, but we defer such a discussion to Chaps. 10 and 11. In Chap. 10, we will recall some basics of matrix valued statistics and matrix valued stochastic processes, which will give us a better understanding of the Wishart process. Furthermore, the Wishart process will turn out to be solvable, in the sense of Chap. 9, hence we firstly develop these powerful techniques and subsequently apply them to the Wishart process.

We conclude this section by recalling the additivity property for Wishart processes, which is analogous to Proposition 3.1.3. Further properties known to hold for squared Bessel processes can be extended to Wishart processes, but such a discussion is deferred to Chap. 11.

Proposition 3.2.1 *If* $\boldsymbol{S} = \{\boldsymbol{S}_t,\ t \geq 0\}$ *and* $\boldsymbol{U} = \{\boldsymbol{U}_t,\ t \geq 0\}$ *are two independent Wishart processes* $WIS_p(\boldsymbol{S}_0, n, \boldsymbol{0}, \boldsymbol{I})$ *and* $WIS_p(\boldsymbol{U}_0, m, \boldsymbol{0}, \boldsymbol{I})$, *respectively, then the sum* $\boldsymbol{S} + \boldsymbol{U} = \{\boldsymbol{S}_t + \boldsymbol{U}_t,\ t \geq 0\}$ *is a Wishart process* $WIS_p(\boldsymbol{S}_0 + \boldsymbol{U}_0, n + m, \boldsymbol{0}, \boldsymbol{I})$.

The proof is presented in Sect. 2.5 of Bru (1991).

3.3 Selected Applications to Finance

The aim of this section is to introduce two models: the stylized *minimal market model* (MMM) and the *transformed constant elasticity of variance* (TCEV) model. Both models we consider to be potential models for the GOP. In addition to providing realistic dynamics for the GOP, these models also turn out to be highly tractable.

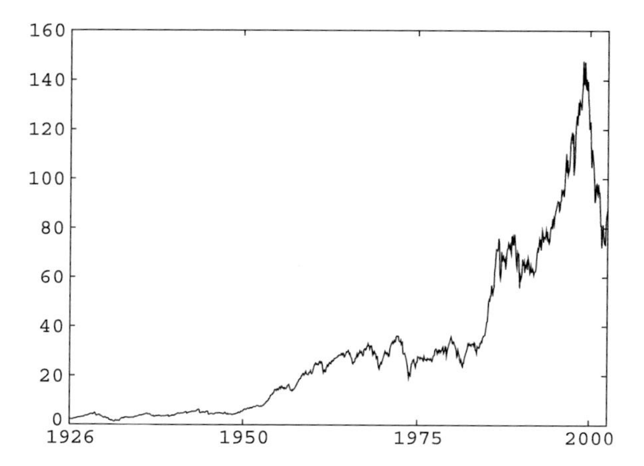

Fig. 3.3.4 Discounted S&P 500

This book focuses on the tractability aspect of models. In particular, we aim to illustrate that for many functionals relevant to finance, the underlying stochastic processes, when they are related to the squared Bessel process, provide models with closed-form solutions.

For an economic justification of these rather realistic models, we refer the reader to Platen and Heath (2010), and also to Baldeaux et al. (2011c). Here, we only recall two pivotal characteristics of these models, which contribute to their realistic representation of the real world dynamics of the GOP:

- volatility should be modeled via a mean-reverting process, which is negatively correlated with the GOP;
- the benchmarked savings account should be modeled as a strict local martingale.

Both points are easily illustrated pictorially: we use the discounted S&P 500 total return index (S&P 500) observed in US dollars from 1926 until 2004 as a proxy for the GOP. This index starts at $\bar{S}_0^{\delta_*} = 2.3$ in January 1926 and has been reconstructed from monthly data provided by Global Financial Data, see Fig. 3.3.4. In Fig. 3.3.5, we show the volatility of this index, which clearly fluctuates around a long-run average. Furthermore, the so-called *leverage effect*, see Black (1976), which expresses a negative correlation between an index and its volatility is a recognized and accepted phenomenon. The two models presented in this section both accommodate a mean-reverting volatility process and the leverage effect.

Secondly, by discussing the two models we challenge the risk-neutral approach to finance by questioning the suitability of the assumption on the existence of an equivalent risk neutral probability measure. In Fig. 3.3.6, we show the candidate Radon-Nikodym derivative for the putative risk neutral measure, which is the benchmarked savings account or the inverse of the discounted index shown in Fig. 3.3.4 if the index is interpreted as the GOP. Clearly, this index appears to trend downwards. This observation conflicts with the classical requirement for the existence of a risk neutral probability measure, where this process should be a martingale. Consequently, we focus on models for which an equivalent risk neutral probability measure does not

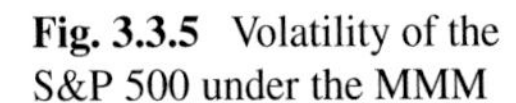

Fig. 3.3.5 Volatility of the S&P 500 under the MMM

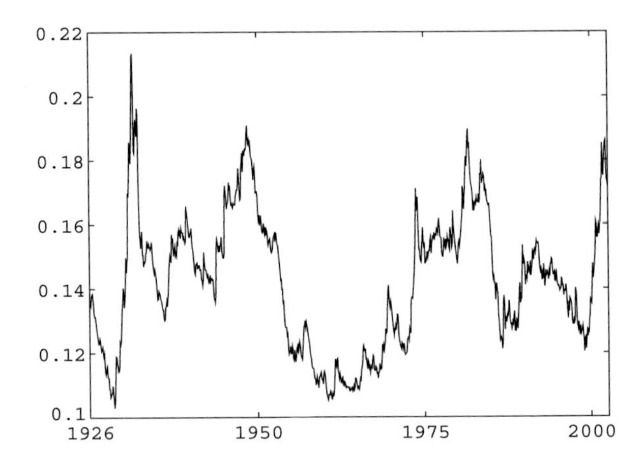

Fig. 3.3.6 Candidate Radon-Nikodym derivative for the putative risk neutral probability measure

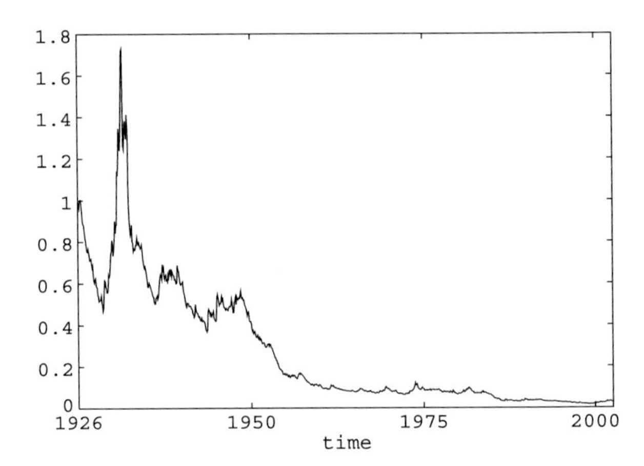

exist. The risk neutral approach breaks down in this context, however, the benchmark approach can still be applied and results in tractable expressions for realistic financial models as will be discussed in this book. It is the aim of this section to point out that even though risk neutral pricing is not applicable, realistic models can still be designed which yield similarly tractable expressions as, for instance, the Black-Scholes model.

3.3.1 Stylized Minimal Market Model

In this subsection, we recall the stylized *minimal market model* (MMM) and demonstrate that it can be used to obtain tractable expressions for some important derivatives. For more information on the stylized MMM, we refer the reader to Platen and Heath (2010).

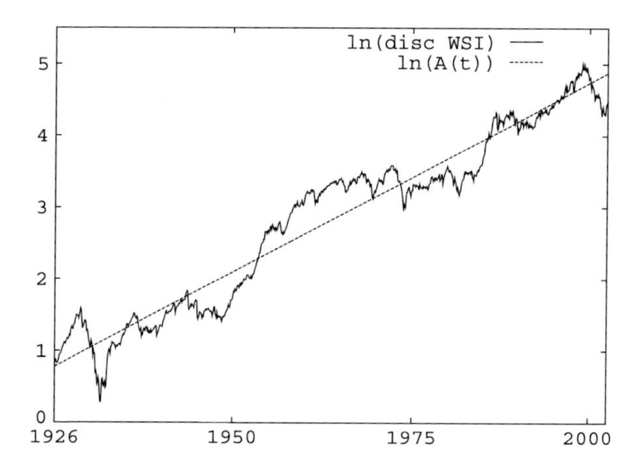

Fig. 3.3.7 Logarithm of discounted S&P 500 and trend line

Under the stylized MMM, we model the discounted GOP $\bar{S}^{\delta_*} = \{\bar{S}_t^{\delta_*},\ t \geq 0\}$ as follows:

$$d\bar{S}_t^{\delta_*} = \alpha_t^{\delta_*} dt + \sqrt{\bar{S}_t^{\delta_*} \alpha_t^{\delta_*}}\, dW_t, \quad t \geq 0.$$

Using the relation

$$X_{\varphi(t)} = \bar{S}_t^{\delta_*},$$

we find that the discounted GOP is a squared Bessel process of dimension four,

$$dX_{\varphi(t)} = 4\, d\varphi(t) + 2\sqrt{X_{\varphi(t)}}\, dW_{\varphi(t)},$$

where

$$dW_{\varphi(t)} = \sqrt{\frac{\alpha_t^{\delta_*}}{4}}\, dW_t.$$

The drift $\alpha_t^{\delta_*}$ of the discounted GOP is modeled as an exponential function of time,

$$\alpha_t^{\delta_*} = \alpha_0 \exp\{\eta t\}.$$

Such a choice seems reasonable. In Fig. 3.3.7 we show the natural logarithm of the discounted S&P 500 index, which is approximately linear. To emphasize this observation we include also the trend line.

The transformed time $\varphi(t)$ is given by the integral of $\alpha_t^{\delta_*}$, i.e.

$$\varphi(t) = \frac{1}{4} \int_0^t \alpha_s^{\delta_*}\, ds = \frac{\alpha_0}{4\eta} \big(\exp\{\eta t\} - 1\big).$$

It is clear that once α_0 and η are known, the MMM is fully parameterized. From the Itô formula it follows immediately that the quadratic variation of the square-root of the discounted GOP satisfies

$$\big[\sqrt{\bar{S}^{\delta_*}}\big]_t = \frac{1}{4} \int_0^t \alpha_s^{\delta_*}\, ds = \varphi(t) - \varphi(0). \tag{3.3.8}$$

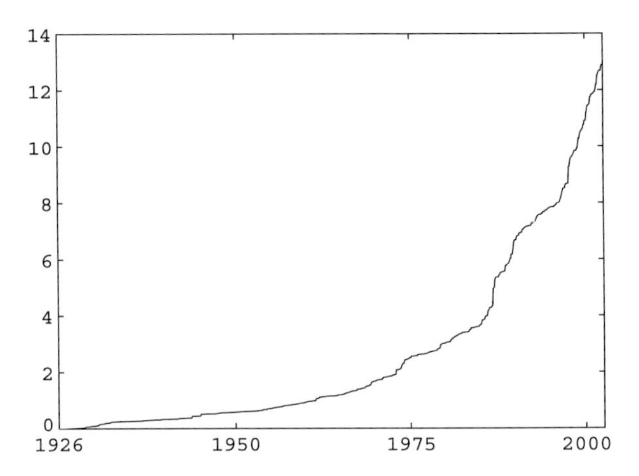

Fig. 3.3.8 Empirical quadratic variation of the square root of the discounted S& P 500

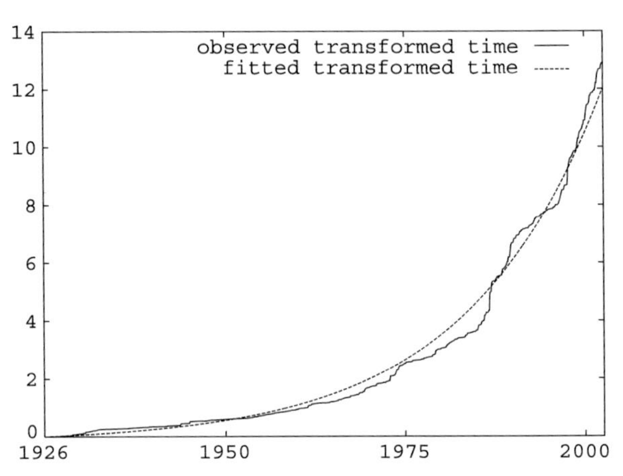

Fig. 3.3.9 Fitted and observed transformed time

The expression (3.3.8) is very useful, as it allows for an easy estimation of α_0 and η. In particular, we show in Fig. 3.3.8 the empirical quadratic variation of the square-root of the discounted GOP. In Fig. 3.3.9, we show the empirical quadratic variation of the square-root of the discounted GOP and the fitted transformed time $\varphi(t)$. We point out that the chosen parameters seem to provide a good fit.

The *normalized GOP process* $Y = \{Y_t,\ t \geq 0\}$ is defined via the ratio

$$Y_t = \frac{\bar{S}_t^{\delta_*}}{\alpha_t^{\delta_*}}, \tag{3.3.9}$$

for $t \geq 0$, which by an application of the Itô formula yields,

$$dY_t = (1 - \eta Y_t)\, dt + \sqrt{Y_t}\, dW_t,$$

for $t \geq 0$ with

$$Y_0 = \frac{\bar{S}_0^{\delta_*}}{\alpha_0^{\delta_*}},$$

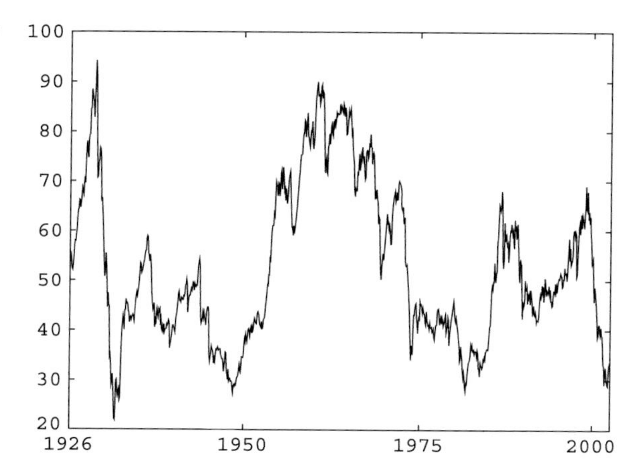

Fig. 3.3.10 Normalized GOP

a square-root process of dimension four. This is a consequence of Proposition 3.1.5, the time-change relationship between a square-root process and a squared Bessel process. We plot the normalized GOP corresponding to the S&P 500 index in Fig. 3.3.10.

The normalized GOP is directly related to the volatility of the GOP: under the MMM, volatility is given by

$$|\theta_t| = \frac{1}{\sqrt{Y_t}}.$$

The squared volatility can be shown to satisfy the SDE

$$d|\theta_t|^2 = d\left(\frac{1}{Y_t}\right) = |\theta_t|^2 \eta \, dt - \left(|\theta_t|^2\right)^{\frac{3}{2}} dW_t,$$

where $t \geq 0$. The diffusion coefficient has the power $3/2$, i.e. we recover the $3/2$ process, see Sect. 3.1. Such a model was suggested in Platen (1997) for modeling the squared volatility of a market index. Furthermore, under the MMM the squared volatility satisfies

$$|\theta_t|^2 = \frac{1}{Y_t},$$

and hence has an *inverse gamma density* as its stationary density.

3.3.2 Standard European Options on the Index Under the MMM

Having investigated the MMM, we now turn to pricing. The conventional approach entails changing from the real-world probability measure to an equivalent martingale measure and consequently to value derivatives. However, an equivalent martingale measure does not exist under the MMM. It turns out that investigating the existence of an equivalent martingale measure is equivalent to pricing a zero coupon using the real world pricing formula (1.3.19). From the real world pricing formula, it is clear that even when assuming a constant short rate, a zero coupon bond is still an index derivative and under the MMM a stochastic quantity. To emphasize this, we assume that the short rate process $r = \{r_t,\ t \geq 0\}$ is constant, i.e. we set $r_t = r$ and note that

$$S_t^{\delta_*} = \bar{S}_t^{\delta_*} S_t^0,$$

where $S_t^0 = \exp\{rt\}$. From the real world pricing formula, we obtain the following price for a zero coupon bond at time t, which matures at $T > t$,

$$\begin{aligned} P_T(t) &= S_t^{\delta_*} E\left(\frac{1}{S_T^{\delta_*}} \,\Big|\, \mathcal{A}_t\right) \\ &= \exp\{-r(T-t)\} E\left(\frac{\bar{S}_t^{\delta_*}}{\bar{S}_T^{\delta_*}} \,\Big|\, \mathcal{A}_t\right). \end{aligned} \tag{3.3.10}$$

Since the discounted GOP is modeled via a time transformed squared Bessel process of dimension $\delta = 4$, computing (3.3.10) amounts to a computation involving the non-central χ^2 distribution. In Sect. 13.1, we will summarize useful properties of this distribution, which facilitate the computations in this section. We introduce the following notation, which will be used throughout this subsection

$$\Psi(x; \nu, \lambda) := P\big(\chi_\nu^2(\lambda) \leq x\big),$$

where $\chi_\nu^2(\lambda)$ denotes a non-central χ^2 random variable with ν degrees of freedom and non-centrality λ. Furthermore, we set

$$\lambda(t, \bar{S}) := \frac{\bar{S}}{\Delta\varphi(t)}, \qquad x(t) := \frac{K/S_T^0}{\Delta\varphi(t)},$$

with

$$\Delta\varphi(t) := \varphi(T) - \varphi(t).$$

The next computation relies on (13.1.3), and we point out that this particular identity is important in the derivation of derivative prices under the MMM. The following presentation follows Hulley and Platen (2012) closely:

$$\begin{aligned} E\left(\frac{\bar{S}_t^{\delta_*}}{\bar{S}_T^{\delta_*}} \,\Big|\, \mathcal{A}_t\right) &= E\left(\frac{\bar{S}_t^{\delta_*}/\Delta\varphi(t)}{\bar{S}_T^{\delta_*}/\Delta\varphi(t)} \,\Big|\, \mathcal{A}_t\right) \\ &= E\left(\frac{\lambda(t, \bar{S}_t^{\delta_*})}{\chi_4^2(\lambda(t, \bar{S}_t^{\delta_*}))} \,\Big|\, \mathcal{A}_t\right) \\ &= 1 - \exp\left\{-\frac{\lambda(t, \bar{S}_t^{\delta_*})}{2}\right\}. \end{aligned}$$

The candidate Radon-Nikodym derivative process $\Lambda = \{\Lambda_t,\ t \geq 0\}$ is given by

$$\Lambda_t = \frac{\bar{S}_0^{\delta_*}}{\bar{S}_t^{\delta_*}}.$$

Hence

$$E(\Lambda_T \mid \mathcal{A}_0) = E\left(\frac{\bar{S}_0^{\delta_*}}{\bar{S}_T^{\delta_*}} \,\Big|\, \mathcal{A}_0\right) = 1 - \exp\left\{-\frac{\lambda(t, \bar{S}_0^{\delta_*})}{2}\right\} < 1,$$

and we conclude that an equivalent martingale measure does not exist. Consequently, risk neutral pricing is not applicable, but real world pricing is.

Now, we turn to call options. In particular, we recall that

$$S_T^{\delta_*} = S_T^0 \bar{S}_T^{\delta_*} = S_T^0 X_{\varphi(T)}.$$

We employ the real world pricing formula to obtain

$$C_{T,K}(t) := S_t^{\delta_*} E\left(\frac{(S_T^{\delta_*} - K)^+}{S_T^{\delta_*}} \,\Big|\, \mathcal{A}_t\right).$$

As in Sect. 2.3, we find it convenient to split the problem of pricing a call option into the computation of an asset-or-nothing binary and a bond-or-nothing binary. We firstly compute the asset-or-nothing binary:

$$\begin{aligned}
A_{T,K}(t) &= S_t^{\delta_*} E\left(\frac{S_T^{\delta_*}}{S_T^{\delta_*}} \mathbf{1}_{S_T^{\delta_*} > K} \,\Big|\, \mathcal{A}_t\right) \\
&= S_t^{\delta_*} E(\mathbf{1}_{S_T^0 X_{\varphi(T)} > K} \mid \mathcal{A}_t) \\
&= S_t^{\delta_*} E(\mathbf{1}_{\chi_4^2(\lambda(t, \bar{S}_t^{\delta_*})) > x(t)} \mid \mathcal{A}_t) \\
&= S_t^{\delta_*} P\big(\chi_4^2\big(\lambda\big(t, \bar{S}_t^{\delta_*}\big)\big) > x(t) \,\big|\, \mathcal{A}_t\big) \\
&= S_t^{\delta_*}\big(1 - \Psi\big(x(t); 4, \lambda\big(t, \bar{S}_t^{\delta_*}\big)\big)\big).
\end{aligned}$$

Similarly, we compute the bond-or-nothing binary:

$$\begin{aligned}
B_{T,K}(t) &= S_t^{\delta_*} E\left(\frac{K}{S_T^{\delta_*}} \mathbf{1}_{S_T^{\delta_*} > K} \,\Big|\, \mathcal{A}_t\right) \\
&= K E\left(\frac{S_t^{\delta_*}}{S_T^{\delta_*}} \mathbf{1}_{S_T^{\delta_*} > K} \,\Big|\, \mathcal{A}_t\right) \\
&= K \exp\big\{-r(T-t)\big\} E\left(\frac{\bar{S}_t^{\delta_*}}{\bar{S}_T^{\delta_*}} \mathbf{1}_{\bar{S}_T^{\delta_*} > K/S_T^0} \,\Big|\, \mathcal{A}_t\right) \\
&= K \exp\big\{-r(T-t)\big\} E\left(\frac{\lambda(t, \bar{S}_t^{\delta_*})}{\chi_4^2(\lambda(t, \bar{S}_t^{\delta_*}))} \mathbf{1}_{\chi_4^2(\lambda(t, \bar{S}_t^{\delta_*})) > x(t)} \,\Big|\, \mathcal{A}_t\right) \\
&= K \exp\big\{-r(T-t)\big\} P\big(\chi_0^2\big(\lambda\big(t, \bar{S}_t^{\delta_*}\big)\big) > x(t)\big) \\
&= K \exp\big\{-r(T-t)\big\}\big(1 - \Psi\big(x(t); 0, \lambda\big(t, \bar{S}_t^{\delta_*}\big)\big)\big),
\end{aligned}$$

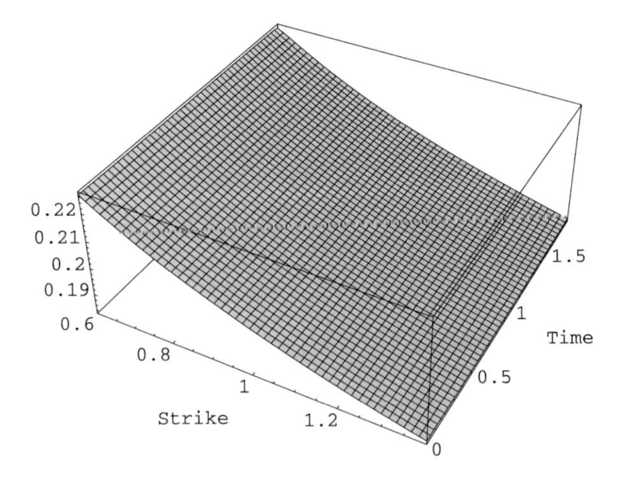

Fig. 3.3.11 Implied volatilities for fair European call prices under the MMM

where we used (13.1.2). Combining the asset-or-nothing binary and the bond-or-nothing binary, we arrive at the price of a European call option:

$$\begin{aligned} c_{T,K}(t) &= S_t^{\delta_*}\left(1 - \Psi\left(x(t); 4, \lambda\left(t, \bar{S}_t^{\delta_*}\right)\right)\right) \\ &\quad - K \exp\{-r(T-t)\}\left(1 - \Psi\left(x(t); 0, \lambda\left(t, \bar{S}_t^{\delta_*}\right)\right)\right). \end{aligned}$$

In Fig. 3.3.11, we show an implied volatility surface for European call options on the GOP under the MMM. Note that the short rate has here to be adjusted for the given maturity according to the fair zero coupon bond price, see e.g. Platen and Heath (2010).

Using the same technique, we arrive at the price of a European put option,

$$\begin{aligned} p_{T,K}(t) &= S_t^{\delta_*} E\left(\frac{(K - S_T^{\delta_*})^+}{S_T^{\delta_*}} \,\middle|\, \mathcal{A}_t\right) \\ &= K \exp\{-r(T-t)\}\left(\Psi\left(x(t); 0, \lambda\left(t, \bar{S}_t^{\delta_*}\right)\right) - \exp\left\{-\frac{1}{2}\lambda\left(t, \bar{S}_t^{\delta_*}\right)\right\}\right) \\ &\quad - S_t^{\delta_*}\Psi\left(x(t); 4, \lambda\left(t, \bar{S}_t^{\delta_*}\right)\right). \end{aligned}$$

It is easily checked that the European call price, the European put price and the zero coupon bond satisfy the well-known put-call parity,

$$p_{T,K}(t) = c_{T,K}(t) - S_t^{\delta_*} + K P_T(t).$$

Note that in this put-call parity relation we use the fair zero coupon bond.

We now focus on *binary options* on an index, i.e. derivatives which pay off a fixed amount in the event that the index exceeds a particular level K at maturity. The real world pricing formula (1.3.19) yields the following formula for the price of a *binary European call option*, which recovers the result from Proposition 5.16 in Hulley (2009):

$$
\begin{aligned}
BC_{T,K}(t) &= S_t^{\delta_*} E\left(\mathbf{1}_{S_T^{\delta_*} > K} \frac{1}{S_T^{\delta_*}} \,\Big|\, \mathcal{A}_t \right) \\
&= S_t^{\delta_*} E\left(\mathbf{1}_{S_T^0 X_{\varphi(T)} > K} \frac{1}{S_T^0 X_{\varphi(T)}} \,\Big|\, \mathcal{A}_t \right) \\
&= \exp\{-r(T-t)\} E\left(\mathbf{1}_{\chi_4^2(\lambda(t, \bar{S}_t^{\delta_*})) > x(t)} \frac{\lambda(t, \bar{S}_t^{\delta_*})}{\chi_4^2(\lambda(t, \bar{S}_t^{\delta_*}))} \,\Big|\, \mathcal{A}_t \right) \\
&= \exp\{-r(T-t)\}\left(1 - \Psi\left(x(t), 0, \lambda\left(t, \bar{S}_t^{\delta_*}\right)\right)\right).
\end{aligned}
$$

Similarly, we can obtain the price of a *binary European put option*, which recovers the result from Proposition 5.17 in Hulley (2009):

$$
\begin{aligned}
BP_{T,K}(t) &= S_t^{\delta_*} E\left(\mathbf{1}_{S_T^{\delta_*} \le K} \frac{1}{S_T^{\delta_*}} \,\Big|\, \mathcal{A}_t \right) \\
&= \exp\{-r(T-t)\}\left(\Psi\left(x(t); 0, \lambda\left(t, \bar{S}_t^{\delta_*}\right)\right) - \exp\left\{ -\frac{1}{2}\lambda\left(t, \bar{S}_t^{\delta_*}\right) \right\} \right).
\end{aligned}
$$

Finally, we remark that the binary European call and the binary European put option sum up to a zero coupon bond:

$$
BC_{T,K}(t) + BP_{T,K}(t) = P_T(t).
$$

3.3.3 Rebates and Barrier Options Under the MMM

We now discuss the pricing of some *path-dependent options*, in particular rebates and barrier options. The presentation is based on Hulley (2009), see also Hulley and Platen (2008). It uses the theory of time-homogeneous scalar diffusions, which can be found e.g. in Borodin and Salminen (2002), Chap. 2, and Rogers et al. (2000), Sect. V.7. The techniques differ from the ones previously discussed in this book. Hence we focus on presenting the results in this subsection and present a brief introduction to this theory in Chap. 16. We remark though that when dealing with one-dimensional diffusions, this theory is powerful and can generate tractable solutions to problems arising in finance and elsewhere, as we now demonstrate.

As in Sect. 2.3, we study a *rebate* written on a market index, again interpreted as the GOP. We recall that a rebate pays one unit of the domestic currency as soon as the index hits a certain level, should this occur before a contracted expiry date, $T > 0$. Again the trigger level is assumed to be a deterministic barrier, $Z_t := z \exp\{rt\}$, for some $z > 0$. As demonstrated in Sect. 2.3, the valuation of rebates is a study of suitable hitting times, which we now define:

$$
\sigma_{z,t} := \inf\left\{u > 0 \colon S_{t+u}^{\delta_*} = Z_{t+u}\right\}
$$

and

$$
\tau_z := \inf\{u > 0 \colon X_u = z\},
$$

where S^{δ_*} denotes the GOP and X the squared Bessel process of dimension four. For rebates and barrier options, we find it convenient to let X be a squared Bessel process of dimension four, started at $\exp\{-rt\}S_t^{\delta_*}$. Consequently,

$$\bar{S}^{\delta_*}_{t+u} \stackrel{d}{=} X_{\varphi_t(u)},$$

for $u \geq 0$, and we have

$$\sigma_{z,t} = \inf\{u > 0\colon X_{\varphi_t(u)} = z\},$$

where we use the notation

$$\varphi_t(u) = \varphi(u+t) - \varphi(t) = \frac{\alpha}{4\eta}\exp\{\eta t\}\big(\exp\{\eta u\} - 1\big).$$

This allows us to realize the following equality in distribution

$$\sigma_{z,t} = \inf\{u > 0\colon X_{\varphi_t(u)} = z\} \stackrel{d}{=} \varphi_t^{-1}(\tau_z) = \frac{1}{\eta}\ln\left(1 + \frac{4\eta}{\alpha}\exp\{-\eta t\}\tau_z\right).$$

We can now easily compute the price of a rebate under the MMM. From the real world pricing formula, (1.3.19), we obtain,

$$\begin{aligned}
R_{\infty,z}(t) &= S_t^{\delta_*} E\left(\frac{1}{S^{\delta_*}_{t+\sigma_{z,t}}} \,\middle|\, \mathcal{A}_t\right) = \frac{\exp\{-rt\}S_t^{\delta_*}}{z} E\big(\exp\{-r\sigma_{z,t}\} \,\big|\, \mathcal{A}_t\big) \\
&= \frac{\exp\{-rt\}S_t^{\delta_*}}{z} E\big(\exp\{-r\varphi_t^{-1}(\tau_z)\} \,\big|\, \mathcal{A}_t\big) \\
&= \frac{\exp\{-rt\}S_t^{\delta_*}}{z} E\left(\left(1 + \frac{4\eta}{\alpha}\exp\{-\eta t\}\tau_z\right)^{-r/\eta}\right) \\
&= \begin{cases} \frac{x}{z}\frac{1}{\Gamma(r/\eta)}\int_0^\infty e^{-s} s^{r/\eta - 1} \frac{\psi_{4\eta/\alpha e^{-\eta t} s}(x)}{\psi_{4\eta/\alpha e^{-\eta t} s}(z)}\, ds & \text{for } x \leq z \\ \frac{x}{z}\frac{1}{\Gamma(r/\eta)}\int_0^\infty e^{-s} s^{r/\eta - 1} \frac{\phi_{4\eta/\alpha e^{-\eta t} s}(x)}{\phi_{4\eta/\alpha e^{-\eta t} s}(z)}\, ds & \text{for } x \geq z, \end{cases}
\end{aligned}$$

for all $t \geq 0$, where we used Proposition 16.3.3 and $x = \exp\{-rt\}S_t^{\delta_*}$. The formula employs the functions ψ and ϕ, which are given by

$$\psi_\alpha(x) = x^{\frac{2-\delta}{4}} I_{\frac{\delta-2}{2}}(\sqrt{2\alpha x}), \tag{3.3.11}$$

where δ refers to the dimension of the squared Bessel process under consideration, in this case $\delta = 4$. Also, we have

$$\phi_\alpha(x) = x^{\frac{2-\delta}{4}} K_{\frac{\delta-2}{2}}(\sqrt{2\alpha x}), \tag{3.3.12}$$

where $\alpha > 0$, $x \geq 0$, and $I_\nu(\cdot)$ and $K_\nu(\cdot)$ denote the modified Bessel functions of the first and second kind, respectively. The functions $\psi_\alpha(\cdot)$ and $\phi_\alpha(\cdot)$ arise naturally when studying one-dimensional diffusions, they are related to the Laplace transform of the transition density of the diffusion under consideration, in this case the squared Bessel process of dimension four. Clearly, pricing a rebate with $T = \infty$ necessarily involves a one-dimensional integration, which can be performed using the techniques to be described in Chap. 12. The change of variables $\exp\{-s\} \mapsto u$ turns this expression into an integral over a bounded interval, which can facilitate the computation from a numerical point of view,

$$R_{\infty,z}(t) = \begin{cases} \frac{x}{z}\frac{1}{\Gamma(r/\eta)}\int_0^1 (-\ln u)^{r/\eta-1}\frac{\psi_{-4\eta/\alpha e^{-\eta t}\ln u}(x)}{\psi_{-4\eta/\alpha e^{-\eta t}\ln u}(z)}\,du & \text{for } x \le z \\ \frac{x}{z}\frac{1}{\Gamma(r/\eta)}\int_0^1 (-\ln u)^{r/\eta-1}\frac{\phi_{-4\eta/\alpha e^{-\eta t}\ln u}(x)}{\phi_{-4\eta/\alpha e^{-\eta t}\ln u}(z)}\,du & \text{for } x \ge z, \end{cases}$$

where $x = \exp\{-rt\}S_t^{\delta_*}$.

We now turn to a rebate with finite maturity $T < \infty$. Computing this expression, the Laplace transform in Eq. (16.3.11) in Chap. 16 is used:

$$\begin{aligned} R_{T,z}(t) &= S_t^{\delta_*} E\left(\frac{\mathbf{1}_{t+\sigma_{z,t}\le T}}{S_{t+\sigma_{z,t}}^{\delta_*}} \,\Big|\, \mathcal{A}_t\right) = \frac{e^{-rt}S_t^{\delta_*}}{z} E\left(\mathbf{1}_{\sigma_{z,t}\le T-t} e^{-r\sigma_{z,t}} \,\big|\, \mathcal{A}_t\right) \\ &= \frac{e^{-rt}S_t^{\delta_*}}{z} E\left(\mathbf{1}_{\varphi_t^{-1}(\tau_z)\le T-t} e^{-r\varphi_t^{-1}(\tau_z)} \,\big|\, \mathcal{A}_t\right) \\ &= \frac{e^{-rt}S_t^{\delta_*}}{z} E\left(\mathbf{1}_{\tau_z\le\varphi(T-t)}\left(1+\frac{4\eta}{\alpha}\exp\{-\eta t\}\tau_z\right)^{-r/\eta} \,\Big|\, \mathcal{A}_t\right). \end{aligned}$$

We now compute the Laplace transform of the price of the rebate with respect to the transformed time to maturity

$$L_\beta\big(R_{T,z}(t)\big) = \begin{cases} \frac{x}{z}\frac{1}{\beta\Gamma(r/\eta)}\int_0^\infty e^{-s}s^{r/\eta-1}\frac{\psi_{\beta+4\eta/\alpha e^{-\eta t}s}(x)}{\psi_{\beta+4\eta/\alpha e^{-\eta t}s}(z)}\,ds & \text{for } x \le z \\ \frac{x}{z}\frac{1}{\beta\Gamma(r/\eta)}\int_0^\infty e^{-s}s^{r/\eta-1}\frac{\phi_{\beta+4\eta/\alpha e^{-\eta t}s}(x)}{\phi_{\beta+4\eta/\alpha e^{-\eta t}s}(z)}\,ds & \text{for } x \ge z \end{cases} \tag{3.3.13}$$

for all $\beta > 0$, where we used Proposition 16.3.3 and $x = e^{-rt}S_t^{\delta_*}$. Recall that ψ_α was defined in (3.3.11) and ϕ_α in (3.3.12). Pricing a finite maturity rebate involves two numerical procedures: the integral needs to be evaluated, e.g. using the techniques from Chap. 12, and consequently the Laplace transform needs to be inverted, e.g. using the methodology from Sect. 13.5. For the numerical integration, performing a change of variables, $e^{-s} \mapsto u$ allows us to write

$$L_\beta\big(R_{T,z}(t)\big) = \begin{cases} \frac{x}{z}\frac{1}{\beta\Gamma(r/\eta)}\int_0^1 (-\ln u)^{r/\eta-1}\frac{-\psi_{\beta+4\eta/\alpha e^{-\eta t}\ln u}(x)}{-\psi_{\beta+4\eta/\alpha e^{-\eta t}\ln u}(z)}\,du & \text{for } x \le z \\ \frac{x}{z}\frac{1}{\beta\Gamma(r/\eta)}\int_0^1 (-\ln u)^{r/\eta-1}\frac{-\phi_{\beta+4\eta/\alpha e^{-\eta t}\ln u}(x)}{-\phi_{\beta+4\eta/\alpha e^{-\eta t}\ln u}(z)}\,du & \text{for } x \ge z. \end{cases}$$

In Fig. 3.3.12 we show the surface of the pricing function $R_{10,50}(\cdot)$ for a rebate with maturity $T = 10$ years and reference level $z = 50$, $\alpha = 1$, $\eta = 0.05$ and $r = 0.04$, as a function of t and $S_t^{\delta_*}$.

We take the same approach to a *barrier option* as in Sect. 2.3 and recall Lemma 2.1.9, which allows us to compute $\tilde{q}_z(\cdot,\cdot,\cdot)$, the transition density of X killed at z with respect to the speed measure. For barrier options under the MMM,

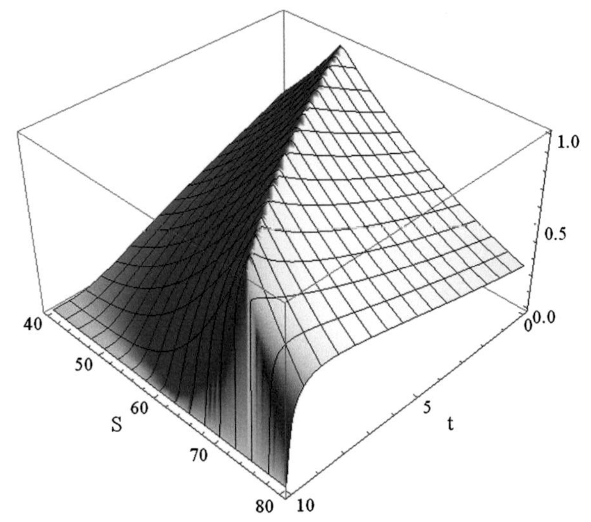

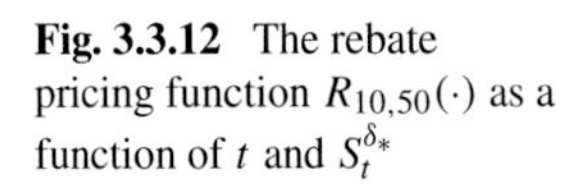
Fig. 3.3.12 The rebate pricing function $R_{10,50}(\cdot)$ as a function of t and $S_t^{\delta_*}$

we find it useful to work with the Laplace transform $\tilde{G}_\alpha^z(x, y)$, the Laplace transform of $\tilde{q}_z(t, x, y)$ with respect to time,

$$\tilde{G}_\alpha^z(x, y) := \int_0^\infty e^{-\alpha t} \tilde{q}_z(t, x, y)\, dt.$$

We point out that in Chap. 16, we provide explicit formulas for $\tilde{G}_\alpha(x, y)$, which we employ here, see Eq. (16.3.6) in Chap. 16. In particular, we focus on a European call with strike K on an index, which is knocked out if the index breaches the same deterministic barrier that was considered in the context of rebates.

$$\begin{aligned}
C_{T,K}^{out}(t) &= S_t^{\delta_*} E\left(\mathbf{1}_{t+\sigma_{z,t} > T} \frac{(S_T^{\delta_*} - K)^+}{S_T^{\delta_*}} \,\middle|\, \mathcal{A}_t \right) \\
&= S_t^{\delta_*} E\left(\mathbf{1}_{\sigma_{z,t} > T-t} \left(1 - \frac{K}{S_T^{\delta_*}} \right)^+ \,\middle|\, \mathcal{A}_t \right) \\
&= S_t^{\delta_*} E\left(\mathbf{1}_{\tau_z > \varphi_t(T-t)} \left(1 - \frac{\exp\{-rT\}K}{X_{\varphi_t(T-t)}} \right)^+ \,\middle|\, \mathcal{A}_t \right) \\
&= S_t^{\delta_*} \int_{e^{-rT}K}^\infty \left(1 - \frac{e^{-rT}K}{y} \right) \tilde{q}_z\big(\varphi_t(T-t), e^{-rt} S_t^{\delta_*}, y\big) m(y)\, dy \\
&= \frac{S_t^{\delta_*}}{2} \int_\kappa^\infty (y - \kappa) \tilde{q}_z\big(\varphi_t(T-t), x, y\big)\, dy,
\end{aligned}$$

where we used $x = \exp\{-rt\} S_t^{\delta_*}$, $\kappa = \exp\{-rT\}K$. In this computation, we used the fact that the speed measure of a squared Bessel process of dimension four satisfies $m(y) = \frac{y}{2}$. We now compute the Laplace transform with respect to the transformed time to maturity, which yields,

$$L_\beta\big(C^{out}_{T,K}(t)\big) = \int_0^\infty e^{-\beta u}\left(\frac{S_t^{\delta_*}}{2}\int_\kappa^\infty (y-\kappa)\tilde{q}_z(u,x,y)\,dy\right)du \quad (3.3.14)$$

$$= \frac{S_t^{\delta_*}}{2}\int_\kappa^\infty (y-\kappa)L_\beta\big(\tilde{q}_z(u,x,y)\big)\,dy$$

$$= \frac{S_t^{\delta_*}}{2}\int_\kappa^\infty (y-\kappa)\tilde{G}^z_\beta(x,y)\,dy, \quad (3.3.15)$$

for all $\beta > 0$. As in Sect. 2.3, we find it convenient to analyze two cases separately: the up-and-out call, where $S_t^{\delta_*} \leq Z_t \iff x \leq z$ and the down-and-out call, for which $S_t^{\delta_*} \geq Z_t \iff x \geq z$. It is clear from Eq. (16.3.6) in Chap. 16, that $x \leq z$ implies that $\tilde{G}^z_\beta(x,y) = 0$ for all $y \geq z$. Hence Eq. (3.3.14) yields

$$L_\beta\big(C^{out}_{T,K}(t)\big) = S_t^{\delta_*}\int_\kappa^{\kappa\vee x}(y-\kappa)\psi_\beta(y)\left(\phi_\beta(x) - \frac{\phi_\beta(z)}{\psi_\beta(z)}\psi_\beta(x)\right)dy$$

$$+ S_t^{\delta_*}\int_{\kappa\vee x}^{\kappa\vee z}(y-\kappa)\psi_\beta(x)\left(\phi_\beta(y) - \frac{\phi_\beta(z)}{\psi_\beta(z)}\psi_\beta(y)\right)dy,$$

if $x \leq z$. We point out that since $w_\beta = 1/2$ for the squared Bessel process of dimension four, the factor $1/2$ from (3.3.14) has disappeared. We now focus on the down-and-out call,

$$L_\beta\big(C^{out}_{T,K}(t)\big) = S_t^{\delta_*}\int_{\kappa\vee z}^{\kappa\vee x}(y-\kappa)\left(\psi_\beta(y) - \frac{\psi_\beta(z)}{\phi_\beta(z)}\phi_\beta(y)\right)\phi_\beta(x)\,dy$$

$$+ S_t^{\delta_*}\int_{\kappa\vee x}^{\infty}(y-\kappa)\left(\psi_\beta(x) - \frac{\psi_\beta(z)}{\phi_\beta(z)}\phi_\beta(x)\right)\phi_\beta(y)\,dy,$$

for $x \geq z$. Again, the factor $\frac{1}{2}$ disappears, due to the fact that $w_\beta = \frac{1}{2}$ for squared Bessel processes of dimension four.

Finally, we point out that pricing the barrier option again involves two numerical procedures: firstly, the integrals need to be computed numerically, e.g. using the techniques from Chap. 12, and subsequently the Laplace transform needs to be inverted, e.g. using the methods from Sect. 13.5.

3.3.4 Exchange Options

We now turn to exchange options, which entitle the owner to exchange one asset for another. In particular, we extend the financial market discussed in this section, which so far consists of the numéraire portfolio and the savings account, by adding two risky securities, whose price processes are denoted by S^a and S^b. This problem is interesting for two reasons: firstly because the dimensionality of the problem is two and not one, as for the other contracts considered in this section, and secondly, because the closed-form solution is expressed in terms of the so-called extended doubly non-central beta distribution, see Sect. 13.3 in Chap. 13. Random variables following the doubly non-central beta distribution arise when one considers ratios of two non-central χ^2-distributed random variables. Moreover, in Hulley (2009) the doubly non-central beta distribution was extended to also allow for one of the non-central χ^2-distributed random variables to have zero degrees of freedom. This extension is an important contribution, since non-central χ^2-random variables with zero degrees of freedom naturally arise in the context of the MMM.

In particular, for $\nu > 0$ and $\lambda_1, \lambda_2 > 0$, we define

$$\beta_{0,\nu}(\lambda_1, \lambda_2) := \frac{\chi_0^2(\lambda_1)}{\chi_0^2(\lambda_1) + \chi_\nu^2(\lambda_2)}, \tag{3.3.16}$$

where $\beta_{0,\nu}(\lambda_1, \lambda_2)$ is an extended non-central beta distributed random variable with degrees of freedom 0 and ν and non-centrality parameters λ_1 and λ_2. Furthermore, $\chi_\nu^2(\lambda)$ denotes a non-central χ^2-distributed random variable with $\nu \geq 0$ degrees of freedom and non-centrality parameter λ. An algorithm to compute the extended non-central beta distribution was presented in Hulley (2009) and is recalled in Sect. 13.4.

We now discuss the financial market in more detail. Having introduced the additional risky securities S^a and S^b, we recall that the market index is the numéraire portfolio under the assumptions of the MMM. Hence the benchmarked prices $\hat{S}^a := S^a / S^{\delta_*}$ and $\hat{S}^b := S^b / S^{\delta_*}$ of the two additional securities must be supermartingales. From the pricing of the zero coupon bond performed in this section, it is known that an inverted squared Bessel process of dimension four is a supermartingale, hence we may model the benchmarked securities as follows:

$$\hat{S}_a := \frac{1}{X^a_{\varphi^a}}, \quad \text{and} \quad \hat{S}_b := \frac{1}{X^b_{\varphi^b}},$$

where X^a and X^b are independent squared Bessel processes of dimension four, which we assume to be given. We use deterministic time-transforms φ^a, φ^b, which, as in Hulley (2009), we do not specify further, and we refer to Platen and Heath (2010), Sect. 14.4 for this approach to modeling the prices of two risky assets S^a and S^b.

The payoff of the exchange option is given by $(S_T^a - S_T^b)^+$. Consequently, the real world pricing formula yields

$$M(t) = S_t^{\delta_*} E\left(\frac{(S_T^a - S_T^b)^+}{S_T^{\delta_*}} \,\middle|\, \mathcal{A}_t \right).$$

We introduce some auxiliary notation,

$$\lambda_a\left(t, S_t^{\delta_*}, S_t^a\right) := \frac{S_t^{\delta_*} / S_t^a}{\Delta\varphi^a(t)},$$

$$\lambda_b\left(t, S_t^{\delta_*}, S_t^b\right) := \frac{S_t^{\delta_*} / S_t^b}{\Delta\varphi_b(t)},$$

where as before

$$\Delta\varphi_a(t) := \varphi_a(T) - \varphi_a(t),$$

$$\Delta\varphi_b(t) := \varphi_b(T) - \varphi_b(t).$$

We recall that $\chi_\nu^2(\lambda)$ denotes a non-central χ^2-distributed random variable with ν degrees of freedom and non-centrality parameter λ and we use $p(\cdot, \nu, \lambda)$ to denote the corresponding probability density function, and $\beta_{0,\nu}(\lambda_1, \lambda_2)$ denotes the extended doubly non-central beta distributed random variable with degrees of freedom 0 and ν and non-centrality parameters λ_1 and λ_2. Now we price the exchange option, where we closely follow Proposition 5.18 in Hulley (2009):

$$
\begin{aligned}
M(t) &= S_t^{\delta_*} E\left(\frac{(S_T^a - S_T^b)^+}{S_T^{\delta_*}} \,\middle|\, \mathcal{A}_t \right) \\
&= S_t^{\delta_*} E\big(\mathbf{1}_{S_T^a \geq S_T^b} \hat{S}_T^a \mid \mathcal{A}_t \big) \\
&\quad - S_t^{\delta_*} E\big(\mathbf{1}_{S_T^a \geq S_T^b} \hat{S}_T^b \mid \mathcal{A}_t \big) \\
&= S_t^a E\left(\mathbf{1}_{\hat{S}_T^a \geq \hat{S}_T^b} \frac{\hat{S}_T^a}{\hat{S}_t^a} \,\middle|\, \mathcal{A}_t \right) \\
&\quad - S_t^b E\left(\mathbf{1}_{\hat{S}_T^a \geq \hat{S}_T^b} \frac{\hat{S}_T^b}{\hat{S}_t^b} \,\middle|\, \mathcal{A}_t \right) \\
&= S_t^a E\left(\mathbf{1}_{X_{\varphi_a(T)}^a \leq X_{\varphi_b(T)}^b} \frac{S_t^{\delta_*}/S_t^a}{X_{\varphi_a(T)}^a} \,\middle|\, \mathcal{A}_t \right) \\
&\quad - S_t^b E\left(\mathbf{1}_{X_{\varphi_a(T)}^a \leq X_{\varphi_b(T)}^b} \frac{S_t^{\delta_*}/S_t^b}{X_{\varphi_b(T)}^b} \,\middle|\, \mathcal{A}_t \right) \\
&= S_t^a E\left(\mathbf{1}_{\Delta\varphi_a(t)\chi_4^2(\lambda(t, S_t^{\delta_*}, S_t^a)) \leq \Delta\varphi_b(t)\chi_4^2(\lambda_b(t, S_t^{\delta_*}, S_t^b))} \frac{\lambda_a(t, S_t^{\delta_*}, S_t^a)}{\chi_4^2(\lambda_a(t, S_t^{\delta_*}, S_t^a))} \,\middle|\, \mathcal{A}_t \right) \\
&\quad - S_t^b \\
&\quad \times E\left(\mathbf{1}_{\Delta\varphi_a(t)\chi_a^2(\lambda_a(t, S_t^{\delta_*}, S_t^a)) \leq \Delta\varphi_b(t)\chi_4^2(\lambda_b(t, S_t^{\delta_*}, S_t^b))} \frac{\lambda_b(t, S_t^{\delta_*}, S_t^b)}{\chi_4^2(\lambda_b(t, S_t^{\delta_*}, S_t^b))} \,\middle|\, \mathcal{A}_t \right) \\
&= S_t^a \int_0^\infty p\big(\zeta, 4, \lambda_b\big(t, S_t^{\delta_*}, S_t^b\big)\big) \\
&\quad \times \int_0^{\frac{\Delta\varphi_b(t)}{\Delta\varphi_a(t)}\zeta} \frac{\lambda_a(t, S_t^{\delta_*}, S_t^a)}{\xi} p\big(\xi, 4, \lambda_a\big(t, S_t^{\delta_*}, S_t^a\big)\big)\, d\xi\, d\zeta \\
&\quad - S_t^b \int_0^\infty \frac{\lambda_b(t, S_t^{\delta_*}, S_t^b)}{\zeta} p\big(\zeta, 4, \lambda_b\big(t, S_t^{\delta_*}, S_t^b\big)\big) \\
&\quad \times \int_0^{\frac{\Delta\varphi_b(t)}{\Delta\varphi_a(t)}\zeta} p\big(\xi, 4, \lambda_a\big(t, S_t^{\delta_*}, S_t^a\big)\big)\, d\xi\, d\zeta.
\end{aligned}
$$

We now use (13.1.2) to obtain

$$
M(t) = S_t^a \int_0^\infty p\big(\zeta, 4, \lambda_b\big(t, S_t^{\delta_*}, S_t^b\big)\big) \int_0^{\frac{\Delta\varphi_b(t)}{\Delta\varphi_a(t)}\zeta} p\big(\xi, 0, \lambda_a\big(t, S_t^{\delta_*}, S_t^a\big)\, d\xi\, d\zeta
$$

$$
\begin{aligned}
&- S_t^b \int_0^\infty p\big(\zeta, 0, \lambda_b\big(t, S_t^{\delta_*}, S_t^b\big)\big) \int_0^{\frac{\Delta\varphi_b(t)}{\Delta\varphi_a(t)}\zeta} p\big(\xi, 4, \lambda_a\big(t, S_t^{\delta_*}, S_t^a\big)\big)\, d\xi\, d\zeta \\
&= S_t^a P\bigg(0 < \chi_0^2\big(\lambda_a\big(t, S_t^{\delta_*}, S_t^a\big)\big) \le \frac{\Delta\varphi_b(t)}{\Delta\varphi_a(t)} \chi_4^2\big(\lambda_b\big(t, S_t^{\delta_*}, S_t^b\big)\big)\bigg) \\
&\quad - S_t^b P\bigg(\chi_4^2\big(\lambda_a\big(t, S_t^{\delta_*}, S_t^a\big)\big) \le \frac{\Delta\varphi_b(t)}{\Delta\varphi_a(t)} \chi_0^2\big(\lambda_b\big(t, S_t^{\delta_*}, S_t^b\big)\big)\bigg) \\
&= S_t^a P\bigg(0 < \beta_{0,4}\big(\lambda_a\big(t, S_t^{\delta_*}, S_t^a\big), \lambda_b\big(t, S_t^{\delta_*}, S_t^b\big)\big) \le \frac{\Delta\varphi_b(t)}{\Delta\varphi_a(t) + \Delta\varphi_b(t)}\bigg) \\
&\quad - S_t^b P\bigg(\beta_{0,4}\big(\lambda_b\big(t, S_t^{\delta_*}, S_t^b\big), \lambda_a\big(t, S_t^{\delta_*}, S_t^a\big)\big) > \frac{\Delta\varphi_a(t)}{\Delta\varphi_a(t) + \Delta\varphi_b(t)}\bigg).
\end{aligned}
$$

3.3.5 Transformed Constant Elasticity of Variance Model

In this subsection, we discuss the *transformed constant elasticity of variance (TCEV) model*, which was introduced in Baldeaux et al. (2011c) as a generalization of the MMM. The TCEV was motivated by the *modified constant elasticity of variance model*, see Chap. 12 in Platen and Heath (2010), Heath and Platen (2002), and also Cox (1996). In particular, on a filtered probability space $(\Omega, \mathcal{A}, \underline{\mathcal{A}}, P)$ with the filtration $\underline{\mathcal{A}} = (\mathcal{A}_t)_{t\ge 0}$ satisfying the usual conditions, see Karatzas and Shreve (1991), we introduce a Brownian motion $W = \{W_t,\ t \ge 0\}$. The savings account discounted GOP $\bar{S}_t^{\delta_*}$ satisfies the following SDE:

$$
d\bar{S}_t^{\delta_*} = \big(\alpha_t^{\delta_*}\big)^{2-2a} \big(\bar{S}_t^{\delta_*}\big)^{2a-1} dt + \big(\alpha_t^{\delta_*}\big)^{1-a} \big(\bar{S}_t^{\delta_*}\big)^{a} dW_t, \qquad (3.3.17)
$$

where $\bar{S}_0^{\delta_*} > 0$ is the initial value, and as for the MMM, $\alpha_t^{\delta_*}$ satisfies

$$
\alpha_t^{\delta_*} = \alpha_0^{\delta_*} \exp\{\eta t\},
$$

where $\alpha_0^{\delta_*} > 0$ and $\eta > 0$. We observe the following:

- for $a = 1$, we recover geometric Brownian motion;
- for $a = \frac{1}{2}$, we recover the MMM.

The second observation motivates us to look for a connection between the TCEV model and the squared Bessel process, which was also observed in Sect. 3.1 for the CEV process. The advantage of such a link is that we can exploit the tractability of the squared Bessel process. In the following, we focus on the case $a \in (0, 1)$, and recall Proposition 8.1 from Baldeaux et al. (2011c).

Proposition 3.3.1 *The process $\bar{S}^{\delta_*} = \{\bar{S}_t^{\delta_*},\ t \ge 0\}$ satisfies the following equality in distribution*:

$$
\bar{S}_t^{\delta_*} \overset{d}{=} X_{\varphi(t)}^{\frac{1}{2-2a}},
$$

where X is a squared Bessel process of dimension $\nu = \frac{3-2a}{1-a}$ and

$$\varphi(t) = \frac{(1-a)(\alpha_0^{\delta_*})^{2-2a} c^2 (\exp\{2(1-a)\eta t\} - 1)}{2\eta}.$$

In Baldeaux et al. (2011c), this model was parameterized via non-parametric kernel based estimation techniques, see e.g. Florens-Zmirou (1993), Jacod (2000), Soulier (1998). We now turn to derivatives pricing.

3.3.6 *Standard European Options on the Index Under the TCEV Model*

In this subsection, we follow Baldeaux et al. (2011c) and price standard derivatives on a market index, which we interpret as the GOP. As in the previous subsection, we point out that even using a constant short rate process $r = \{r_t,\ t \geq 0\}$, where $r_t = r_0$, $t \geq 0$, and r_0 is a fixed constant, zero coupon bonds are index derivatives under the benchmark approach and typically stochastic. Furthermore, the price of a zero coupon bond can be used to confirm if an equivalent martingale measure exists or not. Using a constant short rate, and setting $S_t^0 = \exp\{rt\}$, we now compute the price of a zero coupon bond. We alert the reader to Lemma 13.1.1, where useful identities pertaining to the χ^2-distribution are presented. Using these relations, tractable expressions for standard derivatives under the TCEV model are easily derived, where we employ the notation

$$\Delta\varphi(t) = \varphi(T) - \varphi(t), \qquad \lambda\big(t, \bar{S}_t^{\delta_*}\big) = \frac{(\bar{S}_t^{\delta_*})^{2(1-a)}}{\Delta\varphi(t)},$$

$$x(t) = \frac{(K/S_t^0)^{2(1-a)}}{\Delta\varphi(t)},$$

and use $p(\cdot, \nu, \lambda)$ to denote the probability density function of a non-central χ^2-distributed random variable with ν degrees of freedom and non-centrality parameter λ and $\Psi(\cdot, \nu, \lambda)$ denotes the corresponding probability distribution function. From the real world pricing formula (1.3.19), we have, where $\delta = \frac{3-2a}{1-a}$ denotes the dimensionality of the squared Bessel process,

$$\begin{aligned}
P_T(t) &= S_t^{\delta_*} E\left(\frac{1}{S_T^{\delta_*}} \,\middle|\, \mathcal{A}_t\right) \\
&= \frac{S_t^0 \bar{S}_t^{\delta_*}}{S_T^0} E\left(\frac{1}{X_{\varphi(T)}^{\frac{1}{2-2a}}} \,\middle|\, \mathcal{A}_t\right) \\
&= \frac{S_t^0 \bar{S}_t^{\delta_*}}{S_T^0} \int_0^\infty \frac{1}{(\Delta\varphi(t) y)^{\frac{1}{2-2a}}} p\big(y, \nu, \lambda\big(t, \bar{S}_t^{\delta_*}\big)\big)\, dy \\
&= \frac{S_t^0}{S_T^0} \int_0^\infty \left(\frac{\lambda(t, \bar{S}_t^{\delta_*})}{y}\right)^{\frac{1}{2-2a}} p\big(y, \nu, \lambda\big(t, \bar{S}_t^{\delta_*}\big)\big)\, dy.
\end{aligned}$$

We now use (13.1.8) to conclude,

$$
\begin{aligned}
P_T(t) &= \frac{S_t^0}{S_T^0} \int_0^\infty p\big(\lambda(t, \bar{S}_t^{\delta_*}), \nu, y\big)\, dy \\
&= \frac{S_t^0}{S_T^0} \Psi\left(\lambda(t, \bar{S}_t^{\delta_*}), \frac{1}{1-a}, 0\right),
\end{aligned}
$$

where the last equality used (13.1.9). As in the previous subsection discussing the MMM, the above calculation allows us to confirm that an equivalent martingale measure does not exist for the TCEV model. The candidate Radon-Nikodym derivative process $\Lambda = \{\Lambda_t,\, t \geq 0\}$ is again given by

$$
\Lambda_t = \frac{\bar{S}_0^{\delta_*}}{\bar{S}_t^{\delta_*}}
$$

and we compute

$$
E(\Lambda_T \mid \mathcal{A}_0) = E\left(\frac{\bar{S}_0^{\delta_*}}{\bar{S}_T^{\delta_*}} \,\bigg|\, \mathcal{A}_0\right) = \Psi\left(\lambda(t, \bar{S}_t^{\delta_*}), \frac{1}{1-a}, 0\right) < 1.
$$

This calculation allows us to conclude that an equivalent martingale measure does not exist and that risk neutral pricing is not applicable. We hence continue to employ real world pricing, now turning to call options.

The real world pricing formula (1.3.19) yields the following price for a call option:

$$
\begin{aligned}
c_{T,K}(t) &= S_t^{\delta_*} E\left(\frac{(S_T^{\delta_*} - K)^+}{S_T^{\delta_*}} \,\bigg|\, \mathcal{A}_t\right) \\
&= S_t^{\delta_*} E\left(\left(1 - \frac{K}{S_T^{\delta_*}}\right)^+ \,\bigg|\, \mathcal{A}_t\right) \\
&= S_t^{\delta_*} \int_{x(t)}^\infty \left(1 - \left(\frac{x(t)}{y}\right)^{\frac{1}{2-2a}}\right) p\big(y, \nu, \lambda(t, \bar{S}_t^{\delta_*})\big)\, dy \\
&= S_t^{\delta_*} \big(1 - \Psi\big(x(t), \nu, \lambda(t, \bar{S}_t^{\delta_*})\big)\big) \\
&\quad - S_t^{\delta_*} \int_{x(t)}^\infty \left(\frac{x(t)}{y}\right)^{\frac{1}{2-2a}} p\big(y, \nu, \lambda(t, \bar{S}_t^{\delta_*})\big)\, dy.
\end{aligned}
$$

We now compute,

$$
\begin{aligned}
&S_t^{\delta_*} \int_{x(t)}^\infty \left(\frac{x(t)}{y}\right)^{\frac{1}{2-2a}} p\big(y, \nu, \lambda(t, \bar{S}_t^{\delta_*})\big)\, dy \\
&= \frac{K B_t}{B_T} \int_{x(t)}^\infty \left(\frac{\lambda(t, \bar{S}_t^{\delta_*})}{y}\right)^{\frac{1}{2-2a}} p\big(y, \nu, \lambda(t, \bar{S}_t^{\delta_*})\big)\, dy \\
&= \frac{K B_t}{B_T} \Psi\left(\lambda(t, \bar{S}_t^{\delta_*}), \frac{1}{1-a}, x(t)\right),
\end{aligned}
$$

where we used (13.1.8) and (13.1.10). Hence we arrive at

$$c_{T,K}(t) = S_t^{\delta_*}\big(1 - \Psi\big(x(t), \nu, \lambda\big(t, \bar{S}_t^{\delta_*}\big)\big)\big) - \frac{K S_t^0}{S_T^0}\Psi\left(\lambda\big(t, \bar{S}_t^{\delta_*}\big), \frac{1}{1-a}, x(t)\right).$$

We now price put options:

$$\begin{aligned}
p_{T,K}(t) &= S_t^{\delta_*} E\left(\frac{(K - S_T^{\delta_*})^+}{S_T^{\delta_*}} \,\Big|\, \mathcal{A}_t\right) \\
&= S_t^{\delta_*} E\left(\left(\frac{K}{S_T^{\delta_*}} - 1\right)^+ \,\Big|\, \mathcal{A}_t\right) \\
&= S_t^{\delta_*} \int_0^{x(t)} \left(\frac{x(t)}{y}\right)^{\frac{1}{2-2a}} p\big(y, \nu, \lambda\big(t, \bar{S}_t^{\delta_*}\big)\big)\, dy \\
&\quad - S_t^{\delta_*} \int_0^{x(t)} p\big(y, \nu, \lambda\big(t, \bar{S}_t^{\delta_*}\big)\big)\, dy \\
&= S_t^{\delta_*} \left(\frac{x(t)}{\lambda(t, \bar{S}_t^{\delta_*})}\right)^{\frac{1}{2-2a}} \int_0^{x(t)} \left(\frac{\lambda(t, \bar{S}_t^{\delta_*})}{y}\right)^{\frac{1}{2-2a}} p\big(y, \nu, \lambda\big(t, \bar{S}_t^{\delta_*}\big)\big)\, dy \\
&\quad - S_t^{\delta_*} \int_0^{x(t)} p\big(y, \nu, \lambda\big(t, \bar{S}_t^{\delta_*}\big)\big)\, dy \\
&= \frac{K S_t^0}{S_T^0}\left(\Psi\left(\lambda\big(t, \bar{S}_t^{\delta_*}\big), \frac{1}{1-a}, 0\right) - \Psi\left(\lambda\big(t, \bar{S}_t^{\delta_*}\big), \frac{1}{1-a}, x(t)\right)\right) \\
&\quad - S_t^{\delta_*}\Psi\big(x(t), \nu, \lambda\big(t, \bar{S}_t^{\delta_*}\big)\big).
\end{aligned}$$

The computed prices of the zero coupon bond, the call option and the put option satisfy the put-call parity,

$$p_{T,K}(t) = c_{T,K}(t) - S_t^{\delta_*} + K P_T(t).$$

We also look at binary call and put options. From the real world pricing formula (1.3.19), we obtain the following price for a binary call option

$$\begin{aligned}
BC_{T,K}(t) &= S_t^{\delta_*} E\left(\mathbf{1}_{S_T^{\delta_*} > K} \frac{1}{S_T^{\delta_*}} \,\Big|\, \mathcal{A}_t\right) \\
&= \frac{S_t^0}{S_T^0} E\left(\mathbf{1}_{X_{\varphi(T)}^{\frac{1}{2-2a}} > K/S_T^0} \frac{X_{\varphi(t)}^{\frac{1}{2-2a}}}{X_{\varphi(T)}^{\frac{1}{2-2a}}} \,\Big|\, \mathcal{A}_t\right) \\
&= \frac{S_t^0}{S_T^0} \int_{x(t)}^{\infty} \left(\frac{\lambda(t, \bar{S}_t^{\delta_*})}{y}\right)^{\frac{1}{2-2a}} p\big(y, \nu, \lambda\big(t, \bar{S}_t^{\delta_*}\big)\big)\, dy \\
&= \frac{S_t^0}{S_T^0} \int_{x(t)}^{\infty} p\big(\lambda\big(t, \bar{S}_t^{\delta_*}\big), \nu, y\big)\, dy \\
&= \exp\{-r(T-t)\}\Psi\left(\lambda\big(t, \bar{S}_t^{\delta_*}\big), \frac{1}{1-a}, x(t)\right),
\end{aligned}$$

where we used (13.1.8) and (13.1.10). Regarding binary put options, we compute

$$
\begin{aligned}
BP_{T,K}(t) &= S_t^{\delta_*} E\left(\mathbf{1}_{S_T^{\delta_*} \le K} \frac{1}{S_T^{\delta_*}} \,\Big|\, \mathcal{A}_t\right) \\
&= \frac{S_t^0}{S_T^0} E\left(\mathbf{1}_{X_{\varphi(T)}^{\frac{1}{2-2a}} \le K/S_T^0} \frac{X_{\varphi(t)}^{\frac{1}{2-2a}}}{X_{\varphi(T)}^{\frac{1}{2-2a}}} \,\Big|\, \mathcal{A}_t\right) \\
&= \frac{S_t^0}{S_T^0} \int_0^{x(t)} \left(\frac{\lambda(t, \bar{S}_t^{\delta_*})}{y}\right)^{\frac{1}{2-2a}} p\big(y, \nu, \lambda(t, \bar{S}_t^{\delta_*})\big)\, dy \\
&= \frac{S_t^0}{S_T^0} \int_0^{x(t)} p\big(\lambda(t, \bar{S}_t^{\delta_*}), \nu, y\big)\, dy \\
&= \frac{S_t^0}{S_T^0} \left(\Psi\left(\lambda(t, \bar{S}_t^{\delta_*}), \frac{1}{1-a}, 0\right) - \Psi\left(\lambda(t, \bar{S}_t^{\delta_*}), \frac{1}{1-a}, x(t)\right)\right).
\end{aligned}
$$

We again confirm that

$$
BC_{T,K}(t) + BP_{T,K}(t) = P_T(t),
$$

as should be expected.

3.3.7 *Rebates and Barrier Options Under the TCEV Model*

We now consider *path-dependent options* under the TCEV model, namely rebates and barrier options. The aim of this subsection is to show that the link with the squared Bessel process can again be exploited to price these options.

As in the preceding subsection, we recall that a rebate on a market index, interpreted as the GOP, pays one unit of the domestic currency as soon as the index hits a certain level, should this occur before $T > 0$. The trigger is, as before, assumed to be a deterministic barrier, $Z_t := z\exp\{rt\}$, $z > 0$. We, therefore, define two hitting times:

$$
\sigma_{z,t} = \inf\big\{u > 0\colon\ S_{t+u}^{\delta_*} = Z_{t+u}\big\}
$$

and

$$
\tau_z = \inf\{u > 0\colon\ X_u = z\},
$$

where S^{δ_*} denotes the GOP and X is a squared Bessel process of dimension $\delta = \frac{3-2a}{1-a}$. As for the MMM, we find it convenient to introduce X as a squared Bessel process of dimension δ, started at $\tilde{x} = x^{2-2a}$, where $x = \exp\{-rt\}S_t^{\delta_*}$. Since

$$
\bar{S}_{t+u}^{\delta_*} \stackrel{d}{=} X_{\varphi_t(u)}^{\frac{1}{2-2a}},
$$

we have

$$\begin{aligned}\sigma_{z,t} &= \inf\{u>0\colon S^{\delta_*}_{t+u}=Z_{t+u}\}\\ &= \inf\{X_{\varphi_t(u)}=\tilde{z}\},\end{aligned}$$

where $\tilde{z}=z^{2-2a}$ and

$$\begin{aligned}\varphi_t(u) &= \varphi(u+t)-\varphi(t)\\ &= \frac{(1-a)(\alpha_0^{\delta_*})^{2-2a}c^2}{2\eta}\exp\{2(1-a)\eta t\}\big(\exp\{2(1-a)\eta u\}-1\big)\\ &= \frac{\tilde{\alpha}_0}{4\eta}\exp\{\tilde{\eta}t\}\big(\exp\{\tilde{\eta}u\}-1\big),\end{aligned}$$

where $\tilde{\alpha}_0$ and $\tilde{\eta}$ are defined in the last equality and introduced for convenience. We hence obtain

$$\sigma_{z,t}=\inf\{u>0\colon X_{\varphi_t(u)}=\tilde{z}\}\stackrel{d}{=}\varphi_t^{-1}(\tau_{\tilde{z}})=\frac{1}{\tilde{\eta}}\ln\bigg(\frac{4\eta}{\tilde{\alpha}_0}\exp\{-\tilde{\eta}t\}\tau_{\tilde{z}}+1\bigg).$$

The real world pricing formula (1.3.19) yields

$$\begin{aligned}R_{\infty,z}(t) &= S^{\delta_*}_t E\bigg(\frac{1}{S^{\delta_*}_{t+\sigma_{z,t}}}\,\bigg|\,\mathcal{A}_t\bigg)\\ &= S^{\delta_*}_t E\bigg(\frac{1}{Z_{t+\sigma_{z,t}}}\,\bigg|\,\mathcal{A}_t\bigg)\\ &= \frac{S^{\delta_*}_t\exp\{-rt\}}{z}E\big(\exp\{-r\sigma_{z,t}\}\,\big|\,\mathcal{A}_t\big)\\ &= \frac{x}{z}E\big(\exp\{-r\varphi_t^{-1}(\tau_{\tilde{z}})\}\,\big|\,\mathcal{A}_t\big)\\ &= \frac{x}{z}E\bigg(\bigg(\frac{4\eta}{\tilde{\alpha}_0}\exp\{-\tilde{\eta}t\}\tau_{\tilde{z}}+1\bigg)^{-r/\tilde{\eta}}\,\bigg|\,\mathcal{A}_t\bigg)\\ &= \begin{cases}\frac{x}{z\Gamma(\frac{r}{\tilde{\eta}})}\int_0^\infty\exp\{-s\}s^{r/\tilde{\eta}-1}\frac{\psi_{4\eta/\tilde{\alpha}_0\exp\{-\tilde{\eta}t\}s}(\tilde{x})}{\psi_{4\eta/\tilde{\alpha}_0\exp\{-\tilde{\eta}t\}s}(\tilde{z})}\,ds & \text{for } \tilde{x}\le\tilde{z}\\ \frac{x}{z\Gamma(\frac{r}{\tilde{\eta}})}\int_0^\infty\exp\{-s\}s^{r/\tilde{\eta}-1}\frac{\phi_{4\eta/\tilde{\alpha}_0\exp\{-\tilde{\eta}t\}s}(\tilde{x})}{\phi_{4\eta/\tilde{\alpha}_0\exp\{-\tilde{\eta}t\}s}(\tilde{z})}\,ds & \text{for } \tilde{x}\ge\tilde{z},\end{cases}\end{aligned}$$

where we used Proposition 16.3.3, $x=S^{\delta_*}_t\exp\{-rt\}$, $\tilde{x}=x^{2-2a}=X_0$, and $\tilde{z}=z^{2-2a}$.

We now study finite maturity rebates. From the real world pricing formula (1.3.19) it follows

$$\begin{aligned}R_{T,z}(t) &= S^{\delta_*}_t E\bigg(\frac{\mathbf{1}_{t+\sigma_{z,t}\le T}}{S^{\delta_*}_{t+\sigma_{z,t}}}\,\bigg|\,\mathcal{A}_t\bigg)\\ &= S^{\delta_*}_t E\bigg(\frac{\mathbf{1}_{\sigma_{z,t}\le T-t}}{Z_{t+\sigma_{z,t}}}\,\bigg|\,\mathcal{A}_t\bigg)\\ &= \frac{x}{z}E\big(\mathbf{1}_{\sigma_{z,t}\le T-t}\exp\{-r\sigma_{z,t}\}\,\big|\,\mathcal{A}_t\big)\end{aligned}$$

$$= \frac{x}{z} E\big(\mathbf{1}_{\varphi_t^{-1}(\tau_{\tilde{z}}) \le T-t} \exp\{-r\varphi_t^{-1}(\tau_{\tilde{z}})\} \mid \mathcal{A}_t\big)$$

$$= \frac{x}{z} E\left(\mathbf{1}_{\tau_{\tilde{z}} \le \varphi_t(T-t)} \left(1 + \frac{4\eta}{\tilde{\alpha}_0} \exp\{-\tilde{\eta} t\}\tau_{\tilde{z}}\right)^{-r/\tilde{\eta}} \,\middle|\, \mathcal{A}_t\right).$$

We now compute the Laplace transform of the price, with respect to the transformed time to maturity, where we employ (16.3.11) from Chap. 16:

$$L_\beta\big(R_{T,z}(t)\big)$$
$$= \begin{cases} \frac{x}{z\beta\Gamma(r/\tilde{\eta})} \int_0^\infty \exp\{-s\} s^{r/\tilde{\eta}-1} \frac{\psi_{\beta+4\eta/\tilde{\alpha}_0 \exp\{-\tilde{\eta}t\}s}(\tilde{x})}{\psi_{\beta+4\eta/\tilde{\alpha}_0 \exp\{-\tilde{\eta}t\}s}(\tilde{z})}\, ds & \text{for } \tilde{x} \le \tilde{z} \\ \frac{x}{z\beta\Gamma(r/\tilde{\eta})} \int_0^\infty \exp\{-s\} s^{r/\tilde{\eta}-1} \frac{\phi_{\beta+4\eta/\tilde{\alpha}_0 \exp\{-\tilde{\eta}t\}s}(\tilde{x})}{\phi_{\beta+4\eta/\tilde{\alpha}_0 \exp\{-\tilde{\eta}t\}s}(\tilde{z})}\, ds & \text{for } \tilde{x} \ge \tilde{z}, \end{cases}$$

where $x = \exp\{-rt\} S_t^{\delta_*}$ and $\tilde{x} = x^{2-2a} = X_{\varphi(t)}$.

Now we consider barrier options and recall that $\tilde{q}_z(\cdot,\cdot,\cdot)$ denotes the transition density of X killed at z with respect to the speed measure. The Laplace transform of $\tilde{q}_z(t,x,y)$, with respect to t, is denoted by

$$\tilde{G}_\alpha^z(x,y) = \int_0^\infty \exp\{-\alpha t\} \tilde{q}_z(t,x,y)\, dt.$$

We recall that explicit formulas for $\tilde{G}_\alpha^z(x,y)$ are given in Chap. 16, see (16.3.6). Recalling that $x(t) = \frac{(K/S_t^0)^{2(1-a)}}{\varphi_t(T-t)}$, where $\varphi_t(T-t) = \varphi(T) - \varphi(t)$, we have

$$C_{T,K}^{out}(t) = S_t^{\delta_*} E\left(\mathbf{1}_{t+\sigma_{z,t}>T} \frac{(S_T^{\delta_*} - K)^+}{S_T^{\delta_*}} \,\middle|\, \mathcal{A}_t\right)$$

$$= S_t^{\delta_*} E\left(\mathbf{1}_{\sigma_{z,t}>T-t} \left(1 - \frac{K}{S_T^{\delta_*}}\right)^+ \,\middle|\, \mathcal{A}_t\right)$$

$$= S_t^{\delta_*} E\left(\mathbf{1}_{\tau_{\tilde{z}}>\varphi_t(T-t)} \left(1 - \frac{K/S_T^0}{X_{\varphi_t(T-t)}^{\frac{1}{2-2a}}}\right)^+ \,\middle|\, \mathcal{A}_t\right)$$

$$= S_t^{\delta_*} \int_{x(t)}^\infty \left(1 - \left(\frac{x(t)}{y}\right)^{\frac{1}{2-2a}}\right)^+ \tilde{q}_{\tilde{z}}\big(\varphi_t(T-t), \tilde{x}, y\big) m(y)\, dy,$$

where $m(\cdot)$ denotes the speed measure of X. For a squared Bessel process with index $\nu = \frac{\delta}{2} - 1 \neq 0$, $m(y) = y^\nu/(2|\nu|)$. Hence, one has

$$C_{T,K}^{out}(t) = S_t^{\delta_*} \int_{x(t)}^\infty \big(y^{\frac{1}{2-2a}} - \big(x(t)\big)^{\frac{1}{2-2a}}\big) \frac{\tilde{q}_{\tilde{z}}(\varphi_t(T-t), \tilde{x}, y)}{2|\nu|}\, dy.$$

As before, we now compute the Laplace transform with respect to the transformed time to maturity, which yields:

$$L_\beta\big(C_{T,K}^{out}(t)\big)$$
$$= \int_0^\infty \exp\{-\beta u\} \left(\frac{S_t^{\delta_*}}{2|\nu|} \int_{x(t)}^\infty \big(y^{\frac{1}{2-2a}} - x^{\frac{1}{2-2a}}(t)\big) \tilde{q}_{\tilde{z}}(u, \tilde{x}, y)\, dy\right) du$$

$$= \frac{S_t^{\delta_*}}{2|\nu|} \int_{x(t)}^{\infty} \big(y^{\frac{1}{2-2a}} - x^{\frac{1}{2-2a}}(t)\big) L_\beta\big(\tilde{q}_{\tilde{z}}(u, \tilde{x}, y)\big)\, dy$$

$$= \frac{S_t^{\delta_*}}{2|\nu|} \int_{x(t)}^{\infty} \big(y^{\frac{1}{2-2a}} - x^{\frac{1}{2-2a}}(t)\big) \tilde{G}_\beta^{\tilde{z}}(\tilde{x}, y)\, dy, \tag{3.3.18}$$

for all $\beta > 0$. Again, we focus separately on up-and-out calls, where $S_t^{\delta_*} \le Z_t \iff x \le z \iff \tilde{x} \le \tilde{z}$ and down-and-out calls, where $S_t^{\delta_*} \ge Z_t \iff x \ge z \iff \tilde{x} \ge \tilde{z}$. It is clear from Eq. (16.3.6) in Chap. 16, that $\tilde{x} \le \tilde{z}$ implies that $\tilde{G}_\beta^{\tilde{z}}(\tilde{x}, y) = 0$ for $y \ge \tilde{z}$. Hence

$$\begin{aligned}
&L_\beta\big(C_{T,K}^{out}(t)\big) \\
&= S_t^{\delta_*} \int_{x(t)}^{x(t)\vee\tilde{x}} \big(y^{\frac{1}{2-2a}} - x^{\frac{1}{2-2a}}(t)\big) \psi_\beta(y) \left(\phi_\beta(\tilde{x}) - \frac{\phi_\beta(\tilde{z})}{\psi_\beta(\tilde{z})} \psi_\beta(\tilde{x}) \right) dy \\
&\quad + S_t^{\delta_*} \int_{x(t)\vee\tilde{x}}^{x(t)\vee\tilde{z}} \big(y^{\frac{1}{2-2a}} - x^{\frac{1}{2-2a}}(t)\big) \psi_\beta(\tilde{x}) \left(\phi_\beta(y) - \frac{\phi_\beta(\tilde{z})}{\psi_\beta(\tilde{z})} \psi_\beta(y) \right) dy
\end{aligned}$$

if $x \le z$. We note that $w_\beta = 1/2|\nu|$ for the squared Bessel process of index $\nu = \frac{\delta}{2} - 1 \neq 0$. Hence the factor $1/2|\nu|$ from (3.3.18) disappeared. For the down-and-out call, one has

$$\begin{aligned}
&L_\beta\big(C_{T,K}^{out}(t)\big) \\
&= S_t^{\delta_*} \int_{x(t)\vee\tilde{z}}^{x(t)\vee\tilde{x}} \big(y^{\frac{1}{2-2a}} - x^{\frac{1}{2-2a}}(t)\big) \left(\psi_\beta(y) - \frac{\psi_\beta(\tilde{z})}{\phi_\beta(\tilde{z})} \phi_\beta(y) \right) \phi_\beta(\tilde{x})\, dy \\
&\quad + S_t^{\delta_*} \int_{x(t)\vee\tilde{x}}^{\infty} \big(y^{\frac{1}{2-2a}} - x^{\frac{1}{2-2a}}(t)\big) \left(\psi_\beta(\tilde{x}) - \frac{\psi_\beta(\tilde{z})}{\phi_\beta(\tilde{z})} \phi_\beta(\tilde{x}) \right) \phi_\beta(y)\, dy,
\end{aligned}$$

if $x \ge z$. Again, the factor $1/2|\nu|$ from (3.3.18) disappears, since $w_\beta = 1/2|\nu|$ for the squared Bessel process of index $\nu = \frac{\delta}{2} - 1 \neq 0$.

3.3.8 Exchange Options

In this subsection, we study exchange options, which entitle the owner to exchange one asset for another. As before, we add to the financial market two risky securities, whose price processes are denoted by S^a and S^b. We recall that the benchmarked price processes $\hat{S}^a := S^a/S^{\delta_*}$ and $\hat{S}^b := S^b/S^{\delta_*}$ must be supermartingales. As discussed above, in the TCEV model, if $X_{\varphi_a}^a$ is a squared Bessel process of dimension $\delta_1 = \frac{3-2c_1}{1-c_1}$, and $X_{\varphi_b}^b$ is a squared Bessel process of dimension $\delta_2 = \frac{3-2c_2}{1-c_2}$, then

$$\hat{S}^a := \frac{1}{(X_{\varphi_a}^a)^{\frac{1}{2-2c_1}}} \quad \text{and} \quad \hat{S}^b := \frac{1}{(X_{\varphi_b}^b)^{\frac{1}{2-2c_2}}}$$

are supermartingales, where $c_1 \in (0,1)$ and $c_2 \in (0,1)$. Again, φ^a and φ^b denote deterministic time transforms, which we do not need to specify further, see Hulley (2009).

The payoff of the exchange option is, as before, given by $(S_T^a - S_T^b)^+$, and the real world pricing formula (1.3.19) yields:

$$M(t) = S_t^{\delta_*} E\left(\frac{(S_T^a - S_T^b)^+}{S_T^{\delta_*}} \,\middle|\, \mathcal{A}_t \right).$$

We use the notation

$$\lambda_a := \lambda_a\big(t, S_t^{\delta_*}, S_t^a\big) = \frac{(S_t^{\delta_*}/S_t^a)^{2-2c_1}}{\Delta\varphi_a(t)},$$

$$\lambda_b := \lambda_b\big(t, S_t^{\delta_*}, S_t^b\big) = \frac{(S_t^{\delta_*}/S_t^b)^{2-2c_2}}{\Delta\varphi_b(t)},$$

where $\Delta\varphi_k(t) = \varphi_k(T) - \varphi_k(t)$, $k \in \{a, b\}$, and we recall that $\chi_\nu^2(\lambda)$ denotes a non-central χ^2-distributed random variable with ν degrees of freedom and non-centrality parameter λ and $\Psi(\cdot, \nu, \lambda)$ the corresponding probability distribution function. Hence

$$\begin{aligned}
&S_t^{\delta_*} E\left(\frac{(S_T^a - S_T^b)^+}{S_T^{\delta_*}} \,\middle|\, \mathcal{A}_t \right) \\
&= S_t^{\delta_*} E\left(\frac{1}{(X^a_{\varphi_a(T)})^{\frac{1}{2-2c_1}}} \mathbf{1}_{S_T^a \geq S_T^b} \,\middle|\, \mathcal{A}_t \right) \\
&\quad - S_t^{\delta_*} E\left(\frac{1}{(X^b_{\varphi_b(T)})^{\frac{1}{2-2c_2}}} \,\middle|\, \mathcal{A}_t \right) \\
&= S_t^a E\left(\mathbf{1}_{\hat{S}_T^a \geq \hat{S}_T^b} \left(\frac{\lambda_a}{\chi^2_{\delta_1}(\lambda_a)} \right)^{\frac{1}{2-2c_1}} \,\middle|\, \mathcal{A}_t \right) \\
&\quad - S_t^b E\left(\mathbf{1}_{\hat{S}_T^a \geq \hat{S}_T^b} \left(\frac{\lambda_b}{\chi^2_{\delta_2}(\lambda_b)} \right)^{\frac{1}{2-2c_2}} \,\middle|\, \mathcal{A}_t \right) \\
&= S_t^a E\left(\mathbf{1}_{\Delta\varphi_a(t)\chi^2_{\delta_1}(\lambda_a) \leq (\Delta\varphi_b(t)\chi^2_{\delta_2}(\lambda_b))^{\frac{2-2c_1}{2-2c_2}}} \left(\frac{\lambda_a}{\chi^2_{\delta_1}(\lambda_a)} \right)^{\frac{1}{2-2c_1}} \,\middle|\, \mathcal{A}_t \right) \\
&\quad - S_t^b E\left(\mathbf{1}_{\Delta\varphi_a(t)\chi^2_{\delta_1}(\lambda_a) \leq (\Delta\varphi_b(t)\chi^2_{\delta_2}(\lambda_b))^{\frac{2-2c_1}{2-2c_2}}} \left(\frac{\lambda_b}{\chi^2_{\delta_2}(\lambda_b)} \right)^{\frac{1}{2-2c_2}} \,\middle|\, \mathcal{A}_t \right) \\
&= S_t^a \int_0^\infty p(\zeta, \delta_2, \lambda_b) \int_0^{\frac{(\Delta\varphi_b(t)\zeta)^{\frac{2-2c_1}{2-2c_2}}}{\Delta\varphi_a(t)}} p(\xi, \delta_1, \lambda_a) \left(\frac{\lambda_a}{\xi} \right)^{\frac{1}{2-2c_1}} d\xi\, d\zeta \\
&\quad - S_t^b \int_0^\infty p(\zeta, \delta_2, \lambda_b) \left(\frac{\lambda_b}{\zeta} \right)^{\frac{1}{2-2c_2}} \int_0^{\frac{(\Delta\varphi_b(t)\zeta)^{\frac{2-2c_1}{2-2c_2}}}{\Delta\varphi_a(t)}} p(\xi, \delta_1, \lambda_a)\, d\xi\, d\zeta.
\end{aligned}$$

Regarding the first term, we have

$$S_t^a \int_0^\infty p(\zeta,\delta_2,\lambda_b) \int_0^{\frac{(\Delta\varphi_b(t)\zeta)^{\frac{2-2c_1}{2-2c_2}}}{\Delta\varphi_a(t)}} p(\xi,\delta_1,\lambda_a)\left(\frac{\lambda_a}{\xi}\right)^{\frac{1}{2-2c_1}} d\xi\, d\zeta$$

$$= S_t^a \int_0^\infty p(\zeta,\delta_2,\lambda_b) \int_0^{\frac{(\Delta\varphi_b(t)\zeta)^{\frac{2-2c_1}{2-2c_2}}}{\Delta\varphi_a(t)}} p(\lambda_a,\delta_1,\xi)\, d\xi\, d\zeta$$

$$= S_t^a \int_0^\infty p(\zeta,\delta_2,\lambda_b)\left(\Psi(\lambda_a,\delta_1-2,0) - \Psi\left(\lambda_a,\delta_1-2,\frac{(\Delta\varphi_b(t)\zeta)^{\frac{2-2c_1}{2-2c_2}}}{\Delta\varphi_a(t)}\right)\right) d\zeta$$

$$= S_t^a\left(\Psi(\lambda_a,\delta_1-2,0) - P\left(\chi^2_{\delta_1-2}\left(\frac{(\Delta\varphi_b(t)\chi^2_{\delta_2}(\lambda_b))^{\frac{2-2c_1}{2-2c_2}}}{\Delta\varphi_a(t)}\right) \le \lambda_a\right)\right).$$

Regarding the second term, it follows using (13.1.8) and (13.1.10) that

$$S_t^b \int_0^\infty p(\zeta,\delta_2,\lambda_b)\left(\frac{\lambda_b}{\zeta}\right)^{\frac{1}{2-2c_2}} \int_0^{\frac{(\Delta\varphi_b(t)\zeta)^{\frac{2-2c_1}{2-2c_2}}}{\Delta\varphi_a(t)}} p(\xi,\delta_1,\lambda_a)\, d\xi\, d\zeta$$

$$= S_t^b \int_0^\infty p(\lambda_b,\delta_2,\zeta) \int_0^{\frac{(\Delta\varphi_b(t)\zeta)^{\frac{2-2c_1}{2-2c_2}}}{\Delta\varphi_a(t)}} p(\xi,\delta_1,\lambda_a)\, d\xi\, d\zeta$$

$$= S_t^b \int_0^\infty p(\xi,\delta_1,\lambda_a) \int_{\frac{(\Delta\varphi_a(t)\xi)^{\frac{2-2c_2}{2-2c_1}}}{\Delta\varphi_b(t)}}^\infty p(\lambda_b,\delta_2,\zeta)\, d\zeta\, d\xi$$

$$= S_t^b \int_0^\infty p(\xi,\delta_1,\lambda_a)\Psi\left(\lambda_b,\delta_2-2,\frac{(\Delta\varphi_a(t)\xi)^{\frac{2-2c_2}{2-2c_1}}}{\Delta\varphi_b(t)}\right) d\xi$$

$$= S_t^b P\left(\chi^2_{\delta_2-2}\left(\frac{(\Delta\varphi_a(t)\chi^2_{\delta_1}(\lambda_a))^{\frac{2-2c_2}{2-2c_1}}}{\Delta\varphi_b(t)}\right) \le \lambda_b\right).$$

Finally, we get

$$\begin{aligned} &M(t) \\ &= S_t^a\left(\Psi(\lambda_a,\delta_1-2,0) - P\left(\chi^2_{\delta_1-2}\left(\frac{(\Delta\varphi_b(t)\chi^2_{\delta_2}(\lambda_b))^{\frac{2-2c_1}{2-2c_2}}}{\Delta\varphi_a(t)}\right) \le \lambda_a\right)\right) \\ &\quad - S_t^b P\left(\chi^2_{\delta_2-2}\left(\frac{(\Delta\varphi_a(t)\chi^2_{\delta_1}(\lambda_a))^{\frac{2-2c_2}{2-2c_1}}}{\Delta\varphi_b(t)}\right) \le \lambda_b\right). \end{aligned}$$

Chapter 4
Lie Symmetry Group Methods

A basis for the availability of explicit formulas for derivative prices under the Black-Scholes Model (BSM) and the quadratic models, which we discussed in the previous sections, is the explicitly available transition density for these models. Therefore, it is important to find systematically further diffusion dynamics with explicit transition densities. In this chapter, we show how to obtain transforms, usually Laplace and Fourier transforms, of transition densities of various diffusions beyond the ones we have already studied. Our approach is based on Lie symmetry methods, and has been developed by Craddock and collaborators, see Craddock and Platen (2004), Craddock and Lennox (2007, 2009), Craddock (2009), and Craddock and Dooley (2010). The following motivation follows closely Sect. 2 in Craddock and Lennox (2007). All concepts referred to in this motivation are explained in Sect. 4.2 in more detail. Readers interested in the technical details of Lie symmetry analysis are referred to Bluman and Kumei (1989), and Olver (1993).

4.1 Motivation for Lie Symmetry Methods for Diffusions

We consider the following partial differential equation (PDE), which for our purposes will typically be the Kolmogorov forward or backward equation for a diffusion, or a PDE resulting from the Feynman-Kac formula, see Sect. 15.8 in Chap. 15:

$$u_t = P\big(x, u^{(n)}\big) \quad x \in \Omega \subseteq \Re. \tag{4.1.1}$$

Here $P(\cdot,\cdot)$ is a differential operator, x and t are independent variables, and u is the dependent variable, and n denotes the number of derivatives $u^{(1)}, u^{(2)}, \dots, u^{(n)}$ in x, we typically have $n = 2$. Lie's method, see e.g. Olver (1993), allows us to find vector fields

$$\boldsymbol{v} = \xi(x,t,u)\partial_x + \tau(x,t,u)\partial_t + \phi(x,t,u)\partial_u,$$

which generate one parameter Lie groups that preserve solutions of (4.1.1). It is common in the area to denote the action of $\boldsymbol{v}$ on solutions $u(x,t)$ of (4.1.1) by

$$\rho\big(\exp\{\epsilon \boldsymbol{v}\}\big)u(x,t) = \sigma(x,t;\epsilon)u\big(a_1(x,t;\epsilon), a_2(x,t;\epsilon)\big) \tag{4.1.2}$$

J. Baldeaux, E. Platen, *Functionals of Multidimensional Diffusions with Applications to Finance*, Bocconi & Springer Series 5, DOI 10.1007/978-3-319-00747-2_4,

for some functions σ, a_1, and a_2, where ϵ is the parameter of the group, σ is referred to as the multiplier, and a_1 and a_2 are changes of variables of the symmetry. For the applications we have in mind, ϵ and σ are of crucial importance, ϵ will play the role of the transform parameter of the Fourier or Laplace transform and σ will usually be the Fourier or Laplace transform of the transition density. Following Craddock and Lennox (2007) or Craddock et al. (2009), we assume that (4.1.1) has a fundamental solution $p(x,y,t)$. For this book, it suffices to recall that we can express a solution $u(x,t)$ of the PDE (4.1.1), subject to the initial condition $u(x,0) = f(x)$, in the form

$$u(x,t) = \int_{\Omega} f(y)p(x,y,t)\,dy, \tag{4.1.3}$$

where $p(x,y,t)$ is a fundamental solution of (4.1.1). The key idea of the transform method is to connect (4.1.2) and (4.1.3). Now consider a stationary, i.e. a time-independent solution, say $u_0(x)$. Of course, (4.1.2) yields

$$\rho\big(\exp\{\epsilon \boldsymbol{v}\}\big)u_0(x) = \sigma(x,t;\epsilon)u_0\big(a_1(x,t;\varepsilon)\big),$$

which also solves the initial value problem. We now set $t = 0$ and use (4.1.2) and (4.1.3) to obtain

$$\int_{\Omega} \sigma(y,0;\epsilon)u_0\big(a_1(y,0;\epsilon)\big)p(x,y,t)dy = \sigma(x,t;\epsilon)u_0\big(a_1(x,t;\epsilon)\big). \tag{4.1.4}$$

Since σ, u_0, and a_1 are known functions, we have a family of integral equations for $p(x,y,t)$. In Sect. 4.3, we will discuss as an example the one-dimensional Black-Scholes PDE, which can be reduced to the one-dimensional heat equation

$$u_t = \frac{1}{2}\sigma^2 u_{xx}. \tag{4.1.5}$$

We will show that if $u(x,t)$ solves (4.1.5), then for ϵ sufficiently small, so does

$$\tilde{u}(z,t) = \exp\left\{\frac{\epsilon t^2}{2\sigma^2} - \frac{z\epsilon}{\sigma^2}\right\}u(z - t\epsilon, t).$$

Taking $u_0 = 1$, (4.1.4) gives

$$\int_{-\infty}^{\infty} \exp\left\{-\frac{y\epsilon}{\sigma^2}\right\}p(x,y,t)\,dy = \exp\left\{\frac{\epsilon^2 t}{2\sigma^2} - \frac{x\epsilon}{\sigma^2}\right\}.$$

Setting $a = -\frac{\epsilon}{\sigma^2}$, we get

$$\int_{-\infty}^{\infty} \exp\{ay\}p(x,y,t)\,dy = \exp\left\{\frac{a^2\sigma^2 t}{2} + ax\right\}. \tag{4.1.6}$$

We recognize that (4.1.6) is the moment generating function of the Gaussian distribution. So $p(x,y,t)$ is the Gaussian density with mean x and variance $\sigma^2 t$. We alert the reader to the fact that ϵ plays the role of the transform parameter and $\sigma(\cdot,\cdot;\cdot)$ corresponds to the moment generating function. Finally, we recall a remark from Craddock (2009), namely the fact that Laplace and Fourier transforms can be readily obtained through Lie algebra computations, seems to suggest a relationship between Lie symmetry analysis and harmonic analysis. We refer the reader to Olver

(1993), where this relationship is explored in more detail. It allows one to find transition densities for several other diffusions beyond those previously mentioned in this book.

We now recall some concepts of Lie symmetry analysis, which shall allow the reader to appreciate some of the ideas behind the results subsequently presented. This listing of facts is not aimed at being a rigorous introduction to the topic. The reader interested in studying this method in more detail is referred to Bluman and Kumei (1989), and Olver (1993).

4.2 Some Basic Facts of Lie Symmetry Analysis

The structure of this section is as follows: in Sect. 4.2.1, we introduce symmetry groups of PDEs, and in Sect. 4.2.2 we discuss Lie groups, their associated Lie algebras, vector fields and finally Lie algebras of vector fields. Finally, in Sect. 4.2.3, we discuss prolongations, which allow us to link the two concepts, i.e. they allow us to determine when the Lie group generated by a particular vector field is the symmetry group of a PDE. This crucial link was originally established by *Lie*, see Lie (1881). Finally, we point out that this section follows closely Chap. 1 of the book manuscript (Craddock 2013).

4.2.1 Symmetry Groups of PDEs

In this chapter, we consider single differential equations of order n in m variables on a simply connected subset $\Omega \subseteq \Re^m$. We denote the PDE, as is common in the literature on Lie symmetry analysis, by

$$P\big(\boldsymbol{x}, D^{\boldsymbol{\alpha}} u\big) = 0,$$

where P is a differential operator on $\Omega \times \Re$,

$$D^{\boldsymbol{\alpha}} u = \frac{\partial^{|\boldsymbol{\alpha}|} u}{\partial x_1^{\alpha_1} \dots \partial x_m^{\partial_m}},$$

$\boldsymbol{\alpha} = (\alpha_1, \dots, \alpha_m)$, is a multi-index, $\alpha_i \in \mathcal{N}$, $i \in \{1, \dots, m\}$ and $|\boldsymbol{\alpha}| = \alpha_1 + \dots + \alpha_m$.

Definition 4.2.1 A symmetry group of a differential equation is a group of transformations acting on the independent and dependent variables of the system such that it maps solutions of the equation to other solutions. To be more precise, let $\mathcal{H}_P$ denote the space of all solutions of the PDE

$$P\big(\boldsymbol{x}, D^{\boldsymbol{\alpha}} u\big) = 0.$$

A symmetry $\mathcal{S}$ is a mapping of $\mathcal{H}_P$ into itself, i.e. $\mathcal{S} : \mathcal{H}_P \to \mathcal{H}_P$. Thus if $u \in \mathcal{H}_P$, then we must have $\mathcal{S}u \in \mathcal{H}_P$.

To give the reader a feel for symmetries, we present a trivial example.

Example 4.2.2 For the one-dimensional heat equation,

$$u_t = u_{xx}, \tag{4.2.7}$$

it is well-known, that for an ϵ sufficiently small, $u(x+\epsilon, t)$ and $u(x, t+\epsilon)$ are also solutions of (4.2.7). Hence the mappings $\rho(\epsilon)u(x,t) = u(x+\epsilon, t)$ and $\pi(\epsilon)u(x,t) = u(x, t+\epsilon)$ are symmetries.

The reason for introducing symmetry analysis is that some symmetry groups can map trivial solutions, in some cases $u = 1$, to interesting solutions, such as the transition density of a Brownian motion with drift, as we will illustrate in Sect. 4.3. How does one come up with these interesting symmetry groups? The answer will be given in Sect. 4.2.3, where we present theorems due to Lie and Olver, which essentially give us a mechanical procedure for calculating symmetry groups mapping trivial solutions to interesting ones. The key observation, due to Lie, is that under certain conditions, which he showed to be necessary and sufficient, see Theorem 4.2.11, an analytical problem can be reduced to an algebraic problem. Calculation of symmetry groups can be reduced to the problem of computing vector fields, see Sect. 4.2.3. By exponentiating this vector field, see Sect. 4.3 for an illustration, one can compute the symmetry groups of the PDE under consideration, as it is generated by the given vector field. Next, we introduce Lie groups, Lie algebras, vector fields and Lie algebras of vector fields, see Hall (2003).

4.2.2 Lie Groups

Many Lie groups, though not all, can be realized as matrix groups, which are closed subgroups of the general linear group. We need the definition of matrix Lie groups when introducing Lie algebras.

Definition 4.2.3 The general linear group $GL(n, \Re)$ is the group of all $n \times n$ invertible matrices with entries in $\Re$, and $GL(n, \mathbf{C})$ is the group of all $n \times n$ invertible matrices with entries in $\mathbf{C}$.

We can now define a matrix Lie group. In the next definition, convergence of matrices is understood componentwise, i.e. $\{\boldsymbol{A}_n\}_{n=1}^{\infty}$ converges to $\boldsymbol{A}$, if each entry in $\boldsymbol{A}_n$ converges to the corresponding entry of $\boldsymbol{A}$, where convergence is understood in the sense of convergence of sequences in $\Re$ or $\mathbf{C}$.

Definition 4.2.4 A matrix Lie group of dimension n is a closed subgroup of the general linear group $GL(n, \Re)$ or $GL(n, \mathbf{C})$.

In this book, we restrict ourselves to matrix Lie groups. Next, we introduce Lie algebras. In Definition 4.2.5, we will show how to associate a Lie algebra with a given matrix Lie group. Consequently, analytical problems on the matrix Lie group amount to algebraic problems on the associated Lie algebra. For the purpose of this chapter, it is important to realize that when computing the symmetry group of a PDE, we actually compute its Lie algebra. The latter can be achieved using a mechanical procedure, due to Lie. We illustrate this in Sect. 4.3, but it should suffice as a motivation for introducing Lie algebras. We now introduce Lie algebras and Lie brackets.

Definition 4.2.5 For a given matrix Lie group G of dimension n, the Lie algebra of G consists of the vector space of $n \times n$ matrices $\boldsymbol{A}$, for which the matrix exponential

$$\exp\{\boldsymbol{A}\} = \boldsymbol{I} + \boldsymbol{A} + \frac{\boldsymbol{A}^2}{2} + \cdots = \sum_{k=0}^{\infty} \frac{\boldsymbol{A}^k}{k!} \in G. \tag{4.2.8}$$

In other words, the Lie algebra of G is defined to be

$$\boldsymbol{g} := \big\{\boldsymbol{A} \in \mathcal{M}_n \colon \exp\{\boldsymbol{A}\} \in G\big\},$$

recalling that $\mathcal{M}_n$ is the space of $n \times n$ matrices. For $\boldsymbol{A}, \boldsymbol{B} \in \boldsymbol{g}$, the Lie bracket is $[\boldsymbol{A}, \boldsymbol{B}] := \boldsymbol{AB} - \boldsymbol{BA}$.

The commutator

$$[\boldsymbol{A}, \boldsymbol{B}] = \boldsymbol{AB} - \boldsymbol{BA}$$

of two square matrices $\boldsymbol{A}$, $\boldsymbol{B}$ is known as the Lie bracket of $\boldsymbol{A}$ and $\boldsymbol{B}$ and plays a central role in the theory of Lie algebras. For a proof of the following theorem we refer to Hall (2003).

Theorem 4.2.6 *The Lie algebra of a matrix Lie group is a vector space which is closed under Lie brackets.*

Every finite dimensional Lie algebra can be realized in terms of first order differential operators. So far, we have considered Lie algebras which are vector spaces of matrices. However, equivalently, one can also introduce Lie algebras as vector spaces of first order differential operators,

$$\boldsymbol{v}(f) = \sum_{k=1}^{m} \xi_k(\boldsymbol{x}, u)\partial_{x_k} + \phi(\boldsymbol{x}, u)\partial_u, \tag{4.2.9}$$

where ∂_x denotes the partial derivative with respect to x. We now define a Lie algebra of vector fields. We also include the closure under the Lie bracket as part of the definition.

Definition 4.2.7 Consider a collection of n linearly independent vector fields $\mathcal{V} = \{\boldsymbol{v}_1, \ldots, \boldsymbol{v}_n\}$. Define the Lie bracket of $\boldsymbol{v}_i$ and $\boldsymbol{v}_k$ by

$$[\boldsymbol{v}_i, \boldsymbol{v}_k]f = \boldsymbol{v}_i(\boldsymbol{v}_k f) - \boldsymbol{v}_k(\boldsymbol{v}_i f).$$

Suppose further that for any vectors $\boldsymbol{v}_i, \boldsymbol{v}_j$, $1 \le i \le j \le n$, we have

$$[\boldsymbol{v}_i, \boldsymbol{v}_j] = \sum_{k=1}^{n} c_{i,j}^k \boldsymbol{v}_k,$$

for some constants $c_{i,j}^k$. Denote the linear span of $\mathcal{V}$ by $\mathcal{G}$. Then $\mathcal{G}$ is an n-dimensional Lie algebra, with basis vectors $\{\boldsymbol{v}_1, \ldots, \boldsymbol{v}_n\}$. The numbers $c_{i,j}^k$ are known as the structure constants of the Lie algebra. The Lie bracket of $\boldsymbol{v}$ and $\boldsymbol{w}$ satisfies

$$\begin{aligned}
&[\boldsymbol{v}, a\boldsymbol{w} + bz] = a[\boldsymbol{v}, \boldsymbol{w}] + b[\boldsymbol{v}, z] \\
&[a\boldsymbol{v} + b\boldsymbol{w}, z] = a[\boldsymbol{v}, z] + b[\boldsymbol{w}, z] \\
&[\boldsymbol{v}, \boldsymbol{w}] = -[\boldsymbol{w}, \boldsymbol{v}] \\
&\big[\boldsymbol{v}, [\boldsymbol{w}, z]\big] + \big[z, [\boldsymbol{v}, \boldsymbol{w}]\big] + \big[\boldsymbol{w}, [z, \boldsymbol{v}]\big] = 0.
\end{aligned}$$

The last identity is referred to as *Jacobi's identity*.

We mentioned earlier that instead of dealing with analytical problems on the Lie group, we instead choose to deal with algebraic problems on the Lie algebra. Lie algebras generate Lie groups via the exponential map. For the matrix Lie groups we previously discussed, this exponential map is simply the matrix exponential, see (4.2.8). For Lie algebras of vector fields, this exponential map is the Lie series, which we now introduce.

Every vector field $\boldsymbol{v}$, see (4.2.9), whose coefficients ξ, ϕ are sufficiently well-behaved, generates a one-parameter Lie group. The local Lie group will be called the *flow* of $\boldsymbol{v}$, denoted $\exp\{\epsilon\boldsymbol{v}\}$, where ϵ is the notation for the parameter of the group. The notation $\exp\{\epsilon\boldsymbol{v}\}$ is motivated by noting that the action generated by $\boldsymbol{v}$ can be obtained by summing the so-called Lie series,

$$f\big(\exp\{\epsilon\boldsymbol{v}\}\boldsymbol{x}\big) = f(\boldsymbol{x}) + \epsilon\boldsymbol{v}(f) + \frac{\epsilon^2}{2!}\boldsymbol{v}^2(f) + \cdots = \sum_{n=0}^{\infty} \frac{\epsilon^n}{n!}\boldsymbol{v}^n(f), \tag{4.2.10}$$

where $\boldsymbol{v}$ is tangent to a manifold M and $f \in C^\infty(M)$. Compared with the matrix exponential, the map sending $\boldsymbol{v}$ to $\exp\{\boldsymbol{v}\}$ can be understood as the exponential map for vector fields. Of course, the issue of convergence of the series in (4.2.10) needs to be addressed. In this chapter, we simply assume that it converges, at least for ϵ sufficiently small.

In principle, we could use the Lie series to determine how the group generated by $\boldsymbol{v}$ acts on the function u. However, the Lie series does not turn out to be a useful computational tool. Instead, we will use the following theorem, which appeared in Olver (1993), Chap. 1, to determine how the group generated by $\boldsymbol{v}$ acts on u and the independent variables $\boldsymbol{x}$.

Theorem 4.2.8 *Suppose that the vector field $\boldsymbol{v}$ is tangent to the smooth manifold $X \times U \subseteq \Re^m \times \Re$ and takes the form (4.2.9). Let the group of transformations generated by $\boldsymbol{v}$ act on the point $(\boldsymbol{x}, u) \in X \times U$. If we denote the transformed variables* *by* $(\tilde{\boldsymbol{x}}, \tilde{u})$, *then the new variables satisfy the following system of ordinary differential equations*

$$\frac{d\tilde{x}_k}{d\epsilon} = \xi_k(\tilde{\boldsymbol{x}}, \tilde{u}), \quad k = 1, \dots, m$$
$$\frac{d\tilde{u}}{d\epsilon} = \phi(\tilde{\boldsymbol{x}}, \tilde{u}),$$

with the initial data $\tilde{x}_k(0) = x_k$, $k = 1, \dots, m$, *and* $\tilde{u}(0) = u$.

Intuitively, one can think of the group generated by $\boldsymbol{v}$ as acting on the graph of u, denoted by $(\boldsymbol{x}, u)$, transforming it in some manner. The transformed graph is obtained from Theorem 4.2.8, and hence we write

$$\exp\{\epsilon \boldsymbol{v}\}(\boldsymbol{x}, u) = (\tilde{\boldsymbol{x}}, \tilde{u}). \tag{4.2.11}$$

Equation (4.2.11) describes how the one-parameter group generated by $\boldsymbol{v}$ acts on solutions of the PDE. However, we are primarily interested in knowing when the group generated by $\boldsymbol{v}$ is a symmetry group, i.e. when does it map solutions of (4.1.1) to solutions of (4.1.1).

This question will be answered in the next subsection, the crucial ingredient being prolongations.

4.2.3 Lie's Prolongation Algorithm

We remark that a detailed discussion of Lie symmetries requires substantial technical tools from areas such as differential geometry and the theory of jet bundles. On the other hand, the computation of Lie symmetries is a mechanical procedure, thanks to Theorem 4.2.11 presented in this subsection. Deliberately, we present the computation of Lie symmetries as early as possible. Readers interested in the technical background are referred to Olver (1993).

Firstly, we need to understand how a symmetry group acts on the independent variables $\boldsymbol{x}$, the solution u, and the derivatives of u, since we deal with a PDE. For concreteness, let $X \subseteq \Re^m$ and $U \subseteq \Re$ be smooth manifolds and consider a PDE

$$P\big(\boldsymbol{x}, D^\alpha u\big) = 0, \tag{4.2.12}$$

where $\boldsymbol{x} \in X$ and $u \in U$. For a solution u of (4.2.12), we denote its graph by

$$\Gamma_u = \big\{\big(\boldsymbol{x}, u(\boldsymbol{x})\big) \colon \boldsymbol{x} \in X\big\},$$

which is obtained from u as $\boldsymbol{x}$ takes values in X. We find it convenient to abbreviate $(\boldsymbol{x}, u(\boldsymbol{x}))$ by $(\boldsymbol{x}, u)$. Clearly, when a symmetry group G of (4.2.12) acts on u, we obtain a new solution $\tilde{u}(\tilde{\boldsymbol{x}})$, the graph of which we denote by $\Gamma_{\tilde{u}}$. However, we also need to understand how a symmetry group G affects the distribution of u. This is where prolongations come into play: the n-th prolongation of G, $\mathrm{pr}^n\, G$, extends the action of G to $(\boldsymbol{x}, u)$ and also all derivatives of u, up to order n, as the following definition formally states.

Definition 4.2.9 To determine $\text{pr}^n\, G$, let $\mathcal{D}^n$ be the mapping

$$\mathcal{D}^n : (\boldsymbol{x}, u) \to (\boldsymbol{x}, u, u_{x_1}, \ldots, u_{x_m,\ldots x_m}) = \big(\boldsymbol{x}, u^{(n)}\big).$$

Then the n-th prolongation must satisfy

$$\mathcal{D}^n \circ G = \text{pr}^n\, G \circ \mathcal{D}^n.$$

Intuitively speaking, Definition 4.2.9 says that applying the symmetry group G to $(\boldsymbol{x}, u)$ and subsequently differentiating has the same effect as differentiating first and acting on the derivatives with the prolongation of G. Finally, the n-th prolongation of the symmetry group G has an infinitesimal generator, $\text{pr}^n\, \boldsymbol{v}$. A geometrical interpretation of $\text{pr}^n\, \boldsymbol{v}$ can be found in Olver (1993). We are now in a position to introduce the n-th prolongation of a vector field.

Definition 4.2.10 Given a vector field $\boldsymbol{v}$ with corresponding one-parameter group $\exp\{\epsilon \boldsymbol{v}\}$, we define the n-th prolongation of $\boldsymbol{v}$, $\text{pr}^n\, \boldsymbol{v}$, to be the infinitesimal generator of the corresponding one-parameter group $\text{pr}^n[\exp\{\epsilon \boldsymbol{v}\}]$,

$$\text{pr}^n\, \boldsymbol{v}\,\Big|_{(\boldsymbol{x},u^{(n)})} = \frac{d}{d\epsilon}\bigg|_{\epsilon=0} \text{pr}^n\big[\exp\{\epsilon \boldsymbol{v}\}\big]\big(\boldsymbol{x}, u^{(n)}\big).$$

The next result tells us when a one-parameter Lie group generated by $\boldsymbol{v}$ is a symmetry group of a PDE, which is the central result of Lie symmetry analysis.

Theorem 4.2.11 (Lie) *Let*

$$P\big(\boldsymbol{x}, D^{\alpha} u\big) = 0, \tag{4.2.13}$$

be an n-th order partial differential equation. Let $\boldsymbol{v}$ be a vector field of the form

$$\boldsymbol{v} = \sum_{i=1}^{p} \xi^i(\boldsymbol{x}, u)\partial_{x^i} + \phi(\boldsymbol{x}, u)\partial_u.$$

Then $\boldsymbol{v}$ generates a one parameter group of symmetries of (4.2.13) *if and only if*

$$\text{pr}^n\, \boldsymbol{v}\big[P\big(\boldsymbol{x}, D^{\alpha} u\big)\big] = 0 \tag{4.2.14}$$

whenever $P(\boldsymbol{x}, D^{\alpha} u) = 0$.

Next, we give an explicit formula for $\text{pr}^n\, \boldsymbol{v}$, which is due to Olver (1993).

Theorem 4.2.12 (Olver) *Let*

$$\boldsymbol{v} = \sum_{i=1}^{p} \xi^i(\boldsymbol{x}, u)\partial_{x^i} + \phi(\boldsymbol{x}, u)\partial_u$$

be a vector field defined on an open subset $M \subset X \times U$. The n-th prolongation of $\boldsymbol{v}$ is the vector field

$$\text{pr}^n\, \boldsymbol{v} = \boldsymbol{v} + \sum_{\boldsymbol{J}} \phi^{\boldsymbol{J}}\big(\boldsymbol{x}, u^{(n)}\big)\partial_{u_{\boldsymbol{J}}}$$

defined on the corresponding jet space $M^{(n)} \subset X \times U^{(n)}$, *the second summation being over all (unordered) multi-indices* $\boldsymbol{J} = (j_1, \ldots, j_k)$, *with* $1 \leq j_k \leq p$, $1 \leq k \leq n$. *The coefficient functions* $\phi^{\boldsymbol{J}}$ *of* $\mathrm{pr}^n\, \boldsymbol{v}$ *are given by the following formula*:

$$\phi^{\boldsymbol{J}}\left(\boldsymbol{x}, u^{(n)}\right) = D_{\boldsymbol{J}}\left(\phi - \sum_{i=1}^{p} \xi^i u_i\right) + \sum_{i=1}^{p} \xi^i u_{\boldsymbol{J},i},$$

where $u_i = \frac{\partial u}{\partial x^i}$, *and* $u_{\boldsymbol{J},i} = \frac{\partial u_{\boldsymbol{J}}}{\partial x^i}$, *and* $D_{\boldsymbol{J}}$ *is the total differentiation operator.*

We remark that vector fields satisfying (4.2.14) are referred to as *infinitesimal symmetries*. The next theorem states that the set of all infinitesimal symmetries of a PDE forms a Lie algebra.

Theorem 4.2.13 *Let*

$$P\left(\boldsymbol{x}, D^{\alpha} u\right) = 0, \tag{4.2.15}$$

be an n*-th order partial differential equation. Let the set of all infinitesimal generators of symmetries of* (4.2.15) *be* $\mathbf{g}$. *Then* $\mathbf{g}$ *is a Lie algebra.*

From a practical point of view, as we will illustrate in Sect. 4.3, we apply Theorem 4.2.11 to a given PDE, utilizing the formula for $\mathrm{pr}^n\, \boldsymbol{v}$ given in Theorem 4.2.12. The result is a set of determining equations for the coefficients ξ and ϕ in (4.2.9). These can often be solved by inspection. However, we also recall at this point our earlier remark that the process of computing the infinitesimal symmetries is mechanical. This begs the question if software packages exist, which allow the user to perform these calculations. This is in fact the case, and we refer the reader to Baumann (1998) and Cantwell (2002). These packages could be, for instance, employed to verify the results we present in the next section.

4.3 An Example: The One-Dimensional Black-Scholes PDE

In this section, we illustrate the computation of Lie symmetry groups using the Black-Scholes PDE. However, we remark that via a well-known change of variables, we essentially reduce the problem to the computation of symmetries of the heat equation, the canonical example of Lie symmetry analysis, assuming the underlying asset follows geometric Brownian motion.

It is well-known, see e.g. Black and Scholes (1973), Merton (1973), that for a suitable payoff function $H(.)$, the option pricing formula $V(.,.)$ satisfies the Black-Scholes PDE

$$\frac{\partial V(t,S)}{\partial t} + rS\frac{\partial V(t,S)}{\partial S} + \frac{1}{2}\sigma^2 S^2 \frac{\partial^2 V(t,S)}{\partial S^2} - rV(t,S) = 0, \tag{4.3.16}$$

for $t \in [0, T)$ and $S \in (0, \infty)$, with terminal condition $V(T, S) = H(S)$. We perform the following change of variables: $\tau = T - t$, and obtain the PDE

$$-\frac{\partial V}{\partial \tau} + \frac{1}{2}\sigma^2 S^2 \frac{\partial^2 V}{\partial S^2} + rS\frac{\partial V}{\partial S} - rV = 0,$$

where, for convenience, we drop the explicit dependence of V on τ and S. Consequently, we set $u = V\exp(r\tau)$, which yields the PDE

$$-\frac{\partial u}{\partial \tau} + \frac{1}{2}\sigma^2 S^2 \frac{\partial^2 u}{\partial S^2} + rS\frac{\partial u}{\partial S} = 0$$

and finally, we set

$$z = \ln S + \left(r - \frac{\sigma^2}{2}\right)\tau,$$

to yield

$$-\frac{\partial u}{\partial \tau} + \frac{1}{2}\sigma^2 \frac{\partial^2 u}{\partial z^2} = 0. \tag{4.3.17}$$

Equation (4.3.17) is of second order. Therefore, we employ the second prolongation, and we have $p = 2$, i.e. we deal with two independent parameters, z and τ. Theorem 4.2.11 yields that $\boldsymbol{v}$ generates symmetries of (4.3.17), if and only if

$$\mathrm{pr}^2\,\boldsymbol{v}\left[\frac{1}{2}\sigma^2 u_{zz} - u_\tau\right] = 0$$

whenever $u_\tau = \frac{1}{2}\sigma^2 u_{zz}$, where

$$\boldsymbol{v} = \xi^1\frac{\partial}{\partial z} + \xi^2\frac{\partial}{\partial \tau} + \phi\frac{\partial}{\partial u}.$$

From Theorem 4.2.12 we obtain the second prolongation

$$\mathrm{pr}^2\,\boldsymbol{v} = \boldsymbol{v} + \phi^z\partial_{u_z} + \phi^\tau\partial_{u_\tau} + \phi^{zz}\partial_{u_{zz}} + \phi^{z\tau}\partial_{u_{z\tau}} + \phi^\tau\partial_{u_{\tau\tau}},$$

and apply it to Eq. (4.3.17) to obtain

$$\mathrm{pr}^2\,\boldsymbol{v}\left[\frac{1}{2}\sigma^2 u_{zz} - u_\tau\right] = -\phi^\tau + \frac{1}{2}\sigma^2\phi^{zz}. \tag{4.3.18}$$

Consequently, we need to compute ϕ^τ and ϕ^{zz}. From Theorem 4.2.12 we obtain

$$\begin{aligned}\phi^\tau &= D_\tau\big(\phi - \xi^1 u_z - \xi^2 u_\tau\big) + \xi^1 u_{z\tau} + \xi^2 u_{\tau\tau}\\ &= \phi_\tau + \phi_u u_\tau - \xi^1_\tau u_z - \xi^1_u u_z u_\tau - \xi^2_\tau u_\tau - \xi^2_u (u_\tau)^2,\end{aligned}$$

and

$$\phi^{zz} = D_{zz}\big(\phi - \xi^1 u_z - \xi^2 u_\tau\big) + \xi^1 u_{zzz} + \xi^2 \xi_{zz\tau}.$$

Table 4.3.1 Coefficients for the constant, u, and the derivatives of u

Term	LHS	RHS
1	ϕ_τ	$\frac{1}{2}\sigma^2\phi_{zz}$
u_z	$-\xi^1_\tau$	$(\sigma^2\phi_{zu} - \frac{1}{2}\sigma^2\xi^1_{zz})$
u_τ	$(\phi_u - \xi^2_\tau)$	$(\phi_u - 2\xi^1_z) - \frac{1}{2}\sigma^2\xi^2_{zz}$
$u_z u_\tau$	$-\xi^1_u$	$-3\xi^1_u - \sigma^2\xi^2_{zu}$
u^2_τ	$-\xi^2_u$	$-\xi^2_u$
u^2_z	0	$\frac{1}{2}\sigma^2(\phi_{uu} - 2\xi^1_{zu})$
$u_{z\tau}$	0	$-\sigma^2\xi^2_z$
$u_z u_{z\tau}$	0	$-\sigma^2\xi^2_u$
u^3_z	0	$-\frac{1}{2}\sigma^2\xi^1_{uu}$
$u^2_z u_\tau$	0	$-\frac{1}{2}\sigma^2\xi^2_{uu}$

It can be shown that this equals

$$\begin{aligned}
\phi^{zz} &= \phi_{zz} + \left(2\phi_{zu} - \xi^1_{zz}\right)u_z + \left(\phi_{uu} - 2\xi^1_{zu}\right)(u_z)^2 \\
&\quad + \left(\phi_u - 2\xi^1_z\right)u_{zz} - 3\xi^1_u u_z u_{zz} - 2\xi^2_{zu} u_z u_\tau \\
&\quad - 2\xi^2_z u_{z\tau} - 2\xi^2_u u_z u_{z\tau} - \xi^1_{uu} u^3_z - \xi^2_{zz} u_\tau - \xi^2_{uu}(u_z)^2 u_\tau \\
&\quad - \xi^2_u u_{zz} u_\tau .
\end{aligned}$$

Equality (4.3.18) and replacing

$$u_{zz} = \frac{2}{\sigma^2}u_\tau ,$$

yields

$$\begin{aligned}
&\phi_\tau + \phi_u u_\tau - \xi^1_\tau u_z - \xi^1_u u_z u_\tau - \xi^2_\tau u_\tau - \xi^2_u (u_\tau)^2 \\
&\quad = \frac{1}{2}\sigma^2\biggl[\phi_{zz} + \left(2\phi_{zu} - \xi^1_{zz}\right)u_z + \left(\phi_{uu} - 2\xi^1_{zu}\right)(u_z)^2 \\
&\qquad + \left(\phi_u - 2\xi^1_z\right)\left(\frac{2u_\tau}{\sigma^2}\right) - 3\xi^1_u u_z\left(\frac{2u_\tau}{\sigma^2}\right) - 2\xi^2_{zu} u_z u_\tau - 2\xi^2_z u_{z\tau} \\
&\qquad - 2\xi^2_u u_z u_{z\tau} - \xi^1_{uu} u^3_z - \xi^2_{zz} u_\tau - \xi^2_{uu}(u_z)^2 u_\tau - \xi^2_u \frac{2}{\sigma^2}(u_\tau)^2\biggr].
\end{aligned}$$

Solving for ξ^1, ξ^2, ϕ, we equate coefficients of the partial derivatives of u, which results in the following table, where the coefficients of the terms in the first column are shown in columns two and three, depending on whether they appear on the left or right hand side of the equation.

As mentioned in Sect. 4.2.3, we now use Table 4.3.1 to obtain equations for the coefficient functions ξ^1, ξ^2 and ϕ. From the terms $u_{z\tau}$ and $u_z u_{z\tau}$, we note that ξ^2 only depends on τ, i.e. $\xi^2 = \xi^2(\tau)$. Consequently, $\xi^2_{zu} = 0$, and the term $u_z u_\tau$ yields $\xi^1_u = 0$, hence ξ^1 is a function of z and τ, i.e. $\xi^1 = \xi^1(z, \tau)$ and hence $\xi^1_{zu} = 0$. We make use of this conclusion when considering the coefficient of u^2_z, which gives $\phi_{uu} = 0$, hence

$$\phi(z, \tau, u) = \alpha(z, \tau)u + \beta(z, \tau),$$

where $\alpha(z,\tau)$ and $\beta(z,\tau)$ are functions of z and τ only, and we obtain

$$\begin{aligned}\phi_z &= \alpha_z u + \beta_z\\ \phi_{zz} &= \alpha_{zz} u + \beta_{zz}\\ \phi_{zu} &= \alpha_z\\ \phi_\tau &= \alpha_\tau u + \beta_\tau\\ \phi_{\tau u} &= \alpha_\tau.\end{aligned}$$

From the u_z term, we obtain

$$\xi^1_\tau = \frac{1}{2}\sigma^2\xi^1_{zz} - \sigma^2\phi_{zu} \tag{4.3.19}$$

and the u_τ term yields

$$\xi^1_z = \frac{1}{2}\xi^2_\tau, \tag{4.3.20}$$

as $\xi^2_{zz} = 0$. We hence have

$$\xi^1 = \frac{1}{2}\xi^2_\tau z + A(\tau), \tag{4.3.21}$$

where A is a function of τ. We now use (4.3.19) to get

$$\frac{1}{2}z\xi^2_{\tau\tau} + A_\tau = \xi^1_\tau = \frac{1}{2}\sigma^2\xi^1_{zz} - \sigma^2\alpha_z = -\sigma^2\alpha_z,$$

as $\xi^1_{zz} = 0$, from (4.3.20). Hence, one has

$$\begin{aligned}\alpha_z &= -\frac{1}{2\sigma^2}z\xi^2_{\tau\tau} - \frac{1}{\sigma^2}A_\tau,\\ \alpha_{zz} &= -\frac{1}{2\sigma^2}\xi^2_{\tau\tau}\end{aligned} \tag{4.3.22}$$

and

$$\alpha = -\frac{1}{4\sigma^2}z^2\xi^2_{\tau\tau} - \frac{1}{\sigma^2}zA_\tau + B(\tau),$$

where B is a function of τ, and

$$\alpha_\tau = -\frac{z^2}{4\sigma^2}\xi^2_{\tau\tau\tau} - \frac{z}{\sigma^2}A_{\tau\tau} + B_\tau. \tag{4.3.23}$$

Regarding the constant term 1,

$$\phi_\tau = \frac{1}{2}\sigma^2\phi_{zz}$$

yields

$$\alpha_\tau u + \beta_\tau = \frac{1}{2}\sigma^2(\alpha_{zz}u + \beta_{zz})$$

and hence

$$\alpha_\tau = \frac{1}{2}\sigma^2\alpha_{zz},$$
$$\beta_\tau = \frac{1}{2}\sigma^2\beta_{zz}. \tag{4.3.24}$$

Using (4.3.22), (4.3.23), and (4.3.24), we have

$$-\frac{1}{4\sigma^2}z^2\xi^2_{\tau\tau\tau} - \frac{1}{\sigma^2}zA_{\tau\tau} + B_\tau = \frac{1}{2}\sigma^2\left(-\frac{1}{2\sigma^2}\xi^2_{\tau\tau}\right) = -\frac{1}{4}\xi^2_{\tau\tau}.$$

Comparing the coefficients of the function u and its derivatives, we obtain the following system of equations

$$\xi^2_{\tau\tau\tau} = 0$$
$$A_{\tau\tau} = 0$$
$$B_\tau = -\frac{1}{4}\xi^2_{\tau\tau}.$$

We now introduce the notation

$$\xi^2_{\tau\tau} = C_1,$$
$$\xi^2_\tau = C_1\tau + C_2,$$
$$\xi^2 = \frac{1}{2}C_1\tau^2 + C_2\tau + C_3,$$

where C_1, C_2, and C_3 are constants. Similarly,

$$A_{\tau\tau} = 0,$$

and

$$A(\tau) = C_4\tau + C_5,$$

where C_4 and C_5 are constants. Using

$$B_\tau = -\frac{1}{4}\xi^2_{\tau\tau} = -\frac{1}{4}C_1,$$

we get

$$B(\tau) = -\frac{1}{4}C_1\tau + C_6,$$

where C_6 is a constant. Recall that one has

$$\xi^1 = \frac{z}{2}\xi^2_\tau + A(\tau),$$

and hence

$$\xi^1(z,\tau) = \frac{1}{2}C_1z\tau + \frac{1}{2}C_2z + C_4\tau + C_5.$$

Also, we have

$$\begin{aligned}\phi &= \alpha u + \beta \\ &= \left(-\frac{z^2}{4\sigma^2}\xi^2_{\tau\tau} - \frac{zA_\tau}{\sigma^2} + B(\tau)\right)u + \beta \\ &= \left(C_1\left(-\frac{z^2}{4\sigma^2} - \frac{\tau}{4}\right) - \frac{C_4 z}{\sigma^2} + C_6\right)u + \beta.\end{aligned}$$

Lastly, we get

$$\xi^2(\tau) = \frac{1}{2}C_1\tau^2 + C_2\tau + C_3.$$

Finally, we can obtain our vector fields from

$$\begin{aligned}\boldsymbol{v} &= \xi^1\partial_z + \xi^2\partial_\tau + \phi\partial_u \\ &= \left(\frac{1}{2}C_1 z\tau + \frac{1}{2}C_2 z + C_4\tau + C_5\right)\partial_z \\ &\quad + \left(\frac{1}{2}C_1\tau^2 + C_2\tau + C_3\right)\partial_\tau \\ &\quad + \left[\left(C_1\left(-\frac{z^2}{4\sigma^2} - \frac{\tau}{4}\right) + C_4\left(-\frac{z}{\sigma^2}\right) + C_6\right)u + \beta\right]\partial_u \\ &= C_1\left(\frac{1}{2}z\tau\partial_z + \frac{1}{2}\tau^2\partial_\tau + \left(-\frac{z^2}{4\sigma^2} - \frac{\tau}{4}\right)u\partial_u\right) \\ &\quad + C_2\left(\frac{1}{2}z\partial_z + \tau\partial_\tau\right) \\ &\quad + C_3\partial_\tau + C_4\left(\tau\partial_z - \frac{zu}{\sigma^2}\partial_u\right) \\ &\quad + C_5\partial_z + C_6 u\partial_u + \beta\partial_u.\end{aligned}$$

Consequently, the resulting Lie algebra is spanned by the six vector fields:

$$\begin{aligned}\boldsymbol{v}_1 &= \partial_z \\ \boldsymbol{v}_2 &= \partial_\tau \\ \boldsymbol{v}_3 &= u\partial_u \\ \boldsymbol{v}_4 &= \frac{1}{2}z\partial_z + \tau\partial_\tau \\ \boldsymbol{v}_5 &= \tau\partial_z - \frac{zu}{\sigma^2}\partial_u \\ \boldsymbol{v}_6 &= \frac{1}{2}z\tau\partial_z + \frac{1}{2}\tau^2\partial_\tau - \left(\frac{z^2}{4\sigma^2} + \frac{\tau}{4}\right)u\partial_u\end{aligned}$$

with

$$\boldsymbol{v}_\beta = \beta\partial_u,$$

where $\boldsymbol{v}_\beta$ is an infinite-dimensional sub-algebra. We now exponentiate vector field $\boldsymbol{v}_5$. In particular, from Theorem 4.2.8, we need to solve the following system of ODEs

$$\frac{d\tilde{z}}{d\epsilon} = \tilde{\tau}, \quad \tilde{z}(0) = z,$$
$$\frac{d\tilde{\tau}}{d\epsilon} = 0, \quad \tilde{\tau}(0) = \tau,$$
$$\frac{d\tilde{u}}{d\epsilon} = -\frac{\tilde{z}\tilde{u}}{\sigma^2}, \quad \tilde{u}(0) = u,$$

hence

$$\tilde{\tau} = \tau,$$
$$\tilde{z} = \tau\epsilon + z,$$
$$\tilde{u} = u \exp\left\{-\frac{\tau\epsilon^2}{2\sigma^2} - \frac{z\epsilon}{\sigma^2}\right\},$$

so that

$$\tilde{u} = \exp\left\{\frac{\tilde{\tau}\epsilon^2}{2\sigma^2} - \frac{\tilde{z}\epsilon}{\sigma^2}\right\} u(\tilde{z} - \tilde{\tau}\epsilon, \tilde{\tau}). \tag{4.3.25}$$

Recall that we write $\rho(\exp\{\epsilon\boldsymbol{v}\})u(x,t)$ for the action of the symmetry group generated by $\boldsymbol{v}$ on a solution u. For example, for $\boldsymbol{v}_5$, we have

$$\rho\big(\exp\{\epsilon\boldsymbol{v}_5\}\big) = \exp\left\{\frac{\tau\epsilon^2}{2\sigma^2} - \frac{z\epsilon}{\sigma^2}\right\} u(z - \tau\epsilon, \tau). \tag{4.3.26}$$

The interpretation, as mentioned before, is that if $u(x,t)$ solves (4.3.17), then so does

$$\tilde{u}(\tilde{x}, \tilde{t}) = \rho\big(\exp\{\epsilon\boldsymbol{v}_5\}\big),$$

at least for sufficiently small ϵ.

We now illustrate how to obtain transforms of fundamental solutions for the case of the Black-Scholes PDE. This approach was first introduced in Craddock and Platen (2004), and subsequently developed in Craddock and Lennox (2007), Craddock and Lennox (2009), and Craddock (2009). In general, we have

$$\rho\big(\exp\{\epsilon\boldsymbol{v}\}\big)u(z,\tau) = \sigma(z,\tau;\epsilon)u\big(a_1(z,\tau;\epsilon), a_2(z,\tau;\epsilon)\big), \tag{4.3.27}$$

for a multiplier $\sigma(\cdot, \cdot\, ;\cdot)$, where $a_1(\cdot, \cdot\, ;\cdot)$ and $a_2(\cdot, \cdot\, ;\cdot)$ are the changes of variables of the symmetry. Suppose a fundamental solution $p(z, y, \tau)$ has been obtained. Clearly,

$$u(z,\tau) = \int_{\Re} f(y)p(z,y,\tau)\,dy \tag{4.3.28}$$

solves the initial value problem with

$$u(z, 0) = f(z).$$

Following the transform approach, we aim to connect (4.3.27) and (4.3.28). Consider a stationary solution $u = u_0(z)$, then

$$\rho\big(\exp\{\epsilon \boldsymbol{v}\}\big)u_0(z) = \sigma(z,\tau;\epsilon)u_0\big(a_1(z,\tau;\epsilon)\big),$$

which also solves the initial value problem. Set $\tau = 0$ and use Eq. (4.3.28) to obtain

$$\int_{\Re} \sigma(y,0;\epsilon)u_0\big(a_1(y,0;\epsilon)\big)p(z,y,\tau)\,dy = \sigma(z,\tau;\epsilon)u_0\big(a_1(z,\tau;\epsilon)\big). \tag{4.3.29}$$

Consider (4.3.27) again, from Eq. (4.3.26) we consequently have

$$\sigma(z,\tau;\epsilon) = \exp\left\{\frac{\tau\epsilon^2}{2\sigma^2} - \frac{z\epsilon}{\sigma^2}\right\}$$
$$a_1(z,\tau;\epsilon) = z - \tau\epsilon.$$

Clearly, $u = 1$ solves the PDE (4.3.17), so we have from Eq. (4.3.29)

$$\int_{\Re} \exp\left\{-\frac{y\epsilon}{\sigma^2}\right\}p(z,y,\tau)\,dy = \exp\left\{\frac{\tau\epsilon^2}{2\sigma^2} - \frac{z\epsilon}{\sigma^2}\right\}.$$

Setting $b = -\frac{\epsilon}{\sigma^2}$, we get

$$\int_{\Re} \exp\{yb\}p(z,y,\tau)\,dy = \exp\left\{\frac{\tau b^2\sigma^2}{2} + zb\right\}. \tag{4.3.30}$$

But (4.3.30) is clearly the moment generating function of a normal random variable with mean z and variance $\sigma^2\tau$, so it follows that

$$p(z,y,\tau) = \frac{1}{\sqrt{2\pi\sigma^2\tau}}\exp\left\{-\frac{1}{2\sigma^2\tau}(y-z)^2\right\}.$$

Now changing variables back to Black-Scholes model parameters,

$$\tau = T - t,$$
$$z = \ln S + \left(r - \frac{1}{2}\sigma^2\right)\tau,$$

we get

$$p(t,S;T,y) = \frac{1}{\sqrt{2\pi\sigma^2(T-t)}}\exp\left\{-\frac{(y - (\ln S + (r - \frac{1}{2}\sigma^2)(T-t)))^2}{2\sigma^2(T-t)}\right\}.$$

Transforming variables $y = \ln x$ we obtain

$$p(t,S;T,x) = \frac{1}{x\sqrt{2\pi\sigma^2(T-t)}}\exp\left\{-\frac{(\ln x - (\ln S + (r - \frac{1}{2}\sigma^2)(T-t)))^2}{2\sigma^2(T-t)}\right\},$$

which is the probability density function of a lognormal random variable.

To conclude the illustration, we price a call option, for which the payoff is

$$V(T,S) = H(S) = (S-K)^+,$$

for $K > 0$. Recall that we set $u = V\exp\{r\tau\}$, hence we have

$$\begin{aligned} V(t,S) &= u(S,t)\exp\{r(T-t)\} \\ &= \int_{\Re^+} \exp\{-r(T-t)\}(x-K)^+ p(t,S;T,x)\,dx \\ &= SN(d_1) - K\exp\{-r(T-t)\}N(d_2), \end{aligned}$$

where

$$d_1 = \frac{\ln(\frac{S}{K}) + (r + \frac{1}{2}\sigma)(T-t)}{\sigma\sqrt{T-t}},$$

and

$$d_2 = d_1 - \sigma\sqrt{T-t},$$

which is the standard Black-Scholes option pricing formula.

Finally, we remark that from a financial modeling point of view, integrating the payoff function against the probability density, the fundamental solution yields the price of the option as expected. The above approach is very useful for financial applications, as for any payoff function, given the probability density, we have to solve an integration problem. In this case, the transform was the moment generating function. We will subsequently also recover other transforms, however, if we can invert the transform, if necessary numerically, we only have a one-dimensional integration problem to solve, which can be accomplished using methods to be presented in Chap. 13. For additional examples demonstrating how Lie symmetry methods can be used to solve option pricing problems, we also alert the reader to Caister et al. (2010) and the references therein.

4.4 Results on Transforms of Fundamental Solutions

In this section, we discuss how to obtain transforms of fundamental solutions of the equation

$$u_t = \sigma x^\gamma u_{xx} + f(x)u_x - g(x)u, \quad x \geq 0. \tag{4.4.31}$$

We point out that the equation

$$u_t = \sigma u_{xx} + f(x)u_x - g(x)u, \quad x \in \Re, \tag{4.4.32}$$

is studied in Craddock and Lennox (2009), where we direct the interested reader.

It is straightforward to motivate the study of (4.4.31). For example, setting $\gamma = 1$, then

$$u(x,t) = E\left(\exp\left\{-\int_0^t g(X_s)\,ds\right\}\varphi(X_t)\right),$$

where $u(x,0) = \varphi(x)$ and

$$dX_t = f(X_t)\,dt + \sqrt{2\sigma X_t}\,dW_t,$$

by the Feynman-Kac formula, see Sect. 15.8 in Chap. 15. Consequently, obtaining the fundamental solution of (4.4.31) means that we can compute important functionals of corresponding diffusions.

We remark that the PDE (4.4.31), for $\gamma = 1$, $g(x) = 0$, was studied in Craddock and Platen (2004), with emphasis on obtaining transition densities of important stochastic processes, but also discovering transition densities of stochastic processes that had not been studied before. In Craddock and Lennox (2007), the case $g(x) = \mu x^r$ and an arbitrary γ was considered. Furthermore, besides obtaining explicit transition densities, the authors also explicitly connected the fundamental solutions obtained to other important functionals, such as bond prices, see also Sect. 5.5. This is interesting, as it means that integrals of fundamental solutions have important applications, and not only the fundamental solutions when integrated against a probability density. This line of research was continued in Craddock and Lennox (2009), where Laplace transforms of joint densities of important functionals of diffusions were obtained from (4.4.31). Finally, Craddock improved on the results from Craddock and Lennox (2009) and also studied the PDE (4.4.32). In particular, it has been shown how to obtain generalized Laplace transforms of the fundamental solutions of (4.4.31) and how to obtain Fourier transforms of the fundamental solutions of (4.4.32).

The structure of the remainder of this chapter and the next chapter is as follows: in Sect. 4.4.1, we collect theorems from Craddock (2009), giving the opportunity to compute transforms of fundamental solutions. We illustrate the procedure via examples in Chap. 5. The examples are grouped based on applications. In particular, transition densities are derived, where we follow Craddock and Platen (2004), and Laplace transforms of transition densities are obtained. For the convenience of the reader, the results presented in this section, and additional results which can be obtained via the same method are collected systematically in Sects. 5.3 and 5.4.

4.4.1 Transforms of One-Dimensional Solutions

In this subsection, we show how to obtain transforms of fundamental solutions of Eq. (4.4.31).

As illustrated in Sect. 4.3, it is possible to construct fundamental solutions from trivial solutions. In particular, these trivial solutions will typically be independent of time, whereas fundamental solutions are not. Hence symmetries resulting in trivial transformations in t are not sufficient. Other trivial symmetries include scalings, i.e. for a constant c, if $u(x,t)$ solves (4.4.31), so does $cu(x,t)$.

The next theorem, which is Proposition 2.1 in Craddock and Lennox (2009), produces conditions on the drift f, under which (4.4.31) has nontrivial symmetries. The result was, in slightly simpler form, established in Craddock and Platen (2004), and our proof follows Craddock and Platen (2004).

Theorem 4.4.1 *If $\gamma \neq 2$, the PDE*

$$u_t = \sigma x^\gamma u_{xx} + f(x)u_x - g(x)u, \quad x \geq 0,\ \sigma > 0 \tag{4.4.33}$$

has a nontrivial Lie symmetry group if and only if h, *which is given by* $h(x) = x^{1-\gamma} f(x)$, *satisfies one of the following families of drift equations*

$$\sigma x h' - \sigma h + \frac{1}{2}h^2 + 2\sigma x^{2-\gamma} g(x) = 2\sigma A x^{2-\gamma} + B, \tag{4.4.34}$$

$$\sigma x h' - \sigma h + \frac{1}{2}h^2 + 2\sigma x^{2-\gamma} g(x) = \frac{A x^{4-2\gamma}}{2(2-\gamma)^2} + \frac{B x^{2-\gamma}}{2-\gamma} + C, \tag{4.4.35}$$

$$\sigma x h' - \sigma h + \frac{1}{2}h^2 + 2\sigma x^{2-\gamma} g(x) = \frac{A x^{4-2\gamma}}{2(2-\gamma)^2} + \frac{B x^{3-\frac{3}{2}\gamma}}{3-\frac{3}{2}\gamma} + \frac{C x^{2-\gamma}}{2-\gamma} - \kappa, \tag{4.4.36}$$

with $\kappa = \frac{\gamma}{8}(\gamma - 4)\sigma^2$.

Proof 1. Since the PDE (4.4.33) is of second order, we use the second prolongation of $\boldsymbol{v}$,

$$\mathrm{pr}^2\, \boldsymbol{v} = \boldsymbol{v} + \phi^x \frac{\partial}{\partial u_x} + \phi^t \frac{\partial}{\partial u_t} + \phi^{xx} \frac{\partial}{\partial u_{xx}} + \phi^{xt} \frac{\partial}{\partial u_{xt}} + \phi^{tt} \frac{\partial}{\partial u_{tt}}.$$

Applying the second prolongation of $\boldsymbol{v}$ to (4.4.33), we obtain

$$\begin{aligned}&\mathrm{pr}^2\, \boldsymbol{v}\big[u_t - \sigma x^\gamma u_{xx} - f u_x + g u\big] \\ &\quad = \xi^1\big(-\sigma\gamma x^{\gamma-1} u_{xx} - f_x u_x + g_x u\big) + \phi g - \phi^x f + \phi^t - \phi^{xx}\sigma x^\gamma.\end{aligned}$$

Hence Theorem 4.2.11 yields that whenever u satisfies (4.4.33), we have

$$\phi^t = \xi^1\big(\sigma\gamma x^{\gamma-1} u_{xx} + f_x u_x - g_x u\big) - \phi g + \phi^x f + \phi^{xx}\sigma x^\gamma. \tag{4.4.37}$$

We now compute ϕ^t, ϕ^x and ϕ^{xx} using Theorem 4.2.12, where we note that ξ^1 and ξ^2 must be independent of u, ξ^2 must, furthermore, be independent of x, and ϕ must be linear in u, see also the proof of Proposition 2.1 in Craddock and Lennox (2009). Taking these considerations into account, we obtain

$$\phi^t = \phi_t + \phi_u u_t - \xi^1_t u_x - \xi^2_t u_t,$$

and

$$\phi^x = \phi_x + \phi_u u_x - \xi^1_x u_x$$

and finally

$$\phi^{xx} = \phi_{xx} + 2\phi_{xu} u_x + \phi_u u_{xx} - 2\xi^1_x u_{xx} - \xi^1_{xx} u_x.$$

Substituting the formulas for ϕ^t, ϕ^x, ϕ^{xx} into (4.4.37), we obtain

$$\begin{aligned}&\phi_t + \phi_u u_t - \xi^1_t u_x - \xi^2_t u_t \\ &\quad = \xi^1\big(\sigma\gamma x^{\gamma-1} u_{xx} + f_x u_x - g_x u\big) \\ &\qquad - \phi g + \big(\phi_x + \phi_u u_x - \xi^1_x u_x\big) f \\ &\qquad + \big(\phi_{xx} + (2\phi_{xu} - \xi^1_{xx}) u_x + (\phi_u - 2\xi^1_x) u_{xx}\big)\sigma x^\gamma,\end{aligned}$$

Table 4.4.2 Coefficients for the constant, u, and the derivatives of u

Term	LHS	RHS
1	ϕ_t	$-\phi g + \phi_{xx}\sigma x^\gamma + \phi_x f$
u	$-(\phi_u - \xi^2_t)g$	$-\xi^1 g_x$
u_x	$-\xi^1_t + (\phi_u - \xi^2_t)f$	$\xi^1 f_x + (\phi_u - \xi^1_x)f + (2\phi_{xu} - \xi^1_{xx})\sigma x^\gamma$
u_{xx}	$(\phi_u - \xi^2_t)\sigma x^\gamma$	$\xi^1 \sigma\gamma x^{\gamma-1} + (\phi_u - 2\xi^1_x)\sigma x^\gamma$

and replacing

$$u_t = \sigma x^\gamma u_{xx} + f(x)u_x - g(x)u,$$

we get

$$\begin{aligned}
&\phi_t + (\phi_u - \xi^2_t)(\sigma x^\gamma u_{xx} + f u_x - gu) - \xi^1_t u_x \\
&\quad = \xi^1(\sigma\gamma x^{\gamma-1}u_{xx} + f_x u_x - g_x u) \\
&\qquad - \phi g + (\phi_x + \phi_u u_x - \xi^1_x u_x)f + \phi_{xx}\sigma x^\gamma \\
&\qquad + (2\phi_{xu} - \xi^1_{xx})\sigma x^\gamma u_x + (\phi_u - 2\xi^1_x)u_{xx}\sigma x^\gamma .
\end{aligned}$$

2. As in Sect. 4.3, we use Table 4.4.2 to obtain determining equations for ξ^1, ξ^2 and ϕ, which will be in terms of f and g.

The equation resulting from u_{xx} yields the following ODE for ξ^1:

$$-\xi^2_t \sigma x^\gamma = \xi^1 \sigma\gamma x^{\gamma-1} - 2\xi^1_x \sigma x^\gamma ,$$

from which we obtain

$$\xi^1 = \frac{\xi^2_t x}{2-\gamma} + \rho(t)x^{\gamma/2}, \tag{4.4.38}$$

where $\rho(t)$ is a function of t only. Consequently, one has

$$\xi^1_t = \frac{\xi^2_{tt} x}{2-\gamma} + \rho_t x^{\gamma/2} \tag{4.4.39}$$

$$\xi^1_x = \frac{\xi^2_t}{2-\gamma} + \frac{\gamma}{2}x^{\gamma/2-1}\rho, \tag{4.4.40}$$

$$\xi^1_{xx} = \frac{\gamma}{2}\left(\frac{\gamma}{2} - 1\right)\rho x^{\gamma/2-2}. \tag{4.4.41}$$

Recalling that ϕ is linear in u, we obtain

$$\phi = \alpha(x,t)u + \beta(x,t), \tag{4.4.42}$$

where α and β are functions of x and t only, and hence

$$\phi_t = \alpha_t u + \beta_t \tag{4.4.43}$$

$$\phi_x = \alpha_x u + \beta_x \tag{4.4.44}$$

$$\phi_{xx} = \alpha_{xx} u + \beta_{xx} \tag{4.4.45}$$

$$\phi_u = \alpha.$$

We can now use the equation resulting from u to obtain

$$-\alpha g = -\xi^1 g_x - \xi_t^2 g, \tag{4.4.46}$$

and from the equation resulting from the constant term 1 we obtain, substituting (4.4.43), (4.4.42), (4.4.45), and (4.4.44),

$$\alpha_t u + \beta_t = -(\alpha u + \beta)g + (\alpha_{xx} u + \beta_{xx})\sigma x^\gamma + (\alpha_x u + \beta_x) f.$$

Equating the coefficients of the constant and u, we obtain

$$\alpha_t = -\alpha g + \alpha_{xx}\sigma x^\gamma + \alpha_x f \tag{4.4.47}$$

and

$$\beta_t = -\beta g + \beta_{xx}\sigma x^\gamma + \beta_x f. \tag{4.4.48}$$

Regarding β, we can only say that it is an arbitrary solution of (4.4.33). Now, we use the equation resulting from u_x,

$$-\xi_t^1 - \xi_t^2 f = \xi^1 f_x - \xi_x^1 f + \left(2\alpha_x - \xi_{xx}^1\right)\sigma x^\gamma,$$

which yields, substituting (4.4.39), (4.4.40), and (4.4.41),

$$\begin{aligned}\alpha_x = &-\frac{\xi_{tt}^2 x^{1-\gamma}}{2\sigma(2-\gamma)} - \frac{\rho_t x^{-\gamma/2}}{2\sigma} - \frac{\xi_t^2}{2\sigma(2-\gamma)}\frac{d}{dx}\left(f x^{1-\gamma}\right)\\ &-\frac{\rho}{2\sigma}\frac{d}{dx}\left(f x^{-\gamma/2}\right) + \frac{\gamma}{4}\rho(\gamma/2-1)x^{\gamma/2-2},\end{aligned} \tag{4.4.49}$$

and the second derivative with respect to x,

$$\begin{aligned}\alpha_{xx} = &-\frac{\xi_{tt}^2(1-\gamma)x^{-\gamma}}{2\sigma(2-\gamma)} - \frac{\rho_t(-\gamma/2)x^{-\gamma/2-1}}{2\sigma} - \frac{\xi_t^2}{2\sigma(2-\gamma)}\frac{d^2}{dx^2}\left(f x^{1-\gamma}\right)\\ &-\frac{\rho}{2\sigma}\frac{d^2}{dx^2}\left(f x^{-\gamma/2}\right) + \frac{\gamma}{4}\rho(\gamma/2-1)(\gamma/2-2)x^{\gamma/2-3},\end{aligned} \tag{4.4.50}$$

and also

$$\begin{aligned}\alpha = &-\frac{\xi_{tt}^2 x^{2-\gamma}}{2\sigma(2-\gamma)^2} - \frac{\rho_t x^{1-\gamma/2}}{\sigma(2-\gamma)} - \frac{\xi_t^2}{2\sigma(2-\gamma)}\left(f x^{1-\gamma}\right)\\ &-\frac{\rho}{2\sigma}\left(f x^{-\gamma/2}\right) + \frac{\gamma}{4}\rho x^{\gamma/2-1} + \eta(t),\end{aligned} \tag{4.4.51}$$

where $\eta(t)$ is a function only depending on t and lastly

$$\begin{aligned}\alpha_t = &-\frac{\xi_{ttt}^2 x^{2-\gamma}}{2\sigma(2-\gamma)^2} - \frac{\rho_{tt} x^{1-\gamma/2}}{\sigma(2-\gamma)} - \frac{\xi_{tt}^2}{2\sigma(2-\gamma)}\left(f x^{1-\gamma}\right)\\ &-\frac{\rho_t}{2\sigma}\left(f x^{-\gamma/2}\right) + \frac{\gamma}{4}\rho_t x^{\gamma/2-1} + \eta_t.\end{aligned} \tag{4.4.52}$$

3. We now use (4.4.46), substitute (4.4.52), (4.4.50), and (4.4.49), and perform the obvious cancellations to obtain

$$
\begin{aligned}
&-\frac{\xi^2_{ttt}x^{2-\gamma}}{2\sigma(2-\gamma)^2}-\frac{\rho_{tt}x^{1-\gamma/2}}{\sigma(2-\gamma)}+\eta_t\\
&=-\left(\frac{\xi^2_t x}{2-\gamma}+\rho x^{\gamma/2}\right)g_x-\xi^2_t g-\frac{\xi^2_{tt}(1-\gamma)}{2(2-\gamma)}\\
&\quad-\frac{\xi^2_t x^\gamma}{2(2-\gamma)}\frac{d^2}{dx^2}(fx^{1-\gamma})-\frac{\rho}{2}x^\gamma\frac{d^2}{dx^2}(fx^{-\gamma/2})\\
&\quad-\frac{\xi^2_t}{2\sigma(2-\gamma)}\frac{d}{dx}(fx^{1-\gamma})f-\frac{\rho}{2\sigma}\frac{d}{dx}(fx^{-\gamma/2})f+\frac{\gamma}{4}\rho(\gamma/2-1)x^{\gamma/2-2}f\\
&\quad+\frac{\gamma\sigma}{4}\rho(\gamma/2-1)(\gamma/2-2)x^{3\gamma/2-3}.
\end{aligned}
$$

Setting

$$
Lf=\sigma x^\gamma\left(\frac{x^{1-\gamma}f(x)}{2\sigma(2-\gamma)}\right)''+f(x)\left(\frac{x^{1-\gamma}f(x)}{2\sigma(2-\gamma)}\right)'+g(x)+\frac{xg'(x)}{2-\gamma},
$$

and

$$
Kf=\sigma x^\gamma\left(\frac{\gamma}{4}x^{\gamma/2-1}-\frac{x^{-\gamma/2}}{2\sigma}f(x)\right)''+f\left(\frac{\gamma}{4}x^{\gamma/2-1}-\frac{x^{-\gamma/2}}{2\sigma}f\right)'-x^{\gamma/2}g_x,
$$

we obtain

$$
-\frac{\xi^2_{ttt}x^{2-\gamma}}{2\sigma(2-\gamma)^2}-\frac{\rho_{tt}x^{1-\gamma/2}}{\sigma(2-\gamma)}+\eta_t=-\frac{(1-\gamma)}{2(2-\gamma)}\xi^2_{tt}-Lf\xi^2_t+Kf\rho. \qquad (4.4.53)
$$

4. We now proceed as follows: making an assumption on Lf, which will be (4.4.34), (4.4.35), or (4.4.36), we deduce properties of Kf. Subsequently, as in Sect. 4.3, we solve for ξ^1, ξ^2, and ϕ, and confirm that we obtain nontrivial symmetries. We start with (4.4.34) and (4.4.35):

$$
Lf=Ax^{2-\gamma}+B,
$$

to facilitate the analysis, we employ the notation $h(x)=x^{1-\gamma}f(x)$, and consequently obtain

$$
Lf=\frac{\sigma x^\gamma h''}{2\sigma(2-\gamma)}+\frac{hx^{\gamma-1}h'}{2\sigma(2-\gamma)}+g+\frac{g'x}{2-\gamma}=Ax^{2-\gamma}+B. \qquad (4.4.54)
$$

Hence, we obtain

$$
\frac{\sigma xh''}{2\sigma(2-\gamma)}+\frac{hh'}{2\sigma(2-\gamma)}+gx^{1-\gamma}+\frac{g'x^{2-\gamma}}{2-\gamma}=Ax^{3-2\gamma}+Bx^{1-\gamma}
$$

and using integration by parts yields

$$
\sigma xh'-\sigma h+\frac{h^2}{2}+gx^{2-\gamma}2\sigma=Ax^{4-2\gamma}\sigma+2\sigma Bx^{2-\gamma}+C.
$$

Using $h(x)=x^{1-\gamma}f(x)$, we also obtain

$$
Kf=\sigma x^\gamma\left(\frac{\gamma}{4}x^{\gamma/2-1}-h\frac{x^{\gamma/2-1}}{2\sigma}\right)''+hx^{\gamma-1}\left(\frac{\gamma}{4}x^{\gamma/2-1}-\frac{hx^{\gamma/2-1}}{2\sigma}\right)'-x^{\gamma/2}g_x,
$$

which yields

$$\begin{aligned}Kf &= \frac{\sigma\gamma}{4}(\gamma/2-1)(\gamma/2-2)x^{3\gamma/2-3}\\&\quad -\frac{h''x^{3\gamma/2-1}}{2} - h'(\gamma/2-1)x^{3\gamma/2-2}\\&\quad -\frac{h(\gamma/2-1)(\gamma/2-2)}{2}x^{3\gamma/2-3} + \frac{\gamma}{4}(\gamma/2-1)hx^{3\gamma/2-3}\\&\quad -\frac{h'hx^{3\gamma/2-2}}{2\sigma} - \frac{h^2(\gamma/2-1)x^{3\gamma/2-3}}{2\sigma} - x^{\gamma/2}g_x.\end{aligned}$$

Linking Lf to Kf, we notice that

$$\begin{aligned}&-\frac{h''x^{3\gamma/2-1}}{2} - \frac{h'hx^{3\gamma/2-2}}{2\sigma}\\&\quad = \left(\frac{xh''}{2(2-\gamma)} + \frac{hh'}{2\sigma(2-\gamma)}\right)(-1)(2-\gamma)x^{3\gamma/2-2},\end{aligned}$$

and also

$$\begin{aligned}&-h'(\gamma/2-1)x^{3\gamma/2-2} + h(\gamma/2-1)x^{3\gamma/2-3}\\&\qquad - h^2\frac{(\gamma/2-1)x^{3\gamma/2-3}}{2\sigma}\\&\quad = \left(\sigma xh' - \sigma h + \frac{h^2}{2}\right)\frac{(-1)x^{3\gamma/2-3}(\gamma/2-1)}{\sigma}\end{aligned}$$

and lastly

$$\begin{aligned}-x^{\gamma/2}g_x &= \left(gx^{1-\gamma} + \frac{g'x^{2-\gamma}}{2-\gamma}\right)(-1)(2-\gamma)x^{3\gamma/2-2}\\&\quad + \left(gx^{2-\gamma}2\sigma\right)\frac{(-1)}{\sigma}x^{3\gamma/2-3}(\gamma/2-1).\end{aligned}$$

Consequently, one has

$$\begin{aligned}Kf &= \frac{\sigma\gamma}{4}(\gamma/2-1)(\gamma/2-2)x^{3\gamma/2-3}\\&\quad + \left(Ax^{3-2\gamma} + Bx^{1-\gamma}\right)(-1)(2-\gamma)x^{3\gamma/2-2}\\&\quad + \left(Ax^{4-2\gamma}\sigma + 2\sigma Bx^{2-\gamma} + C\right)\frac{(-1)}{\sigma}x^{3\gamma/2-3}(\gamma/2-1)\\&= (\gamma/2-1)\left(\frac{\sigma\gamma}{4}(\gamma/2-2) - \frac{C}{\sigma}\right)x^{3\gamma/2-3} + A(\gamma/2-1)x^{1-\gamma/2}.\end{aligned}\tag{4.4.55}$$

Table 4.4.3 Coefficients for the constant, x, and powers of x

Term	LHS	RHS
1	η_t	$-\frac{(1-\gamma)}{2(2-\gamma)}\xi^2_{tt} - B\xi^2_t$
$x^{1-\gamma/2}$	$-\frac{\rho_{tt}}{\sigma(2-\gamma)}$	0
$x^{2-\gamma}$	$-\frac{\xi^2_{ttt}}{2\sigma(2-\gamma)^2}$	0
$x^{3\gamma/2-3}$	0	$\rho(\frac{\sigma\gamma}{4}(\frac{\gamma}{2}-1)(\frac{\gamma}{2}-2) - C\frac{(\frac{\gamma}{2}-1)}{\sigma})$

We now substitute (4.4.54) and (4.4.55) into (4.4.53) to obtain

$$\begin{aligned}
&-\frac{x^{2-\gamma}}{2\sigma(2-\gamma)^2}\xi^2_{ttt} - \frac{x^{1-\gamma/2}}{\sigma(2-\gamma)}\rho_{tt} + \eta_t \\
&\quad = -\frac{(1-\gamma)}{2(2-\gamma)}\xi^2_{tt} - \xi^2_t\left(Ax^{2-\gamma} + B\right) \\
&\qquad + \rho\left(\frac{\sigma\gamma}{4}(\gamma/2-1)(\gamma/2-2) - C\frac{(\gamma/2-1)}{\sigma}\right)x^{3\gamma/2-3} \\
&\qquad + A\left(\frac{\gamma}{2}-1\right)x^{1-\gamma/2}\rho.
\end{aligned} \tag{4.4.56}$$

5. As mentioned in Craddock and Platen (2004) and Craddock and Lennox (2009), the cases $A = 0$ and $A \neq 0$ should be treated separately, as they result in different Lie algebras. We also alert the reader to the observation that $A = 0$ corresponds to the first Riccati equation (4.4.34), whereas $A \neq 0$ corresponds to the second Riccati equation, (4.4.35). We firstly consider the case $A = 0$, in which case (4.4.56) collapses to

$$\begin{aligned}
&-\frac{x^{2-\gamma}}{2\sigma(2-\gamma)^2}\xi^2_{ttt} - \frac{x^{1-\frac{\gamma}{2}}\rho_{tt}}{\sigma(2-\gamma)} + \eta_t \\
&\quad = -\frac{(1-\gamma)}{2(2-\gamma)}\xi^2_{tt} - B\xi^2_t \\
&\qquad + \rho\left(\frac{\sigma\gamma}{4}(\gamma/2-1)(\gamma/2-2) - C\frac{(\gamma/2-1)}{\sigma}\right)x^{3\gamma/2-3}.
\end{aligned}$$

Also, recall that $\kappa = \frac{\gamma}{8}(\gamma-4)\sigma^2$. Hence, if $C = \kappa$, the coefficient of ρ is 0 for all ρ. For now we consider the case $C \neq \kappa$ and discuss the case $C = \kappa$ below. We compare coefficients of the powers of x and Table 4.4.3 produces the determining equations for ξ^1, ξ^2, and ϕ.

From the $x^{3\gamma/2-3}$ term, we obtain $\rho = 0$. Using the $x^{2-\gamma}$ term, we get that ξ^2 is a quadratic function in t, hence

$$\xi^2 = \frac{1}{2}C_1t^2 + C_2t + C_3. \tag{4.4.57}$$

Consequently, from the constant term 1, we obtain

$$\eta = -\frac{1}{2}C_1Bt^2 + \left(-\frac{(1-\gamma)C_1}{2(2-\gamma)} - C_2B\right)t + C_4. \tag{4.4.58}$$

Substituting (4.4.57) and (4.4.58) into (4.4.51) , we get

$$\alpha = -\frac{C_1 x^{2-\gamma}}{2\sigma(2-\gamma)^2} - \frac{x^{1-\gamma}}{2\sigma(2-\gamma)} f(C_1 t + C_2) + \left(-\frac{1}{2}C_1 B t^2 + \left(-\frac{(1-\gamma)C_1}{2(2-\gamma)} - C_2 B\right)t + C_4\right).$$

Lastly, one obtains from Eq. (4.4.38)

$$\xi^1 = \frac{C_1 x t}{2-\gamma} + \frac{C_2 x}{2-\gamma}.$$

We now recall that we are looking for vector fields of the form

$$\boldsymbol{v} = \xi^1 \frac{\partial}{\partial x} + \xi^2 \frac{\partial}{\partial t} + \phi \frac{\partial}{\partial u},$$

which upon substitution yields

$$\begin{aligned}
\boldsymbol{v} = &\left(\frac{C_1 x t}{2-\gamma} + \frac{C_2 x}{2-\gamma}\right)\frac{\partial}{\partial x} + \left(\frac{1}{2}C_1 t^2 + C_2 t + C_3\right)\frac{\partial}{\partial t} \\
&+ \left(C_1\left(-\frac{x^{2-\gamma}}{2\sigma(2-\gamma)^2} - \frac{x^{1-\gamma}}{2\sigma(2-\gamma)} t f\right) - C_2 \frac{x^{1-\gamma} f}{2\sigma(2-\gamma)}\right. \\
&\left. - \frac{1}{2}C_1 B t^2 + \left(-\frac{1-\gamma}{2(2-\gamma)}C_1 - C_2 B\right)t + C_4\right) u \frac{\partial}{\partial u} \\
&+ \beta \frac{\partial}{\partial u}.
\end{aligned}$$

Hence a basis for the Lie algebra is

$$\begin{aligned}
\boldsymbol{v}_1 &= \frac{xt}{2-\gamma}\frac{\partial}{\partial x} + \frac{1}{2}t^2 \frac{\partial}{\partial t} \\
&\quad + \left(-\frac{x^{2-\gamma}}{2\sigma(2-\gamma)^2} - \frac{x^{1-\gamma} t f}{2\sigma(2-\gamma)} - \frac{Bt^2}{2} - \frac{(1-\gamma)t}{2(2-\gamma)}\right) u \frac{\partial}{\partial u} \\
\boldsymbol{v}_2 &= \frac{x}{2-\gamma}\frac{\partial}{\partial x} + t\frac{\partial}{\partial t} - \left(\frac{x^{1-\gamma} f}{2\sigma(2-\gamma)} + Bt\right) u \frac{\partial}{\partial u} \\
\boldsymbol{v}_3 &= \frac{\partial}{\partial t} \\
\boldsymbol{v}_4 &= u\frac{\partial}{\partial u}.
\end{aligned}$$

6. There are also vector fields of the form $\boldsymbol{v}_\beta = \beta\frac{\partial}{\partial u}$, β being an arbitrary solution of (4.4.48), so the Lie algebra is infinite-dimensional. However, it is clear that symmetries exist which transform solutions which are constant in t to those that are nonconstant in t. The case where $A = 0$, $C = \kappa$, can be handled similarly to the previous case.

In particular, it can be checked that

$$\xi^2 = \frac{1}{2}C_1 t^2 + C_2 t + C_3,$$
$$\eta = -\frac{(1-\gamma)}{2(2-\gamma)}C_1 t - \frac{C_1 B t^2}{2} - C_2 B t + C_4$$
$$\rho = C_5 t + C_6,$$

as we deduce from the $x^{1-\gamma/2}$ term that ρ can be at most linear in t. Furthermore,

$$\xi^1 = \frac{C_1 x t}{2-\gamma} + \frac{C_2 x}{2-\gamma} + C_5 x^{\gamma/2} t + C_6 x^{\gamma/2},$$

and

$$\begin{aligned}\alpha = &-\frac{C_1 x^{2-\gamma}}{2\sigma(2-\gamma)^2} - \frac{C_5 x^{1-\gamma/2}}{\sigma(2-\gamma)}\\ &+\left(\frac{\gamma}{4}x^{\gamma/2-1} - \frac{x^{-\gamma/2} f}{2\sigma}\right)(C_5 t + C_6)\\ &-\frac{x^{1-\gamma}}{2\sigma(2-\gamma)} f(C_1 t + C_2) - \frac{(1-\gamma)C_1 t}{2(2-\gamma)}\\ &-\frac{C_1 B t^2}{2} - C_2 B t + C_4.\end{aligned}$$

In this case, the vector field

$$\boldsymbol{v} = \xi^1 \frac{\partial}{\partial x} + \xi^2 \frac{\partial}{\partial t} + \phi \frac{\partial}{\partial u}$$

becomes

$$\begin{aligned}\boldsymbol{v} = &\left(\frac{C_1 x t}{2-\gamma} + \frac{C_2 x}{2-\gamma} + C_5 x^{\gamma/2} t + C_6 x^{\gamma/2}\right)\frac{\partial}{\partial x}\\ &+\left(\frac{1}{2}C_1 t^2 + C_2 t + C_3\right)\frac{\partial}{\partial t}\\ &+\left(-\frac{C_1 x^{2-\gamma}}{2\sigma(2-\gamma)^2} - \frac{C_5 x^{1-\gamma/2}}{\sigma(2-\gamma)}\right.\\ &+\left(\frac{\gamma}{4}x^{\gamma/2-1} - \frac{x^{-\gamma/2} f}{2\sigma}\right)(C_5 t + C_6)\\ &-\frac{C_1 x^{1-\gamma}}{2\sigma(2-\gamma)} t f - \frac{C_2 x^{1-\gamma}}{2\sigma(2-\gamma)} f\\ &-\frac{(1-\gamma)C_1 t}{2(2-\gamma)} - \frac{C_1 B t^2}{2}\\ &\left. - C_2 B t + C_4\right) u \frac{\partial}{\partial u}\\ &+\beta \frac{\partial}{\partial u}.\end{aligned}$$

Table 4.4.4 Coefficients for the constant, x and powers of x

Term	LHS	RHS
1	η_t	$-\frac{(1-\gamma)}{2(2-\gamma)}\xi_{tt}^2 - \frac{B}{2\sigma(2-\gamma)}\xi_t^2$
$x^{1-\gamma/2}$	$-\frac{\rho_{tt}}{\sigma(2-\gamma)}$	$\frac{A(\frac{\gamma}{2}-1)}{2(2-\gamma)^2\sigma}\rho$
$x^{2-\gamma}$	$-\frac{\xi_{ttt}^2}{2\sigma(2-\gamma)^2}$	$-\frac{A}{2\sigma(2-\gamma)^2}\xi_t^2$
$x^{3\gamma/2-3}$	0	$\rho(\frac{\sigma\gamma}{4}(\frac{\gamma}{2}-1)(\frac{\gamma}{2}-2) - C\frac{(\frac{\gamma}{2}-1)}{\sigma})$

This means that we obtain the infinitesimal symmetries

$$\begin{aligned}
\boldsymbol{v}_1 &= \frac{xt}{2-\gamma}\frac{\partial}{\partial x} + \frac{1}{2}t^2\frac{\partial}{\partial t} \\
&\quad - \left(\frac{x^{2-\gamma}}{2\sigma(2-\gamma)^2} + \frac{x^{1-\gamma}tf}{2\sigma(2-\gamma)} + \frac{(1-\gamma)t}{2(2-\gamma)} + \frac{Bt^2}{2}\right)u\frac{\partial}{\partial u} \\
\boldsymbol{v}_2 &= \frac{x}{2-\gamma}\frac{\partial}{\partial x} + t\frac{\partial}{\partial t} - \left(\frac{x^{1-\gamma}f}{2\sigma(2-\gamma)} + Bt\right)u\frac{\partial}{\partial u} \\
\boldsymbol{v}_3 &= \frac{\partial}{\partial t} \\
\boldsymbol{v}_4 &= u\frac{\partial}{\partial u} \\
\boldsymbol{v}_5 &= x^{\gamma/2}t\frac{\partial}{\partial x} - \left(\frac{x^{1-\gamma/2}}{\sigma(2-\gamma)} - \left(\frac{\gamma x^{\gamma/2-1}}{4} - \frac{x^{-\gamma/2}f}{2\sigma}\right)t\right)u\frac{\partial}{\partial u} \\
\boldsymbol{v}_6 &= x^{\gamma/2}\frac{\partial}{\partial x} + \left(\frac{\gamma x^{\gamma/2-1}}{4} - \frac{x^{-\gamma/2}f}{2\sigma}\right)u\frac{\partial}{\partial u},
\end{aligned} \tag{4.4.59}$$

and there is an infinitesimal symmetry $\boldsymbol{v}_\beta = \beta\frac{\partial}{\partial u}$, making the Lie algebra infinite-dimensional.

7. We now consider the case $A \neq 0$, which corresponds to the second Riccati equation, (4.4.35). For convenience, compare (4.4.34) and (4.4.35), we relabel the constants, $\sigma A \to \frac{A}{2(2-\gamma)^2}$, $2\sigma B \to \frac{B}{(2-\gamma)}$, which results in Table 4.4.4.

Now we assume that $C = \kappa$, which is the more involved case, however, we recover the $C \neq \kappa$ result by setting $C_5 = C_6 = 0$ in the following. We obtain from the $x^{2-\gamma}$ term

$$\begin{aligned}
\xi^2 &= \frac{C_1\exp\{\sqrt{A}t\}}{\sqrt{A}} - \frac{C_2\exp\{-\sqrt{A}t\}}{\sqrt{A}} + C_3 \\
\xi_t^2 &= C_1\exp\{\sqrt{A}t\} + C_2\exp\{-\sqrt{A}t\} \\
\xi_{tt}^2 &= C_1\sqrt{A}\exp\{\sqrt{A}t\} - C_2\sqrt{A}\exp\{-\sqrt{A}t\} \\
\eta &= C_1\left(-\frac{(1-\gamma)}{2(2-\gamma)} - \frac{B}{\sqrt{A}2\sigma(2-\gamma)}\right)\exp\{\sqrt{A}t\} \\
&\quad + C_2\left(\frac{B}{\sqrt{A}2\sigma(2-\gamma)} - \frac{(1-\gamma)}{2(2-\gamma)}\right)\exp\{-\sqrt{A}t\} + C_4
\end{aligned}$$

$$\begin{aligned}
\rho &= C_5 \exp\left\{\frac{\sqrt{A}}{2}t\right\} + C_6 \exp\left\{-\frac{\sqrt{A}}{2}t\right\} \\
\xi^1 &= \frac{x}{2-\gamma}(C_1 \exp\{\sqrt{A}t\} + C_2 \exp\{-\sqrt{A}t\}) \\
&\quad + x^{\gamma/2}\left(C_5 \exp\left\{\frac{\sqrt{A}}{2}t\right\} + C_6 \exp\left\{-\frac{\sqrt{A}}{2}t\right\}\right)
\end{aligned}$$

and

$$\begin{aligned}
\alpha &= -\frac{x^{2-\gamma}}{2\sigma(2-\gamma)^2}(C_1\sqrt{A}\exp\{\sqrt{A}t\} - C_2\sqrt{A}\exp\{-\sqrt{A}t\}) \\
&\quad - \frac{x^{1-\gamma/2}}{\sigma(2-\gamma)}\left(C_5\frac{\sqrt{A}}{2}\exp\left\{\frac{\sqrt{A}}{2}t\right\} + C_6\left(-\frac{\sqrt{A}}{2}\right)\exp\left\{-\frac{\sqrt{A}}{2}t\right\}\right) \\
&\quad + \left(\frac{\gamma}{4}x^{\gamma/2-1} - \frac{x^{-\gamma/2}f}{2\sigma}\right)\left(C_5 \exp\left\{\frac{\sqrt{A}}{2}t\right\} + C_6 \exp\left\{-\frac{\sqrt{A}}{2}t\right\}\right) \\
&\quad - \frac{x^{1-\gamma}}{2\sigma(2-\gamma)}f(C_1 \exp\{\sqrt{A}t\} + C_2 \exp\{-\sqrt{A}t\}) \\
&\quad + C_1\left(-\frac{(1-\gamma)}{2(2-\gamma)} - \frac{B}{\sqrt{A}2\sigma(2-\gamma)}\right)\exp\{\sqrt{A}t\} \\
&\quad + C_2\left(\frac{B}{\sqrt{A}2\sigma(2-\gamma)} - \frac{(1-\gamma)}{2(2-\gamma)}\right)\exp\{-\sqrt{A}t\} + C_4.
\end{aligned}$$

Finally, we look for symmetries of the form

$$\boldsymbol{v} = \xi^1\frac{\partial}{\partial x} + \xi^2\frac{\partial}{\partial t} + \phi\frac{\partial}{\partial u},$$

and obtain the following basis for the infinite-dimensional Lie algebra,

$$\begin{aligned}
\boldsymbol{v}_1 &= \frac{x}{2-\gamma}\exp\{\sqrt{A}t\}\frac{\partial}{\partial x} + \frac{\exp\{\sqrt{A}t\}}{\sqrt{A}}\frac{\partial}{\partial t} \\
&\quad - \left(\frac{x^{2-\gamma}}{2\sigma(2-\gamma)^2}\sqrt{A} + \frac{x^{1-\gamma}}{2\sigma(2-\gamma)}f\right. \\
&\quad \left. + \frac{B}{\sqrt{A}2\sigma(2-\gamma)} + \frac{(1-\gamma)}{2(2-\gamma)}\right)\exp\{\sqrt{A}t\}u\frac{\partial}{\partial u} \\
\boldsymbol{v}_2 &= \frac{x}{2-\gamma}\exp\{-\sqrt{A}t\}\frac{\partial}{\partial x} - \frac{\exp\{-\sqrt{A}t\}}{\sqrt{A}}\frac{\partial}{\partial t} \\
&\quad + \left(\frac{\sqrt{A}x^{2-\gamma}}{2\sigma(2-\gamma)^2} - \frac{x^{1-\gamma}}{2\sigma(2-\gamma)}f\right. \\
&\quad \left. + \frac{B}{\sqrt{A}2\sigma(2-\gamma)} - \frac{(1-\gamma)}{2(2-\gamma)}\right)\exp\{-\sqrt{A}t\}u\frac{\partial}{\partial u} \\
\boldsymbol{v}_3 &= \frac{\partial}{\partial t}
\end{aligned}$$

$$\begin{aligned}
\boldsymbol{v}_4 &= u\frac{\partial}{\partial u}\\
\boldsymbol{v}_5 &= x^{\gamma/2}\exp\left\{\frac{\sqrt{A}}{2}t\right\}\frac{\partial}{\partial x}\\
&\quad+\left(-\frac{x^{1-\gamma/2}\sqrt{A}}{2\sigma(2-\gamma)}+\frac{\gamma x^{\gamma/2-1}}{4}-\frac{x^{-\gamma/2}f}{2\sigma}\right)\exp\left\{\frac{\sqrt{A}}{2}t\right\}u\frac{\partial}{\partial u}\\
\boldsymbol{v}_6 &= \exp\left\{-\frac{\sqrt{A}}{2}t\right\}x^{\gamma/2}\frac{\partial}{\partial x}\\
&\quad+\left(\frac{x^{1-\gamma/2}\sqrt{A}}{2\sigma(2-\gamma)}+\frac{\gamma x^{\gamma/2-1}}{4}-\frac{x^{-\gamma/2}f}{2\sigma}\right)\exp\left\{-\frac{\sqrt{A}}{2}t\right\}u\frac{\partial}{\partial u}
\end{aligned}$$

and, of course, $\boldsymbol{v}_\beta = \beta\frac{\partial}{\partial u}$, where β is an arbitrary function satisfying (4.4.48).

For the (4.4.36) case, we refer the reader to Craddock and Lennox (2009), Proposition 2.1.

Finally, we consider the case where Lf satisfies neither (4.4.34), (4.4.35), nor (4.4.36). In that case, it is clear from Eq. (4.4.53) that $\xi_t^2 = 0$, hence ξ^2 is a constant function and we do not obtain a symmetry group transforming solutions which are constant in time to solutions which are not. □

We now present symmetries for the Riccati equations (4.4.34), (4.4.35), and (4.4.36). All of these symmetries will be generalized Laplace transforms. We need the following lemma, which appeared as Lemma 1.1 in Craddock (2009).

Lemma 4.4.2 *Suppose we have a linear PDE*

$$P(\boldsymbol{x}, D^{\boldsymbol{\alpha}})u = \sum_{|\boldsymbol{\alpha}|\le n} a_{\boldsymbol{\alpha}}(\boldsymbol{x})D^{\boldsymbol{\alpha}}u, \quad \boldsymbol{x}\in\Omega\subseteq\Re^m, \tag{4.4.60}$$

where $\boldsymbol{\alpha} = (\alpha_1,\ldots,\alpha_m,)$, $\alpha_i\in\mathcal{N}$, $|\boldsymbol{\alpha}| = \alpha_1+\cdots+\alpha_m$ *and* $D^{\boldsymbol{\alpha}} = \frac{\partial^{|\boldsymbol{\alpha}|}}{\partial_{x_1}^{\alpha_1}\ldots\partial_{x_m}^{\alpha_m}}$. *For a continuous one parameter family of solutions* $U_\epsilon(\boldsymbol{x})$ *of* (4.4.60), *for* $\epsilon\in I\subseteq\Re$, *where* I *is an interval containing* 0, *we have that for a function* $\varphi: I\to\Re$ *with sufficiently rapid decay,*

$$u(\boldsymbol{x}) = \int_I \varphi(\epsilon)U_\epsilon(\boldsymbol{x})\,d\epsilon$$

solves (4.4.60). *If the PDE is time dependent and* $U_\epsilon(\boldsymbol{x},t)$ *is the family of symmetry solutions, then*

$$u(\boldsymbol{x},t) = \int_I \varphi(\epsilon)U_\epsilon(\boldsymbol{x},t)\,d\epsilon$$

and $u(\boldsymbol{x},0) = \int_I\varphi(\epsilon)U_\epsilon(\boldsymbol{x},0)\,d\epsilon$ *solve* (4.4.60) *and so does* $\frac{d^nU_\epsilon(\boldsymbol{x})}{d\epsilon^n}$, *for all* $n = 0,1,2,\ldots$.

We now prove the following result, see Theorem 3.1 in Craddock and Lennox (2009).

Theorem 4.4.3 *Suppose $\gamma \neq 2$ and $h(x) = x^{1-\gamma} f(x)$ is a solution of the Riccati equation*

$$\sigma x h' - \sigma h + \frac{1}{2}h^2 + 2\sigma x^{2-\gamma} g(x) = 2\sigma A x^{2-\gamma} + B.$$

Then the PDE (4.4.31) has a symmetry of the form

$$\begin{aligned}\overline{U}_\epsilon(x,t) = {} & \frac{1}{(1+4\epsilon t)^{\frac{1-\gamma}{2-\gamma}}} \exp\left\{\frac{-4\epsilon(x^{2-\gamma} + A\sigma(2-\gamma)^2 t^2)}{\sigma(2-\gamma)^2(1+4\epsilon t)}\right\} \\ & \times \exp\left\{\frac{1}{2\sigma}\left(F\left(\frac{x}{(1+4\epsilon t)^{\frac{2}{2-\gamma}}}\right) - F(x)\right)\right\} \\ & \times u\left(\frac{x}{(1+4\epsilon t)^{\frac{2}{2-\gamma}}}, \frac{t}{1+4\epsilon t}\right), \end{aligned} \tag{4.4.61}$$

where $F'(x) = f(x)/x^\gamma$ and u is a solution of the PDE (4.4.31). That is, for ϵ sufficiently small, U_ϵ is a solution of (4.4.31) whenever u is. If $u(x,t) = u_0(x)$ with u_0 an analytic, stationary solution, then there is a fundamental solution $p(x,y,t)$ of (4.4.31) such that

$$\int_0^\infty \exp\{-\lambda y^{2-\gamma}\} u_0(y) p(x,y,t)\, dy = U_\lambda(x,t).$$

Here $U_\lambda(x,t) = \overline{U}_{\frac{1}{4}\sigma(2-\gamma)^2\lambda}(x,t)$. Further, if $u_0 = 1$, then $\int_0^\infty p(x,y,t)\, dy = 1$.

Proof 1. We recall from the proof of Theorem 4.4.1, that there is a symmetry $\boldsymbol{v}$, obtained by multiplying the right hand side of Eq. (4.4.59) by 8,

$$\begin{aligned}\boldsymbol{v} = {} & \frac{8xt}{2-\gamma}\frac{\partial}{\partial x} + 4t^2 \frac{\partial}{\partial t} \\ & - \left(\frac{4x^{2-\gamma}}{\sigma(2-\gamma)^2} + \frac{4x^{1-\gamma} t f}{\sigma(2-\gamma)} + \beta t + 4At^2\right) u \frac{\partial}{\partial u},\end{aligned}$$

where we set $\beta = \frac{4(1-\gamma)}{2-\gamma}$. Exponentiating the symmetry we obtain

$$\begin{aligned} & \frac{d\tilde{x}}{d\epsilon} = \frac{8\tilde{x}\tilde{t}}{2-\gamma}, \quad \tilde{x}(0) = x, \\ & \frac{d\tilde{t}}{d\epsilon} = 4\tilde{t}^2, \quad \tilde{t}(0) = t, \\ & \frac{d\tilde{u}}{d\epsilon} = -\left(\frac{4\tilde{x}^{2-\gamma}}{\sigma(2-\gamma)^2} + \frac{4\tilde{x}^{1-\gamma}\tilde{t} f(\tilde{x})}{\sigma(2-\gamma)} + \beta\tilde{t} + 4A\tilde{t}^2\right)\tilde{u}, \quad \tilde{u}(0) = u. \end{aligned} \tag{4.4.62}$$

From this system of equations, we obtain

$$\begin{aligned} \tilde{t} &= \frac{t}{1-4\epsilon t}, \\ t &= \frac{\tilde{t}}{1+4\epsilon\tilde{t}}, \end{aligned} \tag{4.4.63}$$

and

$$\begin{aligned} \tilde{x} &= x(1-4\epsilon t)^{-\frac{2}{2-\gamma}} \\ x &= \tilde{x}(1-4\epsilon t)^{\frac{2}{2-\gamma}}. \end{aligned} \tag{4.4.64}$$

Substituting (4.4.63) and (4.4.64) into (4.4.62), we obtain

$$\begin{aligned} \frac{d\tilde{u}}{\tilde{u}} = -\Bigg(&\frac{4x^{2-\gamma}(1-4\epsilon t)^{-2}}{\sigma(2-\gamma)^2} \\ &+ \frac{4x^{1-\gamma}(1-4\epsilon t)^{\frac{-4+3\gamma}{2-\gamma}} tf(x(1-4\epsilon t)^{\frac{-2}{2-\gamma}})}{\sigma(2-\gamma)} \\ &+ \beta t(1-4\epsilon t)^{-1} + 4At^2(1-4\epsilon t)^{-2}\Bigg) d\epsilon, \end{aligned}$$

and hence

$$\begin{aligned} \ln \tilde{u} = &-\frac{x^{2-\gamma}(1-4\epsilon t)^{-1}}{\sigma(2-\gamma)^2 t} \\ &- \frac{4}{\sigma(2-\gamma)} \int x^{1-\gamma}(1-4\epsilon t)^{\frac{-4+3\gamma}{2-\gamma}} tf\big(x(1-4\epsilon t)^{-\frac{2}{2-\gamma}}\big)\, d\epsilon \\ &+ \frac{\beta t \ln(1-4\epsilon t)}{4t} - At^2\frac{(1-4\epsilon t)^{-1}}{t} + C. \end{aligned}$$

Regarding the integral

$$-\frac{4}{\sigma(2-\gamma)} \int x^{1-\gamma}(1-4\epsilon t)^{\frac{-4+3\gamma}{2-\gamma}} tf\big(x(1-4\epsilon t)^{-\frac{2}{2-\gamma}}\big)\, d\epsilon,$$

we use the change of variables, $z = x(1-4\epsilon t)^{\frac{-2}{2-\gamma}}$, to obtain

$$\begin{aligned} &\frac{4}{\sigma(2-\gamma)} \int x^{1-\gamma}(1-4\epsilon t)^{\frac{-4+3\gamma}{2-\gamma}} tf\big(x(1-4\epsilon t)^{-\frac{2}{2-\gamma}}\big)\, d\epsilon \\ &= \frac{1}{2\sigma}\int z^{-\gamma} f(z)\, dz \\ &= \frac{1}{2\sigma} F(z), \end{aligned}$$

where we recalled that $F'(z) = \frac{f(z)}{z^{\gamma}}$. Consequently,

$$\begin{aligned} \tilde{u} = &\exp\left\{-\frac{x^{2-\gamma}}{\sigma(2-\gamma)^2 t(1-4\epsilon t)}\right\} \\ &\times \exp\left\{-\frac{1}{2\sigma}F\big(x(1-4\epsilon t)^{\frac{-2}{2-\gamma}}\big)\right\} \\ &\times (1-4\epsilon t)^{\beta/4} \exp\left\{-\frac{At}{1-4\epsilon t}\right\} \exp\{C\}. \end{aligned}$$

Also, setting $\epsilon = 0$ we have

$$u = \exp\left\{-\frac{x^{2-\gamma}}{\sigma(2-\gamma)^2 t}\right\}\exp\left\{-\frac{1}{2\sigma}F(x)\right\}\exp\{-At\}\exp\{C\},$$

which yields

$$\begin{aligned}\tilde{u} &= u\exp\left\{-\frac{x^{2-\gamma}((1-4\epsilon t)^{-1}-1)}{\sigma(2-\gamma)^2 t}\right\}\\ &\times \exp\left\{-\frac{(F(x(1-4\epsilon t)^{\frac{-2}{2-\gamma}})-F(x))}{2\sigma}\right\}\\ &\times \exp\left\{-At\left((1-4\epsilon t)^{-1}-1\right)\right\}(1-4\epsilon t)^{\beta/4}.\end{aligned}$$

Finally, changing back to the new parameters $\tilde{x}$, $\tilde{t}$, we obtain

$$\begin{aligned}\tilde{u} &= u\exp\left\{-\frac{\tilde{x}^{2-\gamma}(1+4\epsilon\tilde{t})^{-1}4\epsilon}{\sigma(2-\gamma)^2}\right\}\\ &\times \exp\left\{\frac{1}{2\sigma}\left(F\left(\frac{\tilde{x}}{(1+4\epsilon\tilde{t})^{2/(2-\gamma)}}\right)-F(\tilde{x})\right)\right\}\\ &\times \exp\left\{-\frac{4A\epsilon\tilde{t}^2}{1+4\epsilon\tilde{t}}\right\}(1+4\epsilon\tilde{t})^{-\beta/4},\end{aligned}$$

and, recalling $\beta = \frac{4(1-\gamma)}{2-\gamma}$, yields

$$\begin{aligned}\tilde{u}(\tilde{x},\tilde{t}) &= u\left(\frac{\tilde{x}}{(1+4\epsilon\tilde{t})^{\frac{2}{2-\gamma}}}, \frac{\tilde{t}}{1+4\epsilon\tilde{t}}\right)\frac{1}{(1+4\epsilon\tilde{t})^{\frac{1-\gamma}{2-\gamma}}}\\ &\times \exp\left\{\frac{-4\epsilon}{\sigma(2-\gamma)^2(1+4\epsilon\tilde{t})}\left(\tilde{x}^{2-\gamma}+A\tilde{t}^2\sigma(2-\gamma)^2\right)\right\}\\ &\times \exp\left\{\frac{1}{2\sigma}\left(F\left(\frac{\tilde{x}}{(1+4\epsilon\tilde{t})^{2/(2-\gamma)}}\right)-F(\tilde{x})\right)\right\}.\end{aligned}$$

Of course, the distinction between old and new parameters now becomes redundant, and so we have

$$\begin{aligned}\overline{U}_\epsilon(x,t) &= \frac{1}{(1+4\epsilon t)^{(1-\gamma)/(2-\gamma)}}\\ &\times \exp\left\{\frac{-4\epsilon}{\sigma(2-\gamma)^2(1+4\epsilon t)}\left(x^{2-\gamma}+At^2\sigma(2-\gamma)^2\right)\right\}\\ &\times \exp\left\{\frac{1}{2\sigma}\left(F\left(\frac{x}{(1+4\epsilon t)^{2/(2-\gamma)}}\right)-F(x)\right)\right\}\\ &\times u\left(\frac{x}{(1+4\epsilon t)^{\frac{2}{2-\gamma}}}, \frac{t}{1+4\epsilon t}\right).\end{aligned}$$

We now use the notation

$$U_\lambda(x,t) = \overline{U}_{(1/4)\sigma(2-\gamma)^2\lambda}(x,t).$$

This completes the first part of the proof.

2. We now prove that if $u(x,t) = u_0(x)$ with u_0 an analytic, stationary solution, then there is a fundamental solution $p(x, y, t)$ of (4.4.31) so that

$$\int_0^\infty \exp\{-\lambda y^{2-\gamma}\}u_0(y)p(x, y, t)\,dy = U_\lambda(x,t),$$

recalling that $U_\lambda(x,t) = \overline{U}_{(1/4)\sigma(2-\gamma)^2\lambda}(x,t)$. If p is a fundamental solution of (4.4.31), then we have

$$U_\lambda(x,t) = \int_0^\infty U_\lambda(y, 0)p(x, y, t)\,dy.$$

However, from (4.4.61), we have

$$U_\lambda(x, 0) = \exp\{-\lambda x^{2-\gamma}\}u(x, 0),$$

hence

$$U_\lambda(x,t) = \int_0^\infty \exp\{-\lambda y^{2-\gamma}\}u_0(y)p(x, y, t)\,dy,$$

for a stationary solution $u_0(y)$. Changing variables, i.e. setting $z = y^{2-\gamma}$, we obtain

$$U_\lambda(x,t) = \int_0^\infty \exp\{-\lambda z\}u_0\big(z^{\frac{1}{2-\gamma}}\big)p\big(x, z^{\frac{1}{2-\gamma}}, t\big)z^{\frac{1}{2-\gamma}-1}\frac{1}{2-\gamma}\,dz, \tag{4.4.65}$$

and hence it is shown that $U_\lambda(x,t)$ should be the Laplace transform of the distribution given in (4.4.65). This follows from the forthcoming Proposition 4.4.4. We now show that p is a fundamental solution. We note that if U_λ is a solution of (4.4.31), then, by Lemma 4.4.2, so is

$$u(x,t) = \int_0^\infty U_\lambda(x,t)\varphi(\lambda)\,d\lambda,$$

provided that the test function φ decays sufficiently rapidly against U_λ. Setting $t = 0$,

$$\begin{aligned}
u(x, 0) &= \int_0^\infty U_\lambda(x, 0)\varphi(\lambda)\,d\lambda \\
&= \int_0^\infty \exp\{-\lambda x^{2-\gamma}\}u_0(x)\varphi(\lambda)\,d\lambda \\
&= u_0(x)\int_0^\infty \exp\{-\lambda x^{2-\gamma}\}\varphi(\lambda)\,d\lambda \\
&= u_0(x)\Phi(x),
\end{aligned}$$

where $\Phi(x)$ denotes the generalized Laplace transform of φ. We now integrate $u_0(y)\Phi(y)$ against $p(x, y, t)$ to confirm that $p(x, y, t)$ is a fundamental solution.

In particular, it follows from Fubini's theorem that

$$\begin{aligned}\int_0^\infty u_0(y)\Phi(y)p(x,y,t)\,dy &= \int_0^\infty u_0(y)\int_0^\infty \exp\{-\lambda y^{2-\gamma}\}\varphi(\lambda)\,d\lambda p(x,y,t)\,dy\\ &= \int_0^\infty\int_0^\infty u_0(y)\exp\{-\lambda y^{2-\gamma}\}\varphi(\lambda)p(x,y,t)\,dy\,d\lambda\\ &= \int_0^\infty \varphi(\lambda)\int_0^\infty u_0(y)\exp\{-\lambda y^{2-\gamma}\}p(x,y,t)\,dy\,d\lambda\\ &= \int_0^\infty \varphi(\lambda)U_\lambda(x,t)\,d\lambda\\ &= u(x,t).\end{aligned}$$

But $u(x,t)$ solves Eq. (4.4.31), hence we have shown that integrating initial data $u_0(y)\Phi(y)$ against $p(x,y,t)$ produces a solution of the PDE, with initial data $u_0(y)\Phi(y)$. This establishes that p is a fundamental solution of (4.4.31), completing the second part of the proof.

Regarding the final part of the proof, if $u_0(x)=1$, then

$$U_0(x,t)=1,$$

but

$$U_0(x,t)=\int_0^\infty p(x,y,t)\,dy,$$

which completes the third part of the proof. □

Finally, we present the following proposition, which was employed in the proof of Theorem 4.4.3, see Proposition 3.2 in Craddock and Lennox (2009).

Proposition 4.4.4 *The solution $U_\lambda(x,t)$ in Theorem* 4.4.3 *is the Laplace transform of a distribution.*

The symmetries of the Riccati equations (4.4.35) and (4.4.36) can be obtained using similar arguments. We present the following result, which corresponds to (4.4.35). For the case of (4.4.36), we refer the reader to Theorem 2.8 in Craddock (2009).

Theorem 4.4.5 *Consider the PDE*

$$u_t=\sigma x^\gamma u_{xx}+f(x)u_x-g(x)u,\quad \gamma\neq 2,\ x\geq 0 \tag{4.4.66}$$

and suppose that g and $h(x)=x^{1-\gamma}f(x)$ satisfy

$$\sigma x h'-\sigma h+\frac{1}{2}h^2+2\sigma x^{2-\gamma}g(x)=\frac{A}{2(2-\gamma)^2}x^{4-2\gamma}+\frac{B}{2-\gamma}x^{2-\gamma}+C, \tag{4.4.67}$$

where $A>0$, B and C are arbitrary constants. Let u_0 be a stationary, analytic solution of (4.4.66).

Then (4.4.66) *has a solution of the form*

$$\overline{U}_\epsilon(x,t)$$
$$= u\left(\frac{x}{(1+2\sqrt{A}\epsilon\sinh(\sqrt{A}t)+2A\epsilon^2(\cosh(\sqrt{A}t)-1))^{\frac{1}{2-\gamma}}}\right)$$
$$\times\exp\left\{\frac{-A\epsilon x^{2-\gamma}}{\sigma(2-\gamma)^2}\left(\frac{\cosh(\sqrt{A}t)+\sqrt{A}\epsilon\sinh(\sqrt{A}t)}{1+2\sqrt{A}\epsilon\sinh(\sqrt{A}t)+2A\epsilon^2(\cosh(\sqrt{A}t)-1)}\right)\right\}$$
$$\times\exp\left\{-\frac{Bt}{2\sigma(2-\gamma)}\right\}\left|\frac{\cosh(\frac{\sqrt{A}t}{2})+(1+2\sqrt{A}\epsilon)\sinh(\frac{\sqrt{A}t}{2})}{\cosh(\frac{\sqrt{A}t}{2})-(1-2\sqrt{A}\epsilon)\sinh(\frac{\sqrt{A}t}{2})}\right|^{\frac{B}{2\sigma\sqrt{A}(2-\gamma)}}$$
$$\times\exp\left\{\frac{1}{2\sigma}F\left(\frac{x}{(1+2A\epsilon^2(\cosh(\sqrt{A}t)-1)+2\sqrt{A}\epsilon\sinh(\sqrt{A}t))^{\frac{1}{2-\gamma}}}\right)\right.$$
$$\left.-\frac{F(x)}{2\sigma}\right\}\left(1+2\sqrt{A}\epsilon\sinh(\sqrt{A}t)+2A\epsilon^2\left(\cosh(\sqrt{A}t)-1\right)\right)^{-\frac{c}{2}}, \quad (4.4.68)$$

where $F'(x)=\frac{f(x)}{x^\gamma}$ *and* $c=\frac{(1-\gamma)}{2-\gamma}$. *Furthermore, there exists a fundamental solution* $p(x,y,t)$ *of* (4.4.66) *such that*

$$\int_0^\infty \exp\left\{-\lambda y^{2-\gamma}\right\}u_0(y)p(x,y,t)\,dy = U_\lambda(x,t), \quad (4.4.69)$$

where $U_\lambda(x,t)=\overline{U}_{\frac{\sigma(2-\gamma)^2\lambda}{A}}(x,t)$.

Proof 1. The proof follows that of Theorem 2.5 in Craddock (2009). In the proof of Theorem 4.4.1, we found the following infinitesimal symmetries

$$\begin{aligned}
\mathbf{v}_1 &= \frac{\sqrt{A}x}{2-\gamma}\exp\{\sqrt{A}t\}\partial_x+\exp\{\sqrt{A}t\}\partial_t \\
&\quad -\left(\frac{Ax^{2-\gamma}}{2\sigma(2-\gamma)^2}+\frac{\sqrt{A}x^{1-\gamma}}{2\sigma(2-\gamma)}f(x)+\alpha\right)\exp\{\sqrt{A}t\}u\partial_u \\
\mathbf{v}_2 &= \frac{-\sqrt{A}x}{2-\gamma}\exp\{-\sqrt{A}t\}\partial_x+\exp\{-\sqrt{A}t\}\partial_t \\
&\quad -\left(\frac{Ax^{2-\gamma}}{2\sigma(2-\gamma)^2}-\frac{\sqrt{A}x^{1-\gamma}}{2\sigma(2-\gamma)}f(x)+\beta\right)\exp\{-\sqrt{A}t\}u\partial_u \\
\mathbf{v}_3 &= \partial_t \\
\mathbf{v}_4 &= u\partial_u,
\end{aligned}$$

where $\alpha=\frac{1-\gamma}{2(2-\gamma)}\sqrt{A}+\frac{B}{2\sigma(2-\gamma)}$, $\beta=-\frac{1-\gamma}{2(2-\gamma)}\sqrt{A}+\frac{B}{2\sigma(2-\gamma)}$. We now combine these symmetries as follows

$$\mathbf{v}=\mathbf{v}_1+\mathbf{v}_2-2\mathbf{v}_3+\frac{B}{(2-\gamma)\sigma}\mathbf{v}_4,$$

which yields

$$\begin{aligned}
\boldsymbol{v} &= \frac{2\sqrt{A}x}{2-\gamma}\sinh(\sqrt{A}t)\partial_x + 2\big(\cosh(\sqrt{A}t)-1\big)\partial_t \\
&\quad - \left(\frac{Ax^{2-\gamma}}{\sigma(2-\gamma)^2}\cosh(\sqrt{A}t) + \frac{\sqrt{A}x^{1-\gamma}f}{\sigma(2-\gamma)}\sinh(\sqrt{A}t) + \sqrt{A}c\sinh(\sqrt{A}t)\right. \\
&\quad \left. + \frac{B}{\sigma(2-\gamma)}\big(\cosh(\sqrt{A}t)-1\big)\right)u\partial_u \\
&= \frac{2\sqrt{A}x}{2-\gamma}\sinh(\sqrt{A}t)\partial_x + 2\big(\cosh(\sqrt{A}t)-1\big)\partial_t - \frac{\tilde{g}}{\sigma}u\partial_u,
\end{aligned}$$

where

$$\begin{aligned}
\tilde{g}(x,t) &= \frac{Ax^{2-\gamma}}{(2-\gamma)^2}\cosh(\sqrt{A}t) + \frac{\sqrt{A}x^{1-\gamma}f(x)}{(2-\gamma)}\sinh(\sqrt{A}t) + \sqrt{A}c\sigma\sinh(\sqrt{A}t) \\
&\quad + \frac{B}{(2-\gamma)}\big(\cosh(\sqrt{A}t)-1\big)
\end{aligned}$$

and $c = \frac{1-\gamma}{2-\gamma}$.

2. We solve the following ODEs

$$\begin{aligned}
\frac{d\tilde{x}}{d\epsilon} &= \frac{2\sqrt{A}\tilde{x}}{2-\gamma}\sinh(\sqrt{A}\tilde{t}), \quad \tilde{x}(0)=x \\
\frac{d\tilde{t}}{d\epsilon} &= 2\big(\cosh(\sqrt{A}\tilde{t})-1\big), \quad \tilde{t}(0)=t \\
\frac{d\tilde{u}}{d\epsilon} &= -\frac{\tilde{g}(\tilde{x},\tilde{t})}{\sigma}\tilde{u}, \quad \tilde{u}(0)=u.
\end{aligned}$$

First, we solve the ODE

$$\frac{d\tilde{t}}{d\epsilon} = 2\big(\cosh(\sqrt{A}\tilde{t})-1\big), \quad \tilde{t}(0)=t,$$

to yield

$$\tilde{t} = \frac{1}{\sqrt{A}}\log\left(\frac{\sqrt{A}\epsilon - \sqrt{A}\epsilon e^{\sqrt{A}t} + e^{\sqrt{A}t}}{1+\sqrt{A}\epsilon - \sqrt{A}\epsilon e^{\sqrt{A}t}}\right). \tag{4.4.70}$$

Consequently, we solve

$$\frac{d\tilde{x}}{d\epsilon} = \frac{2\sqrt{A}\tilde{x}}{2-\gamma}\sinh(\sqrt{A}\tilde{t}), \quad \tilde{x}(0)=x,$$

to yield

$$\tilde{x} = x\big(\sqrt{A}e^{\sqrt{A}t}\epsilon(-1+\sqrt{A}\epsilon) + \sqrt{A}\epsilon e^{-\sqrt{A}t}(1+\sqrt{A}\epsilon) + \big(1-2A\epsilon^2\big)\big)^{-\frac{1}{2-\gamma}}. \tag{4.4.71}$$

Finally, we solve

$$\frac{d\tilde{u}}{\tilde{u}} = a_1\tilde{x}^{2-\gamma}\cosh(\sqrt{A}\tilde{t})\,d\epsilon + a_2\big(\cosh(\sqrt{A}\tilde{t}) - 1\big)\,d\epsilon$$
$$+ a_3\sinh(\sqrt{A}\tilde{t})\tilde{x}^{1-\gamma}f(\tilde{x})\,d\epsilon + a_4\sinh(\sqrt{A}\tilde{t})\,d\epsilon, \quad \tilde{u}(0) = u, \tag{4.4.72}$$

where

$$a_1 = -\frac{A}{\sigma(2-\gamma)^2}, \qquad a_2 = -\frac{B}{\sigma(2-\gamma)},$$
$$a_3 = -\frac{\sqrt{A}}{\sigma(2-\gamma)}, \qquad a_4 = -\sqrt{A}c.$$

We integrate the expression on the right hand side of Eq. (4.4.72) term-by-term to yield

$$a_1\int \tilde{x}^{2-\gamma}\cosh(\sqrt{A}\tilde{t})\,d\epsilon$$
$$= \frac{a_1 e^{\sqrt{A}t}(1 + 2\sqrt{A}\epsilon + e^{\sqrt{A}t} - 2\sqrt{A}\epsilon e^{\sqrt{A}t})x^{2-\gamma}}{2\sqrt{A}(-1 + e^{\sqrt{A}t})(-1 - \sqrt{A}\epsilon + \sqrt{A}\epsilon e^{\sqrt{A}t})(-\sqrt{A}\epsilon - e^{\sqrt{A}t} + \sqrt{A}\epsilon e^{\sqrt{A}t})},$$

also

$$a_2\int \big(\cosh(\sqrt{A}\tilde{t}) - 1\big)\,d\epsilon = \frac{a_2}{2}\tilde{t}$$
$$= \frac{a_2}{2\sqrt{A}}\log\left(\frac{\sqrt{A}\epsilon - \sqrt{A}\epsilon e^{\sqrt{A}t} + e^{\sqrt{A}t}}{1 + \sqrt{A}\epsilon - \sqrt{A}\epsilon e^{\sqrt{A}t}}\right),$$

regarding the third term

$$a_3\int \sinh(\sqrt{A}\tilde{t})\tilde{x}^{1-\gamma}f(\tilde{x})\,d\epsilon = a_3\frac{(2-\gamma)}{2\sqrt{A}}F(\tilde{x})$$

and lastly

$$a_4\int \sinh(\sqrt{A}\tilde{t})\,d\epsilon$$
$$= \frac{a_4}{2\sqrt{A}}(2-\gamma)\left(\log(x) + \frac{\sqrt{A}t}{2-\gamma}\right)$$
$$- \frac{a_4}{2\sqrt{A}}\log\big(\sqrt{A}\epsilon + A\epsilon^2 + e^{\sqrt{A}t} - 2A\epsilon^2 e^{\sqrt{A}t} - \sqrt{A}\epsilon e^{2\sqrt{A}t} + A\epsilon^2 e^{2\sqrt{A}t}\big)$$

and hence

$$\log(\tilde{u})$$
$$= C + a_1$$
$$\times \frac{e^{\sqrt{A}t}(1 + 2\sqrt{A}\epsilon + e^{\sqrt{A}t} - 2\sqrt{A}\epsilon e^{\sqrt{A}t})x^{2-\gamma}}{2\sqrt{A}(-1 + e^{\sqrt{A}t})(-1 - \sqrt{A}\epsilon + \sqrt{A}\epsilon e^{\sqrt{A}t})(-\sqrt{A}\epsilon - e^{\sqrt{A}t} + \sqrt{A}\epsilon e^{\sqrt{A}t})}$$

$$+\frac{a_2}{2\sqrt{A}}\log\left(\frac{\sqrt{A}\epsilon-\sqrt{A}\epsilon e^{\sqrt{A}t}+e^{\sqrt{A}t}}{1+\sqrt{A}\epsilon-\sqrt{A}\epsilon e^{\sqrt{A}t}}\right)$$
$$+a_3\frac{(2-\gamma)}{2\sqrt{A}}F(\tilde{x})$$
$$-a_4\frac{\log(\sqrt{A}\epsilon+A\epsilon^2+e^{\sqrt{A}t}-2A\epsilon^2e^{\sqrt{A}t}-\sqrt{A}\epsilon e^{2\sqrt{A}t}+A\epsilon^2e^{2\sqrt{A}t})}{2\sqrt{A}}.$$

From the initial condition, we get

$$\log(u)=C+\frac{a_1(1+e^{\sqrt{A}t})x^{2-\gamma}}{2\sqrt{A}(-1+e^{\sqrt{A}t})}$$
$$+\frac{a_2}{2\sqrt{A}}\log\left(e^{\sqrt{A}t}\right)+a_3\frac{(2-\gamma)}{2\sqrt{A}}F(x)-\frac{a_4}{2}\frac{\log(e^{\sqrt{A}t})}{\sqrt{A}}.$$

Consequently, we obtain the following solution for $\tilde{u}$,

$$\log(\tilde{u})$$
$$=\log(u)+\frac{a_1x^{2-\gamma}}{2\sqrt{A}}$$
$$\times\left(\frac{e^{\sqrt{A}t}(1+2\sqrt{A}\epsilon+e^{\sqrt{A}t}-2\sqrt{A}\epsilon e^{\sqrt{A}t})}{(-1+e^{\sqrt{A}t})(-1-\sqrt{A}\epsilon+\sqrt{A}\epsilon e^{\sqrt{A}t})(-\sqrt{A}\epsilon-e^{\sqrt{A}t}+\sqrt{A}\epsilon e^{\sqrt{A}t})}\right.$$
$$\left.-\frac{(1+e^{\sqrt{A}t})}{(-1+e^{\sqrt{A}t})}\right)+\frac{a_2}{2}\left(\frac{\log(\frac{\sqrt{A}\epsilon-\sqrt{A}\epsilon e^{\sqrt{A}t}+e^{\sqrt{A}t}}{1+\sqrt{A}\epsilon-\sqrt{A}\epsilon e^{\sqrt{A}t}})}{\sqrt{A}}-t\right)$$
$$+\frac{a_3(2-\gamma)}{2\sqrt{A}}\left(F(\tilde{x})-F(x)\right)+\frac{a_4}{2}$$
$$\times\left(-\frac{\log(\sqrt{A}\epsilon+A\epsilon^2+e^{\sqrt{A}t}-2A\epsilon^2e^{\sqrt{A}t}-\sqrt{A}\epsilon e^{2\sqrt{A}t}+A\epsilon^2e^{2\sqrt{A}t})}{\sqrt{A}}+t\right).$$

3. We now express t and x in terms of the new parameters $\tilde{t}$ and $\tilde{x}$,

$$t=\frac{1}{\sqrt{A}}\log\left(\frac{\sqrt{A}\epsilon-(1+\sqrt{A}\epsilon)e^{\sqrt{A}\tilde{t}}}{-\sqrt{A}\epsilon e^{\sqrt{A}\tilde{t}}+\sqrt{A}\epsilon-1}\right),$$

also

$$x=\frac{\tilde{x}}{(1+2\sqrt{A}\epsilon\sinh(\sqrt{A}\tilde{t})+2A\epsilon^2(\cosh(\sqrt{A}\tilde{t})-1))^{\frac{1}{2-\gamma}}}.$$

Substituting, we get for the first term

$$\frac{a_1}{2}\epsilon\tilde{x}^{2-\gamma}\frac{(1+e^{2\sqrt{A}\tilde{t}}+\sqrt{A}\epsilon(-1+e^{2\sqrt{A}\tilde{t}}))}{(1+\sqrt{A}\epsilon(-1+e^{\sqrt{A}\tilde{t}}))(e^{\sqrt{A}\tilde{t}}-\sqrt{A}\epsilon+\sqrt{A}\epsilon e^{\sqrt{A}\tilde{t}})}$$
$$=a_1\epsilon\tilde{x}^{2-\gamma}\frac{\cosh(\sqrt{A}\tilde{t})+\sqrt{A}\epsilon\sinh(\sqrt{A}\tilde{t})}{1+2\sqrt{A}\epsilon\sinh(\sqrt{A}\tilde{t})+2A\epsilon^2(\cosh(\sqrt{A}\tilde{t})-1)},$$

for the second term

$$\frac{a_2}{2\sqrt{A}}\left(\sqrt{A}\tilde{t}-\log\left(\frac{-e^{\sqrt{A}\tilde{t}}-\sqrt{A}\epsilon(-1+e^{\sqrt{A}\tilde{t}})}{-1-\sqrt{A}\epsilon(-1+e^{\sqrt{A}\tilde{t}})}\right)\right)$$

$$=\frac{a_2\tilde{t}}{2}-\frac{a_2}{2\sqrt{A}}\log\left(\frac{\cosh(\frac{\sqrt{A}\tilde{t}}{2})+(1+2\sqrt{A}\epsilon)\sinh(\frac{\sqrt{A}\tilde{t}}{2})}{\cosh(\frac{\sqrt{A}\tilde{t}}{2})-(1-2\sqrt{A}\epsilon)\sinh(\frac{\sqrt{A}\tilde{t}}{2})}\right),$$

for the third term

$$\frac{a_3(2-\gamma)}{2\sqrt{A}}\left(F(\tilde{x})-F\left(\frac{\tilde{x}}{(1+2\sqrt{A}\epsilon\sinh(\sqrt{A}\tilde{t})+2A\epsilon^2(\cosh(\sqrt{A}\tilde{t})-1))^{\frac{1}{2-\gamma}}}\right)\right),$$

and for the fourth term

$$\frac{a_4}{2\sqrt{A}}\log\left(e^{-\sqrt{A}\tilde{t}}\left(1+\sqrt{A}\epsilon\left(-1+e^{\sqrt{A}\tilde{t}}\right)\right)\left(e^{\sqrt{A}\tilde{t}}+\sqrt{A}\epsilon\left(-1+e^{\sqrt{A}\tilde{t}}\right)\right)\right)$$

$$=\frac{a_4}{2\sqrt{A}}\log\left(1+2\sqrt{A}\epsilon\sinh(\sqrt{A}\tilde{t})+2A\epsilon^2\left(\cosh(\sqrt{A}\tilde{t})-1\right)\right).$$

Hence, substituting a_1, a_2, a_3, and a_4, and no longer emphasizing the difference between new and old variables, one has

$$\begin{aligned}
&\overline{U}_\epsilon(x,t)\\
&=u\left(\frac{x}{(1+2\sqrt{A}\epsilon\sinh(\sqrt{A}t)+2A\epsilon^2(\cosh(\sqrt{A}t)-1))^{\frac{1}{2-\gamma}}}\right)\\
&\times\exp\left\{-\frac{A\epsilon x^{2-\gamma}}{\sigma(2-\gamma)^2}\left(\frac{\cosh(\sqrt{A}t)+\sqrt{A}\epsilon\sinh(\sqrt{A}t)}{1+2\sqrt{A}\epsilon\sinh(\sqrt{A}t)+2A\epsilon^2(\cosh(\sqrt{A}t)-1)}\right)\right\}\\
&\times\exp\left\{-\frac{Bt}{2\sigma(2-\gamma)}\right\}\left|\frac{\cosh(\frac{\sqrt{A}t}{2})+(1+2\sqrt{A}\epsilon)\sinh(\frac{\sqrt{A}t}{2})}{\cosh(\frac{\sqrt{A}t}{2})-(1-2\sqrt{A}\epsilon)\sinh(\frac{\sqrt{A}t}{2})}\right|^{\frac{B}{2\sigma(2-\gamma)\sqrt{A}}}\\
&\times\exp\left\{\frac{1}{2\sigma}\left(F\left(\frac{x}{(1+2\sqrt{A}\epsilon\sinh(\sqrt{A}t)+2A\epsilon^2(\cosh(\sqrt{A}t)-1))^{\frac{1}{2-\gamma}}}\right)\right.\right.\\
&\left.\left.-F(x)\right)\right\}\left(1+2\sqrt{A}\epsilon\sinh(\sqrt{A}t)+2A\epsilon^2\left(\cosh(\sqrt{A}t)-1\right)\right)^{-\frac{c}{2}}.
\end{aligned}$$

We change parameters, setting $\epsilon\to\frac{\sigma(2-\gamma)^2\lambda}{A}$ and set

$$U_\lambda(x,t)=\overline{U}_{\frac{\sigma(2-\gamma)^2\lambda}{A}}(x,t),$$

and find that, for a stationary, analytic solution, say $u_0(x)$,

$$U_\lambda(x,0)=u_0(x)e^{-\frac{A\epsilon x^{2-\gamma}}{\sigma(2-\gamma)^2}}=u_0(x)e^{-\lambda x^{2-\gamma}}.$$

Similar to the proof of Theorem 4.4.3, we have

$$\int_0^\infty e^{-\lambda y^{2-\gamma}}u_0(y)p(x,y,t)\,dy=U_\lambda(x,t),$$

for a fundamental solution p, so $U_\lambda(x,t)$ is the generalized Laplace transform of $u_0 p$. For the proof of this fact, we refer to the proof of Theorem 2.5 in Craddock (2009). □

We now recall Corollary 2.6 from Craddock (2009).

Corollary 4.4.6 *Under the assumptions of Theorem 4.4.5, suppose that we have a stationary solution $u_0(x) = 1$. Then the resulting fundamental solution has the property that*

$$\int_0^\infty p(x,y,t)\,dy = 1.$$

Proof From Theorem 4.4.5, we have that

$$U_\lambda(x,t) = \int_0^\infty e^{-\lambda y^{2-\gamma}} p(x,y,t)\,dy.$$

But

$$U_0(x,t) = \left| \frac{\cosh(\frac{\sqrt{A}t}{2}) + \sinh(\frac{\sqrt{A}t}{2})}{\cosh(\frac{\sqrt{A}t}{2}) - \sinh(\frac{\sqrt{A}t}{2})} \right|^{\frac{B}{2\sigma\sqrt{A}(2-\gamma)}} e^{-\frac{Bt}{2\sigma(2-\gamma)}} = 1,$$

which concludes the proof. □

For the case $A < 0$, we direct the reader to Craddock (2009), see Remark 2.7.

Concluding this chapter, we remark that in Sects. 5.3 and 5.4 the results presented in this section, and additional results which can be obtained via the same method, are collected.

Chapter 5
Transition Densities via Lie Symmetry Methods

In this chapter, we discuss how to obtain explicit transition densities and Laplace transforms of joint transition densities for various diffusions using Lie symmetry methods. We begin with a motivating example, and subsequently present two cautionary examples. The chapter continues with transition densities, which could have useful applications in finance or other areas of application, but are new and have therefore not received so far much attention in the literature. It is hoped that this chapter encourages readers to construct their own examples and apply them to problems they encounter. Subsequently, we present Laplace transforms of joint transition densities in Sect. 5.4. Section 5.5 illustrates how Lie symmetry methods can be powerfully combined with probability theory to enlarge the scope of results that can be obtained.

5.1 A Motivating Example

In this section, we firstly present an example, which exemplifies how explicit transition densities can be found via Lie symmetry methods. The squared Bessel process sits at the heart of the developments in Chap. 3, and our motivating example is also based on this process. Consequently, we consider a squared Bessel process of dimension δ, $\delta \geq 2$,

$$dX_t = \delta\, dt + 2\sqrt{X_t}\, dW_t,$$

where $X_0 = x > 0$, whose transition density satisfies the Kolmogorov backward equation

$$u_t = 2x u_{xx} + \delta u_x.$$

Hence in Eq. (4.4.1), we set $\sigma = 2$, $f = \delta$, $g = 0$, and $\gamma = 1$, and in Eq. (4.4.34), we set $h = \delta$, $A = 0$, $B = -2\delta + \frac{\delta^2}{2}$. Now, we employ Theorem 4.4.3 with $u(x,t) = 1$ and $F(x) = \delta \ln x$ to obtain

$$\overline{U}_\epsilon(x,t) = \exp\left\{ -\frac{4\epsilon x}{\sigma(1+4\epsilon t)} \right\} (1+4\epsilon t)^{-\frac{\delta}{\sigma}}, \tag{5.1.1}$$

J. Baldeaux, E. Platen, *Functionals of Multidimensional Diffusions with Applications to Finance*, Bocconi & Springer Series 5, DOI 10.1007/978-3-319-00747-2_5,

where $\sigma = 2$. Setting $\epsilon = \frac{\sigma\lambda}{4}$ in Eq. (5.1.1), we obtain the Laplace transform

$$\begin{aligned} U_\lambda(x,t) &= \int_0^\infty \exp\{-\lambda y\} p(t,x,y)\,dy \\ &= \exp\left\{-\frac{x\lambda}{1+2\lambda t}\right\}(1+2\lambda t)^{-\frac{\delta}{2}}, \end{aligned}$$

which is easily inverted to yield

$$p(t,x,y) = \frac{1}{2t}\left(\frac{x}{y}\right)^{\frac{\nu}{2}} I_\nu\left(\frac{\sqrt{xy}}{t}\right)\exp\left\{-\frac{(x+y)}{2t}\right\}, \tag{5.1.2}$$

where $\nu = \frac{\delta}{2} - 1$ denotes the index of the squared Bessel process. Of course, Eq. (5.1.2) shows the transition density of a squared Bessel process started at time 0 in x for being at time t in y. Recall that I_ν denotes the modified Bessel function of the first kind, and that we plotted this transition density in Fig. 3.1.1. We also show it in Fig. 5.3.1.

5.2 Two Cautionary Examples

The previous example begs the question whether a fundamental solution is necessarily a transition density. Fundamental solutions are known not to be unique, and the following example, which is again based on a squared Bessel process and taken from Craddock and Lennox (2009), shows that a fundamental solution is not necessarily a transition density.

Example 5.2.1 Consider a squared Bessel process of dimension three, $\delta = 3$, the transition density of which satisfies the Kolmogorov backward equation

$$u_t = 2xu_{xx} + 3u_x, \tag{5.2.3}$$

a stationary solution of which is $u_1(x) = 1/\sqrt{x}$. Again, we employ Theorem 4.4.3, to obtain

$$\begin{aligned} &\int_0^\infty \frac{1}{\sqrt{y}} \exp\{-\lambda y\} p(t,x,y)\,dy \\ &\quad = \exp\left\{-\frac{x\lambda}{1+2\lambda t}\right\}(1+2\lambda t)^{-\frac{3}{2}} u_1\left(\frac{x}{(1+2\lambda t)^2}\right) \\ &\quad = \exp\left\{-\frac{x\lambda}{1+2\lambda t}\right\}(1+2\lambda t)^{-\frac{3}{2}} \frac{(1+2\lambda t)}{\sqrt{x}} \\ &\quad = \exp\left\{-\frac{x\lambda}{1+2\lambda t}\right\}(1+2\lambda t)^{-\frac{1}{2}} \frac{1}{\sqrt{x}}, \end{aligned}$$

so that we have

$$p(t,x,y) = \frac{\exp\{-\frac{y+x}{2t}\}\cosh(\frac{\sqrt{xy}}{t})}{\sqrt{2t\pi x}}.$$

We note that

$$\int_0^\infty p(t,x,y)\,dy = \frac{\sqrt{2t}}{\sqrt{\pi x}}\exp\left\{-\frac{x}{2t}\right\} + \operatorname{erf}\left(\frac{\sqrt{x}}{\sqrt{2t}}\right).$$

This fundamental solution does not integrate to 1, and hence is not a transition probability density.

We conclude that not all fundamental solutions are transition probability densities. From Example 5.2.1, it is tempting to deduce that fundamental solutions integrating to 1 are transition probability densities. The next example, which stems from Craddock (2009), see Proposition 2.10, shows that also this conjecture is false. As the preceding two examples, it is again based on a squared Bessel process. The example makes use of the following proposition, Proposition 2.4 in Craddock (2009), which shows how to invert a Laplace transform when studying squared Bessel processes.

Proposition 5.2.2 *For a nonnegative integer n, the following equality holds,*

$$L_y^{-1}\left(\lambda^n \exp\left\{\frac{k}{\lambda}\right\}\right) = \sum_{l=0}^{n} \frac{k^l}{l!}\delta^{n-l}(y) + \left(\frac{k}{y}\right)^{\frac{n+1}{2}} I_{n+1}(2\sqrt{ky}),$$

where L_λ is the Laplace transform, $\delta(y)$ is the Dirac delta function and I_n is a modified Bessel function of the first kind with index n.

We now present the example.

Example 5.2.3 Consider a squared Bessel process of dimension 2δ. The transition density satisfies the Kolmogorov backward equation

$$u_t = 2xu_{xx} + 2\delta u_x. \tag{5.2.4}$$

It is easily verified that the stationary solutions $u_0(x) = 1$ and $u_1(x) = x^{1-\delta}$ satisfy (5.2.4). In Sect. 5.1, it was shown that the stationary solution $u_0(x) = 1$ produces the correct transition density. We will now investigate the fundamental solution produced by $u_1(x) = x^{1-\delta}$. Applying Theorem 4.4.3 with $A = 0$, we obtain

$$U_\lambda(x,t) = x^{1-\delta}\exp\left\{-\frac{\lambda x}{(1+2\lambda t)}\right\}(1+2\lambda t)^{\delta-2},$$

i.e.

$$\begin{aligned} L_\lambda &= \int_0^\infty \exp\{-\lambda y\} u_1(y) q(t,x,y)\,dy \\ &= x^{1-\delta}\exp\left\{-\frac{\lambda x}{1+2\lambda t}\right\}(1+2\lambda t)^{\delta-2}. \end{aligned}$$

We now invert the Laplace transform, which yields

$$u_1(y)q(t,x,y) = x^{1-\delta}\exp\left\{-\frac{x+y}{2t}\right\}(2t)^{\delta-2} L_y^{-1}\left(\lambda^{\delta-2}\exp\left\{\frac{k}{\lambda}\right\}\right),$$

where $k = \frac{x}{(2t)^2}$. We now apply Proposition 5.2.2 to yield

$$q(t,x,y) = (2t)^{-1}\left(\frac{y}{x}\right)^{\frac{\delta-1}{2}} \exp\left\{-\frac{(x+y)}{2t}\right\} I_{\delta-1}\left(\frac{\sqrt{xy}}{t}\right) + (2t)^{\delta-2}\left(\frac{y}{x}\right)^{\delta-1} \exp\left\{-\frac{(x+y)}{2t}\right\} \sum_{l=0}^{\delta-2} \frac{k^l}{l!} \delta^{\delta-2-l}(y).$$

We have

$$\int_0^\infty (2t)^{\delta-2}\left(\frac{y}{x}\right)^{\delta-1} \exp\left\{-\frac{x+y}{2t}\right\} \sum_{l=0}^{\delta-2} \frac{x^l}{(2t)^{2l} l!} \delta^{(\delta-2-l)}(y)\, dy = 0,$$

since the Dirac delta function and their derivatives select the value of the test function $y^{\delta-1}$ and its derivatives at zero. Also, we recognize that

$$(2t)^{-1}\left(\frac{y}{x}\right)^{\frac{\delta-1}{2}} \exp\left\{-\frac{(x+y)}{2t}\right\} I_{\delta-1}\left(\frac{\sqrt{xy}}{t}\right)$$

is the transition density of a squared Bessel process of dimension 2δ, cf. (3.1.4), and hence

$$\int_0^\infty (2t)^{-1}\left(\frac{y}{x}\right)^{\frac{\delta-1}{2}} \exp\left\{-\frac{(x+y)}{2t}\right\} I_{\delta-1}\left(\frac{\sqrt{xy}}{t}\right) dy = 1.$$

Finally, we observe that $U_0(x,t) = u_1(x)$ and

$$U_\lambda(x,t) = \int_0^\infty \exp\{-\lambda y\} u_1(y) q(t,x,y)\, dy,$$

which yields

$$\int_0^\infty u_1(y) q(t,x,y)\, dy = u_1(x),$$

and hence $q(t,x,y)$ is not the transition density.

However, in Craddock (2009), the following useful check for processes satisfying

$$X_t = X_0 + \int_0^t f(X_s)\, ds + \int_0^t \sqrt{2\sigma X_s}\, dW_s$$

was presented, see Proposition 2.11 in Craddock (2009), which we now recall.

Proposition 5.2.4 *Let $X = \{X_t, t \geq 0\}$ be an Itô diffusion which is the unique strong solution of*

$$X_t = X_0 + \int_0^t f(X_s)\, ds + \int_0^T \sqrt{2\sigma X_s}\, dW_s,$$

where $W = \{W_t, t \geq 0\}$ is a standard Wiener process and $X_0 = x > 0$. Suppose further that f is measurable and there exist constants $K > 0$, $a > 0$ such that

$\|f(x)\| \leq K\exp\{ax\}$ for all x. Then there exists a $T > 0$ such that $u(x,t,\lambda) = E(\exp\{-\lambda X_t\})$ is the unique strong solution of the first order PDE

$$\frac{\partial u}{\partial t} + \lambda^2\sigma\frac{\partial u}{\partial \lambda} + \lambda E\big(f(X_t)\exp\{-\lambda X_t\}\big) = 0, \tag{5.2.5}$$

subject to $u(x,0,\lambda) = \exp\{-\lambda x\}$, for $0 \leq t < T$, $\lambda > a$.

Finally, we show that Proposition 5.2.4 can be used to confirm that the fundamental solution $u_0(x) = 1$ produces the correct fundamental solution, see Example 2.3 in Craddock (2009).

Example 5.2.5 For the squared Bessel process $X = \{X_t, t \geq 0\}$ of dimension δ given by the SDE

$$dX_t = \delta\,dt + 2\sqrt{X_t}\,dW_t,$$

where $X_0 = x > 0$, Eq. (5.2.5) yields that $E(\exp\{-\lambda X_t\})$ is the unique solution of the PDE

$$u_t + 2\lambda^2 u_\lambda + \lambda\delta u = 0,$$

where $u(x,0,\lambda) = \exp\{-\lambda x\}$. It can be confirmed that

$$u(x,t,\lambda) = E\big(\exp\{-\lambda X_t\}\big) = \frac{1}{(1+2\lambda t)^{\frac{\delta}{2}}}\exp\left\{-\frac{\lambda x}{1+2\lambda t}\right\},$$

satisfies the PDE and boundary conditions, and coincides with the result produced by the fundamental solution corresponding to $u_0(x) = 1$ in Example 5.2.3.

The above examples indicate that one has to be careful when deciding which fundamental solution yields the desired transition probability density.

5.3 One-Dimensional Examples

In this section, we aim to illustrate how to derive one-dimensional transition densities using the results from Chap. 4. We emphasize that the process of deriving transition densities is mechanical and easily applied to the study of novel stochastic processes. In this regard, we recall examples of transition densities studied in Craddock and Platen (2004) and provide the reader with additional references. It is intended that this section encourages readers to study stochastic processes that are tractable and potentially more suitable to their applications than those processes that have been employed in the past mainly because they were considered to be tractable from a conventional perspective.

We illustrate the derivation of the transition density of the square-root process, where we follow the presentation in Craddock (2009). In particular, we assume that

$$dX_t = (a - bX_t)\,dt + \sqrt{2\sigma X_t}\,dW_t, \tag{5.3.6}$$

where $X_0 = x > 0$ and a, b, and σ are assumed to be positive and $\frac{a}{\sigma} \geq 1$.

Proposition 5.3.1 *The transition density of the process X as specified in the SDE* (5.3.6) *started in x at time* 0 *being in y at time t is given by the explicit formula*

$$p(x,t,y) = \frac{b\exp\{bt(\frac{a}{\sigma}+1)\}}{\sigma(\exp\{bt\}-1)}\left(\frac{y}{x}\right)^{\frac{\nu}{2}}\exp\left\{\frac{-b(x+\exp\{bt\}y)}{\sigma(\exp\{bt\}-1)}\right\} \times I_\nu\left(\frac{b\sqrt{xy}}{\sigma\sinh(\frac{bt}{2})}\right), \quad (5.3.7)$$

where $\nu = \frac{a}{\sigma} - 1 \geq 0$.

Proof We note that the transition density of X satisfies the Kolmogorov backward equation

$$u_t = \sigma x u_{xx} + (a - bx)$$

and that Eq. (4.4.67) is satisfied with

$$h(x) = (a - bx), \qquad g = 0, \qquad \gamma = 1.$$

Hence we employ Theorem 4.4.5 with $\gamma = 1$, $u_0 = 1$, $A = b^2$, $B = -ab$, $C = \frac{1}{2}a^2 - a\sigma$, and $F(x) = a\ln(x) - bx$ to obtain

$$\begin{aligned}
\overline{U}_\epsilon(x,t) &= \exp\left\{\frac{-b^2\epsilon x}{\sigma}\left(\frac{\cosh(bt)+b\epsilon\sinh(bt)}{1+2b\epsilon\sinh(bt)+2b^2\epsilon^2(\cosh(bt)-1)}\right)\right\} \\
&\quad\times\exp\left\{\frac{tab}{2\sigma}\right\}\left|\frac{\cosh(\frac{bt}{2})+(1+2b\epsilon)\sinh(\frac{bt}{2})}{\cosh(\frac{bt}{2})-(1-2b\epsilon)\sinh(\frac{bt}{2})}\right|^{\frac{-ab}{2\sigma b}} \\
&\quad\times\exp\left\{\frac{1}{2\sigma}F\left(\frac{x}{(1+2b^2\epsilon^2(\cosh(bt)-1)+2b\epsilon\sinh(bt))}\right)-\frac{F(x)}{2\sigma}\right\} \\
&= \exp\left\{b(at-2b\epsilon x)\cosh\left(\frac{bt}{2}\right)+at\sinh\left(\frac{bt}{2}\right)+2ab\epsilon t\sinh\left(\frac{bt}{2}\right)\right. \\
&\quad\left.+2b\epsilon x\sinh\left(\frac{bt}{2}\right)\right\}\left(\cosh\left(\frac{bt}{2}\right)+(1+2b\epsilon)\sinh\left(\frac{bt}{2}\right)\right)^{-\frac{a}{\sigma}},
\end{aligned}$$

where the last equality can be shown using MATHEMATICA. We have

$$\begin{aligned}
&\left(\cosh\left(\frac{bt}{2}\right)+(1+2b\epsilon)\sinh\left(\frac{bt}{2}\right)\right)^{-\frac{a}{\sigma}} \\
&\quad= \exp\left\{\frac{bta}{2\sigma}\right\}\left(\exp\{bt\}+b\epsilon\left(\exp\{bt\}-1\right)\right)^{-\frac{a}{\sigma}}.
\end{aligned}$$

Also, it follows that

$$\begin{aligned}
&\exp\left\{b(at-2b\epsilon x)\cosh\left(\frac{bt}{2}\right)+at\sinh\left(\frac{bt}{2}\right)+2ab\epsilon t\sinh\left(\frac{bt}{2}\right)\right. \\
&\qquad\left.+2b\epsilon x\sinh\left(\frac{bt}{2}\right)\right\} \\
&\quad= \exp\left\{\frac{abt}{2\sigma}\right\}\exp\left\{-\frac{b^2\epsilon x}{\sigma(\exp\{bt\}+b\epsilon(\exp\{bt\}-1))}\right\}.
\end{aligned}$$

Now, one obtains

$$\begin{aligned}\overline{U}_\epsilon(x,t) &= \exp\left\{\frac{abt}{2\sigma}\right\}\big(\exp\{bt\} + b\epsilon\big(\exp\{bt\} - 1\big)\big)^{-\frac{a}{\sigma}}\\ &\quad\times \exp\left\{\frac{abt}{2\sigma}\right\}\exp\left\{-\frac{b^2\epsilon x}{\sigma(\exp\{bt\} + b\epsilon(\exp\{bt\} - 1))}\right\}\\ &= \exp\left\{\frac{abt}{\sigma}\right\}\big(\exp\{bt\} + b\epsilon\big(\exp\{bt\} - 1\big)\big)^{-\frac{a}{\sigma}}\\ &\quad\times \exp\left\{-\frac{b^2\epsilon x}{\sigma(\exp\{bt\} + b\epsilon(\exp\{bt\} - 1))}\right\}.\end{aligned}$$

Substituting $\epsilon = \frac{\lambda\sigma}{b^2}$, we get

$$\begin{aligned}\overline{U}_\epsilon(x,t) &= U_\lambda(x,t)\\ &= \int_0^\infty \exp\{-\lambda y\} p(t,x,y)\,dy\\ &= \exp\left\{\frac{abt}{\sigma}\right\}\big(b\exp\{bt\} + \lambda\sigma\big(\exp\{bt\} - 1\big)\big)^{-\frac{a}{\sigma}} b^{\frac{a}{\sigma}}\\ &\quad\times \exp\left\{-\frac{b\lambda x}{(b\exp\{bt\} + \lambda\sigma(\exp\{bt\} - 1))}\right\}.\end{aligned}$$

This Laplace transform can be easily inverted to yield (5.3.7). It can be confirmed via Proposition 5.2.4 that the density in (5.3.7) is the correct transition probability density. □

In Fig. 3.1.2, a plot of the transition density of a square-root process was shown. We now recall some results from Craddock and Platen (2004). In particular, we study generalizations of the squared Bessel process. We focus on the process $X = \{X_t,\ t \geq 0\}$, given by the SDE

$$dX_t = a(X_t)\,dt + \sqrt{2X_t}\,dW_t, \tag{5.3.8}$$

for $t \geq 0$ with $X_0 > 0$. Then, following Craddock and Platen (2004), Platen and Heath (2010), and Platen and Bruti-Liberati (2010), by applying the results of Chap. 4, we distinguish ten cases:

(i) for the constant drift function

$$a(x) = \alpha > 0,$$

we recover the squared Bessel process of dimension $\delta = 2\alpha$ with transition density

$$p(0,x;t,y) = \frac{1}{t}\left(\frac{x}{y}\right)^{\frac{1-\alpha}{2}} I_{\alpha-1}\left(\frac{2\sqrt{xy}}{t}\right)\exp\left\{-\frac{(x+y)}{t}\right\}.$$

Here $I_{\alpha-1}$ is again the modified Bessel function of the first kind with index $\alpha - 1$, see also Eq. (3.1.4) and Fig. 3.1.1

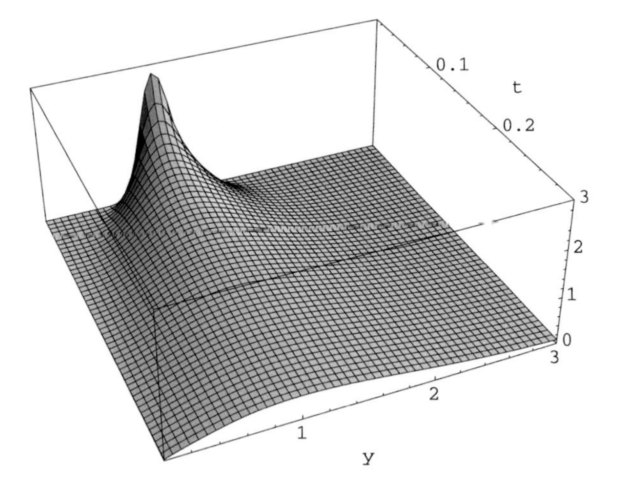

Fig. 5.3.1 Transition density for a squared Bessel process, case (i)

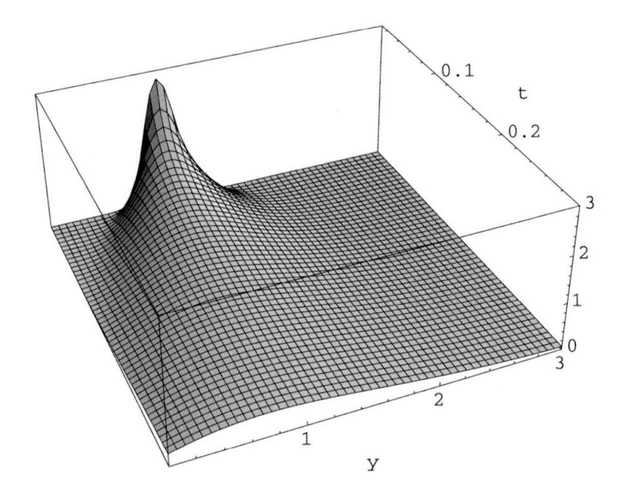

Fig. 5.3.2 Transition density for case (ii)

(ii) setting the drift function to

$$a(x) = \frac{\mu\, x}{1 + \frac{\mu}{2}\, x}$$

for $\mu > 0$, we obtain the transition density

$$p(0, x; t, y) = \frac{\exp\{-\frac{(x+y)}{t}\}}{(1 + \frac{\mu}{2}\, x)t} \left[\left(\sqrt{\frac{x}{y}} + \frac{\mu\, \sqrt{x\, y}}{2} \right) I_1\left(\frac{2\, \sqrt{x\, y}}{t} \right) + t\, \delta(y) \right]$$

with $\delta(\cdot)$ denoting the Dirac delta function. For $y = 0$ one can interpret $\frac{\exp\{-\frac{x}{t}\}}{(1+\frac{\mu}{2}\, x)}$ as the probability of absorption at zero. In Fig. 5.3.2 we show the above transition density for $x = 1$ and $\mu = 1$

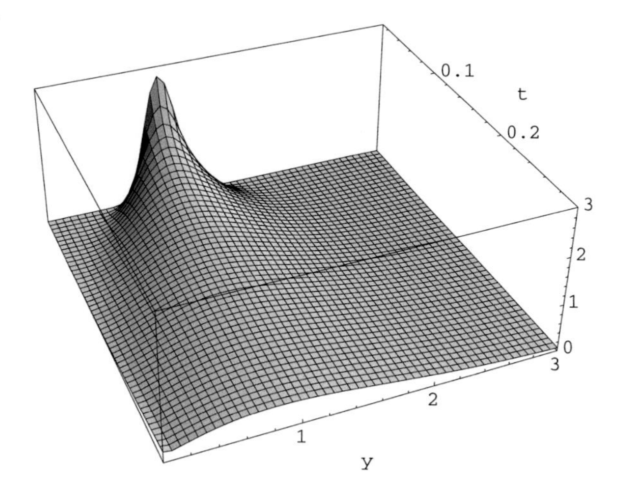

Fig. 5.3.3 Transition density for case (iii)

(iii) the drift function

$$a(x) = \frac{1 + 3\sqrt{x}}{2(1 + \sqrt{x})},$$

results in the transition density

$$p(0, x; t, y) = \frac{\cosh(\frac{2\sqrt{xy}}{t})}{\sqrt{\pi y t}(1 + \sqrt{x})}\left(1 + \sqrt{y}\tanh\left(\frac{2\sqrt{xy}}{t}\right)\right) \times \exp\left\{-\frac{(x + y)}{t}\right\}.$$

In Fig. 5.3.3 we display the corresponding transition density for $x = 1$

(iv) studying the drift function

$$a(x) = 1 + \mu \tanh\left(\mu + \frac{1}{2}\mu \ln(x)\right)$$

for $\mu = \frac{1}{2}\sqrt{\frac{5}{2}}$, we obtain the transition density

$$p(0, x; t, y) = \left(\frac{x}{y}\right)^{\frac{\mu}{2}}\left[I_{-\mu}\left(\frac{2\sqrt{xy}}{t}\right) + e^{2\mu} y^{\mu} I_{\mu}\left(\frac{2\sqrt{xy}}{t}\right)\right] \times \frac{\exp\{-\frac{x+y}{t}\}}{(1 + \exp\{2\mu\}x^{\mu})\,t}. \tag{5.3.9}$$

The shape of the density (5.3.9) for $x = 1$ looks quite similar to that in Fig. 5.3.3

(v) given the drift function

$$a(x) = \frac{1}{2} + \sqrt{x},$$

we obtain the transition density

$$p(0,x;t,y) = \cosh\left(\frac{(t+2\sqrt{x})\sqrt{y}}{t}\right)\frac{\exp\{-\sqrt{x}\}}{\sqrt{\pi\, y\, t}} \times \exp\left\{-\frac{(x+y)}{t} - \frac{t}{4}\right\}. \tag{5.3.10}$$

Also the transition density (5.3.10) for $x = 1$ shows a lot of similarity with that in Fig. 5.3.3

(vi) the drift function

$$a(x) = \frac{1}{2} + \sqrt{x}\tanh(\sqrt{x}),$$

results in the transition density

$$p(0,x;t,y) = \frac{\cosh(\frac{2\sqrt{xy}}{t})}{\sqrt{\pi y t}}\frac{\cosh(\sqrt{y})}{\cosh(\sqrt{x})}\exp\left\{-\frac{(x+y)}{t} - \frac{t}{4}\right\}. \tag{5.3.11}$$

The above transition density (5.3.11) for $x = 1$ has also a similar shape as that in Fig. 5.3.3

(vii) when the drift function satisfies

$$a(x) = \frac{1}{2} + \sqrt{x}\coth(\sqrt{x})$$

the process has the transition density

$$p(0,x;t,y) = \frac{\sinh(\frac{2\sqrt{x\,y}}{t})}{\sqrt{\pi\, y\, t}}\frac{\sinh(\sqrt{y})}{\sinh(\sqrt{x})}\exp\left\{-\frac{(x+y)}{t} - \frac{t}{4}\right\}.$$

This transition density has for $x = 1$ some similarity with that shown in Fig. 5.3.1

(viii) using the drift function

$$a(x) = 1 + \cot\big(\ln(\sqrt{x})\big)$$

for $x \in (\exp\{-2\pi\}, 1)$, then we obtain the real valued transition density

$$p(0,x;t,y) = \frac{\exp\{-\frac{(x+y)}{t}\}}{2\imath t\sin(\ln(\sqrt{x}))}\left(y^{\frac{\imath}{2}} I_\imath\left(\frac{2\sqrt{x\,y}}{t}\right) - y^{-\frac{\imath}{2}} I_{-\imath}\left(\frac{2\sqrt{x\,y}}{t}\right)\right), \tag{5.3.12}$$

where $\imath$ denotes the imaginary unit. We plot in Fig. 5.3.4 the transition density (5.3.12) for $x = \frac{1}{2}$. Note that the process X lives on the bounded interval $(\exp\{-2\pi\}, 1)$

(ix) choosing the drift function

$$a(x) = x\coth\left(\frac{x}{2}\right),$$

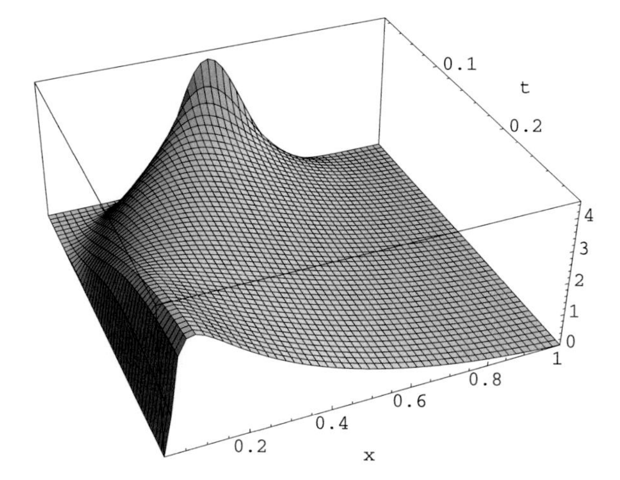

Fig. 5.3.4 Transition density for case (viii)

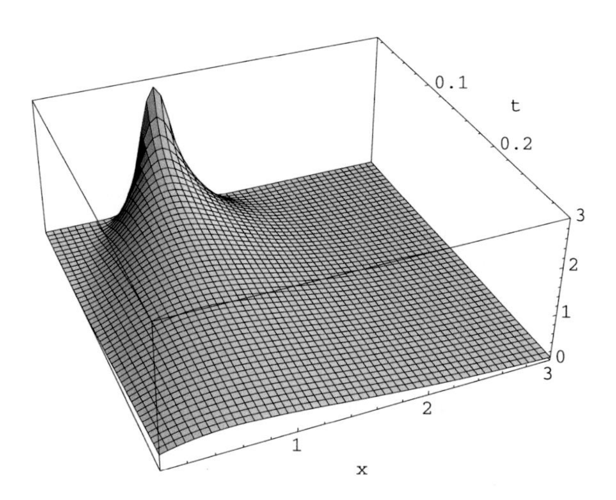

Fig. 5.3.5 Transition density for case (x)

then we obtain the transition density

$$p(0,x;t,y) = \frac{\sinh(\frac{y}{2})}{\sinh(\frac{x}{2})} \exp\left\{-\frac{(x+y)}{2\tanh(\frac{t}{2})}\right\} \times \left[\frac{\exp\{\frac{t}{2}\}}{\exp\{t\}-1}\sqrt{\frac{x}{y}} I_1\left(\frac{\sqrt{x\,y}}{\sinh(\frac{t}{2})}\right) + \delta(y)\right],$$

where $\delta(\cdot)$ is again the Dirac delta function. Figure 5.3.1 displayed a transition density of similar shape

(x) lastly, setting the drift function to

$$a(x) = x\tanh\left(\frac{x}{2}\right)$$

we obtain the transition density

$$p(0,x;t,y) = \frac{\cosh(\frac{y}{2})}{\cosh(\frac{x}{2})} \exp\left\{-\frac{(x+y)}{2\tanh(\frac{t}{2})}\right\}$$
$$\times \left[\frac{\exp\{\frac{t}{2}\}}{\exp\{t\}-1}\sqrt{\frac{x}{y}} I_1\left(\frac{\sqrt{x\,y}}{\sinh(\frac{t}{2})}\right) + \delta(y)\right].$$

We plot in Fig. 5.3.5 the transition density for $x = 1$.

Many of the above diffusion processes are very recent in the literature and essentially discovered in Craddock and Platen (2004). They offer new dynamics ready to be employed in modeling, for instance, in finance.

5.4 Laplace Transforms of Joint Transition Densities

In this section, we present Laplace transforms of the type

$$E\left(\exp\left\{-\lambda X_t - \mu \int_0^t X_s\,ds - \gamma \int_0^t \frac{ds}{X_s}\right\}\right), \tag{5.4.13}$$

for suitable stochastic processes $X = \{X_t,\ t \geq 0\}$. These Laplace transforms have important applications. For example, if X is the independent short rate process and $\lambda = \gamma = 0$ and $\mu = 1$, then Eq. (5.4.13) contributes to the price of a zero coupon bond, see also Sect. 5.5. However, there are many applications beyond interest rate modeling. For instance, in Chap. 6 we will design exact Monte Carlo schemes for stochastic volatility models based on results from this section. In Sect. 8.5.2, we will focus on exact and quasi-Monte Carlo methods for realized variance derivatives, to illustrate further possible applications of the results presented in this section. At the heart of such applications sits the observation that for some tasks, the fundamental solution is sometimes more interesting than its Laplace transform, see Sect. 8.5.2 and Chap. 6. Hence even though we might not always be able to integrate the fundamental solution to calculate the Laplace transform, we may be nevertheless able to calculate and subsequently use the fundamental solution.

We illustrate this type of technique in the following result, see Craddock and Lennox (2009).

Proposition 5.4.1 *Let $X = \{X_t,\ t \geq 0\}$ be a squared Bessel process,*

$$dX_t = \delta\,dt + 2\sqrt{X_t}\,dW_t,$$

where $\delta \geq 2$ and $X_0 = x > 0$. Then the function $u(x,t)$ given by

$$u(x,t) = E\left(\exp\left\{-\lambda X_t - \mu \int_0^t \frac{ds}{X_s}\right\}\right)$$
$$= \exp\{-x/2t\}\left(\frac{x}{2t}\right)^d \frac{\Gamma(\alpha)\,{}_1F_1(\alpha,\beta,x/(2t+4t^2\lambda))}{\Gamma(\beta)(1+2\lambda t)^\alpha},$$

where ${}_1F_1(a,b,z)$ *is Kummer's confluent hypergeometric function, satisfies the PDE*

$$u_t = 2xu_{xx} + \delta u_x - \frac{\mu}{x}u,$$

whose fundamental solution is given by

$$p(t,x,y) = \frac{1}{2t}\left(\frac{x}{y}\right)^{(1-\delta/2)/2} I_{2d+\delta/2-1}\left(\frac{\sqrt{xy}}{t}\right)\exp\left\{-\frac{(x+y)}{2t}\right\}, \tag{5.4.14}$$

where $d = \frac{1}{4}(2-\delta+\sqrt{(\delta-2)^2+8\mu})$, $\alpha = d+\frac{\delta}{2}$, *and* $\beta = 2d+\frac{\delta}{2}$.

Proof The drift function $f(x)=\delta$ satisfies the first Ricatti equation (4.4.34), where $\sigma=2$, $\gamma=1$, $g(x)=\frac{\mu}{x}$, and $A=0$. Choosing the stationary solution $u_0(x)=x^d$, where $d=\frac{1}{4}(2-\delta+\sqrt{(\delta-2)^2+8\mu})$, we obtain from Theorem 4.4.3

$$\overline{U}_\epsilon(x,t) = \exp\left\{-\frac{4\epsilon x}{2(1+4\epsilon t)}\right\}\frac{x^d}{(1+4\epsilon t)^{2d+\frac{\delta}{2}}}.$$

Next, we set $\epsilon = \frac{\sigma\lambda}{4} = \frac{\lambda}{2}$ to obtain

$$\begin{aligned}\overline{U}_\epsilon(x,t) = U_\lambda(x,t) &= \int_0^\infty y^d p(t,x,y)\exp\{-\lambda y\}\,dy\\ &= \frac{x^d}{(1+2\lambda t)^{\frac{\delta}{2}+2d}}\exp\left\{-\frac{\lambda x}{(1+2\lambda t)}\right\}.\end{aligned}$$

Inverting this Laplace transform, we obtain the fundamental solution

$$p(t,x,y) = \frac{1}{2t}\exp\left\{-\frac{x+y}{2t}\right\}\left(\frac{x}{y}\right)^{\frac{1-\delta/2}{2}} I_{\frac{\delta}{2}+2d-1}\left(\frac{\sqrt{xy}}{t}\right).$$

We obtain

$$\int_0^\infty e^{-\lambda y}p(t,x,y)\,dy = \frac{\Gamma(\alpha)}{\Gamma(\beta)}\left(\frac{x}{2t}\right)^d e^{-\frac{x}{2t}}\,{}_1F_1\left(\alpha,\beta,\frac{x}{2t+4t^2\lambda}\right)(1+2\lambda t)^{-\alpha},$$

by integrating the modified Bessel function of the first kind term-by-term. □

We now recall results from Craddock and Lennox (2009), where Eq. (4.4.35) was handled via group invariant solutions. In particular, this approach produced Whittaker transforms of fundamental solutions. Although such integral transforms have known inversion integrals, explicit inversion is usually not possible, as few of these transforms have been computed and tabulated. However, in Craddock (2009), Eqs. (4.4.35) and (4.4.36) were handled via symmetry methods, namely by using the full group of symmetries, see also the proof of Theorem 4.4.5. This approach produces generalized Laplace transforms of the fundamental solutions.

As fundamental solutions will play an important role in Chap. 6, we present both, Laplace transforms and fundamental solutions themselves.

Theorem 5.4.2 *Let $X = \{X_t,\, t \geq 0\}$ be a squared Bessel process where*

$$dX_t = \delta\, dt + 2\sqrt{X_t}\, dW_t,$$

for $\delta \geq 2$ and $X_0 = x > 0$. Then the function $u(x,t)$ given by

$$\begin{aligned} u(x,t) &= E\left(\exp\left\{-\lambda X_t - \frac{b^2}{2}\int_0^t X_s\, ds\right\}\right) \\ &= \frac{\exp\{-(xb/2)(1+2\lambda b^{-1}\coth(bt))/(\coth(bt)+2\lambda b^{-1})\}}{(\cosh(bt)+2\lambda b^{-1}\sinh(bt))^{\delta/2}}. \end{aligned}$$

satisfies the PDE

$$u_t = 2xu_{xx} + \delta u_x - \frac{b^2}{2}xu,$$

whose fundamental solution is given by

$$p(t,x,y) = \frac{b}{2\sinh(bt)}\left(\frac{y}{x}\right)^{\delta/4-1/2}\exp\left\{-\frac{b(x+y)}{2\tanh(bt)}\right\} I_{(\delta-2)/2}\left(\frac{b\sqrt{xy}}{\sinh(bt)}\right).$$

We have the following result pertaining to square-root processes satisfying the SDE,

$$dX_t = (a - bX_t)\, dt + \sqrt{2\sigma X_t}\, dW_t, \tag{5.4.15}$$

where $X_0 = x > 0$.

Proposition 5.4.3 *Let $X = \{X_t,\, t \geq 0\}$ be a square-root process of dimension $\delta = \frac{4a}{2\sigma} \geq 2$, whose dynamics satisfy the SDE (5.4.15). Then the function $u(x,t)$ is given by*

$$\begin{aligned} u(x,t) &= E\left(\exp\left\{-\lambda X_t - \mu\int_0^t \frac{ds}{X_s}\right\}\right) \\ &= \frac{\Gamma(k+\nu/2+1/2)}{\Gamma(\nu+1)}\beta x^{-k}\exp\left\{\frac{b}{2\sigma}\left(at + x - \frac{x}{\tanh(bt/2)}\right)\right\} \\ &\quad \times \frac{e^{\beta^2/(2\alpha)}}{\beta\alpha^k} M_{-k,\nu/2}\left(\frac{\beta^2}{\alpha}\right), \end{aligned}$$

where $\nu = \frac{1}{\sigma}\sqrt{(a-\sigma)^2 + 4\mu\sigma}$, $k = \frac{a}{2\sigma}$, $\alpha = \frac{b}{2\sigma}(1+\coth(\frac{bt}{2})) + \lambda$, $\beta = \frac{b\sqrt{x}}{2\sigma\sinh(\frac{bt}{2})}$, and $M_{s,r}(z)$ denotes the Whittaker function of the first kind. Furthermore, $u(x,t)$ satisfies the PDE

$$u_t = \sigma x u_{xx} + (a - bx)u_x - \frac{\mu}{x}u,$$

whose fundamental solution is given by

$$\begin{aligned} p(t,x,y) &= \frac{b}{2\sigma\sinh(bt/2)}\left(\frac{y}{x}\right)^{a/(2\sigma)-1/2}\exp\left\{\frac{b}{2\sigma}\left(at + (x-y) - \frac{x+y}{\tanh(bt/2)}\right)\right\} \\ &\quad \times I_\nu\left(\frac{b\sqrt{xy}}{\sigma\sinh(bt/2)}\right). \end{aligned} \tag{5.4.16}$$

Finally, we present the Laplace transform of the joint density for

$$\left(X_t, \int_0^t X_s\,ds, \int_0^t \frac{ds}{X_s}\right).$$

In particular, we consider the function

$$u(x,t) = E\left(\exp\left\{-\lambda X_t - (b^2/2)\int_0^t X_s\,ds - \nu\int_0^t \frac{ds}{X_s}\right\}\right),$$

where $X_0 > 0$. We have the following result.

Proposition 5.4.4 *Let $X = \{X_t, t \geq 0\}$ be a squared Bessel process of dimension $\delta \geq 2$. Then*

$$\begin{aligned} u(x,t) &= E\left(\exp\left\{-\lambda X_t - (b^2/2)\int_0^t X_s\,ds - \nu\int_0^t \frac{ds}{X_s}\right\}\right) \\ &= \exp\{-bx/(2\tanh(bt))\}\frac{\Gamma(\alpha)}{\Gamma(\beta)}\frac{b^{a/2}(x\exp\{bt\})^\gamma(\exp\{2bt\}-1)^{-\gamma}}{(\cosh(bt)+(2\lambda/b)\sinh(bt))^\delta} \\ &\quad\times {}_1F_1\left(\alpha, \beta, \frac{b^2x\,\mathrm{csch}(bt)}{2b\cosh(bt)+4\lambda\sinh(bt)}\right), \end{aligned}$$

where $a = \sqrt{(\delta-2)^2+8\nu}$, $\delta = \frac{1}{4}(2+a+\delta)$, $\gamma = \frac{1}{4}(2+a-\delta)$, $\alpha = \frac{1}{4}(a+\delta+2)$, $\beta = \frac{a+2}{2}$, and ${}_1F_1(a,b,z)$ is Kummer's confluent hypergeometric function and csch denotes the hyperbolic cosecant, $\mathrm{csch}(x) = \frac{2\exp\{x\}}{\exp\{2x\}-1}$. Furthermore, $u(x,t)$ satisfies the PDE

$$u_t = 2xu_{xx} + \delta u_x - \frac{b^2}{2}xu - \nu\frac{u}{x},$$

whose fundamental solution is given by

$$\begin{aligned} p(t,x,y) &= \frac{b}{2\sinh(bt)}\exp\big(-b(x+y)/(2\tanh(bt))\big)\left(\frac{y}{x}\right)^{(\delta-2)/4} \\ &\quad\times I_{\sqrt{(\delta-2)^2+8\nu}/2}\left(\frac{b\sqrt{xy}}{\sinh(bt)}\right). \end{aligned}$$

This result provides important access to functionals of squared Bessel processes that have explicit formulas. We point out that Proposition 5.4.4 will be applied in Sects. 5.5, 6.3, and 6.4.

5.5 Bond Pricing in Quadratic Models

So far in this chapter, we have illustrated how Lie symmetry methods can be used to obtain transition densities and Laplace transforms of joint transition densities. This section illustrates that by combining results obtained via Lie symmetry methods with probability theory, the scope of results that can be obtained is increased. We illustrate this using two examples. Firstly, we use Proposition 5.4.4, which provides the Laplace transform of joint transition densities of the squared Bessel process with the change of law result from Pitman and Yor (1982), see Proposition 3.1.6, which connects squared Bessel and square-root processes, to price zero coupon bonds in the Cox, Ingersoll, Ross (CIR) model introduced in Cox et al. (1985). Secondly, we recall from Sect. 3.1, that a $3/2$ process is simply the inverse of a squared Bessel process, and use this observation and Proposition 5.4.3, which deals with square-root processes, to price zero coupon bonds under a $3/2$ process for the short-rate.

We begin with the pricing of a zero coupon bond in the CIR model. Recall that in the CIR model, the short rate is modeled using a square-root process,

$$dr_t = k(\theta - r_t)\,dt + \sigma\sqrt{r_t}\,dW_t, \tag{5.5.17}$$

where $r_0 \geq 0$ and $\frac{4k\theta}{\sigma^2} \geq 2$. Consequently, we are interested in computing

$$E\left(\exp\left\{-\int_t^T r_s\,ds\right\}\,\Big|\,\mathcal{A}_t\right), \tag{5.5.18}$$

where we use $\mathcal{A}_t$ to denote $\mathcal{A}_t = \sigma\{r_s,\ s \leq t\}$. We find it convenient to reduce the pricing problem to the study of Laplace transforms of squared Bessel processes. As discussed in Sect. 3.1, we recall that there are at least two methods for reducing the study of square-root processes to the study of squared Bessel processes. These are transformation of space-time and the change of law, see Propositions 3.1.5 and 3.1.6. As discussed in Sect. 3.1, using the standard change of time technique, we transform (5.5.17) into a square-root process with volatility coefficient 2: we introduce the process $\rho = \{\rho_t, t \geq 0\}$ via $\rho_t = r_{\frac{4t}{\sigma^2}}$, and obtain the following SDE for ρ_t:

$$d\rho_t = (2j\rho_t + \delta)\,dt + 2\sqrt{\rho_t}\,d\tilde{W}_t, \tag{5.5.19}$$

where $\tilde{W} = \{\tilde{W}_t,\ t \geq 0\}$ is a standard Brownian motion, $j = -\frac{2k}{\sigma^2}$, and $\delta = \frac{4k\theta}{\sigma^2}$. We use ${}^{j}P^{n}_{\rho_0}$ to denote the law of ρ, and set $\mathcal{F}_t = \sigma\{\rho_s, s \leq t\}$. Due to the functional dependence of r and ρ, we have $\mathcal{A}_{\frac{4t}{\sigma^2}} = \mathcal{F}_t, t \geq 0$. By Proposition 3.1.6, the following absolute continuity relationship between square-root and squared Bessel processes holds:

$${}^{j}P^{\delta}_{\rho_0}\,|_{\mathcal{F}_t} = \exp\left\{\frac{j}{2}(\rho_t - \rho_0 - \delta t) - \frac{j^2}{2}\int_0^t \rho_s\,ds\right\} P^{\delta}_{\rho_0}\Big|_{\mathcal{F}_t}. \tag{5.5.20}$$

We now use Eq. (5.5.20) to change the pricing problem (5.5.18) into one that can be solved using the Laplace transforms of densities of squared Bessel processes from Sect. 5.4. We point out that this technique will also be used in Chap. 6. The next theorem shows how to derive the well-known bond pricing formula in the CIR model by combining the results from Sect. 5.4, in particular Proposition 5.4.4, with the change of law formula from Pitman and Yor (1982).

Theorem 5.5.1 *Assume that the dynamics of* r_t *are given by* (5.5.17) *and* $\frac{4k\theta}{\sigma^2} \geq 2$. *Then we have the following formula for a zero coupon price at time* t *with maturity date* $T > t$:

$$E\left(\exp\left\{-\int_t^T r_s\, ds\right\} \Bigm| \mathcal{A}_t\right) = A(t,T)\exp\{-B(t,T)r_t\},$$

where

$$A(t,T) = \left(\frac{2h\exp((k+h)(T-t)/2)}{2h+(k+h)(\exp(h(T-t))-1)}\right)^{\frac{2k\theta}{\sigma^2}}$$

$$B(t,T) = \frac{2(\exp((T-t)h)-1)}{2h+(k+h)(\exp((T-t)h)-1)}$$

$$h = \sqrt{k^2+2\sigma^2}.$$

Proof Setting $\tilde{t} := \frac{t\sigma^2}{4}$ and $\tilde{T} := \frac{T\sigma^2}{4}$, we employ Eq. (5.5.20) to obtain

$$\begin{aligned}
&E\left(\exp\left\{-\int_t^T r_s\, ds\right\} \Bigm| \mathcal{A}_t\right)\\
&= E\left(\exp\left\{-\int_t^T \rho_{\frac{s\sigma^2}{4}}\, ds\right\} \Bigm| \mathcal{F}_{\frac{t\sigma^2}{4}}\right)\\
&= E\left(\exp\left\{-\frac{4}{\sigma^2}\int_{\tilde{t}}^{\tilde{T}} \rho_{\tilde{s}}\, d\tilde{s}\right\} \Bigm| \mathcal{F}_{\tilde{t}}\right)\\
&= \tilde{E}\left(\exp\left\{\frac{j}{2}\rho_{\tilde{T}} - \frac{j}{2}\rho_{\tilde{t}} - \frac{j\delta(\tilde{T}-\tilde{t})}{2} - \left(\frac{j^2}{2}+\frac{4}{\sigma^2}\right)\int_{\tilde{t}}^{\tilde{T}} \rho_{\tilde{s}}\, d\tilde{s}\right\} \Bigm| \mathcal{F}_{\tilde{t}}\right),
\end{aligned}$$

where we use E to denote the expectation with respect to ${}^{j}P^{\delta}_{\rho_0}$ and $\tilde{E}$ to denote the expectation with respect to $P^{\delta}_{\rho_0}$. Also, we recall that $\delta = \frac{4k\theta}{\sigma^2}$ and $j = -\frac{2k}{\sigma^2}$. Now we define

$$\frac{b^2}{2} = \frac{j^2}{2} + \frac{4}{\sigma^2}$$

to obtain

$$b = \frac{2}{\sigma^2}\sqrt{k^2+2\sigma^2} = \frac{2}{\sigma^2}h$$

and we also set $\lambda = -\frac{j}{2}$. It now follows from Theorem 5.4.2 that

$$\begin{aligned}
&\tilde{E}\left(\exp\left\{\frac{j}{2}\rho_{\tilde{T}} - \frac{j}{2}\rho_{\tilde{t}} - \frac{j\delta(\tilde{T}-\tilde{t})}{2} - \left(\frac{j^2}{2}+\frac{4}{\sigma^2}\right)\int_{\tilde{t}}^{\tilde{T}} \rho_{\tilde{s}}\, d\tilde{s}\right\} \Bigm| \mathcal{F}_{\tilde{t}}\right)\\
&= \exp\left\{\frac{k^2\theta(T-t)}{\sigma^2} + r_t\frac{k}{\sigma^2}\right\}
\end{aligned}$$

$$\times \frac{\exp\left\{\dfrac{-r_t \dfrac{h(1+2\lambda\frac{\coth(b(\tilde{T}-\tilde{t}))\sigma^2}{2\sqrt{k^2+2\sigma^2}})}{\sigma^2(\coth(b(\tilde{T}-\tilde{t}))+2\frac{\lambda\sigma^2}{2\sqrt{k^2+2\sigma^2}})}}{\coth(b(\tilde{T}-\tilde{t}))+2\lambda\frac{\sigma^2}{2\sqrt{k^2+2\sigma^2}}}\right\}}{(\cosh(b(\tilde{T}-\tilde{t}))+2\lambda b^{-1}\sinh(b(\tilde{T}-\tilde{t})))^{\delta/2}}.$$

It can be checked that

$$\frac{\exp\{\frac{k^2\theta(T-t)}{\sigma^2}\}}{(\cosh(b(\tilde{T}-\tilde{t}))+2\lambda b^{-1}\sinh(b(\tilde{T}-\tilde{t})))^{\delta/2}} = A(t,T).$$

Finally,

$$\exp\left\{-r_t\left(\frac{h}{\sigma^2}\left(\frac{1+\frac{k}{h}\coth(b(\tilde{T}-\tilde{t}))}{\coth(b(\tilde{T}-\tilde{t}))+\frac{k}{h}}\right)-\frac{k}{\sigma^2}\right)\right\} = \exp\{-r_t B(t,T)\},$$

is completing the proof. □

Next, we discuss zero coupon bond pricing in the 3/2 model. However, we point out that these techniques are also useful when studying volatility derivatives, see e.g. Carr and Sun (2007). We recall the 3/2 process from Sect. 3.1, which is given by

$$dr_t = \kappa r_t(\theta - r_t)\,dt + \sigma r_t^{3/2}\,dW_t,$$

where $r_0 > 0$. Consequently, we are interested in computing

$$E\left(\exp\left\{-\int_t^T r_s\,ds\right\}\,\middle|\,\mathcal{A}_t\right),$$

where $\mathcal{A}_t = \sigma\{r_s,\ s \le t\}$. Now, we define $v_t = \frac{1}{r_t}$, and obtain by Itô's formula

$$dv_t = \left(\kappa + \sigma^2 - \kappa\theta v_t\right)dt - \sigma\sqrt{v_t}\,dW_t.$$

Since $v_t = \frac{1}{r_t}$, we have

$$E\left(\exp\left\{-\int_t^T r_s\,ds\right\}\,\middle|\,\mathcal{A}_t\right) = E\left(\exp\left\{-\int_t^T \frac{ds}{v_s}\right\}\,\middle|\,\mathcal{A}_t\right).$$

We now simply use Proposition 5.4.3 to yield

$$\begin{aligned} &E\left(\exp\left\{-\int_t^T r_s\,ds\right\}\,\middle|\,\mathcal{A}_t\right) \\ &= \frac{\Gamma(k+\frac{\nu}{2}+\frac{1}{2})}{\Gamma(\nu+1)}\beta r_t^k \exp\left\{\frac{b}{\sigma^2}\left(a\tau + r_t^{-1} - \frac{r_t^{-1}}{\tanh(b\tau/2)}\right)\right\} \\ &\quad\times \frac{\exp\{\beta^2/(2\alpha)\}}{\beta\alpha^k} M_{-k,\nu/2}\left(\frac{\beta^2}{\alpha}\right), \end{aligned}$$

where $\nu = \frac{2}{\sigma^2}\sqrt{(\kappa + \sigma^2 - \frac{\sigma^2}{2})^2 + 2\sigma^2}$, $k = \frac{\kappa+\sigma^2}{\sigma^2}$, $\alpha = \frac{\kappa\theta}{\sigma^2}(1 + \coth(\frac{\kappa\theta\tau}{2}))$, $\beta = \frac{\kappa\theta v_t^{-\frac{1}{2}}}{\sigma^2 \sinh(\frac{\kappa\theta\tau}{2})}$, $a = \kappa + \sigma^2$, and $b = \kappa\theta$. This result can be shown to match Theorem 3 in Carr and Sun (2007).

Note that similar calculations yield corresponding results for other diffusion processes captured in Chap. 4.

Chapter 6
Exact and Almost Exact Simulation

The aim of this chapter is to discuss the simulation of tractable models, illustrated in the context of stochastic volatility models. For two popular stochastic volatility models, the Heston model, see Heston (1993), and the $3/2$ model, see Heston (1997) and Lewis (2000), we present exact simulation algorithms, where we use results from Sect. 5.4. These techniques are based on the *inverse transform method*, which we firstly recall. Moving to higher dimensions, it seems more difficult to generalize these techniques except for the trivial dependence structure, the independent case. Consequently, we recall almost exact simulation schemes from Platen and Bruti-Liberati (2010), which are applicable in the multidimensional case. Finally, we point out that in Chap. 11 we will discuss advanced multidimensional stochastic volatility models based on the Wishart process, which have been successfully applied to the modeling of stochastic volatility.

We introduce these simulation methods in an equity context, in particular, we concentrate on modeling stocks and stock indices. However, these methods are also applicable in other areas, for example in interest rate modeling: the *stochastic volatility Brace-Gątarek-Musiela model* introduces stochastic volatility processes in the context of the *LIBOR market model*. The techniques discussed in this chapter are also applicable in such a context, see e.g. Chap. 16 in Brace (2008), in particular, Sect. 16.4, which deals with simulation.

6.1 Sampling by Inverse Transform Methods

Conceptually, we simulate the given models, one- and multidimensional models, using the inverse transform method, which was discussed e.g. in Chap. 2 in Platen and Bruti-Liberati (2010). The forthcoming brief description of the inverse transform method follows this discussion closely.

The well-known inverse transform method can be applied for the generation of a continuous random variable Y with given probability distribution function F_Y. From

J. Baldeaux, E. Platen, *Functionals of Multidimensional Diffusions with Applications to Finance*, Bocconi & Springer Series 5, DOI 10.1007/978-3-319-00747-2_6,

a uniformly distributed random variable $0 < U < 1$, we obtain an F_Y distributed random variable $y(U)$ by realizing that

$$U = F_Y\big(y(U)\big), \tag{6.1.1}$$

so that

$$y(U) = F_Y^{-1}(U). \tag{6.1.2}$$

Here F_Y^{-1} denotes the inverse function of F_Y. More generally, one can still set

$$y(U) = \inf\big\{y\colon\ U \leq F_Y(y)\big\} \tag{6.1.3}$$

in the case when F_Y is no longer continuous, where $\inf\{y\colon\ U \leq F_Y(y)\}$ denotes the lower limit of the set $\{y\colon\ U \leq F_Y(y)\}$. If U is a $U(0,1)$ uniformly distributed random variable, then the random variable $y(U)$ in (6.1.2) will be F_Y-distributed. The above calculation in (6.1.2) may need to apply a root finding method, for instance, a Newton method, see Press et al. (2002). Obviously, given an explicit transition distribution function for the solution of a one-dimensional SDE we can sample a trajectory directly from this transition law at given time instants. One simply starts with the initial value, generates the first increment and sequentially the subsequent random increments of the simulated trajectory, using the inverse transform method for the respective transition distributions that emerge.

Also in the case of a two-dimensional SDE we can simulate by sampling from the bivariate transition distribution. We first identify the marginal transition distribution function F_{Y_1} of the first component. Then we use the inverse transform method, as above, for the exact simulation of an outcome of the first component of the two-dimensional random variable based on its marginal distribution function. Afterwards, we exploit the conditional transition distribution function $F_{Y_2|Y_1}$ of the second component Y_2, given the simulated first component Y_1, and use again the inverse transform method to simulate also the second component of the considered SDE. This simulation method is exact as long as the root finding procedure involved can be interpreted as being exact. It exploits a well-known basic result on multivariate distribution functions, see for instance Rao (1973).

It is obvious that this simulation technique can be generalized to the exact simulation of increments of solutions of some d-dimensional SDEs. Based on a given d-variate transition distribution function one needs to find the marginal distribution F_{Y_1} and the conditional distributions $F_{Y_2|Y_1}, F_{Y_3|Y_1,Y_2}, \ldots, F_{Y_d|Y_1,Y_2,\ldots,Y_{d-1}}$. Then the inverse transform method can be applied to each conditional transition distribution function one after the other. This also shows that it is sufficient to characterize explicitly in a model just the marginal and conditional transition distribution functions.

Note also that nonparametrically described transition distribution functions are sufficient for application of the inverse transform method. Of course, explicitly known parametric distributions are preferable for a number of practical reasons. They certainly reduce the complexity of the problem itself by splitting it into a sequence of problems. Finally, we recall that explicit transition densities have already been presented in Chaps. 2, 3, and 5.

Regarding the simulation of stochastic volatility models describing the evolution of a stock or index price, we proceed as follows, assuming a price process with SDE

$$dS_t = \mu S_t \, dt + \sqrt{V_t} S_t \, dW_t,$$

where $V = \{V_t, \, t \geq 0\}$ is a square-root process, see Sect. 3.1, if we deal with the Heston model; or a $3/2$ process, see Sect. 3.1, if we deal with the $3/2$ model. For both models, to obtain a realization of S_t, we firstly simulate V_t, subsequently we simulate $\int_0^t V_s \, ds$ conditional on V_t, and lastly S_t, which, conditional on V_t and $\int_0^t V_s \, ds$, follows a conditional Gaussian distribution. As discussed in Chap. 3, the distribution of V_t is known for the square-root and the $3/2$ process, see Sect. 3.1. Regarding the conditional distribution of $\int_0^t V_s \, ds$, we compute the Laplace transform of $\int_0^t V_s \, ds$, conditional on V_t. Subsequently, the probability distribution is easily recovered by an approach due to Feller, see Feller (1971). Having obtained the conditional probability distribution, the inversion method is applicable. To compute the Laplace transform of $\int_0^t V_s ds$ conditional on V_t, we rely on the results from Sect. 5.4, especially the fundamental solutions. We compute the relevant conditional Laplace transforms in Sect. 6.2, and also compute additional conditional Laplace transforms, such as the Hartman-Watson law for squared Bessel processes. Subsequently, in Sects. 6.3 and 6.4, we show how to apply the results from Sect. 6.2 to the Heston and the $3/2$ model.

6.2 Computing Conditional Laplace Transforms

In this section, we discuss how Laplace transforms of the form

$$E\left(\exp\left\{-\frac{b^2}{2}\int_0^t X_s \, ds - \nu \int_0^t \frac{ds}{X_s}\right\} \Bigg| X_t\right), \tag{6.2.4}$$

where $X = \{X_t, \, t \geq 0\}$ is a one-dimensional diffusion process to be specified below, can be computed using the results from Sect. 5.4. Such Laplace transforms turn out to play important roles in the design of exact simulation methods for stochastic volatility models, as we will show in Sects. 6.3 and 6.4. We point out when computing conditional Laplace transforms of the above form that Lie symmetry methods turn out to be crucial.

Formally, we consider the computation of the functional

$$u(x,t) = E\left(\exp\left\{-\lambda X_t - \frac{b^2}{2}\int_0^t X_s \, ds - \nu \int_0^t \frac{ds}{X_s}\right\}\right),$$

where $X = \{X_t, \, t \geq 0\}$ is such that its drift f satisfies one of the Ricatti equations (4.4.34), (4.4.35), or (4.4.36), and $X_0 = x$. We identify the corresponding PDE for u and denote the fundamental solution by $p(x, y, t)$.

However,

$$u(x,t) = E\left(\exp\left\{-\lambda X_t - \frac{b^2}{2}\int_0^t X_s\,ds - \nu\int_0^t \frac{ds}{X_s}\right\}\right)$$
$$= \int_0^\infty \exp\{-\lambda y\} E\left(\exp\left\{-\frac{b^2}{2}\int_0^t X_s\,ds - \nu\int_0^t \frac{ds}{X_s}\right\}\,\Big|\, X_t = y\right)$$
$$\times\, q(x,y,t)\,dy,$$

where $q(x,y,t)$ denotes the transition density of $X = \{X_t,\, t \geq 0\}$. Since $p(x,y,t)$ is a fundamental solution of the associated PDE we immediately have

$$E\left(\exp\left\{-\frac{b^2}{2}\int_0^t X_s\,ds - \nu\int_0^t \frac{ds}{X_s}\right\}\,\Big|\, X_t = y\right) = \frac{p(t,x,y)}{q(t,x,y)}.$$

Assuming the fundamental solution $p(x,y,t)$ and the transition density $q(x,y,t)$ are available in closed-form, the simple steps presented above outline a systematic approach to computing conditional Laplace transforms. As an illustration, we compute the Hartman-Watson law for squared Bessel processes, see also Jeanblanc et al. (2009), Proposition 6.5.1.1.

Proposition 6.2.1 *Assume that* $\delta \geq 2$, *and that* $X = \{X_t,\, t \geq 0\}$ *is given by the SDE*

$$dX_t = \delta\,dt + 2\sqrt{X_t}\,dW_t,$$

where $X_0 = x > 0$. *Then*

$$E\left(\exp\left\{-\frac{b^2}{2}\int_0^t \frac{ds}{X_s}\right\}\,\Big|\, X_t = y\right) = \frac{I_{\sqrt{b^2+\nu^2}}(\sqrt{xy}/t)}{I_\nu(\sqrt{xy}/t)},$$

where $\nu = \delta/2 - 1$.

Proof The proof follows immediately from Proposition 5.4.1, where the fundamental solution of the PDE

$$u_t = 2xu_{xx} + \delta u_x - \frac{b^2}{2}\frac{u}{x}$$

is given by

$$p(x,y,t) = \frac{1}{2t}\left(\frac{x}{y}\right)^{(1-\delta/2)/2} I_{2d+\frac{\delta}{2}-1}\left(\frac{\sqrt{xy}}{t}\right)\exp\left\{-\frac{(x+y)}{2t}\right\},$$

where $d = \frac{1}{4}(2 - \delta + \sqrt{(\delta-2)^2 + 4b^2})$ and the transition density of the squared Bessel process is of the form

$$q(x,y,t) = \frac{1}{2t}\left(\frac{x}{y}\right)^{(1-\delta/2)/2} I_{\delta/2-1}\left(\frac{\sqrt{xy}}{t}\right)\exp\left\{-\frac{(x+y)}{2t}\right\}.$$

Lastly, note that

$$2d + \frac{n}{2} - 1 = \sqrt{b^2+\nu^2},$$

where $\nu = \frac{\delta}{2} - 1$ is the index of the squared Bessel process, which finishes the proof. □

The next result is due to Pitman and Yor (1982). However, we present an alternative proof which employs Lie symmetry methods.

Proposition 6.2.2 *Assume $\delta \geq 2$, and that $X = \{X_t,\ t \geq 0\}$ is given by*

$$dX_t = \delta\, dt + 2\sqrt{X_t}\, dW_t,$$

$X_0 = x > 0$. *Then*

$$E\left(\exp\left\{-\frac{b^2}{2}\int_0^t X_s\, ds\right\} \,\middle|\, X_t = y\right)$$

$$= \frac{bt}{\sinh(bt)} \exp\left\{\frac{x+y}{2t}\left(1 - bt\coth(bt)\right)\right\} \frac{I_\nu\left(\frac{b\sqrt{xy}}{\sinh(bt)}\right)}{I_\nu\left(\frac{\sqrt{xy}}{t}\right)}.$$

Proof The proof follows along the lines of the proof of Proposition 6.2.1. From Proposition 5.4.2, we have that the fundamental solution of the PDE

$$u_t = 2xu_{xx} + \delta u_x - \frac{b^2}{2}\frac{u}{x},$$

is given by

$$p(x,y,t) = \frac{b}{2\sinh(bt)}\left(\frac{y}{x}\right)^{\frac{\delta/2-1}{2}} \exp\left\{-\frac{b(x+y)}{2\tanh(bt)}\right\} I_{(\delta-2)/2}\left(\frac{b\sqrt{xy}}{\sinh(bt)}\right).$$

Recalling the transition density of the squared Bessel process, the result follows. □

Proposition 6.2.2 plays a crucial role in the Broadie-Kaya exact simulation scheme for the Heston model, see Broadie and Kaya (2006). Consequently, the fundamental solutions presented in Chap. 5 can be used for this stochastic volatility model.

Finally, the following result can be used in the design of an exact simulation scheme for the 3/2 model, which is another stochastic volatility model.

Proposition 6.2.3 *Let $X = \{X_t,\ t \geq 0\}$ be a squared Bessel process of dimension δ, where $\delta \geq 2$. Then*

$$E\left(\exp\left\{-\frac{b^2}{2}\int_0^t X_s\, ds - \mu\int_0^t \frac{ds}{X_s}\right\} \,\middle|\, X_t = y\right)$$

$$= \frac{bt}{\sinh(bt)} \exp\left\{\frac{(x+y)}{2t}\left(1 - tb\coth(bt)\right)\right\} \frac{I_{\sqrt{\nu^2+2\mu}}\left(\frac{b\sqrt{xy}}{\sinh(bt)}\right)}{I_\nu\left(\frac{\sqrt{xy}}{t}\right)}.$$

Proof From Proposition 5.4.4, we have that the fundamental solution of

$$u_t = 2xu_{xx} + \delta u_x - \left(\frac{b^2 x}{2} + \frac{\mu}{x}\right)u$$

is

$$p(x, y, t) = \frac{b}{2\sinh(bt)} \exp\left\{-\frac{b(x+y)}{2\tanh(bt)}\right\}\left(\frac{y}{x}\right)^{\frac{\delta-2}{4}} I_{\sqrt{\nu^2+2\mu}}\left(\frac{b\sqrt{xy}}{\sinh(bt)}\right).$$

Recalling the transition density of the squared Bessel process, the result follows. □

We remind the reader that the fundamental solutions obtained via Lie symmetry methods sit at the heart of the computations of the results, not the Laplace transform of the solutions.

These conditional Laplace transforms are now applied to two stochastic volatility models, the Heston and the 3/2 model.

6.3 Exact Simulation of the Heston Model

In this section, we present the approach proposed by Broadie and Kaya (2006) to simulate the stock price under the Heston model exactly. We recall that the dynamics of the stock price and squared volatility under the Heston model satisfy the SDE,

$$dS_t = \mu S_t\, dt + \rho\sqrt{V_t} S_t\, dB_t + \sqrt{1-\rho^2}\sqrt{V_t} S_t\, dW_t, \tag{6.3.5}$$

$$dV_t = \kappa(\theta - V_t)\, dt + \sigma\sqrt{V_t}\, dB_t, \tag{6.3.6}$$

respectively, where $W = \{W_t,\, t \geq 0\}$ and $B = \{B_t,\, t \geq 0\}$ are independent Brownian motions. Integrating the stock price, we have

$$S_t = S_0 \exp\left\{\mu t - \frac{1}{2}\int_0^t V_s\, ds + \rho\int_0^t \sqrt{V_s}\, dB_s + \sqrt{1-\rho^2}\int_0^t \sqrt{V_s}\, dW_s\right\}.$$

We now integrate the squared volatility or the variance process

$$V_t = V_0 + \kappa\theta t - \kappa\int_0^t V_s\, ds + \sigma\int_0^t \sqrt{V_s}\, dB_s.$$

Hence one obtains

$$\int_0^t \sqrt{V_s}\, dB_s = \frac{V_t - V_0 - \kappa\theta t + \kappa\int_0^t V_s\, ds}{\sigma}. \tag{6.3.7}$$

Consequently, it follows

$$S_t = S_0 \exp\left\{\mu t - \frac{1}{2}\int_0^t V_s\, ds + \frac{\rho}{\sigma}\left(V_t - V_0 - \kappa\theta t + \kappa\int_0^t V_s\, ds\right) + \sqrt{1-\rho^2}\int_0^t \sqrt{V_s}\, dW_s\right\}.$$

We now present the exact simulation algorithm, and subsequently explain the individual steps in detail.

Algorithm 6.1 Exact simulation for the Heston model

1: Generate a sample of V_t given V_0
2: Generate a sample of $\int_0^t V_s\,ds$ given V_t
3: Compute $\int_0^t \sqrt{V_s}\,dB_s$ from (6.3.7) given V_t and $\int_0^t V_s\,ds$
4: Generate a sample from S_t, given $\int_0^t \sqrt{V_s}dB_s$ and $\int_0^t V_s\,ds$

6.3.1 Simulating V_t

In Sect. 5.3, the transition density of the square-root process of dimension δ was derived, see also Sect. 3.1, from which we can obtain the following equality in distribution

$$V_t \stackrel{d}{=} \frac{\sigma^2(1-\exp\{-\kappa t\})}{4\kappa}\chi_\delta^2\left(\frac{4\kappa\exp\{-\kappa t\}}{\sigma^2(1-\exp\{-\kappa t\})}\right),$$

where $\delta = \frac{4\theta\kappa}{\sigma^2}$ and $\chi_\delta^2(\lambda)$ denotes a non-central χ^2 random variable with δ degrees of freedom and non-centrality parameter λ. One way of sampling non-central χ^2 random variables, which was also used in Broadie and Kaya (2006), proceeds as follows: from Johnson et al. (1995), it is known that for $\delta > 1$,

$$\chi_\delta^2(\lambda) = \chi_1^2(\lambda) + \chi_{\delta-1}^2,$$

and hence

$$\chi_\delta^2(\lambda) = (Z+\sqrt{\lambda})^2 + \chi_{\delta-1}^2,$$

where Z is a standard normal random variable independent of $\chi_{\delta-1}^2$. Furthermore, for $\delta > 0$, we have the following equality in distribution:

$$\chi_\delta^2(\lambda) \stackrel{d}{=} \chi_{\delta+2N}^2,$$

where N is a Poisson random variable with mean $\frac{\lambda}{2}$. Since a χ^2-distributed random variable is a special case of a gamma random variable, we can use algorithms to sample from the gamma distribution. Lastly, we remark that in Sect. 13.2, we will present an algorithm to compute the cumulative distribution function of a non-central χ^2 random variable with $\delta \geq 0$ degrees of freedom, and hence we can also sample by inverting the cumulative distribution function as discussed in Sect. 6.1.

6.3.2 Simulating $\int_0^t V_s\,ds$ Given V_t

We point out that the challenging step in Algorithm 6.1 is the simulation of the integrated variance, $\int_0^t V_s\,ds$, conditional on the end point of the integral, V_t. This problem is solved by computing the Laplace transform of $\int_0^t V_s\,ds$, conditional on V_t, by combining a probabilistic result with a result from Sect. 5.4.

The method illustrates that results obtained via Lie symmetry analysis can be powerfully combined with results from probability theory, see also Sect. 5.5 for additional examples. Having obtained the Laplace transform, we use it to compute the characteristic function, which in turn can be used to compute the probability distribution function.

In fact, the approach is similar to the approach presented in Sect. 5.5. We introduce a time-change, i.e. we set $\rho_t = V_{\frac{4t}{\sigma^2}}$ to obtain the following SDE for $\rho = \{\rho_t,\, t \geq 0\}$,

$$d\rho_t = (2j\rho_t + \delta)\,dt + 2\sqrt{\rho_t}\,d\tilde{B}_t, \tag{6.3.8}$$

where $\tilde{B} = \{\tilde{B}_t,\, t \geq 0\}$ is a standard Brownian motion, $j = -\frac{2\kappa}{\sigma^2}$, and $\delta = \frac{4\kappa\theta}{\sigma^2}$. We now recall formula (6.d) from Pitman and Yor (1982), which reads

$$^{j}P^{\delta,t}_{\rho_0 \to y} = \frac{\exp\{-\frac{j^2}{2}\int_0^t \rho_s\,ds\}}{P^{\delta,t}_{\rho_0 \to y}} P^{\delta,t}_{\rho_0 \to y}, \tag{6.3.9}$$

using $^{j}P^{\delta,t}_{\rho_0 \to y}$ to denote the bridge for $\{\rho_s,\, 0 \leq s \leq t\}$ obtained by conditioning $^{j}P^{\delta}_{\rho_0}$ on $\rho_t = y$, where $^{j}P^{\delta}_{\rho_0}$ denotes the law of $\rho = \{\rho_t,\, t \geq 0\}$ started at ρ_0. Equation (6.3.9) is the analogue of Proposition 3.1.6, but for bridge constructions. We are now in a position to prove the following theorem.

Theorem 6.3.1 *Let $V = \{V_t,\, t \geq 0\}$ be given by Eq.* (6.3.6). *Then*

$$\begin{aligned}
&E\left(\exp\left\{-a\int_0^t V_s\,ds\right\}\,\Big|\, V_t\right) \\
&= \frac{\gamma(a)\exp\{-\frac{(\gamma(a)-\kappa)t}{2}\}(1-\exp\{-\kappa t\})}{\kappa(1-\exp\{-\gamma(a)t\})} \\
&\quad \times \exp\left\{\frac{V_0+V_t}{\sigma^2}\left(\frac{\kappa(1+\exp\{-\kappa t\})}{1-\exp\{-\kappa t\}} - \frac{\gamma(a)(1+\exp\{-\gamma(a)t\})}{1-\exp\{-\gamma(a)t\}}\right)\right\} \\
&\quad \times \frac{I_{\frac{\delta}{2}-1}\left(\frac{4\gamma(a)\sqrt{V_0 V_t}}{\sigma^2}\frac{\exp\{-\frac{\gamma(a)t}{2}\}}{(1-\exp\{-\gamma(a)t\})}\right)}{I_{\frac{\delta}{2}-1}\left(\frac{4\kappa\sqrt{V_0 V_t}}{\sigma^2}\frac{\exp\{-\frac{\kappa t}{2}\}}{(1-\exp\{-\kappa t\})}\right)},
\end{aligned}$$

where $\gamma(a) = \sqrt{\kappa^2 + 2\sigma^2 a}$.

Proof The steps of the proof are as follows: in fact, they are similar to the proof of Theorem 5.5.1. We firstly change the volatility coefficient of V_t from σ to 2, using the well-known time-change discussed above. Subsequently, we apply the bridge construction from Eq. (6.3.9), which is analogous to Eq. (5.5.20), to reduce the problem to the computation of conditional Laplace transforms involving a squared Bessel process, which we derived in Sect. 6.2.

As in Sect. 5.5, we set $\rho_t = V_{\frac{4t}{\sigma^2}}$ to obtain the SDE (6.3.8) for $\rho = \{\rho_t,\, t \geq 0\}$.

Recalling Eq. (6.3.9), we compute

$$
\begin{aligned}
& E\left(\exp\left\{-a\int_0^t V_s\,ds\right\}\,\middle|\,V_t\right) \\
&= E\left(\exp\left\{-a\int_0^t \rho_{\frac{\sigma^2 s}{4}}\,ds\right\}\,\middle|\,\rho_{\frac{\sigma^2 t}{4}}\right) \\
&= E\left(\exp\left\{-\frac{4a}{\sigma^2}\int_0^{\frac{\sigma^2 t}{4}} \rho_s\,ds\right\}\,\middle|\,\rho_{\frac{\sigma^2 t}{4}}\right) \\
&= \frac{\tilde{E}(\exp\{-(\frac{j^2}{2}+\frac{4a}{\sigma^2})\int_0^{\frac{\sigma^2 t}{4}} \rho_s\,ds\} \mid \rho_{\frac{\sigma^2 t}{4}})}{\tilde{E}(\exp\{-\frac{j^2}{2}\int_0^{\frac{\sigma^2 t}{4}} \rho_s\,ds\} \mid \rho_{\frac{\sigma^2 t}{4}})},
\end{aligned}
$$

where we use E to denote the expectation with respect to ${}^j P^{\delta,t}_{\rho_0\to y}$, and $\tilde{E}$ the expectation with respect to $P^{\delta,t}_{\rho_0\to y}$. Applying Proposition 6.2.2 to both, the numerator and the denominator, and recalling that $\rho_{\frac{\sigma^2 t}{4}} = V_t$ and $j = -\frac{2\kappa}{\sigma^2}$, the result follows. □

As described in Broadie and Kaya (2006), we now obtain the characteristic function $\Phi(b)$ by setting $a = -\imath b$,

$$
\Phi(b) = E\left(\exp\left\{\imath b\int_0^t V_s\,ds\right\}\,\middle|\,V_t\right).
$$

The probability distribution function can be obtained by Fourier inversion methods, see Feller (1971):

$$
\begin{aligned}
P\left(\int_0^t V_s\,ds \le x\,\middle|\,V_t\right) &= \frac{1}{\pi}\int_{-\infty}^{\infty}\frac{\sin(ux)}{u}\Phi(u)\,du \\
&= \frac{2}{\pi}\int_0^{\infty}\frac{\sin(ux)}{u}\Re\big(\Phi(u)\big)\,du,
\end{aligned}
\tag{6.3.10}
$$

where $\Re(\Phi(u))$ denotes the real part of $\Phi(u)$.

The final integral in Eq. (6.3.10) can be computed numerically and one can then sample by inversion.

6.3.3 Generating S_t

We recall that in Step 3 of Algorithm 6.1, we computed $\int_0^t \sqrt{V_s}\,dB_s$ in terms of V_t and $\int_0^t V_s\,ds$. Due to the independence of $V = \{V_t,\, t \ge 0\}$ and $W = \{W_t,\, t \ge 0\}$, it is clear that $\int_0^t \sqrt{V_s}\,dW_s$ given $\int_0^t V_s\,ds$ follows a normal distribution with mean 0 and variance $\int_0^t V_s\,ds$. Hence $\log(S_t)$ follows a conditionally normal distribution with mean

$$
\log(S_0) + \mu t - \frac{1}{2}\int_0^t V_s\,ds + \frac{\rho}{\sigma}\left(V_t - V_0 - \kappa\theta t + \kappa\int_0^t V_s\,ds\right)
$$

and random variance

$$(1-\rho^2)\int_0^t V_s\,ds.$$

In this way, we can obtain samples of S_t satisfying the dynamics (6.3.5).

6.4 Exact Simulation of the 3/2 Model

It is very useful to have exact simulation algorithms for important models. In this section, we closely follow the approach from Baldeaux (2012a) to simulate exactly the stock price or index under the 3/2 model, see e.g. Carr and Sun (2007), Heston (1997), Itkin and Carr (2010), and Lewis (2000). We remark that this approach is similar to the approach from Broadie and Kaya (2006), which we discussed in the previous section.

The dynamics of the stock price under the 3/2 model are described by the system of SDEs,

$$dS_t = \mu S_t\,dt + \rho\sqrt{V_t}S_t\,dB_t + \sqrt{1-\rho^2}\sqrt{V_t}S_t\,dW_t, \tag{6.4.11}$$

$$dV_t = \kappa V_t(\theta - V_t)\,dt + \sigma(V_t)^{3/2}\,dB_t, \tag{6.4.12}$$

where $B=\{B_t,\ t\ge 0\}$ and $W=\{W_t,\ t\ge 0\}$ are independent Brownian motions. The key observation, as already discussed in Sects. 3.1 and 5.5, is that V_t is the inverse of a square-root process. Defining $X_t=\frac{1}{V_t}$, we obtain

$$dX_t = \left(\kappa+\sigma^2-\kappa\theta X_t\right)dt - \sigma\sqrt{X_t}\,dB_t. \tag{6.4.13}$$

Expressing the stock price in terms of the process $X=\{X_t,\ t\ge 0\}$, we obtain

$$S_t = S_0\exp\left\{\mu t - \frac{1}{2}\int_0^t (X_s)^{-1}\,ds + \rho\int_0^t (\sqrt{X_s})^{-1}\,dB_s + \sqrt{1-\rho^2}\int_0^t (\sqrt{X_s})^{-1}\,dW_s\right\}. \tag{6.4.14}$$

It is useful to study $\log(X_t)$, for which we obtain the following SDE

$$d\log(X_t) = \left(\frac{\kappa+\frac{\sigma^2}{2}}{X_t} - \kappa\theta\right)dt - \sigma(\sqrt{X_t})^{-1}\,dB_t.$$

Hence

$$\log(X_t) = \log(X_0) + \left(\kappa+\frac{\sigma^2}{2}\right)\int_0^t \frac{ds}{X_s} - \kappa\theta t - \sigma\int_0^t (\sqrt{X_s})^{-1}\,dB_s,$$

or equivalently

$$\int_0^t (\sqrt{X_s})^{-1}\,dB_s = \frac{1}{\sigma}\left(\log\left(\frac{X_0}{X_t}\right) + \left(\kappa+\frac{\sigma^2}{2}\right)\int_0^t \frac{ds}{X_s} - \kappa\theta t\right). \tag{6.4.15}$$

Algorithm 6.2 describes how to simulate the stock price given by (6.4.11) exactly.

We now discuss the individual steps of the algorithm. Clearly, Steps (1), (3), and (4) are very similar to steps (1), (3), and (4) of Algorithm 6.1.

Algorithm 6.2 Exact simulation for the 3/2 model

1: Generate a sample of X_t given X_0
2: Generate a sample of $\int_0^t \frac{ds}{X_s}$ given X_t
3: Compute $\int_0^t (\sqrt{X_s})^{-1}\, dB_s$ from (6.4.15) given X_t and $\int_0^t \frac{ds}{X_s}$
4: Generate a sample from S_t, given $\int_0^t (\sqrt{X_s})^{-1}\, dB_s$ and $\int_0^t (X_s)^{-1}\, ds$

6.4.1 Simulating X_t

Since $X = \{X_t,\, t \geq 0\}$ is a square-root process, see Eq. (6.4.13), we can immediately apply the methodology from Sect. 6.3.

6.4.2 Simulating $\int_0^t \frac{ds}{X_s}$ Given X_t

We approach Step (2) of Algorithm 6.2 in the same manner as Step (2) of Algorithm 6.1. However, we end up having to compute a different conditional Laplace transform. Fortunately, the relevant Laplace transform can be computed as shown in Sect. 6.2 using Lie symmetry methods. As before, we change the volatility coefficient of X from σ to 2, using the standard time change, which was also used in Sect. 6.3: we define $\rho_t = X_{\frac{4t}{\sigma^2}}$ to obtain the SDE

$$d\rho_t = (2j\rho_t + \delta)\, dt + 2\sqrt{\rho_t}\, d\tilde{B}_t,$$

where

$$\delta = \frac{4(\kappa + \sigma^2)}{\sigma^2}$$

and $j = -\frac{2\kappa\theta}{\sigma^2}$ and $\tilde{B} = \{\tilde{B}_t,\, t \geq 0\}$ is a standard Brownian motion. Now we use formula (6.3.9) again, but this time to obtain a different conditional Laplace transform in the numerator.

$$\begin{aligned}
&E\left(\exp\left\{-a\int_0^t \frac{ds}{X_s}\right\} \bigg| X_t\right) \\
&= E\left(\exp\left\{-a\int_0^t \frac{ds}{\rho_{\frac{\sigma^2 s}{4}}}\right\} \bigg| \rho_{\frac{\sigma^2 t}{4}}\right) \\
&= E\left(\exp\left\{-\frac{4a}{\sigma^2}\int_0^{\frac{\sigma^2 t}{4}} \frac{ds}{\rho_s}\right\} \bigg| \rho_{\frac{\sigma^2 t}{4}}\right) \\
&= \tilde{E}\left(\exp\left\{-\frac{4a}{\sigma^2}\int_0^{\frac{\sigma^2 t}{4}} \frac{ds}{\rho_s} - \frac{j^2}{2}\int_0^{\frac{\sigma^2 t}{4}} \rho_s\, ds\right\} \bigg| \rho_{\frac{\sigma^2 t}{4}}\right) \Big/ \\
&\quad \tilde{E}\left(\exp\left\{-\frac{j^2}{2}\int_0^{\frac{\sigma^2 t}{4}} \rho_s\, ds\right\} \bigg| \rho_{\frac{\sigma^2 t}{4}}\right),
\end{aligned} \tag{6.4.16}$$

where we use E to denote the expectation with respect to ${}^{j}P^{\delta,t}_{\rho_0\to y}$, and $\tilde{E}$ denotes the expectation with respect to $P^{\delta,t}_{\rho_0\to y}$, see Eq. (6.3.9).

Computing the numerator in (6.4.16) using Proposition 6.2.3 and the denominator using Proposition 6.2.2 yields the following result.

Theorem 6.4.1 *Let X be given by* (6.4.13). *Then*

$$E\left(\exp\left\{-a\int_0^t \frac{ds}{X_s}\right\}\Bigg|\, X_t\right) = \frac{I_{\sqrt{\nu^2+8a/\sigma^2}}\left(-\frac{2\kappa\theta\sqrt{X_t X_0}}{\sigma^2\sinh(-\frac{\kappa\theta t}{2})}\right)}{I_\nu\left(-\frac{2\kappa\theta\sqrt{X_t X_0}}{\sigma^2\sinh(-\frac{\kappa\theta t}{2})}\right)},$$

where $\delta = \frac{4(\kappa+\sigma^2)}{\sigma^2}$ *and* $\nu = \frac{\delta}{2} - 1$.

Consequently, we can proceed as in Sect. 6.3: we compute the Laplace transform using Theorem 6.4.1, compute the probability distribution of $\int_0^t \frac{ds}{X_s}$ conditional on X_t, and sample by inversion.

6.4.3 *Simulating S_t*

As in Sect. 6.3, in Step 3) of Algorithm 6.2, we compute $\int_0^t (\sqrt{X_s})^{-1}\, dB_s$ in terms of X_t and $\int_0^t \frac{ds}{X_s}$. Due to the independence of $X = \{X_t,\, t \geq 0\}$ and $W = \{W_t,\, t \geq 0\}$, it follows that $\int_0^t (\sqrt{X_s})^{-1}\, dW_s$ given $\int_0^t (X_s)^{-1}\, ds$ follows a normal distribution with mean 0 and variance $\int_0^t (X_s)^{-1}\, ds$. Hence $\log(S_t)$ follows a normal distribution with mean

$$\log(S_0) + \mu t - \frac{1}{2}\int_0^t (X_s)^{-1}\, ds + \rho\int_0^t (\sqrt{X_s})^{-1}\, dB_s$$

and variance

$$\left(1-\rho^2\right)\int_0^t (X_s)^{-1}\, ds.$$

6.5 Stochastic Volatility Models with Jumps in the Stock Price

In this section, we extend the model to the case where the stock price process is also subjected to jumps. We follow the presentation in Broadie and Kaya (2006), see also Korn et al. (2010), Sect. 7.2.3, and deal with the Heston model. However, the argument does not rely on the specification of the volatility process, but tells us how to modify the approach from Sects. 6.3 and 6.4 to allow for jumps. Hence the discussion presented in this section also applies to the 3/2 model. The following model was presented in Bates (1996), we also refer the reader to Chap. 5 in Gatheral (2006), where it is referred to as the *SVJ model*,

$$dS_t = S_{t-}\left((r-\lambda\bar{\mu})\,dt + \sqrt{V_t}\left(\rho\, dB_t + \sqrt{1-\rho^2}\, dW_t\right) + (Y_t - 1)\, dN_t\right),$$
$$dV_t = \kappa(\theta - V_t)\,dt + \sigma\sqrt{V_t}\, dB_t, \tag{6.5.17}$$

where $N = \{N_t,\ t \geq 0\}$ is a Poisson process with constant intensity λ. The processes $B = \{B_t,\ t \geq 0\}$ and $W = \{W_t,\ t \geq 0\}$ are independent Brownian motions and independent of the Poisson process, and the jump variables $Y = \{Y_t,\ t \geq 0\}$ are a family of independent random variables all having the same lognormal distribution with mean μ_s and variance σ_s^2. Furthermore,

$$E(Y_t - 1) = \bar{\mu},$$

and hence

$$\mu_s = \log(1+\bar{\mu}) - \frac{1}{2}\sigma_s^2.$$

Integrating the SDE for the stock price (6.5.17), we obtain

$$S_t = \tilde{S}_t \prod_{j=1}^{N_t} \tilde{Y}_j, \tag{6.5.18}$$

where

$$\tilde{S}_t = S_0 \exp\left\{(r-\lambda\bar{\mu})t - \frac{1}{2}\int_0^t V_s\, ds + \rho \int_0^t \sqrt{V_s}\, dB_s + \sqrt{1-\rho^2}\int_0^t \sqrt{V_s}\, dW_s\right\},$$

and $\tilde{Y}_j$, $j = 1, \ldots, N_t$, denotes the size of the j-th jump. As discussed in Broadie and Kaya (2006), Korn et al. (2010), Eq. (6.5.18) motivates the simulation algorithm for the SVJ model: we firstly simulate the diffusion part as in Sect. 6.3 and consequently take care of the jump part, $\prod_{j=1}^{N_t} \tilde{Y}_j$. Algorithm 6.3 is the analogue of Algorithm 6.1 and also appeared in Broadie and Kaya (2006) and in similar form in Korn et al. (2010).

Algorithm 6.3 Exact Simulation Algorithm for the SVJ model

1: Generate a sample of V_t given V_0
2: Generate a sample from the distribution of $\int_0^t V_s\, ds$ given V_t and V_0
3: Recover $\int_0^t \sqrt{V_s}\, dB_s$ from (6.3.6) given V_t, V_0 and $\int_0^t V_s\, ds$
4: Generate $\tilde{S}_t$
5: Generate N_t
6: Generate $\prod_{j=1}^{N_t} \tilde{Y}_j$, given N_t

Since the $\tilde{Y}_j$, $j = 1, \ldots, N_t$, are mutually independent and each follows a log-normal distribution with mean μ_s and variance σ_s^2, it is clear that

$$\sum_{j=1}^{N_t} \log(\tilde{Y}_j) | N_t \sim N\big(N_t \mu_s, N_t \sigma_s^2\big).$$

There are alternative approaches to simulating $\prod_{j=1}^{N_t} \tilde{Y}_j$: in Sect. 3.5 in Glasserman (2004), it was shown how to simulate N_t by simulating the jump times of the Poisson process. Furthermore, as discussed in Broadie and Kaya (2006), given N_t, one can simulate the jump sizes $\tilde{Y}_j$, $j = 1, \ldots, N_t$, individually. However, Algorithm 6.3 results in a problem that is of fixed dimension. More precisely, the dimension of the problem in Algorithm 6.3 is five, i.e. five random numbers are used to obtain a realization of S_t. Having a problem of fixed dimensionality is important when applying quasi-Monte Carlo methods, permitting an effective way of tackling multidimensional problems, see Chap. 12, hence we choose the formulation presented in Algorithm 6.3.

6.6 Stochastic Volatility Models with Simultaneous Jumps in the Volatility Process and the Stock Price

In this section, we briefly extend the SVJ model from Sect. 6.5 to allow for simultaneous jumps in the stock price and the volatility process, the *SVCJ model*. As argued in Gatheral (2006), it is unrealistic to assume that the instantaneous volatility would not jump if the stock price did. Hence the following model, introduced in Duffie et al. (2000), allows for simultaneous jumps in the stock price and the volatility,

$$\begin{aligned} dS_t &= S_{t-}\big((r - \lambda\bar{\mu})\,dt + \sqrt{V_t}\big(\rho\,dB_t + \sqrt{1-\rho^2}\,dW_t\big) + \big(Y_t^s - 1\big)\,dN_t\big), \\ dV_t &= \kappa(\theta - V_t)\,dt + \sigma\sqrt{V_t}\,dB_t + Y^v dN_t, \end{aligned}$$

where $N = \{N_t, t \geq 0\}$ is again a Poisson process with constant intensity λ, $Y^s = \{Y_t^s, t \geq 0\}$ is the relative jump size of the stock price, and $Y^v = \{Y_t^v, t \geq 0\}$ is the jump size of the variance. The magnitudes of the jumps in the stock price and variance processes are dependent, via the parameter ρ_J, in the following way: the distribution of Y_t^v is exponential with mean μ_v and given Y^v, Y^s is lognormally distributed with mean $\mu_s + \rho_J Y^v$ and variance σ_s^2. The parameters μ_s and $\bar{\mu}$ are related via

$$\mu_s = \log\big((1+\bar{\mu})(1-\rho_J\mu_v)\big) - \frac{1}{2}\sigma_s^2,$$

hence only one needs to be specified. Due to the occurrence of jumps in the volatility, we have to modify the previous procedure. Essentially, we simulate the variance and the stock price process at each jump time. Algorithm 6.4 is the analogue of Algorithms 6.1 and 6.3 and we point out that this algorithm also appeared in Broadie and Kaya (2006), see Sect. 6.2.

Algorithm 6.4 Exact Simulation Algorithm for the SVCJ model

1: Simulate the arrival time of the next jump, τ_j.
2: **if** $\tau_j > T$ **then**
3: Set $\tau_j \to T$
4: **end if**
5: Simulate $V_{\tau_j^-}$ and $S_{\tau_j^-}$, using the time step $\Delta t \to \tau_j - t^*$
6: **if** $\tau_j = T$ **then**
7: Go to Step 13
8: **else**
9: Generate Y^v from an exponential distribution with mean μ_v and set

$$V_{\tau_j} \to V_{\tau_j^-} + Y^v.$$

10: **end if**
11: Generate Y^s by sampling from a lognormal distribution with mean $(\mu_s + \rho_J Y^s)$ and variance σ_s^2. Set $S_{\tau_j} \to S_{\tau_j^-} Y^s$.
12: Set $S_{t^*} \to S_{\tau_j}$, $V_{t^*} \to V_{\tau_j}$, $t^* \to \tau_j$ and go to Step 1
13: Set $S_T \to S_{\tau_j^-}$

6.7 Multidimensional Stochastic Volatility Models

In this section, we discuss the extension of the methodology presented in Sects. 6.3 and 6.4 to the multidimensional case. We firstly explain why a generalization of this methodology is not straightforward, which motivates us to consider almost exact simulation schemes, see Platen and Bruti-Liberati (2010), Chap. 2, for more information on this topic. Furthermore, in Chap. 11 we study advanced stochastic volatility models based on Wishart processes.

Consider the following simple case, with SDEs

$$dS_t^1 = \mu^1 S_t^1 \, dt + \sqrt{V_t^1} S_t^1 \, dW_t^1, \tag{6.7.19}$$

$$dS_t^2 = \mu^2 S_t^2 \, dt + \sqrt{V_t^2} S_t^2 \, dW_t^2, \tag{6.7.20}$$

where the two Brownian motions $W^1 = \{W_t^1, t \geq 0\}$ and $W^2 = \{W_t^2, t \geq 0\}$ co-vary, say $d[W^1, W^2]_t = \rho \, dt$. The volatility processes, V^1 and V^2, which can be square-root or 3/2 processes, see Sect. 3.1, are here driven by Brownian motions independent of W^1 and W^2. Of course, S^1 and S^2 can be simulated as discussed in Sects. 6.3 and 6.4, however, S^1 and S^2 are not independent. In particular, given $V_t^j, \int_0^t V_s^j \, ds$, $j = 1, 2$, we have that, for $j = 1, 2$,

$$\log(S_t^j) \sim N(\mu_j, \sigma_j^2),$$

with

$$\mu_j = \log(S_0^j) - \frac{1}{2} \int_0^t V_s^j \, ds$$

and

$$\sigma_j^2 = \int_0^t V_s^j \, ds.$$

Here the conditional covariance is given by

$$\rho \int_0^t \sqrt{V_s^1} \sqrt{V_s^2} \, ds, \tag{6.7.21}$$

where we recall that ρ denotes the correlation between W^1 and W^2. The computation of the integral in Eq. (6.7.21) does not follow immediately from the methods discussed in Sect. 5.4. We hence recall the almost exact simulation methodology from Platen and Bruti-Liberati (2010).

6.7.1 *Matrix Square-Root Processes via Time-Changed Wishart Processes*

In this subsection, we briefly recall from Platen and Bruti-Liberati (2010) how to obtain a matrix square-root process from a time-changed Wishart process. The diagonal elements of this process will play the role of V^1 and V^2 in Eqs. (6.7.19) and (6.7.20). We point out that this discussion is based on the simple Wishart process from Sect. 3.2. Once we fully develop the theory of Wishart processes in Chap. 11, we can employ more advanced stochastic volatility models, as in Da Fonseca et al. (2008c).

Recall from Sect. 3.2 that square-root processes can be obtained by time-changing a squared Bessel process. As in Platen and Bruti-Liberati (2010), we consider the function

$$s_t = s_0 \exp\{ct\},$$

where $s_0 > 0$ and consider the transformed time

$$\varphi(t) = \varphi(0) + \frac{1}{4} \int_0^t \frac{b^2}{s_u} \, du,$$

and compute

$$\varphi(t) = \varphi(0) + \frac{b^2}{4cs_0} \big(1 - \exp\{-ct\}\big).$$

Let $X = \{X_t, \, t \geq 0\}$ denote a squared Bessel process of dimension $\delta > 0$, then we obtain a square-root process $Y = \{Y_t, \, t \geq 0\}$ of the same dimension $\delta > 0$ as follows: setting

$$Y_t = s_t X_{\varphi(t)},$$

we obtain the following dynamics for Y,

$$dY_t = \left(\frac{\delta}{2} b^2 + cY_t\right) dt + b\sqrt{Y_t} \, dU_t,$$

where

$$dU_t = \sqrt{\frac{4s_t}{b^2}}\, dW_{\varphi(t)},$$

and since

$$[U]_t = \int_0^t \frac{4s_z}{b^2}\, d\varphi(z) = t,$$

$U = \{U_t,\ t \geq 0\}$ is a Brownian motion, by Levy's characterization theorem, see Sect. 15.3. This procedure is easily generalized. Recall the Wishart process from Sect. 3.2, so $\boldsymbol{W}_t$ is an $n \times p$ matrix, whose elements are independent scalar Brownian motions and $\boldsymbol{W}_0 = \boldsymbol{C}$ is the initial state matrix. We set

$$\boldsymbol{X}_t = \boldsymbol{W}_t^\top \boldsymbol{W}_t, \qquad \boldsymbol{X}_0 = \boldsymbol{C}^\top \boldsymbol{C},$$

so $\boldsymbol{X} = \{\boldsymbol{X}_t,\ t \geq 0\}$ is a Wishart process $WIS_p(\boldsymbol{X}_0, n, \boldsymbol{0}, \boldsymbol{I}_p)$. Following Platen and Bruti-Liberati (2010), we generalize the idea of time-changing a squared Bessel process to time-changing a Wishart process and set

$$\boldsymbol{\Sigma}_t = s_t \boldsymbol{X}_{\varphi(t)},$$

to obtain the SDE

$$d\boldsymbol{\Sigma}_t = \left(\frac{\delta}{4} b^2 \boldsymbol{I} + c\boldsymbol{\Sigma}_t\right) dt + \frac{b}{2}\left(\sqrt{\boldsymbol{\Sigma}_t}\, d\boldsymbol{U}_t + d\boldsymbol{U}_t^\top \sqrt{\boldsymbol{\Sigma}_t}\right), \tag{6.7.22}$$

for $t \geq 0$, $\boldsymbol{\Sigma}_0 = s_0 \boldsymbol{X}_{\varphi(0)}$, and $d\boldsymbol{U}_t = \sqrt{\frac{4s_t}{b_t^2}}\, d\boldsymbol{W}_{\varphi(t)}$ is the differential of a matrix Wiener process.

6.7.2 Multidimensional Heston Model with Independent Prices

We firstly focus on the case where the volatility process and the Brownian motion driving the stock price are independent. We study the following model

$$d\boldsymbol{S}_t = \boldsymbol{A}_t(\boldsymbol{r}\, dt + \sqrt{\boldsymbol{B}_t}\, d\boldsymbol{W}_t),$$

where $\boldsymbol{S} = \{\boldsymbol{S}_t = (S_t^1, S_t^2, \dots, S_t^d)^\top,\ t \geq 0\}$ is a vector process and $\boldsymbol{A} = \{\boldsymbol{A}_t = [A_t^{i,j}]_{i,j=1}^d,\ t \geq 0\}$ is a diagonal matrix process with elements

$$A_t^{i,j} = \begin{cases} S_t^i & \text{for } i = j \\ 0 & \text{otherwise.} \end{cases} \tag{6.7.23}$$

Additionally, $\boldsymbol{r} = (r_1, r_2, \dots, r_d)^\top$ is a d-dimensional vector and $\boldsymbol{W} = \{\boldsymbol{W}_t = (W_t^1, W_t^2, \dots, W_t^d)^\top,\ t \geq 0\}$ is a d-dimensional vector of correlated Wiener processes. Moreover, $\boldsymbol{B} = \{\boldsymbol{B}_t = [B_t^{i,j}]_{i,j=1}^d,\ t \geq 0\}$ is a matrix process with elements

$$B_t^{i,j} = \begin{cases} \Sigma_t^{i,i} & \text{for } i = j \\ 0 & \text{otherwise.} \end{cases} \tag{6.7.24}$$

Note that $\boldsymbol{B}$ is the generalization of V in the one-dimensional case. Here, the matrix process $\boldsymbol{\Sigma} = \{\boldsymbol{\Sigma}_t = [\Sigma_t^{i,j}]_{i,j=1}^d, \, t \geq 0\}$ is a matrix square-root process given by the SDE (6.7.22). Therefore, $\boldsymbol{B}_t$ can be constructed from the diagonal elements of $\boldsymbol{\Sigma}_t$. Recall that these elements $\Sigma_t^{1,1}, \Sigma_t^{2,2}, \dots, \Sigma_t^{d,d}$ form square-root processes and that, for simplicity, we assumed that $\boldsymbol{B}$ is independent of $\boldsymbol{W}$.

We illustrate the simulation in a two-dimensional example. The corresponding two-dimensional SDE for the two prices can be represented as

$$
\begin{aligned}
dS_t^1 &= S_t^1 r_1 \, dt + S_t^1 \sqrt{\Sigma_t^{1,1}} \, d\tilde{W}_t^1, \\
dS_t^2 &= S_t^2 r_2 \, dt + S_t^2 \sqrt{\Sigma_t^{2,2}} \big[\varrho \, d\tilde{W}_t^1 + \sqrt{1-\varrho^2} \, d\tilde{W}_t^2\big],
\end{aligned}
$$

where $t \geq 0$. Here, $\Sigma^{1,1}$ and $\Sigma^{2,2}$ are diagonal elements of the 2×2 matrix given by (6.7.22) and $\tilde{W}^1$ and $\tilde{W}^2$ are independent Wiener processes. The logarithmic transformation $\boldsymbol{X}_t = \log(\boldsymbol{S}_t)$ yields the following SDE

$$
\begin{aligned}
dX_t^1 &= \left(r_1 - \frac{1}{2}\Sigma_t^{1,1}\right) dt + \sqrt{\Sigma_t^{1,1}} \, dW_t^1, \\
dX_t^2 &= \left(r_2 - \frac{1}{2}\Sigma_t^{2,2}\right) dt + \sqrt{\Sigma_t^{2,2}} \big[\varrho \, d\tilde{W}_t^1 + \sqrt{1-\varrho^2} \, d\tilde{W}_t^2\big],
\end{aligned}
$$

for $t \geq 0$. This results in the following representations:

$$
\begin{aligned}
X_{t_{i+1}}^1 &= X_{t_i}^1 + r_1(t_{i+1} - t_i) - \frac{1}{2}\int_{t_i}^{t_{i+1}} \Sigma_u^{1,1} \, du + \int_{t_i}^{t_{i+1}} \sqrt{\Sigma_u^{1,1}} \, d\tilde{W}_u^1, \\
X_{t_{i+1}}^2 &= X_{t_i}^2 + r_2(t_{i+1} - t_i) - \frac{1}{2}\int_{t_i}^{t_{i+1}} \Sigma_u^{2,2} \, du + \varrho \int_{t_i}^{t_{i+1}} \sqrt{\Sigma_u^{2,2}} \, d\tilde{W}_u^1 \\
&\quad + \sqrt{1-\varrho^2} \int_{t_i}^{t_{i+1}} \sqrt{\Sigma_u^{2,2}} \, d\tilde{W}_u^2.
\end{aligned}
$$

We approximate the integral $\int_{t_i}^{t_{i+1}} \Sigma_u^{j,j} \, du$, $j = 1, 2$, using e.g. the trapezoidal rule. Consequently, we can simulate the model, noting that conditional on $\int_{t_i}^{t_{i+1}} \Sigma_u^{j,j} \, du$ and $X_{t_i}^j$, $j = 1, 2$, we obtain that $X_{t_{i+1}}^j$ follows a normal distribution with mean

$$
X_{t_i}^j + r_j(t_{i+1} - t_i) - \frac{1}{2}\int_{t_i}^{t_{i+1}} \Sigma_u^{j,j} \, du, \quad j = 1, 2,
$$

and variance

$$
\int_{t_i}^{t_{i+1}} \Sigma_u^{j,j} \, du.
$$

Furthermore, $X_{t_{i+1}}^1$ and $X_{t_{i+1}}^2$ have the conditional covariance

$$
\varrho \int_{t_i}^{t_{i+1}} \sqrt{\Sigma_u^{1,1}} \sqrt{\Sigma_u^{2,2}} \, du,
$$

which we approximate, for example, using the trapezoidal rule and the trajectories of $\Sigma^{1,1}$ and $\Sigma^{2,2}$.

6.7.3 Multidimensional Heston Model with Correlated Prices

We now consider a multidimensional version of the Heston model, which allows for correlation of the volatility vector $\boldsymbol{\Sigma}$ with the vector asset price process $\boldsymbol{S}$. We define the generalization by the system of SDEs

$$d\boldsymbol{S}_t = \boldsymbol{A}_t\big(\boldsymbol{r}\,dt + \sqrt{\boldsymbol{B}_t}\big(\boldsymbol{C}\,d\boldsymbol{W}_t^1 + \boldsymbol{D}\,d\boldsymbol{W}_t^2\big)\big),$$
$$d\boldsymbol{\Sigma}_t = (\boldsymbol{a} - \boldsymbol{E}\,\boldsymbol{\Sigma}_t)\,dt + \boldsymbol{F}\sqrt{\boldsymbol{B}_t}\,d\boldsymbol{W}_t^1,$$

for $t \geq 0$. Here, $\boldsymbol{S} = \{\boldsymbol{S}_t = (S_t^1, S_t^2, \ldots, S_t^d)^\top, t \geq 0\}$ and $\boldsymbol{r} = (r_1, r_2, \ldots, r_d)^\top$. The matrix $\boldsymbol{A}_t = [A_t^{i,j}]_{i,j=1}^d$ is given by (6.7.23) and $\boldsymbol{B}_t = [B_t^{i,j}]_{i,j=1}^d$ is a matrix with elements as in (6.7.24). Additionally, $\boldsymbol{C} = [C^{i,j}]_{i,j=1}^d$ is a diagonal matrix with elements

$$C^{i,j} = \begin{cases} \varrho_i & \text{for } i = j \\ 0 & \text{otherwise,} \end{cases}$$

and $\boldsymbol{D} = [D^{i,j}]_{i,j=1}^d$ is a diagonal matrix with elements

$$D^{i,j} = \begin{cases} \sqrt{1-\varrho_i^2} & \text{for } i = j \\ 0 & \text{otherwise,} \end{cases}$$

where $\varrho_i \in [-1, 1]$, $i \in \{1, 2, \ldots, d\}$. Moreover, $\boldsymbol{\Sigma} = \{\boldsymbol{\Sigma}_t = (\Sigma_t^{1,1}, \Sigma_t^{2,2}, \ldots, \Sigma_t^{d,d})^\top, t \in [0, \infty)\}$ and $\boldsymbol{a} = (a_1, a_2, \ldots, a_d)^\top$. The matrix $\boldsymbol{E} = [E^{i,j}]_{i,j=1}^d$ is a diagonal matrix with elements

$$E^{i,j} = \begin{cases} b_i & \text{for } i = j \\ 0 & \text{otherwise,} \end{cases}$$

and $\boldsymbol{F} = [F^{i,j}]_{i,j=1}^d$ is a diagonal matrix with elements

$$F^{i,j} = \begin{cases} \sigma_i & \text{for } i = j \\ 0 & \text{otherwise.} \end{cases}$$

Furthermore, $\boldsymbol{W}^1 = \{\boldsymbol{W}_t^1 = (W_t^{1,1}, W_t^{1,2}, \ldots, W_t^{1,d})^\top, t \geq 0\}$ is a vector of independent Wiener processes and $\boldsymbol{W}^2 = \{\boldsymbol{W}_t^2 = (W_t^{2,1}, W_t^{2,2}, \ldots, W_t^{2,d})^\top, t \geq 0\}$ is a vector of correlated Wiener processes which are independent of $\boldsymbol{W}^1$. In two dimensions, the model looks as follows:

$$d\Sigma_t^{1,1} = \big(a_1 - b_1\Sigma_t^{1,1}\big)\,dt + \sigma_1\sqrt{\Sigma_t^{1,1}}\,dW_t^{1,1},$$
$$d\Sigma_t^{2,2} = \big(a_2 - b_2\Sigma_t^{2,2}\big)\,dt + \sigma_2\sqrt{\Sigma_t^{2,2}}\,dW_t^{1,2},$$

for $t \geq 0$. The two-dimensional asset price process is given by

$$dS_t^1 = r_1 S_t^1\,dt + S_t^1\sqrt{\Sigma_t^{1,1}}\Big(\varrho_1\,dW_t^{1,1} + \sqrt{1-\varrho_1^2}\,dW_t^{2,1}\Big),$$
$$dS_t^2 = r_2 S_t^2\,dt + S_t^2\sqrt{\Sigma_t^{2,2}}\Big(\varrho_2\,dW_t^{1,2} + \sqrt{1-\varrho_2^2}\,dW_t^{2,2}\Big),$$

for $t \geq 0$. Hence we can simulate $\Sigma^{1,1}$ and $\Sigma^{2,2}$ via the non-central χ^2-distribution, see Sect. 3.1, or the elements of a matrix square-root process. We can now generate samples of the logarithm of the stock price, $\boldsymbol{X}_t = \log(\boldsymbol{S}_t)$, using the representation

$$
\begin{aligned}
X^1_{t_{i+1}} &= X^1_{t_i} + r_1(t_{i+1} - t_i) + \frac{\varrho_1}{\sigma_1}\big(\Sigma^{1,1}_{t_{i+1}} - \Sigma^{1,1}_{t_i} - a_1(t_{i+1} - t_i)\big) \\
&\quad + \left(\frac{\varrho_1 b_1}{\sigma_1} - \frac{1}{2}\right) \int_{t_i}^{t_{i+1}} \Sigma^{1,1}_u \, du + \sqrt{1 - \varrho_1^2} \int_{t_i}^{t_{i+1}} \sqrt{\Sigma^{1,1}_u} \, dW^{2,1}_u, \\
X^2_{t_{i+1}} &= X^2_{t_i} + r_2(t_{i+1} - t_i) + \frac{\varrho_2}{\sigma_2}\big(\Sigma^{2,2}_{t_{i+1}} - \Sigma^{2,2}_{t_i} - a_2(t_{i+1} - t_i)\big) \\
&\quad + \left(\frac{\varrho_2 b_2}{\sigma_2} - \frac{1}{2}\right) \int_{t_i}^{t_{i+1}} \Sigma^{2,2}_u \, du + \sqrt{1 - \varrho_2^2} \int_{t_i}^{t_{i+1}} \sqrt{\Sigma^{2,2}_u} \, dW^{2,2}_u.
\end{aligned}
$$

Hence we approximate $\int_{t_i}^{t_{i+1}} \Sigma^{j,j}_u \, du$, $j = 1, 2$, using e.g. the trapezoidal rule. We recall that given

$$
\Sigma^{j,j}_{t_{i+1}}, \Sigma^{j,j}_{t_i}, \int_{t_i}^{t_{i+1}} \Sigma^{j,j}_u \, du, X^j_{t_i}, \quad j = 1, 2,
$$

the random variables $X^j_{t_{i+1}}$, $j = 1, 2$, are conditionally Gaussian with mean

$$
\begin{aligned}
&X^j_{t_i} + r_j(t_{i+1} - t_i) + \frac{\varrho_j}{\sigma_j}\big(\Sigma^{j,j}_{t_{i+1}} - \Sigma^{j,j}_{t_i} - a_j(t_{i+1} - t_i)\big) \\
&\quad + \left(\frac{\varrho_j b_j}{\sigma_j} - \frac{1}{2}\right) \int_{t_i}^{t_{i+1}} \Sigma^{j,j}_u \, du
\end{aligned}
$$

and variance

$$
\left(1 - \varrho_j^2\right) \int_{t_i}^{t_{i+1}} \Sigma^{j,j}_u \, du.
$$

Lastly, if $d[W^{2,1}, W^{2,2}]_t = \rho \, dt$, then the covariance between $X^1_{t_{i+1}}$ and $X^2_{t_{i+1}}$, conditional on $\Sigma^{j,j}_{t_{i+1}}, \Sigma^{j,j}_{t_i}, \int_{t_i}^{t_{i+1}} \Sigma^{j,j}_u \, du, X^j_{t_i}$, $j = 1, 2$, is

$$
\rho \sqrt{1 - \varrho_1^2} \sqrt{1 - \varrho_2^2} \int_{t_i}^{t_{i+1}} \sqrt{\Sigma^{1,1}_u} \sqrt{\Sigma^{2,2}_u} \, du.
$$

Concluding the chapter we mention that in Chap. 11 we will introduce another Heston model based on the Wishart process.

Chapter 7
Affine Diffusion Processes on the Euclidean Space

Affine processes have been applied to problems in finance due to their ability to capture some stylized facts of financial time series, but also due to their computational tractability. They are characterized by the fact that their associated characteristic function is exponentially affine in the state variables. Consequently, this characteristic function is also referred to as the affine transform.

In this book, we present two approaches to affine processes: the first, presented in this chapter, is concerned with admissibility, which studies whether the affine transform is well-defined. References dealing with this question include Duffie and Kan (1996), Duffie et al. (2003), Filipović and Mayerhofer (2009), and Cuchiero et al. (2011). The second approach is referred to as the Grasselli-Tebaldi approach, see Grasselli and Tebaldi (2008). It studies the question of solvability, i.e. whether the affine transform is analytically tractable. Of course, solvability is a stronger condition than admissibility, since in order to compute the affine transform, it has to exist. We emphasize that the question of solvability is closer related to the spirit of this book than admissibility. However, for completeness, we discuss both approaches.

We alert the reader to the fact that traditionally, affine processes were studied on the Euclidean state space and positive factors take values in $(\Re^+)^m$, see e.g. Filipović and Mayerhofer (2009). In Gouriéroux and Sufana (2004b), it was noted that when studying matrix-valued processes, positive factors can take values as elements in positive semidefinite matrices. This allows one to model more general dependence structures, which we will explore in Chap. 11. Finally, we remark that the domain where factors take values was generalized to symmetric cones in Grasselli and Tebaldi (2008).

We proceed as follows: in the current chapter and Chap. 8, we discuss admissibility and in Chap. 9 the solvability of affine processes, taking values in the Euclidean state space. Later in Chap. 11, we will discuss the solvability of certain matrix-valued processes, the Wishart processes.

In the first, more theoretical part of the current chapter, we provide the definition of affine processes on the Euclidean space and discuss the admissibility of affine processes, i.e. their existence and uniqueness. In particular, we discuss necessary and sufficient conditions to guarantee the existence of the affine transform. Given

J. Baldeaux, E. Platen, *Functionals of Multidimensional Diffusions with Applications to Finance*, Bocconi & Springer Series 5, DOI 10.1007/978-3-319-00747-2_7,

this theoretical background, we will then apply affine processes to problems arising in mathematical finance. In Sect. 7.2, we present recipes, which can be used to arrive at a pricing formula in the classical risk-neutral setting. Finally, in Sect. 7.3, we discuss pricing using affine diffusions under the benchmark approach. In Chap. 8, the recipes presented in this chapter, for the classical risk-neutral setting and the benchmark approach, are applied to concrete examples from finance.

7.1 Theoretical Background on Affine Diffusions

In this section, we recall parts of the theory of affine processes. The discussion is based on Filipović and Mayerhofer (2009). For further details, the interested reader is referred also to Filipović (2009).

7.1.1 Definition of Affine Processes

We start our discussion of *affine processes* on a general Euclidean state space and subsequently consider a specific state space, referred to as *canonical state space* in Filipović and Mayerhofer (2009). For a given dimension $d \geq 1$ and a closed state space $\mathcal{X} \subset \Re^d$ with non-empty interior, we consider for a process $\boldsymbol{X} = \boldsymbol{X}^{\boldsymbol{x}} = \{\boldsymbol{X}_t,\, t \in [0, T]\}$ the following SDE

$$d\boldsymbol{X}_t = \boldsymbol{b}(\boldsymbol{X}_t)\, dt + \boldsymbol{\rho}(\boldsymbol{X}_t)\, d\boldsymbol{W}_t, \tag{7.1.1}$$

$\boldsymbol{X}_0 = \boldsymbol{x}$, where $\boldsymbol{b} : \mathcal{X} \to \Re^d$ is assumed to be continuous. Here, $\boldsymbol{\rho} : \mathcal{X} \to \Re^{d \times d}$ is assumed to be measurable so that the diffusion matrix

$$\boldsymbol{a}(\boldsymbol{x}) = \boldsymbol{\rho}(\boldsymbol{x})\boldsymbol{\rho}(\boldsymbol{x})^\top \tag{7.1.2}$$

is continuous in $\boldsymbol{x} \in \mathcal{X}$. Furthermore, $\boldsymbol{W}$ denotes a d-dimensional Brownian motion defined on a filtered probability space $(\Omega, \mathcal{A}, \underline{\mathcal{A}}, P)$, and we assume throughout this chapter that for each $\boldsymbol{x} \in \mathcal{X}$, there exists a unique solution $\boldsymbol{X} = \boldsymbol{X}^{\boldsymbol{x}}$ to (7.1.1). We denote by $\mathbf{C}$ the set of complex numbers and by $\imath$ the imaginary unit.

We now define an affine process.

Definition 7.1.1 The process $\boldsymbol{X}$ is affine, if the $\mathcal{A}_t$-conditional characteristic function of $\boldsymbol{X}_T$ is exponential affine in $\boldsymbol{X}_t$, for all $t \leq T$. This means that there exist $\mathbf{C}$- and $\mathbf{C}^d$-valued functions $\phi(t, \boldsymbol{u})$ and $\boldsymbol{\psi}(t, \boldsymbol{u})$, respectively, with jointly continuous t-derivatives such that $\boldsymbol{X} = \boldsymbol{X}^{\boldsymbol{x}}$ satisfies the conditional characteristic function

$$E\big(e^{\boldsymbol{u}^\top \boldsymbol{X}_T} \bigm| \mathcal{A}_t\big) = e^{\phi(T-t,\boldsymbol{u}) + \boldsymbol{\psi}(T-t,\boldsymbol{u})^\top \boldsymbol{X}_t} \tag{7.1.3}$$

for all $\boldsymbol{u} \in \imath\Re^d$, $t \leq T$ and $\boldsymbol{x} \in \mathcal{X}$.

The functions ϕ and $\boldsymbol{\psi}$ will play a crucial role in the further discussion. For now, we remind the reader that since $E(e^{\boldsymbol{u}^\top \boldsymbol{X}_T} \mid \mathcal{A}_t)$ is bounded by one, for $\boldsymbol{u} \in \imath\Re^d$, the real part of the exponent $\phi(T-t, \boldsymbol{u}) + \boldsymbol{\psi}(T-t, \boldsymbol{u})^\top \boldsymbol{X}_t$ in (7.1.3) has to be negative. We remark that $\phi(t, \boldsymbol{u})$ and $\boldsymbol{\psi}(t, \boldsymbol{u})$ for $t \geq 0$ and $\boldsymbol{u} \in \imath\Re^d$ are uniquely determined by (7.1.3), and satisfy, in particular, the initial conditions $\phi(0, \boldsymbol{u}) = 0$ and $\boldsymbol{\psi}(0, \boldsymbol{u}) = \boldsymbol{u}$. In the next theorem, we present necessary and sufficient conditions on the diffusion matrix $\boldsymbol{a}(\boldsymbol{x})$ and the drift $\boldsymbol{b}(\boldsymbol{x})$ for $\boldsymbol{X}$ to be affine. We refer the reader for the proof to Theorem 2.2 in Filipović and Mayerhofer (2009).

Theorem 7.1.2 *Suppose $\boldsymbol{X}$ is affine. Then the diffusion matrix $\boldsymbol{a}(\boldsymbol{x})$ and drift $\boldsymbol{b}(\boldsymbol{x})$ are affine in $\boldsymbol{x}$, i.e.*

$$\begin{aligned} \boldsymbol{a}(\boldsymbol{x}) &= \boldsymbol{a} + \sum_{i=1}^d x_i \boldsymbol{\alpha}_i \\ \boldsymbol{b}(\boldsymbol{x}) &= \boldsymbol{b} + \sum_{i=1}^d x_i \boldsymbol{\beta}_i = \boldsymbol{b} + \boldsymbol{B}\boldsymbol{x} \end{aligned} \tag{7.1.4}$$

for some $d \times d$-matrices $\boldsymbol{a}$ and $\boldsymbol{\alpha}_i$, and d-vectors $\boldsymbol{b}$ and $\boldsymbol{\beta}_i$, where we denote by

$$\boldsymbol{B} = (\boldsymbol{\beta}_1, \ldots, \boldsymbol{\beta}_d)$$

the $d \times d$-matrix with i-th column vector $\boldsymbol{\beta}_i$, $1 \leq i \leq d$. Moreover, ϕ and $\boldsymbol{\psi} = (\psi_1, \ldots, \psi_d)^\top$ solve the system of Riccati equations

$$\begin{aligned} \partial_t \phi(t, \boldsymbol{u}) &= \frac{1}{2} \boldsymbol{\psi}(t, \boldsymbol{u})^\top \boldsymbol{a} \boldsymbol{\psi}(t, \boldsymbol{u}) + \boldsymbol{b}^\top \boldsymbol{\psi}(t, \boldsymbol{u}) \\ \phi(0, \boldsymbol{u}) &= 0 \\ \partial_t \psi_i(t, \boldsymbol{u}) &= \frac{1}{2} \boldsymbol{\psi}(t, \boldsymbol{u})^\top \boldsymbol{\alpha}_i \boldsymbol{\psi}(t, \boldsymbol{u}) + \boldsymbol{\beta}_i^\top \boldsymbol{\psi}(t, \boldsymbol{u}), \quad 1 \leq i \leq d \\ \psi(0, \boldsymbol{u}) &= \boldsymbol{u}. \end{aligned} \tag{7.1.5}$$

In particular, ϕ is determined by $\boldsymbol{\psi}$ via simple integration:

$$\phi(t, \boldsymbol{u}) = \int_0^t \left(\frac{1}{2} \boldsymbol{\psi}(s, \boldsymbol{u})^\top \boldsymbol{a} \boldsymbol{\psi}(s, \boldsymbol{u}) + \boldsymbol{b}^\top \boldsymbol{\psi}(s, \boldsymbol{u}) \right) ds.$$

Conversely, suppose the diffusion matrix $\boldsymbol{a}(\boldsymbol{x})$ and drift $\boldsymbol{b}(\boldsymbol{x})$ are affine of the form (7.1.4) *and suppose there exists a solution $(\phi, \boldsymbol{\psi})$ of the Riccati equations* (7.1.5) *such that $\phi(t, \boldsymbol{u}) + \boldsymbol{\psi}(t, \boldsymbol{u})^\top \boldsymbol{x}$ has negative real part for all $t \geq 0$, $\boldsymbol{u} \in \imath\Re^d$ and $\boldsymbol{x} \in \mathcal{X}$. Then $\boldsymbol{X}$ is affine with conditional characteristic function* (7.1.3).

As mentioned above, the functions ϕ and $\boldsymbol{\psi}$ play a crucial role in the study of affine processes. To improve our understanding of them, we provide the following lemma, see also Filipović (2009), Lemma 10.1. However, as we subsequently fix a state space, its statement can be sharpened, see the forthcoming Theorem 7.1.5.

Lemma 7.1.3 *Let $\boldsymbol{a}$ and $\boldsymbol{\alpha}_i$ be real $d \times d$-matrices, and $\boldsymbol{b}$ and $\boldsymbol{\beta}_i$ be real d-vectors, $1 \le i \le d$. The letter K can represent either $\Re$ or $\mathbf{C}$.*

- *For every $\boldsymbol{u} \in K^d$, there exists some $t_+(\boldsymbol{u}) \in (0, \infty]$ such that there exists a unique solution $(\phi(., \boldsymbol{u}), \boldsymbol{\psi}(., \boldsymbol{u})) : [0, t_+(\boldsymbol{u})) \to K \times K^d$ of the Riccati equations (7.1.5). In particular, $t_+(\mathbf{0}) = \infty$.*
- *The domain*

$$\mathcal{D}_K = \left\{(t, \boldsymbol{u}) \in \Re^+ \times K^d \,\middle|\, t < t_+(\boldsymbol{u})\right\}$$

is open in $\Re^+ \times K^d$ and maximal in the sense that for all $\boldsymbol{u} \in K^d$ either $t_+(\boldsymbol{u}) = \infty$ or $\lim_{t \uparrow t_+(\boldsymbol{u})} \|\boldsymbol{\psi}(t, \boldsymbol{u})\| = \infty$, respectively.
- *For every $t \in \Re^+$, the t-section*

$$\mathcal{D}_K(t) = \left\{\boldsymbol{u} \in K^d \,\middle|\, (t, \boldsymbol{u}) \in \mathcal{D}_K\right\}$$

is an open neighborhood of $\mathbf{0}$ in K^d. Moreover, $\mathcal{D}_K(0) = K^d$ and $\mathcal{D}_K(t_1) \supseteq \mathcal{D}_K(t_2)$ for $0 \le t_1 \le t_2$.
- *ϕ and $\boldsymbol{\psi}$ are analytic functions on $\mathcal{D}_K$.*
- *$\mathcal{D}_\Re = \mathcal{D}_\mathbf{C} \cap (\Re^+ \times \Re^d)$.*

We shall call $\mathcal{D}_K$ the maximal domain for the Riccati equation.

Regarding the conditions in Theorem 7.1.2, we note the following interplay between the parameters $\boldsymbol{a}, \boldsymbol{\alpha}_i, \boldsymbol{b}, \boldsymbol{\beta}_i$ in (7.1.4) and the state space $\mathcal{X}$:

- $\boldsymbol{a}, \boldsymbol{\alpha}_i, \boldsymbol{b}, \boldsymbol{\beta}_i$ must be such that $\boldsymbol{X}$ does not leave the set $\mathcal{X}$, and
- $\boldsymbol{a}, \boldsymbol{\alpha}_i$ must be such that $\boldsymbol{a} + \sum_i^d x_i \boldsymbol{\alpha}_i$ is symmetric and positive semi-definite for all $\boldsymbol{x} \in \mathcal{X}$.

Following Filipović and Mayerhofer (2009), we now assume that the state space under consideration is of the form

$$\mathcal{X} = \left(\Re^+\right)^m \times \Re^n \tag{7.1.6}$$

for some integers $m, n \ge 0$ with $m + n = d$. This type of state space is sufficient for the applications we discuss in this chapter and the following chapter. For a discussion on this issue and references to other state spaces, we refer the reader to Remark 3.1 in Filipović and Mayerhofer (2009) and Chaps. 10 and 11 of this book.

For the state space given in (7.1.6), we now present necessary and sufficient conditions on the functions $a(\boldsymbol{x})$ and $b(\boldsymbol{x})$ for $\boldsymbol{X}$ to be affine. Following Filipović and Mayerhofer (2009), we find the following notation helpful: we consider the index sets:

$$I = \{1, \ldots, m\} \quad \text{and} \quad J = \{m + 1, \ldots, m + n\}.$$

For any vector $\boldsymbol{\mu}$, matrix $\boldsymbol{\nu}$, and index sets M, N, we define by

$$\boldsymbol{\mu}_M = (\mu_i)_{i \in M}, \qquad \boldsymbol{\nu}_{MN} = (\nu_{ij})_{i \in M, j \in N}$$

the respective sub-vector and sub-matrix. The following result represents Theorem 3.2 in Filipović and Mayerhofer (2009).

Theorem 7.1.4 *The process $\boldsymbol{X}$ on the canonical state space $(\Re^+)^m \times \Re^n$ is affine if and only if $\boldsymbol{a}(\boldsymbol{x})$ and $\boldsymbol{b}(\boldsymbol{x})$ are of the form* (7.1.4) *for parameters $\boldsymbol{a}$, $\boldsymbol{\alpha}_i$, $\boldsymbol{b}$, $\boldsymbol{\beta}_i$, which are admissible in the following sense*:

$$
\begin{aligned}
&\boldsymbol{a}, \boldsymbol{\alpha}_i \text{ are symmetric positive semidefinite},\\
\boldsymbol{a}_{II} &= 0 \quad \left(\text{and thus } \boldsymbol{a}_{IJ} = \boldsymbol{a}_{JI}^\top = 0\right),\\
\boldsymbol{\alpha}_j &= 0 \quad \text{for all } j \in J,\\
\alpha_{i,kl} = \alpha_{i,lk} &= 0 \quad \text{for } k \in I \setminus \{i\}, \text{ for all } 1 \le i, l \le d,\\
\boldsymbol{b} &\in \left(\Re^+\right)^m \times \Re^n,\\
\boldsymbol{B}_{IJ} &= \boldsymbol{0},\\
&\boldsymbol{B}_{II} \text{ has nonnegative off-diagonal elements.}
\end{aligned}
\tag{7.1.7}
$$

In this case, the corresponding system of Riccati equations (7.1.5) *simplifies to*

$$
\begin{aligned}
\partial_t \phi(t, \boldsymbol{u}) &= \frac{1}{2} \boldsymbol{\psi}_J(t, \boldsymbol{u})^\top \boldsymbol{a}_{JJ} \boldsymbol{\psi}_J(t, \boldsymbol{u}) + \boldsymbol{b}^\top \boldsymbol{\psi}(t, \boldsymbol{u}),\\
\phi(0, \boldsymbol{u}) &= 0,\\
\partial_t \psi_i(t, \boldsymbol{u}) &= \frac{1}{2} \boldsymbol{\psi}(t, \boldsymbol{u})^\top \boldsymbol{\alpha}_i \boldsymbol{\psi}(t, \boldsymbol{u}) + \boldsymbol{\beta}_i^\top \boldsymbol{\psi}(t, \boldsymbol{u}), \quad i \in I,\\
\partial_t \boldsymbol{\psi}_J(t, \boldsymbol{u}) &= \boldsymbol{B}_{JJ}^\top \boldsymbol{\psi}_J(t, \boldsymbol{u}),\\
\psi(0, \boldsymbol{u}) &= \boldsymbol{u},
\end{aligned}
\tag{7.1.8}
$$

and there exists a unique global solution $(\phi(., \boldsymbol{u}), \boldsymbol{\psi}(., \boldsymbol{u})) : \Re^+ \to \mathbf{C}^- \times (\mathbf{C}^-)^m \times \imath\Re^n$ for all initial values $\boldsymbol{u} \in (\mathbf{C}^-)^m \times \imath\Re^n$. In particular, the equation for $\boldsymbol{\psi}_J$ forms an autonomous linear system with unique global solution $\boldsymbol{\psi}_J(t, \boldsymbol{u}) = \exp\{\boldsymbol{B}_{JJ}^\top t\}\boldsymbol{u}_J$ for all $\boldsymbol{u}_J \in \mathbf{C}^n$.

We point out that the admissibility conditions (7.1.7) are well illustrated in Filipović and Mayerhofer (2009). We now go back to Definition 7.1.1 and examine it in the light of the canonical state space. In particular, we wish to extend both, the definition of the functions ϕ and $\boldsymbol{\psi}$ beyond $\boldsymbol{u} \in \imath\Re^d$, but also the validity of the affine transform. The next result, Theorem 3.3 in Filipović and Mayerhofer (2009), shows that this is possible.

Theorem 7.1.5 *Suppose $\boldsymbol{X}$ is affine with admissible parameters as given in* (7.1.7). *Let $\tau > 0$. Then*:

- $\mathcal{S}(\mathcal{D}_\Re(\tau)) \subset \mathcal{D}_{\mathbf{C}}(\tau)$;
- $\mathcal{D}_\Re(\tau) = M(\tau)$ *where*

$$
M(\tau) = \left\{\boldsymbol{u} \in \Re^d \;\middle|\; E\left(e^{\boldsymbol{u}^\top \boldsymbol{X}^{\boldsymbol{x}}(\tau)}\right) < \infty \text{ for all } \boldsymbol{x} \in \left(\Re^+\right)^m \times \Re^n\right\};
$$

- $\mathcal{D}_\Re(\tau)$ *and* $\mathcal{D}_\Re$ *are convex sets; moreover, for all* $0 \le t \le T$ *and* $\boldsymbol{x} \in (\Re^+)^m \times \Re^n$;
- (7.1.3) *holds for all* $\boldsymbol{u} \in \mathcal{S}(\mathcal{D}_\Re(T - t))$;
- (7.1.3) *holds for all* $\boldsymbol{u} \in (\mathbf{C}^-)^m \times \imath\Re^n$;
- $M(t) \supseteq M(T)$.

It is clear that extending the validity of the affine transform is a crucial result for applications, in particular, for option pricing. Before discussing applications, we finally address the following technical issue: we concentrate on an affine process $\boldsymbol{X}$ on the canonical state space $(\Re^+)^m \times \Re^n$ with admissible parameters $\boldsymbol{a}, \boldsymbol{\alpha}_i, \boldsymbol{b}$, and $\boldsymbol{\beta}_i$, which means that for any $\boldsymbol{x} \in (\Re^+)^m \times \Re^n$, the process $\boldsymbol{X} = \boldsymbol{X}^{\boldsymbol{x}}$ satisfies

$$d\boldsymbol{X}_t = (\boldsymbol{b} + \boldsymbol{B}\boldsymbol{X}_t)\,dt + \boldsymbol{\rho}(\boldsymbol{X}_t)\,d\boldsymbol{W}_t, \tag{7.1.9}$$

$X_0 = \boldsymbol{x}$ and $\boldsymbol{\rho}(\boldsymbol{x})\boldsymbol{\rho}(\boldsymbol{x})^\top = \boldsymbol{a} + \sum_{i \in I} x_i \boldsymbol{\alpha}_i$. So far, the entire discussion in this chapter was based on the premise that there exists a unique solution $\boldsymbol{X} = \boldsymbol{X}^{\boldsymbol{x}}$ to (7.1.9). However, if $\boldsymbol{a}(\boldsymbol{x}) = \boldsymbol{\rho}(\boldsymbol{x})\boldsymbol{\rho}(\boldsymbol{x})^\top$ is affine, then $\boldsymbol{\rho}(\boldsymbol{x})$ cannot be Lipschitz continuous in $\boldsymbol{x}$, in general, which means that the existence and uniqueness of a solution to (7.1.9) has to be investigated. Following Sect. 8 in Filipović and Mayerhofer (2009), we show how $\boldsymbol{X}$ can always be realized as a unique solution of the SDE (7.1.9) on the canonical state space $\mathcal{X} = (\Re^+)^m \times \Re^n$ and for particular choices of $\boldsymbol{\rho}(\boldsymbol{x})$.

We now proceed as follows: we illustrate that even though the law of an affine process is uniquely determined by its characteristics, it can be realized by infinitely many variants of the SDE, by replacing $\boldsymbol{\rho}(\boldsymbol{x})$ in (7.1.9) by $\boldsymbol{\rho}(\boldsymbol{x})\boldsymbol{D}$, for any orthogonal matrix $\boldsymbol{D}$. Consequently, we present an algorithm, which produces a canonical choice of $\boldsymbol{\rho}(\boldsymbol{x})$, and establishes that the resulting SDE admits a unique solution, the law of which is uniquely determined by $\boldsymbol{a}, \boldsymbol{\alpha}_i, \boldsymbol{b}, \boldsymbol{\beta}_i$, but is independent of the choice of $\boldsymbol{\rho}$.

Firstly, we show that the law of an SDE is uniquely determined by $\boldsymbol{a}$, $\boldsymbol{\alpha}_i$, $\boldsymbol{b}$, $\boldsymbol{\beta}_i$, but can be realized via an infinite number of SDEs. Firstly, note that for any orthogonal $d \times d$-matrix $\boldsymbol{D}$, the function $\boldsymbol{\rho}(\boldsymbol{x})\boldsymbol{D}$ results in the same diffusion matrix as $\boldsymbol{\rho}(\boldsymbol{x})$, since $\boldsymbol{\rho}(\boldsymbol{x})\boldsymbol{D}\boldsymbol{D}^\top\boldsymbol{\rho}(\boldsymbol{x})^\top = \boldsymbol{\rho}(\boldsymbol{x})\boldsymbol{\rho}(\boldsymbol{x})^\top$. However, from Theorem 7.1.4, it is known that given admissible parameters $\boldsymbol{a}$, $\boldsymbol{\alpha}_i$,$\boldsymbol{b}$, $\boldsymbol{\beta}_i$, the functions $(\phi(\cdot,\boldsymbol{u}), \boldsymbol{\psi}(\cdot,\boldsymbol{u})) : \Re^+ \to \mathbf{C}^- \times (\mathbf{C}^-)^m \times \imath\Re^n$ are uniquely determined as solutions of the Riccati equations (7.1.8), for all $\boldsymbol{u} \in (\mathbf{C}^-)^m \times \imath\Re^n$. These, in turn, uniquely determine the law of the process $\boldsymbol{X}$. Indeed, for any $0 \le t_1 < t_2$ and $\boldsymbol{u}_1, \boldsymbol{u}_2 \in (\mathbf{C}^-)^m \times \imath\Re^n$, we iterate the affine transform (7.1.3) to obtain

$$\begin{aligned} &E\left(e^{\boldsymbol{u}_1^\top \boldsymbol{X}_{t_1} + \boldsymbol{u}_2^\top \boldsymbol{X}_{t_2}}\right) \\ &\quad = e^{\phi(t_2-t_1,\boldsymbol{u}_2)+\phi(t_1,\boldsymbol{u}_1+\boldsymbol{\psi}(t_2-t_1,\boldsymbol{u}_2))+\boldsymbol{\psi}(t_1,\boldsymbol{u}_1+\boldsymbol{\psi}(t_2-t_1,\boldsymbol{u}_2))^\top \boldsymbol{x}}. \end{aligned}$$

We conclude that the joint distribution of $(\boldsymbol{X}_{t_1}, \boldsymbol{X}_{t_2})$ is uniquely determined by the functions ϕ and $\boldsymbol{\psi}$. Iterating this argument, one concludes that every finite-dimensional distribution, and thus the law of $\boldsymbol{X}$, is uniquely determined by the parameters $\boldsymbol{a}$, $\boldsymbol{\alpha}_i$, $\boldsymbol{b}$, and $\boldsymbol{\beta}_i$. Consequently, the law of an affine process $\boldsymbol{X}$, while uniquely determined by its characteristics (7.1.4), can be realized by infinitely many variants of the SDE (7.1.9), by replacing $\boldsymbol{\rho}(\boldsymbol{x})$ by $\boldsymbol{\rho}(\boldsymbol{x})\boldsymbol{D}$, for an orthogonal $d \times d$ matrix $\boldsymbol{D}$. We now recall the canonical choice of $\boldsymbol{\rho}(\boldsymbol{x})$ presented in Filipović and Mayerhofer (2009):

- from Lemma 7.1 in Filipović and Mayerhofer (2009), it follows that every affine process $\boldsymbol{X}$ on $(\Re^+)^m \times \Re^n$ can be written as $\boldsymbol{X} = \boldsymbol{\Lambda}^{-1}\boldsymbol{Y}$ for some invertible

$d \times d$ matrix $\boldsymbol{\Lambda}$ and some affine process $\boldsymbol{Y}$ on $(\Re^+)^m \times \Re^n$ with block-diagonal diffusion matrix. Hence one can focus on $\rho(\boldsymbol{x})$ for which

$$\boldsymbol{\rho}(\boldsymbol{x})\boldsymbol{\rho}(\boldsymbol{x})^\top = \begin{pmatrix} \operatorname{diag}(x_1, \dots, x_q, 0, \dots, 0) & \mathbf{0} \\ \mathbf{0} & \boldsymbol{a} + \sum_{i \in I} x_i \boldsymbol{\alpha}_{i,JJ} \end{pmatrix}$$

for some integer $0 \leq q \leq m$. Since $0 \leq q \leq m$, $\boldsymbol{\rho}(\boldsymbol{x}) \equiv \boldsymbol{\rho}(\boldsymbol{x}_I)$ is a function of $\boldsymbol{x}_I$ only.

- set $\boldsymbol{\rho}_{IJ}(\boldsymbol{x}) \equiv \mathbf{0}$, $\boldsymbol{\rho}_{JI}(\boldsymbol{x}) \equiv \mathbf{0}$, and

$$\boldsymbol{\rho}_{II}(\boldsymbol{x}_I) = \operatorname{diag}(\sqrt{x_1}, \dots, \sqrt{x_q}, 0, \dots, 0).$$

Choose for $\boldsymbol{\rho}_{JJ}(\boldsymbol{x}_I)$ any measurable $n \times n$-matrix-valued function satisfying

$$\boldsymbol{\rho}_{JJ}(\boldsymbol{x}_I)\boldsymbol{\rho}_{JJ}(\boldsymbol{x}_I)^\top = \boldsymbol{a} + \sum_{i \in I} x_i \boldsymbol{\alpha}_{i,JJ},$$

see Sect. 8 in Filipović and Mayerhofer (2009) for a discussion on how to choose such a function.

- consequently, the SDE (7.1.9) now reads

$$\begin{aligned} d\boldsymbol{X}_I &= (\boldsymbol{b}_I + \boldsymbol{B}_{II}\boldsymbol{X}_I)\,dt + \boldsymbol{\rho}_{II}(\boldsymbol{X}_I)\,d\boldsymbol{W}_I, \\ d\boldsymbol{X}_J &= (\boldsymbol{b}_J + \boldsymbol{B}_{JI}\boldsymbol{X}_I + \boldsymbol{B}_{JJ}\boldsymbol{X}_J)\,dt + \boldsymbol{\rho}_{JJ}(\boldsymbol{X}_I)\,d\boldsymbol{W}_J, \\ \boldsymbol{X}_0 &= \boldsymbol{x}. \end{aligned}$$

Lemma 8.2 in Filipović and Mayerhofer (2009) asserts the existence and uniqueness of an $(\Re^+)^m \times \Re^n$-valued weak solution $\boldsymbol{X} = \boldsymbol{X}^{\boldsymbol{x}}$ for any $\boldsymbol{x} \in (\Re^+)^m \times \Re^n$.

We conclude this theoretical part with the following result, which presents Theorem 8.1 in Filipović and Mayerhofer (2009).

Theorem 7.1.6 *Let $\boldsymbol{a}$, $\boldsymbol{\alpha}_i$, $\boldsymbol{b}$, $\boldsymbol{\beta}_i$ be admissible parameters. Then there exists a measurable function $\boldsymbol{\rho} : (\Re^+)^m \times \Re^n \to \Re^{d \times d}$ with $\boldsymbol{\rho}(\boldsymbol{x})\boldsymbol{\rho}(\boldsymbol{x})^\top = \boldsymbol{a} + \sum_{i \in I} x_i \boldsymbol{\alpha}_i$, and such that, for any $\boldsymbol{x} \in (\Re^+)^m \times \Re^n$, there exists a unique $(\Re^+)^m \times \Re^n$-valued solution $\boldsymbol{X} = \boldsymbol{X}^{\boldsymbol{x}}$ of* (7.1.9). *Moreover, the law of $\boldsymbol{X}$ is uniquely determined by $\boldsymbol{a}$, $\boldsymbol{\alpha}_i$, $\boldsymbol{b}$, $\boldsymbol{\beta}_i$ and does not depend on the particular choice of $\boldsymbol{\rho}$.*

Finally, we remark that the above discussion is similar to the forthcoming discussion in Sect. 9.3, which will show how to represent in normal form the affine processes discussed in Dai and Singleton (2000).

7.2 Pricing Using Affine Diffusions

In this section, we present classical approaches to pricing in affine models, following Filipović and Mayerhofer (2009). Here the expectation can be interpreted as being taken under some assumed risk neutral probability measure.

7.2.1 Classical Approaches to Pricing in Affine Models

We firstly present, following Filipović and Mayerhofer (2009), two classical approaches to pricing in affine models. The following assumption sits at the heart of the classical approach.

Assumption 7.2.1 *The process $(r_t)_{t\geq 0}$ is an affine transform of $(\boldsymbol{X}_t)_{t\geq 0}$,*

$$r_t = c + \boldsymbol{\gamma}^\top \boldsymbol{X}_t,$$

where $\boldsymbol{X}$ is affine on the canonical state space $(\Re^+)^m \times \Re^n$ with admissible parameters $\boldsymbol{a}, \boldsymbol{\alpha}_i, \boldsymbol{b}, \boldsymbol{\beta}_i$, where $i \in \{1, \dots, d\}$, given in Eq. (7.1.7) and $c \in \Re$, $\boldsymbol{\gamma} \in \Re^d$.

We are interested in computing expectations of the form

$$\pi(t) = E\big(e^{-\int_t^T r_s ds} f(\boldsymbol{X}_T) \,\big|\, \mathcal{A}_t\big),$$

where we assume that f is such that

$$E\big(e^{-\int_0^T r_s d_s} \big|f(\boldsymbol{X}_T)\big|\big) < \infty.$$

Both classical approaches are based on the following result, see Theorem 4.1 in Filipović and Mayerhofer (2009).

Theorem 7.2.2 *Let $\tau > 0$. The following statements are equivalent:*

- $E(e^{-\int_0^\tau r_s ds}) < \infty$ *for all $\boldsymbol{x} \in (\Re^+)^m \times \Re^n$, where $\boldsymbol{X}_0 = \boldsymbol{x}$.*
- *There exists a unique solution $(\Phi(\cdot, \boldsymbol{u}), \boldsymbol{\Psi}(\cdot, \boldsymbol{u})) : [0, \tau] \to \mathbf{C} \times \mathbf{C}^d$ of*

$$\begin{aligned}
\partial_t \Phi(t, \boldsymbol{u}) &= \frac{1}{2} \boldsymbol{\Psi}_J(t, \boldsymbol{u})^\top \boldsymbol{a}_{JJ} \boldsymbol{\Psi}_J(t, \boldsymbol{u}) + \boldsymbol{b}^\top \boldsymbol{\Psi}(t, \boldsymbol{u}) - c, \\
\Phi(0, \boldsymbol{u}) &= 0, \\
\partial_t \boldsymbol{\Psi}_i(t, \boldsymbol{u}) &= \frac{1}{2} \boldsymbol{\Psi}(t, \boldsymbol{u})^\top \boldsymbol{\alpha}_i \boldsymbol{\Psi}(t, \boldsymbol{u}) + \boldsymbol{\beta}_i^\top \boldsymbol{\Psi}(t, \boldsymbol{u}) - \gamma_i, \quad i \in I \\
\partial_t \boldsymbol{\Psi}_J(t, \boldsymbol{u}) &= \boldsymbol{B}_{JJ}^\top \boldsymbol{\Psi}_J(t, \boldsymbol{u}) - \boldsymbol{\gamma}_J, \\
\Psi(0, \boldsymbol{u}) &= \boldsymbol{u},
\end{aligned} \tag{7.2.1}$$

for $\boldsymbol{u} = 0$.

In either case, there exists an open convex neighborhood U of $\mathbf{0}$ in $\Re^d$ such that the system of Riccati equations (7.2.1) admits a unique solution $(\Phi(\cdot, \boldsymbol{u}), \boldsymbol{\Psi}(\cdot, \boldsymbol{u}))$: $[0, \tau] \to \mathbf{C} \times \mathbf{C}^d$ for all $\boldsymbol{u} \in \mathcal{S}(U)$, and we have

$$E\big(e^{-\int_t^T r_s ds} e^{\boldsymbol{u}^\top \boldsymbol{X}_T} \,\big|\, \mathcal{A}_t\big) = e^{\Phi(T-t, \boldsymbol{u}) + \boldsymbol{\Psi}(T-t, \boldsymbol{u})^\top \boldsymbol{X}_t},$$

for all $\boldsymbol{u} \in \mathcal{S}(U)$, $t \leq T \leq t + \tau$ and $\boldsymbol{x} \in (\Re^+)^m \times \Re^n$.

For the remainder of this section, we assume that one of the conditions of Theorem 7.2.2 is satisfied. In practice, given an affine process, we would firstly have to solve the Riccati equations to make use of Theorem 7.2.2. There are two approaches we can pursue:

7.2.2 Forward-Measure Approach

We define the Radon-Nikodym derivative

$$\Lambda_F = \frac{dP^T}{dP} = \frac{1}{P_T(0) S_T^0},$$

where $P_T(t) = E(e^{-\int_t^T r_s\,ds} \mid \mathcal{A}_t)$ and $S_T^0 = \exp\{\int_0^T r_s\,ds\}$. Here P is interpreted as the assumed risk neutral probability measure and P^T as the T-forward measure with the zero coupon bond $P_T(t)$ as numéraire. We also set

$$\Lambda_F(t) = \left.\frac{dP^T}{dP}\right|_{\mathcal{A}_t} = E(\Lambda_F | \mathcal{A}_t)$$

and note that $\Lambda_F(t)$ is a strictly positive $(\mathcal{A}, P)$-martingale. From Bayes's Theorem, see Sect. 15.8, one has

$$E_{P^T}\big(f(\boldsymbol{X}_T) \mid \mathcal{A}_t\big) = \frac{E(f(\boldsymbol{X}_T) e^{-\int_t^T r_s\,ds} \mid \mathcal{A}_t)}{P_T(t)}.$$

Up to normalization with $E(e^{-\int_t^T r_s\,ds} \mid \mathcal{A}_t)$, calculating

$$\pi(t) = E\big(e^{-\int_t^T r_s\,ds} f(\boldsymbol{X}_T) \mid \mathcal{A}_t\big)$$

amounts to computing $E_{P^T}(f(\boldsymbol{X}_T) \mid \mathcal{A}_t)$. The following result, Corollary 4.2 in Filipović and Mayerhofer (2009), is used to compute the characteristic function of $\boldsymbol{X}_T$ under P^T.

Corollary 7.2.3 *For any maturity $T \leq \tau$, the T-zero coupon bond price at $t \leq T$ is given as*

$$P_T(t) = e^{-A(T-t) - \boldsymbol{B}(T-t)^\top \boldsymbol{X}_t},$$

where we denote

$$A(t) = -\Phi(t, \boldsymbol{0}), \qquad \boldsymbol{B}(t) = -\boldsymbol{\Psi}(t, \boldsymbol{0}).$$

Moreover, for $t \leq T \leq S \leq \tau$, the $\mathcal{A}_t$-conditional characteristic function of $\boldsymbol{X}_T$ under the S-forward measure P^S is given by

$$E_{P^S}\big(e^{\boldsymbol{u}^\top \boldsymbol{X}_T} \mid \mathcal{A}_t\big) = \frac{e^{-A(S-T) + \boldsymbol{\Phi}(T-t, \boldsymbol{u} - \boldsymbol{B}(S-T)) + \boldsymbol{\Psi}(T-t, \boldsymbol{u} - \boldsymbol{B}(S-T))^\top \boldsymbol{X}_t}}{P_S(t)}, \tag{7.2.2}$$

for all $\boldsymbol{u} \in \mathcal{S}(U + \boldsymbol{B}(S-T))$, where U is the neighborhood of $\boldsymbol{0}$ in $\Re^d$ from Theorem 7.2.2.

Consequently, using (7.2.2), we either recognize the characteristic function, or we invert it numerically. The resulting distribution is denoted by $q(t, T, d\boldsymbol{x})$ and we can finally compute

$$\pi(t) = P_T(t) \int_{(\Re^+)^m \times \Re^n} f(\boldsymbol{x}) q(t, T, d\boldsymbol{x}).$$

7.2.3 Applying the Fubini Theorem

For this second approach, we assume that the payoff function f admits the representation

$$f(\boldsymbol{x}) = \int_{\Re^d} e^{(\boldsymbol{u}+\imath \boldsymbol{y})^\top \boldsymbol{x}} \tilde{f}(\boldsymbol{y})\, d\boldsymbol{y}, \tag{7.2.3}$$

for some integrable function $\tilde{f} : \Re^d \to \mathbf{C}$ and some constant $\boldsymbol{u} \in U$. Then we may be able to apply Fubini's theorem to change the order of integration, which gives

$$\begin{aligned}
\pi(t) &= E\big(e^{-\int_t^T r_s\, ds} f(\boldsymbol{X}_T) \,\big|\, \mathcal{A}_t\big) \\
&= E\left(e^{-\int_t^T r_s\, ds} \int_{\Re^d} e^{(\boldsymbol{u}+\imath \boldsymbol{y})^\top \boldsymbol{X}_T} \tilde{f}(\boldsymbol{y})\, d\boldsymbol{y} \,\middle|\, \mathcal{A}_t\right) \\
&= \int_{\Re^d} E\big(e^{-\int_t^T r_s\, ds + (\boldsymbol{u}+\imath \boldsymbol{y})^\top \boldsymbol{X}_T} \,\big|\, \mathcal{A}_t\big) \tilde{f}(\boldsymbol{y})\, d\boldsymbol{y} \\
&= \int_{\Re^d} e^{\Phi(T-t, \boldsymbol{u}+\imath \boldsymbol{y}) + \boldsymbol{\Psi}(T-t, \boldsymbol{u}+\imath \boldsymbol{y})^\top \boldsymbol{X}_t} \tilde{f}(\boldsymbol{y})\, d\boldsymbol{y},
\end{aligned} \tag{7.2.4}$$

where we used Theorem 7.2.2 to arrive at (7.2.4). The next lemma, Lemma 4.3 in Filipović and Mayerhofer (2009), shows that $\tilde{f}$ can be found by Fourier transformation.

Lemma 7.2.4 *Let $f : \Re^d \to \mathbf{C}$ be a measurable function and $\boldsymbol{u} \in \Re^d$ be such that the function $h(\boldsymbol{x}) = e^{-\boldsymbol{u}^\top \boldsymbol{x}} f(\boldsymbol{x})$ and its Fourier transform*

$$\hat{h}(\boldsymbol{y}) = \int_{\Re^d} h(\boldsymbol{x}) e^{-\imath \boldsymbol{y}^\top \boldsymbol{x}}\, d\boldsymbol{x}$$

are integrable on $\Re^d$. Then (7.2.3) *holds for almost all $\boldsymbol{x} \in \Re^d$ for*

$$\tilde{f}(\boldsymbol{x}) = \frac{1}{(2\pi)^d} \hat{h}(\boldsymbol{x}).$$

Moreover, the right-hand side of (7.2.3) *is continuous in $\boldsymbol{x}$. Hence, if f is continuous then* (7.2.3) *holds for all $\boldsymbol{x} \in \Re^d$.*

Finally, we summarize the classical approach to pricing in affine models, where we are interested in computing

$$E\big(e^{-\int_t^T g(\boldsymbol{X}_s)\, ds} f(\boldsymbol{X}_T) \,\big|\, \mathcal{A}_t\big)$$

under an assumed risk neutral probability measure:

1. postulate that $g(\boldsymbol{x})$ is an affine function of $\boldsymbol{x}$,
$$g(\boldsymbol{x}) = c + \boldsymbol{\gamma}^\top \boldsymbol{x};$$
2. solve the system of Riccati equations, (7.2.1);
3. • identify the law of $\boldsymbol{X}_T$ under the forward measure, either by inspection or numerical inversion;
   • represent f as in (7.2.3);
4. compute the resulting integral numerically.

7.3 Pricing Using Affine Diffusions Under the Benchmark Approach

The aim of this section is to demonstrate how to modify the classical approaches to pricing in affine models presented in Sects. 7.2.2 and 7.2.3, to be applicable under the benchmark approach. Recall that under the forward-measure approach, Sect. 7.2.2, the distribution of the state variables under the forward measure played the key role. Under the benchmark approach, we can price under the real world probability measure, but replace the distribution with suitable fundamental solutions, see also Chap. 5. As this approach was developed in Craddock and Platen (2004, 2009), and Craddock and Lennox (2009), we refer to it as the *Craddock-Lennox-Platen approach*. Regarding the approach from Sect. 7.2.3, we point out that explicit formulas for the affine transform sit at the heart of this approach. When dealing with the benchmark approach, we find that *benchmarked Laplace transforms* naturally arise. Furthermore, using recent results from Chan and Platen (2011) and Lennox (2011), for the models employed under the benchmark approach, these benchmarked Laplace transforms are often available in closed-form. We conclude this chapter by discussing how to employ *forward measures under the benchmark approach*.

7.3.1 *Craddock-Lennox-Platen Approach to the One-Dimensional Problem*

We now place ourselves in a one-dimensional setup and rely on results presented in Craddock and Lennox (2009). We are interested in computing

$$E\big(e^{-\int_0^T g(X_s)} f(X_T)\big),$$

where X is a one-dimensional affine process started at x, and P is interpreted as the real world probability measure. Essentially, we wish to address two questions in this subsection:

(i) is there a more general approach to the formulated problem than the steps outlined above, in particular, having to solve a different set of Riccati equations for every affine process?
(ii) can we choose functions which are more general than the affine functions presented under Assumption 7.2.1?

It is clear that both approaches presented in the previous section require us to solve the Riccati equations. Therefore, the classical approach cannot be used to obtain an affirmative answer to (i). Regarding (ii), it is clear that both classical approaches crucially relied on Theorem 7.2.2, which presented a closed form solution for

$$E\big(e^{-\int_0^T r_s\,ds} e^{\boldsymbol{u}^\top \boldsymbol{X}_T}\big).$$

This result exploited the fact that the process r is an affine transform of $\boldsymbol{X}$.

We now show how to obtain an affirmative answer to the questions (i) and (ii) by using the *Craddock-Lennox-Platen* approach, see e.g. Craddock and Platen (2004, 2009), Craddock and Lennox (2009). For the remainder of this subsection, we set $m = 1$, $n = 0$, and hence $d = 1$. We have the following corollary to Theorem 7.1.4.

Corollary 7.3.1 *The process X on $\Re^+$ is affine if and only if $a(x)$ and $b(x)$ are affine of the form given in Theorem 7.1.4 for parameters a, α, b, β, which are admissible in the sense that*

$$a = 0, \qquad \alpha \geq 0,$$

hence $a(x) = \alpha x$, $x \in \Re^+$, $b \in \Re^+$,

$$b(x) = b + \beta x.$$

We now introduce the functional

$$u(x,t) = E\left(e^{-\int_0^t g(X_s)\,ds} f(X_t)\right),$$

where $X_0 = x$ and use the Feynman-Kac formula, see Sect. 15.8, to obtain the following Cauchy problem:

$$\begin{aligned} u_t &= (b + \beta x)u_x + \frac{1}{2}\alpha x u_{xx} - g(x)u, \\ u(x,0) &= f(x). \end{aligned} \tag{7.3.5}$$

The following result is from Craddock and Lennox (2009), see Theorem 4.4.1 in Chap. 4 of this book.

Corollary 7.3.2 *The PDE* (7.3.5) *has a non-trivial Lie symmetry group if and only if g is a solution of one of the following families of drift equations*

$$Lf = Ax + B, \tag{7.3.6}$$

or

$$Lf = Ax + Bx^{1/2} - \frac{3}{8}\left(\frac{1}{2}\alpha\right)^2, \tag{7.3.7}$$

where

$$Lf = \frac{b(x)}{\alpha}\beta + g(x) + xg'(x). \tag{7.3.8}$$

In this section, we focus on Eq. (7.3.6). We remark that it is trivial to solve (7.3.6) and (7.3.7) for g.

Corollary 7.3.3 *If $g(x) = \tilde{A}x + \tilde{B} + \frac{\tilde{C}}{x}$, then* (7.3.6) *is satisfied. Similarly, if $g(x) = \tilde{A}x + \tilde{B}x^{1/2} + (-\frac{3}{8}(\frac{1}{2}\alpha)^2 - \frac{b\beta}{\alpha}) + \frac{\tilde{C}}{x}$, then* (7.3.7) *is satisfied.*

Proof Substituting $b(x) = b + \beta x$, we have

$$\frac{b\beta}{\alpha} + \frac{\beta^2 x}{\alpha} + g(x) + xg'(x) = Ax + B, \quad \text{for } x > 0,$$

and we recognize the first-order ordinary differential equation

$$g'(x) + \frac{g(x)}{x} = \left(A - \frac{\beta^2}{\alpha}\right) + \left(B - \frac{b\beta}{\alpha}\right)\frac{1}{x}.$$

The integrating factor is simply x, and we obtain

$$g(x) = \frac{x(A - \frac{\beta^2}{\alpha})}{2} + \left(B - \frac{b\beta}{\alpha}\right) + \frac{C}{x}.$$

The second part of the proof can be completed analogously. □

We conclude that using the Craddock-Lennox-Platen approach we can handle a function $g(x)$ of the form

$$g(x) = \tilde{A}x + \tilde{B} + \frac{\tilde{C}}{x}.$$

Next, we address part (i). In particular, we point out that for the cases

$$g(x) = x\mu, \tag{7.3.9}$$

$$g(x) = \frac{\mu}{x}, \tag{7.3.10}$$

$$g(x) = \frac{\nu}{x} + \mu x, \tag{7.3.11}$$

fundamental solutions to the Cauchy problem are given in Craddock and Lennox (2009), which we will recall below. Consequently, for g given by (7.3.9), (7.3.10), or (7.3.11), computing

$$E\left(e^{-\int_0^T g(X_s)\,ds} f(X_T)\right)$$

amounts to computing the following one-dimensional integral numerically

$$\int_0^\infty f(y)p(x, y, T)\,dy,$$

where $p(x, y, T)$ denotes the fundamental solution of (7.3.5), and x is the starting point of the affine process. We now recall these fundamental solutions.

Corollary 7.3.4 *Let $g(x) = \mu x$, where $\mu > 0$, then there is a fundamental solution of the PDE* (7.3.5) *of the form*

$$p(x, y, t) = \frac{\sqrt{Axy}e^{-(F(x)-F(y))/(2\sigma)}}{2\sigma \sinh(\sqrt{A}t/2)} \exp\left\{-\frac{Bt}{2\sigma} - \frac{\sqrt{A}(x+y)}{2\sigma \tanh(\sqrt{A}t/2)}\right\}$$
$$\times \left(C_1(y)I_\nu\left(\frac{\sqrt{Ayx}}{\sigma \sinh(\sqrt{A}t/2)}\right) + C_2(y)I_{-\nu}\left(\frac{\sqrt{Ayx}}{\sigma \sinh(\sqrt{A}t/2)}\right)\right),$$

where $\nu = \frac{\sqrt{\sigma^2+2C}}{\sigma}$, $F'(x) = \frac{b(x)}{x}$, *and we interpret* $I_{-\nu}$ *to be* $K_\nu(z)$ *if* ν *is an integer, where* $\sigma = \frac{\alpha}{2}$, $A = \beta^2 + 2\alpha\mu$, $B = b\beta$, *and* $C = (b^2 - \alpha b)$.

The following results were also covered by our Propositions 5.4.3 and 5.4.4.

Corollary 7.3.5 *Suppose* $g(x) = \frac{\mu}{x}$, *where* $\mu > 0$, *then there is a fundamental solution to the PDE* (7.3.5) *of the form*

$$p(x,y,t) = \frac{\sqrt{A}e^{(F(y)-F(x))/(2\sigma)}}{2\sigma \sinh(\sqrt{A}t/2)} \frac{\sqrt{x}}{\sqrt{y}} \exp\left\{-\frac{Bt}{2\sigma} - \frac{\sqrt{A}(x+y)}{2\sigma \tanh(\sqrt{A}t/2)}\right\} \times \left(C_1(y) I_\nu\left(\frac{\sqrt{Axy}}{\sigma \sinh(\sqrt{A}t/2)}\right) + C_2(y) I_{-\nu}\left(\frac{\sqrt{Axy}}{\sigma \sinh(\sqrt{A}t/2)}\right)\right),$$

in which $F'(x) = b(x)/x$ *and* $\nu = \frac{\sqrt{2C+4\mu\sigma+\sigma^2}}{\sigma}$, *and we interpret* $I_{-\nu}(z)$ *to be* $K_\nu(z)$ *if* ν *is an integer,* $\sigma = \frac{\alpha}{2}$, $A = \beta^2$, $B = b\beta$, *and* $C = \frac{b^2-\alpha b}{2}$.

Let us provide for this case an example.

Example 7.3.6 Setting $b = a_1$, $\beta = -a_2$, $\alpha = 2\sigma$ in Corollary 7.3.5, we obtain the square-root process

$$dX_t = (a_1 - a_2 X_t)\,dt + \sqrt{2\sigma X_t}\,dW_t,$$

started at x. Using the fundamental solution $p(x,y,t)$, we can compute

$$E\left(e^{-\mu \int_0^T \frac{ds}{X_s}} f(X_T)\right) = \int_0^\infty f(y) p(x,y,T)\,dy,$$

which usually has to be evaluated numerically. However, for the case $f(x) = e^{-\lambda x}$, we obtain the explicit formula

$$E\left(e^{-\lambda X_t - \mu \int_0^t \frac{ds}{X_s}}\right) = \frac{\Gamma(k+\nu/2+1/2)}{\Gamma(\nu+1)} \beta x^{-k} \times \exp\left\{\frac{a_2}{2\sigma}\left(a_1 t + x - \frac{x}{\tanh(a_2 t/2)}\right)\right\} \times \frac{1}{\beta \alpha^k} e^{\beta^2/(2\alpha)} M_{-k,\nu/2}\left(\frac{\beta^2}{\alpha}\right),$$

where $M_{s,r}(z)$ is the Whittaker function of the first kind, $\nu = \frac{1}{\sigma}\sqrt{(a_1-\sigma)^2 + 4\mu\sigma}$, $k = \frac{a_1}{2\sigma}$, $\alpha = \frac{a_2}{2\sigma}(1 + \coth(\frac{a_2 t}{2})) + \lambda$, and $\beta = \frac{a_2\sqrt{x}}{2\sigma \sinh(a_2 t/2)}$.

Finally, we consider the case $g(x) = \frac{\nu}{x} + \mu x$, where $\mu > 0$, $\nu > 0$.

Corollary 7.3.7 *Suppose* $g(x) = \frac{\nu}{x} + \mu x$, $\mu > 0$, $\nu > 0$, *then there is a fundamental solution to the PDE* (7.3.5) *of the form*

$$
\begin{aligned}
p(x, y, t) = {} & \frac{\sqrt{Axy}}{2\sigma \sinh(\sqrt{A}t/2)} e^{-(Bt+\sqrt{A}(x+y)\coth(\sqrt{A}t/2)+F(x)-F(y))/(2\sigma)} \\
& \times \left(C_1(y) I_{\sqrt{\sigma^2+2C}/\sigma} \left(\frac{\sqrt{Axy}}{\sigma \sinh(\sqrt{A}t/2)} \right) \right. \\
& \left. + C_2(y) I_{-\sqrt{\sigma^2+2C}/\sigma} \left(\frac{\sqrt{Axy}}{\sigma \sinh(\sqrt{A}t/2)} \right) \right).
\end{aligned}
$$

As usual, $F'(x) = b(x)/x$, $I_{-\nu}(z) = K_\nu(z)$ *if* ν *is an integer,* $A = \beta^2 + 2\alpha\mu$, $B = b\beta$, $C = \frac{1}{2}(b^2 - \alpha b + 2\alpha\nu)$, *and* $\sigma = \frac{\alpha}{2}$.

Following the discussion in Craddock and Lennox (2009), we usually have to consider the case $C_1 = 1$, $C_2 = 0$.

In conclusion, we have the following procedure for calculating the above type of functional for a one-dimensional affine process using the Craddock-Lennox-Platen approach:

1. postulate that $g(x)$ is given by

$$
g(x) = ax + b + \frac{c}{x};
$$

2. compute numerically the one-dimensional integral

$$
\int_0^\infty f(y) p(x, y, T)\, dy,
$$

using the fundamental solution $p(x, y, T)$ identified in Corollaries 7.3.4, 7.3.5, and 7.3.7.

7.3.2 Benchmarked Laplace Transforms

In this subsection, we discuss benchmarked Laplace transforms. These functionals arise naturally in the context of the benchmark approach when applying the real world pricing formula, as we illustrate using a simple example. For more advanced examples, we refer the reader to Sect. 8.5.

As already discussed in Sect. 7.2.1, affine models have been applied to interest rate modeling. We place ourselves in the stylized MMM, see Sect. 3.3, and assume the following basic model for the short rate:

$$
r_t = \frac{c}{Y_t}, \tag{7.3.12}
$$

for $t \geq 0$, where $c > 0$ and $Y = \{Y_t, t \geq 0\}$ denotes the normalized GOP, see Eq. (3.3.9). Though very simple, the diffusion coefficient of the interest rate has the power $\frac{3}{2}$, which is also a feature of the interest rate models discussed in Ahn and Gao (1999), and Platen (1999). The real world pricing formula (1.3.19) yields the following expression for the time $t=0$ zero coupon bond price:

$$P_T(0) = S_0^{\delta_*} E\left(\frac{1}{S_T^{\delta_*}}\right) = \frac{S_0^{\delta_*}}{\alpha_T^{\delta_*}} E\left(\frac{\exp\{-c\int_0^T \frac{ds}{Y_s}\}}{Y_T}\right).$$

Due to the presence of Y_T in the denominator, we refer to

$$E\left(\frac{\exp\{-c\int_0^T \frac{ds}{Y_s}\}}{Y_T}\right) \tag{7.3.13}$$

as a benchmarked Laplace transform. The following result, see Proposition 8.1 in Chan and Platen (2011) and Proposition 2.0.41 in Lennox (2011), gives us access to many useful benchmarked Laplace transforms. We choose to present this result in generality, and employ the notation from Lennox (2011). We introduce the square-root process $X = \{X_t,\ t \geq 0\}$, where

$$dX_t = (a - bX_t)\,dt + \sqrt{2\sigma X_t}\,dW_t, \tag{7.3.14}$$

and $X_0 = x > 0$.

Proposition 7.3.8 *Assume that* $X = \{X_t,\ t \geq 0\}$ *is given by* (7.3.14) *and that* $\frac{2a}{\sigma} \geq 2$. *Let* $\beta = 1 + m - \alpha + \frac{\nu}{2}$, $m = \frac{1}{2}(\frac{a}{\sigma} - 1)$, *and* $\nu = \frac{1}{\sigma}\sqrt{(a-\sigma)^2 + 4\mu\sigma}$. *Then if* $m > \alpha - \frac{\nu}{2} - 1$,

$$\begin{aligned} &E\left(\exp\left\{-\mu\int_0^t \frac{ds}{X_s}\right\}X_t^{-\alpha}\right) \\ &= \frac{1}{2^\nu x^m}\exp\left\{-\frac{bx}{\sigma(e^{bt}-1)} + bmt\right\}\left(\frac{b\exp\{bt\}}{(e^{bt}-1)\sigma}\right)^{-m+\alpha-\frac{\nu}{2}} \\ &\quad \times \left(\frac{b^2 x}{\sigma^2\sinh^2(\frac{bt}{2})}\right)^{\nu/2}\frac{\Gamma(\beta)}{\Gamma(1+\nu)}\,{}_1F_1\left(\beta, 1+\nu, \frac{bx}{\sigma(e^{bt}-1)}\right). \end{aligned}$$

Setting $a = 1$, $b = \eta$, and $\sigma = \frac{1}{2}$, we can now compute (7.3.13). In Sect. 8.5, we present further examples, which can be calculated using Proposition 7.3.8.

Finally, we present Proposition 2.0.42 from Lennox (2011).

Proposition 7.3.9 *Assume that* $X = \{X_t,\ t \geq 0\}$ *is given by Eq.* (7.3.14) *and that* $\frac{2a}{\sigma} \geq 2$. *Define* $A = b^2 + 4\mu\sigma$, $m = \frac{1}{\sigma}\sqrt{(a-\sigma)^2 + 4\sigma\nu}$, $\beta = \frac{\sqrt{Ax}}{\sigma\sinh(\frac{\sqrt{A}t}{2})}$, *and* $k = \frac{\sqrt{A} + b\tanh(\frac{\sqrt{A}t}{2})}{2\sigma\tanh(\frac{\sqrt{A}t}{2})}$. *Then if* $a > (2\alpha - 3)\sigma$, *for* $\mu > 0$, $\nu \geq 0$,

$$\begin{aligned} &E\left(X_t^{-\alpha}\exp\left\{-\nu\int_0^t \frac{ds}{X_s} - \mu\int_0^t X_s\,ds\right\}\right) \\ &= \frac{\sqrt{A}x^{\frac{1}{2}-\frac{a}{2\sigma}}}{2\sigma\sinh(\frac{\sqrt{A}t}{2})}\left(\frac{\beta}{2}\right)^m \exp\left\{\frac{b(x+at) - \sqrt{A}x\coth(\frac{\sqrt{A}t}{2})}{2\sigma}\right\}k^{-(1+\frac{a}{2\sigma}+\frac{1}{2}+\frac{m}{2}-\alpha)} \\ &\quad \times \frac{\Gamma(1+\frac{a}{2\sigma}+\frac{1}{2}+\frac{m}{2}-\alpha)}{\Gamma(1+m)}\,{}_1F_1\left(1-\alpha+\frac{a}{2\sigma}+\frac{1}{2}+\frac{m}{2}, 1+m, \frac{\beta^2}{4k}\right). \end{aligned}$$

7.3.3 Forward Measures Under the Benchmark Approach

In this subsection, we discuss how to employ the concept of a forward measure under the benchmark approach. Recall from Sect. 3.3, that under the stylized MMM, the benchmarked savings account

$$\frac{\hat{S}_T^{\delta_*}}{\hat{S}_0^{\delta_*}}$$

could not be used as a Radon-Nikodym derivative to introduce an equivalent risk neutral probability measure, as this process is a strict supermartingale. However, under the benchmark approach, benchmarked derivative prices are martingales, see Chap. 1. Hence, the benchmarked zero coupon bond, $\hat{P}_T(t)$, appears as a candidate to define an equivalent probability measure. We introduce the Radon-Nikodym process $\Lambda = \{\Lambda_t,\ t \in [0, T]\}$,

$$\Lambda_t = \frac{\hat{P}_T(t)}{\hat{P}_T(0)}, \qquad t \leq T,$$

where $\hat{P}_T(t) = \frac{P_T(t)}{S_t^{\delta_*}}$ denotes the price of a benchmarked zero coupon bond, and $P_T(t)$ denotes the time t price of a zero coupon bond maturing at T. Clearly, since $\hat{P}_T(t)$ forms a martingale $\Lambda = \{\Lambda_t,\ t \in [0, T]\}$ is a martingale, where $\Lambda_0 = 1$. Consequently, we can use the Radon-Nikodym derivative

$$\Lambda_T = \frac{1}{S_T^{\delta_*}} \frac{1}{E(\frac{1}{S_T^{\delta_*}})},$$

to define a new probability measure, the T-forward measure P^T, via

$$\frac{dP^T}{dP} = \frac{1}{S_T^{\delta_*}} \frac{1}{E(\frac{1}{S_T^{\delta_*}})}.$$

Note that P is here interpreted as the real world probability measure. We hence obtain the following pricing rule for a nonnegative $\mathcal{A}_T$-measurable payoff H_T satisfying

$$E\left(\frac{H_T}{S_T^{\delta_*}}\right) < \infty.$$

Proposition 7.3.10 *The* real world forward price *at time t of an $\mathcal{A}_T$-measurable payoff H_T to be paid at T, where $E(\frac{H_T}{S_T^{\delta_*}}) < \infty$, is given by*

$$E_{P^T}(H_T \mid \mathcal{A}_t) = \frac{E\left(\frac{H_T}{S_T^{\delta_*}} \mid \mathcal{A}_t\right)}{\hat{P}_T(t)}.$$

Proof The proof follows immediately from the real world pricing formula and the Bayes rule, see Sect. 15.8,

$$\begin{aligned} E_{P^T}(H_T \mid \mathcal{A}_t) &= \frac{E(H_T \Lambda_T \mid \mathcal{A}_t)}{E(\Lambda_T \mid \mathcal{A}_t)} \\ &= E\big(\Lambda_T \Lambda_t^{-1} H_T \mid \mathcal{A}_t\big) \\ &= E\left(\frac{H_T}{S_T^{\delta_*}} \,\middle|\, \mathcal{A}_t\right) \frac{1}{\hat{P}_T(t)}. \end{aligned}$$ □

In the case when an equivalent risk neutral probability measure exists, the forward price derived in Sect. 7.2.2 under the classical approach can be shown to coincide with the above forward price. We emphasize that Proposition 7.3.10 does not rely on the assumption of the existence of an equivalent risk neutral probability measure.

The T-forward measure P^T is employed as an auxiliary measure for pricing, hence we are particularly interested in the distribution of random variables under the measure P^T. Using Proposition 7.3.10, we can compute the affine transform of a random variable X under P^T, where we use the real world probability measure P to perform the computation, which is the same approach as employed in Theorem 7.2.2:

$$E_{P^T}\big(\exp\{uX\} \mid \mathcal{A}_t\big) = \frac{S_t^{\delta_*}}{P_T(t)} E\left(\frac{\exp\{uX\}}{S_T^{\delta_*}} \,\middle|\, \mathcal{A}_t\right),$$

where $u \in \mathbf{C}$.

A crucial observation is the following: if we want to apply the methodology from Sect. 7.2.3, see also Sect. 8.6, to a typical problem arising in finance, such as the pricing of index options, we are in fact interested in the affine transform of $\ln(S_T^{\delta_*})$. This results in the following computation:

$$E_{P^T}\big(\exp\{u \ln(S_T^{\delta_*})\} \mid \mathcal{A}_t\big) = \frac{S_t^{\delta_*}}{P_T(t)} E\big((S_T^{\delta_*})^{u-1} \mid \mathcal{A}_t\big). \tag{7.3.15}$$

Under the benchmark approach, we preferably employ realistic, but also tractable models, hence Eq. (7.3.15) is easily computed. We illustrate this in the one-and two-dimensional setting in Sect. 8.6.

Chapter 8
Pricing Using Affine Diffusions

The aim of this chapter is to illustrate how to price derivatives using affine diffusions in the classical risk-neutral setting and under the benchmark approach. In the classical risk-neutral setting, the affine transform plays a crucial role in the pricing of derivatives. In particular, there are essentially two ways in which this transform has been employed:

- the affine transform can be used to determine the law of the vector of random variables under consideration, if necessary numerically;
- the affine transform can be employed together with the Fourier transform.

In this chapter, we first show how to use the affine transform to determine the law of a vector of random variables. Later, we combine this with the Fourier transform. We present the theory, mainly relying on Filipović and Mayerhofer (2009). Subsequently, we illustrate the theory by using two one-dimensional examples.

Under the benchmark approach, we can work under the real world probability measure, using the Craddock-Lennox-Platen approach from Sect. 7.3.1, or benchmarked Laplace transforms, or we can employ the forward measure from Sect. 7.3.3. In Sect. 8.5, we illustrate the usage of benchmarked Laplace transforms, and in Sect. 8.6, we work under the forward measure.

8.1 Theoretical Background

As in Chap. 7, we work on a filtered probability space $(\Omega, \mathcal{A}, \underline{\mathcal{A}}, P)$ and use $\boldsymbol{X}$ to denote an affine process that assumes values in the canonical state space $\mathcal{X} = (\Re^+)^m \times \Re^n$. The dynamics of $\boldsymbol{X}$ are given by

$$d\boldsymbol{X}_t = \boldsymbol{b}(\boldsymbol{X}_t)\,dt + \boldsymbol{\rho}(\boldsymbol{X}_t)\,d\boldsymbol{W}_t, \tag{8.1.1}$$

where $\boldsymbol{X}_0 = \boldsymbol{x}$ and

$$\boldsymbol{\rho}(\boldsymbol{x})\boldsymbol{\rho}(\boldsymbol{x})^\top = \boldsymbol{a}(\boldsymbol{x}).$$

Affine processes are frequently used in the context of short rate models, and we restate Assumption 7.2.1, also to recall the notation used therein.

J. Baldeaux, E. Platen, *Functionals of Multidimensional Diffusions with Applications to Finance*, Bocconi & Springer Series 5, DOI 10.1007/978-3-319-00747-2_8,

Assumption 8.1.1 *The process* $r = \{r_t, t \geq 0\}$ *is an affine transform of* $\boldsymbol{X} = \{\boldsymbol{X}_t, t \geq 0\}$,

$$r_t = c + \boldsymbol{\gamma}^\top \boldsymbol{X}_t,$$

where $\boldsymbol{X}$ *is an affine process on the canonical state space* $(\Re^+)^m \times \Re^n$ *given by Eq.* (8.1.1) *with admissible parameters* $\boldsymbol{a}, \boldsymbol{\alpha}_i, \boldsymbol{b}$, *and* $\boldsymbol{\beta}_i$, *where* $i \in \{1, \dots, d\}$, *given in Eq.* (7.1.7), *and* $c \in \Re$, $\boldsymbol{\gamma} \in \Re^d$.

We are interested in computing conditional expectations of the form

$$\pi(t) = E\left(\exp\left\{ -\int_t^T r_s \, ds \right\} f(\boldsymbol{X}_T) \,\Big|\, \mathcal{A}_t \right) \tag{8.1.2}$$

and hence impose the integrability condition

$$E\left(\exp\left\{ -\int_0^T r_s \, ds \right\} \left| f(\boldsymbol{X}_T) \right| \right) < \infty$$

for the remainder of this chapter. In Eq. (8.1.2), the expectation is taken with respect to the measure P. This refers either to the case when P denotes some assumed equivalent risk-neutral probability measure or the case when P denotes the real world probability measure. In the remainder of the section, we discuss how to compute such discounted Laplace transforms. We recall Theorem 7.2.2, where we assume that the expectation is taken with respect to the measure P, irrespective of whether this refers to an assumed risk-neutral measure or the real world probability measure. We point out that if P corresponds to an assumed risk neutral probability measure and if f is simply the constant one, then the computation of (8.1.2) yields the price at time t of a zero coupon bond maturing at time T. For the remainder of the section, we assume that the conditions of Theorem 7.2.2 are satisfied. We have the following result, see Corollary 4.2 in Filipović and Mayerhofer (2009).

Theorem 8.1.2 *Let* $\tau > 0$ *and assume that the conditions of Theorem* 7.2.2 *are satisfied. Then for any maturity* $T \leq \tau$, *the* T*-zero coupon bond price at* $t \leq T$ *is given as*

$$E\left(\exp\left\{ -\int_t^T r_s \, ds \right\} \,\Big|\, \mathcal{A}_t \right) = \exp\left\{ -A(T-t) - \boldsymbol{B}(T-t)^\top \boldsymbol{X}_t \right\} \tag{8.1.3}$$

where we denote

$$A(t) = -\Phi(t, \boldsymbol{0}), \qquad \boldsymbol{B}(t) = -\boldsymbol{\Psi}(t, \boldsymbol{0}).$$

Moreover, for $t \leq T \leq S \leq \tau$, *the* $\mathcal{A}_t$*-conditional characteristic function of* $\boldsymbol{X}_T$ *is given by*

$$E\left(\exp\left\{ -\int_t^S r_s \, ds + \boldsymbol{u}^\top \boldsymbol{X}_T \right\} \,\Big|\, \mathcal{A}_t \right)$$
$$= e^{-A(S-T) + \Phi(T-t, \boldsymbol{u} - \boldsymbol{B}(S-T)) + \boldsymbol{\Psi}(T-t, \boldsymbol{u} - \boldsymbol{B}(S-T))^\top \boldsymbol{X}_t} \tag{8.1.4}$$

for all $\boldsymbol{u} \in \mathcal{S}(U + B(S-T))$, *where* U *is the neighborhood of* $\boldsymbol{0}$ *in* $\Re^d$ *from Theorem* 7.2.2.

We remark that if P corresponds to the risk-neutral probability measure, then equality (8.1.4) gives the law of $\boldsymbol{X}_T$ under a forward measure P^S, defined via the Radon-Nikodym derivative

$$\Lambda_F = \frac{dP^S}{dP} = \frac{1}{E((S_S^0)^{-1})} \frac{1}{S_S^0},$$

where $S_t^0 = \exp\{\int_0^t r_s\, ds\}$. From Bayes' Theorem, see Sect. 15.8,

$$E_{P^S}\big(\exp\{\boldsymbol{u}^\top \boldsymbol{X}_T\} \mid \mathcal{A}_t\big) = \frac{E(\exp\{-\int_t^S r_s\, ds + \boldsymbol{u}^\top \boldsymbol{X}_T\} \mid \mathcal{A}_t)}{E(\exp\{-\int_t^S r_s\, ds\} \mid \mathcal{A}_t)}. \tag{8.1.5}$$

The expression $E(\exp\{-\int_t^S r_s\, ds\} \mid \mathcal{A}_t)$ was computed in (8.1.3) and

$$E\left(\exp\left\{-\int_t^S r_s\, ds + \boldsymbol{u}^\top \boldsymbol{X}_T\right\} \bigg| \mathcal{A}_t\right)$$

in Eq. (8.1.4). One can now recognize the law of $\boldsymbol{X}_T$ under P_S, or compute it numerically. Finally, we point out that computations using forward measures under the benchmark approach will be performed in Sect. 8.6.

We now illustrate how to apply Theorem 8.1.2. Clearly, this requires the solution of the system of Riccati equations (7.2.1). In some cases, such as the Vasiček and the CIR model, explicit solutions can be found, and we now show how to obtain them.

8.2 One-Dimensional Examples

In this section, we discuss two one-dimensional examples, which feature prominently in the finance literature.

8.2.1 Vasiček Model

The state space of the Vasiček model, see Vasiček (1977), is $\Re$, and we set $r_t = X_t$, so that we consider the one-dimensional affine process

$$dr_t = (b + \beta r_t)\, dt + \sigma\, dW_t, \tag{8.2.6}$$

where $\sigma \geq 0$, $b, \beta \in \Re$. Given this parametrization, the system of Riccati equations (7.2.1) now reads

$$\begin{aligned} \partial_t \Phi(t,u) &= \frac{1}{2}\Psi^2(t,u)\sigma^2 + b\Psi(t,u), \\ \Phi(0,u) &= 0, \\ \partial_t \Psi(t,u) &= \beta\Psi(t,u) - 1, \\ \Psi(0,u) &= u. \end{aligned} \tag{8.2.7}$$

This system is easily solved, in particular, we obtain

$$\Psi(t,u) = \exp\{\beta t\}u - \frac{\exp\{\beta t\}-1}{\beta}$$

and

$$\begin{aligned}\Phi(t,u) = \frac{1}{2}\sigma^2\bigg[&\frac{u^2}{2\beta}\big(\exp\{2\beta t\}-1\big) + \frac{1}{2\beta^3}\big(\exp\{2\beta t\} - 4\exp\{\beta t\} + 3 + 2\beta t\big)\\ &- \frac{u}{\beta^2}\big(\exp\{2\beta t\} - 2\exp\{\beta t\} + 1\big)\bigg]\\ &+ b\bigg[\frac{u}{\beta}\big(\exp\{\beta t\}-1\big) - \frac{\exp\{\beta t\}-1-t\beta}{\beta^2}\bigg],\end{aligned}$$

which holds for all $u \in \mathbf{C}$, and hence (8.1.4) holds for all $u \in \mathbf{C}$. This allows us, via Theorem 8.1.2, to compute

$$E\left(\exp\left\{-\int_t^T r_s\,ds\right\}\,\middle|\,\mathcal{A}_t\right) = \exp\big\{-A(T-t) - B(T-t)r_t\big\},$$

where

$$\begin{aligned}A(t) &= -\Phi(t,0)\\ &= -\frac{b}{\beta^2}\big(1 - \exp\{\beta t\} + \beta t\big) - \frac{\sigma^2}{4\beta^3}\big(3 - 4\exp\{\beta t\} + \exp\{2\beta t\} + 2\beta t\big),\end{aligned}$$

and

$$B(t) = -\Psi(t,0) = \frac{\exp\{\beta t\}-1}{\beta}.$$

Furthermore, we have, by invoking Eq. (8.1.5),

$$\begin{aligned}&E_{P^S}\big(\exp\{ur_T\}\,\big|\,\mathcal{A}_t\big)\\ &= \exp\bigg\{u\bigg(\exp\big\{\beta(T-t)\big\}r_t - \frac{\sigma^2}{2\beta^2}\big(2 - \exp\big\{\beta(S-T)\big\} + \exp\big\{\beta(S+T-2t)\big\}\\ &\quad - 2\exp\big\{\beta(T-t)\big\}\big) - \frac{b}{\beta}\big(1 - \exp\big\{\beta(T-t)\big\}\big)\\ &\quad + \frac{\sigma^2 u}{4\beta}\big(\exp\big\{2\beta(T-t)\big\} - 1\big)\bigg)\bigg\}.\end{aligned}$$

This means, we identify the distribution of r_T under P^S conditional on $\mathcal{A}_t$ as Gaussian with mean

$$\begin{aligned}&\exp\big\{\beta(T-t)\big\}r_t - \frac{\sigma^2}{2\beta^2}\big(2 - \exp\big\{\beta(S-T)\big\}\\ &\quad + \exp\big\{\beta(S+T-2t)\big\} - 2\exp\big\{\beta(T-t)\big\}\big)\\ &\quad - \frac{b}{\beta}\big(1 - \exp\big\{\beta(T-t)\big\}\big)\end{aligned}$$

and variance

$$\frac{\sigma^2\beta}{2}\big(\exp\{2\beta(T-t)\}-1\big). \tag{8.2.8}$$

For the special case $S=T$, i.e. $P^T=P^S$, this distribution reduces to the well-known law of a Gaussian random variable with mean

$$\exp\{\beta(T-t)\}r_t-\frac{\sigma^2}{2\beta^2}\big(\exp\{2\beta(T-t)\}-\exp\{\beta(T-t)\}\big)$$
$$-\left(\frac{b}{\beta}+\frac{\sigma^2}{2\beta^2}\right)\big(1-\exp\{\beta(T-t)\}\big)$$

and variance (8.2.8). These results are in line with well-known results on pricing under the Vasiček model, see e.g. Mamon (2004).

8.2.2 CIR Model

We now discuss the CIR model, see Cox et al. (1985), following the presentation in Filipović and Mayerhofer (2009). In this case, the state space is $\Re^+$. We set $r_t=X_t$, and deal with the following model for the short rate

$$dr_t=(b+\beta r_t)\,dt+\sigma\sqrt{r_t}\,dW_t, \tag{8.2.9}$$

where $b,\sigma>0$ and $\beta<0$. The system of Ricatti equations (7.2.1) now reads

$$\begin{aligned}\partial_t\Phi(t,u)&=b\Psi(t,u),\\ \Phi(0,u)&=0,\\ \partial_t\Psi(t,u)&=\frac{1}{2}\sigma^2\Psi^2(t,u)+\beta\Psi(t,u)-1,\\ \Psi(0,u)&=u.\end{aligned} \tag{8.2.10}$$

To solve system (8.2.10), we use the following lemma, which appeared as Lemma 5.2 in Filipović and Mayerhofer (2009).

Lemma 8.2.1 *Consider the Riccati differential equation*

$$\partial_t G=AG^2+BG-C,\quad G(0,u)=u, \tag{8.2.11}$$

where $A,B,C\in\mathbf{C}$ and $u\in\mathbf{C}$, with $A\neq 0$ and $B^2+4AC\in\mathbf{C}\setminus\Re^-$. Let $\sqrt{\cdot}$ denote the analytic extension of the real square root to $\mathbf{C}\setminus\Re^-$, and define $\lambda=\sqrt{B^2+4AC}$.

- *The function*

$$G(t,u)=-\frac{2C(\exp\{\lambda t\}-1)-(\lambda(\exp\{\lambda t\}+1)+B(\exp\{\lambda t\}-1))u}{\lambda(\exp\{\lambda t\}+1)-B(\exp\{\lambda t\}-1)-2A(\exp\{\lambda t\}-1)u}$$

is the unique solution of (8.2.11) *on its maximal interval of existence* $[0,t_+(u))$.

Moreover,

$$\int_0^t G(s,u)\,ds$$
$$= \frac{1}{A}\log\left(\frac{2\lambda\exp\{\frac{\lambda-B}{2}t\}}{\lambda(\exp\{\lambda t\}+1)-B(\exp\{\lambda t\}-1)-2A(\exp\{\lambda t\}-1)u}\right). \tag{8.2.12}$$

- *If, in addition,* $A>0$, $B\in\Re$, $\Re(C)\geq 0$ *and* $u\in\mathbf{C}^-$, *then* $t_+(u)=\infty$ *and* $G(t,u)$ *is* $\mathbf{C}^-$*-valued.*

Invoking Lemma 8.2.1, we conclude that $A=\frac{1}{2}\sigma^2$, $B=\beta$, $C=1$, $\lambda=\sqrt{\beta^2+2\sigma^2}$ and

$$\begin{aligned}\Psi(t,u) &= -\frac{2(\exp\{\lambda t\}-1)-(\lambda(\exp\{\lambda t\}+1)+\beta(\exp\{\lambda t\}-1))u}{\lambda(\exp\{\lambda t\}+1)-\beta(\exp\{\lambda t\}-1)-\sigma^2(\exp\{\lambda t\}-1)u}\\ &= -\frac{L_1(t)-L_2(t)u}{L_3(t)-L_4(t)u},\end{aligned}$$

where

$$\begin{aligned}L_1(t) &= 2\big(\exp\{\lambda t\}-1\big)\\ L_2(t) &= \lambda\big(\exp\{\lambda t\}+1\big)+\beta\big(\exp\{\lambda t\}-1\big)\\ L_3(t) &= \lambda\big(\exp\{\lambda t\}+1\big)-\beta\big(\exp\{\lambda t\}-1\big)\\ L_4(t) &= \sigma^2\big(\exp\{\lambda t\}-1\big)\end{aligned}$$

and

$$\begin{aligned}\Phi(t,u) &= \frac{2b}{\sigma^2}\log\left(\frac{2\lambda\exp\{\frac{\lambda-\beta}{2}t\}}{\lambda(\exp\{\lambda t\}+1)-\beta(\exp\{\lambda t\}-1)-\sigma^2(\exp\{\lambda t\}-1)u}\right)\\ &= \frac{2b}{\sigma^2}\log\left(\frac{L_5(t)}{L_3(t)-L_4(t)u}\right),\end{aligned}$$

i.e. we set

$$L_5(t)=2\lambda\exp\left\{\frac{\lambda-\beta}{2}t\right\},$$

where $(\Phi(\cdot,u),\Psi(\cdot,u)):\Re^+\to\mathbf{C}^-\times\mathbf{C}^-$ and (8.1.4) holds for all $u\in\mathbf{C}^-$ and $t\leq T$. As an application of the above result, we can obtain from Theorem 8.1.2

$$E\left(\exp\left\{-\int_t^T r_s\,ds\right\}\,\middle|\,\mathcal{A}_t\right)=\exp\big\{-A(T-t)-B(T-t)r_t\big\},$$

where

$$A(t)=-\Phi(t,0)=\frac{2b}{\sigma^2}\log\left(\frac{L_3(t)}{L_5(t)}\right)$$

and

$$B(t)=-\Psi(t,0)=\frac{L_1(t)}{L_3(t)},$$

which is the same result as the one derived in Sect. 5.5 using Lie symmetry methods.

We can also compute the law of r_T under P^S, conditional on $\mathcal{A}_t$. Applying Eq. (8.1.5), this gives

$$\begin{aligned}
&E_{P^S}\big(\exp\{ur_T\}\,\big|\,\mathcal{A}_t\big)\\
&= \frac{\exp\{-A(S-T)+\Phi(T-t,u-B(S-T))+\Psi(T-t,u-B(S-T))r_t\}}{P_S(t)}\\
&= \exp\big\{-A(S-T)+\Phi\big(T-t,u-B(S-T)\big)+\Psi\big(T-t,u-B(S-T)\big)r_t\\
&\quad +A(S-t)+B(S-t)r_t\big\}\\
&= \left(\frac{L_5(S-T)L_5(T-t)L_3(S-t)}{L_3(S-T)(L_3(T-t)-L_4(T-t)(u-B(S-T)))L_5(S-t)}\right)^{\frac{2b}{\sigma^2}}\\
&\quad \times \exp\left\{r_t\left(\frac{L_1(S-t)}{L_3(S-t)}-\frac{L_1(T-t)-L_2(T-t)(u-B(S-T))}{L_3(T-t)-L_4(T-t)(u-B(S-T))}\right)\right\}.
\end{aligned}$$

It can be confirmed that

$$\frac{L_5(S-T)L_5(T-t)L_3(S-t)}{L_3(S-T)(L_3(T-t)-L_4(T-t)(u-B(S-T)))L_5(S-t)} = \frac{1}{1-C_1(t,T,S)u}$$

and also that

$$\begin{aligned}
&\frac{L_1(S-t)}{L_3(S-t)}-\frac{L_1(T-t)-L_2(T-t)(u-B(S-T))}{L_3(T-t)-L_4(T-t)(u-B(S-T))}\\
&\quad = -C_2(t,T,S)+\frac{C_2(t,T,S)}{1-C_1(t,T,S)u},
\end{aligned}$$

where

$$\begin{aligned}
C_1(t,T,S) &= \frac{L_3(S-T)L_4(T-t)}{2\lambda L_3(S-t)} \quad \text{and}\\
C_2(t,T,S) &= \frac{L_2(T-t)}{L_4(T-t)}-\frac{L_1(S-t)}{L_3(S-t)}.
\end{aligned}$$

To identify the distribution of r_T under P^S conditional on $\mathcal{A}_t$, we recall the following well-known result, which in this form appeared as Lemma 5.1 in Filipović and Mayerhofer (2009), see also Sects. 3.1 and 13.1.

Lemma 8.2.2 *The non-central χ^2-distribution with $\delta > 0$ degrees of freedom and non-centrality parameter $\lambda > 0$ has the density function*

$$p(x,\delta,\lambda) = \frac{1}{2}\exp\left\{-\frac{x+\lambda}{2}\right\}\left(\frac{x}{\lambda}\right)^{\frac{\delta}{4}-\frac{1}{2}} I_{\frac{\delta}{2}-\frac{1}{2}}(\sqrt{\lambda x}), \quad x \geq 0$$

and characteristic function

$$\int_{\Re^+}\exp\{ux\}p(x,\delta,\lambda)\,dx = \frac{\exp\{\frac{\lambda u}{1-2u}\}}{(1-2u)^{\frac{\delta}{2}}}, \quad u \in \mathbf{C}^-.$$

Here $I_\nu(x) = \sum_{j\geq 0}\frac{1}{j!\Gamma(j+\nu+1)}(\frac{x}{2})^{2j+\nu}$ denotes the modified Bessel function of the first kind of order $\nu > -1$, see e.g. Abramowitz and Stegun (1972), *Sect.* 9.6.

Using Lemma 8.2.2, we conclude that under P^S, the random variable $\frac{2r_T}{C_1(t,T,S)}$, conditional on $\mathcal{A}_t$, follows a non-central χ^2-distribution with $\frac{4b}{\sigma^2}$ degrees of freedom and non-centrality parameter $2C_2(t,T,S)r_t$. These results are consistent with well-known pricing formulas under the CIR model.

8.3 Fourier Transform Approach

We recall that the methodology in the previous section relied on using the characteristic function to identify the law of $\boldsymbol{X}_T$, either by inspection or numerical inversion. The approach presented in the current section also uses the characteristic function, but in a different manner. We follow the approach presented in Filipović (2009), where the following economic interpretation was presented.

We start with the economic interpretation and later present the approach in a rigorous fashion. Its applications to some examples will conclude the section. Essentially, we express the payoff function $f(\boldsymbol{x})$ as follows

$$f(\boldsymbol{x}) = \int_{\Re^q} \exp\{(\boldsymbol{v} + \imath \boldsymbol{L}\boldsymbol{\lambda})^\top \boldsymbol{x}\} \tilde{f}(\boldsymbol{\lambda})\, d\boldsymbol{\lambda}, \quad d\boldsymbol{x}\text{-a.s.},$$

where $\tilde{f}(\boldsymbol{\lambda})$ denotes an integrable function. Economically, this means that we set up a static hedge using claims with complex payoffs $\exp\{(\boldsymbol{v}+\imath\boldsymbol{L}\boldsymbol{\lambda})^\top\boldsymbol{x}\}$, each weighted by $\tilde{f}(\boldsymbol{\lambda})$. The linearity of pricing rules ensures that the price of the claim with payoff $f(\boldsymbol{x})$ is given by the weighted average of the prices of the claims with payoffs $\exp\{(\boldsymbol{v}+\imath\boldsymbol{L}\boldsymbol{\lambda})^\top\boldsymbol{x}\}$, each weighted by $\tilde{f}(\boldsymbol{\lambda})$. The following theorem, which appeared as Theorem 10.5 in Filipović (2009), makes this argument rigorous.

Theorem 8.3.1 *Suppose either condition* (i) *or* (ii) *of Theorem* 7.2.2 *is met for some* $\tau \geq T$, *and let* $\mathcal{D}_{\Re}(T)$ *denote the maximal domain for the system of Riccati equations* (7.2.1). *Assume that* f *satisfies*

$$f(\boldsymbol{x}) = \int_{\Re^q} \exp\{(\boldsymbol{v} + \imath \boldsymbol{L}\boldsymbol{\lambda})^\top \boldsymbol{x}\} \tilde{f}(\boldsymbol{\lambda})\, d\boldsymbol{\lambda}, \quad d\boldsymbol{x}\text{-a.s.}, \tag{8.3.13}$$

for some $\boldsymbol{v} \in \mathcal{D}_{\Re}(T)$ *and* $d \times q$ *matrix* $\boldsymbol{L}$, *and some integrable function* $\tilde{f} : \Re^q \to \mathbf{C}$, *for some positive integer* $q \leq d$. *Then the price* (8.1.2) *is well defined and given by the formula*

$$\pi(t) = \int_{\Re^q} \exp\{\Phi(T-t, \boldsymbol{v}+\imath\boldsymbol{L}\boldsymbol{\lambda}) + \boldsymbol{\Psi}(T-t, \boldsymbol{v}+\imath\boldsymbol{L}\boldsymbol{\lambda})^\top \boldsymbol{X}_t\} \tilde{f}(\boldsymbol{\lambda})\, d\boldsymbol{\lambda}. \tag{8.3.14}$$

If f is continuous in $\boldsymbol{x}$, then (8.3.13) holds for all $\boldsymbol{x}$, which follows since the right-hand side of (8.3.14) is continuous in $\boldsymbol{x}$, by the Riemann-Lebesgue theorem.

Of course, the applicability of Theorem 8.3.1 depends on how easy it is to come up with a representation of the form (8.3.13). Following Filipović (2009), we can find some examples useful for finance. We refer also to Sect. 8.4 for a more constructive approach.

8.3.1 Examples of Fourier Decompositions

Following Filipović (2009) and Hurst and Zhou (2010), we discuss European call and put options, exchange options, and spread options. For the proofs of the following results, we refer the reader to Filipović (2009).

Lemma 8.3.2 *Let $K > 0$. For any $y \in \Re$ the following identities hold*:

$$\frac{1}{2\pi}\int_{\Re} \exp\{(w+\imath\lambda)y\}\frac{K^{-(w-1+\imath\lambda)}}{(w+\imath\lambda)(w-1+\imath\lambda)}\,d\lambda = \begin{cases} (K-e^y)^+ & \text{if } w<0 \\ (e^y-K)^+ - e^y & \text{if } 0<w<1, \\ (e^y-K)^+ & \text{if } w>1. \end{cases}$$

Clearly, the case $0<w<1$ also equals $(K-e^y)^+ - K$.

By setting $K = e^z$ in Lemma 8.3.2, we obtain the payoff of an exchange option.

Corollary 8.3.3 *For any $y, z \in \Re$ the following identities hold*:

$$\frac{1}{2\pi}\int_{\Re}\frac{\exp\{(w+\imath\lambda)y-(w-1+\imath\lambda)z\}}{(w+\imath\lambda)(w-1+\imath\lambda)}\,d\lambda = \begin{cases} (e^y-e^z)^+ & \text{if } w>1, \\ (e^y-e^z)^+ - e^y & \text{if } 0<w<1. \end{cases}$$

Lastly, we discuss the payoff of a spread-option.

Lemma 8.3.4 *Let $\mathbf{w} = (w_1, w_2)^\top \in \Re^2$ be such that $w_2 < 0$ and $w_1 + w_2 > 1$. Then for any $\mathbf{y} = (y_1, y_2)^\top \in \Re^2$ the following identity holds*:

$$\left(e^{y_1}-e^{y_2}-1\right)^+(2\pi)^2 = \int_{\Re^2}\exp\{(\mathbf{w}+\imath\boldsymbol{\lambda})^\top \mathbf{y}\} \times \frac{\Gamma(w_1+w_2-1+\imath(\lambda_1+\lambda_2))\Gamma(-w_2-\imath\lambda_2)}{\Gamma(w_1+1+\imath\lambda_1)}\,d\lambda_1\,d\lambda_2,$$

where the gamma function $\Gamma(z) = \int_0^\infty t^{-1+z}e^{-t}dt$ is defined for all complex z with $\Re(z) > 0$.

8.4 A Special Class of Payoff Functions

Following Filipović (2009), we point out that for a special class of payoff functions, we can apply both approaches, the one from Sect. 8.3 and the one from Sect. 8.1. For particular payoff functions, we can compute $\tilde{f}$, as needed for the Fourier transform approach from Sect. 8.3, but one can also compute the relevant densities. The following theorem is Theorem 10.6 in Filipović (2009).

Theorem 8.4.1 *Suppose either condition* (i) *or* (ii) *of Theorem* 7.2.2 *is met for some* $\tau \geq T$, *and let* $\mathcal{D}_{\Re}$ *denote the maximal domain for the system of Riccati equations* (7.2.1). *Assume that* f *is of the form*

$$f(\boldsymbol{x}) = e^{\boldsymbol{v}^{\top}\boldsymbol{x}} h\big(\boldsymbol{L}^{\top}\boldsymbol{x}\big)$$

for some $\boldsymbol{v} \in \mathcal{D}_{\Re}(T)$ *and* $d \times q$*-matrix* $\boldsymbol{L}$, *and some integrable function* $h : \Re^q \to \Re$, *for a positive integer* $q \leq d$. *Define the bounded function*

$$\tilde{f}(\boldsymbol{\lambda}) = \frac{1}{(2\pi)^q} \int_{\Re^q} e^{-\imath\boldsymbol{\lambda}^{\top}\boldsymbol{y}} h(\boldsymbol{y})\, d\boldsymbol{y}, \quad \boldsymbol{\lambda} \in \Re^q.$$

- *If* $\tilde{f}$ *is an integrable function in* $\boldsymbol{\lambda} \in \Re^q$, *then the assumptions of Theorem* 8.3.1 *are met.*
- *If* $\boldsymbol{v} = \boldsymbol{L}\boldsymbol{w}$, *for some* $\boldsymbol{w} \in \Re^q$, *and* $e^{\Phi(T-t,\boldsymbol{v}+\imath\boldsymbol{L}\boldsymbol{\lambda})+\boldsymbol{\Psi}(T-t,\boldsymbol{v}+\imath\boldsymbol{L}\boldsymbol{\lambda})^{\top}\boldsymbol{X}_t}$ *is an integrable function in* $\boldsymbol{\lambda} \in \Re^q$, *then the* $\mathcal{A}_t$*-conditional distribution of the* $\Re^q$*-valued random variable* $\boldsymbol{Y} = \boldsymbol{L}^{\top}\boldsymbol{X}_T$ *under the* T*-forward measure* P^T *admits the continuous density function*

$$q(t,T,\boldsymbol{y}) = \frac{1}{(2\pi)^q} \int_{\Re^q} e^{-(\boldsymbol{w}+\imath\boldsymbol{\lambda})^{\top}\boldsymbol{y}} \frac{e^{\Phi(T-t,\boldsymbol{v}+\imath\boldsymbol{L}\boldsymbol{\lambda})+\boldsymbol{\Psi}(T-t,\boldsymbol{v}+\imath\boldsymbol{L}\boldsymbol{\lambda})^{\top}\boldsymbol{X}_t}}{P_T(t)}\, d\boldsymbol{\lambda}.$$

In either case, the integral in (8.3.14) *is well-defined and the pricing formula* (8.3.14) *holds.*

8.5 Pricing Using Benchmarked Laplace Transforms

In this section, we discuss pricing under the benchmark approach using benchmarked Laplace transforms. We have two applications:

- a standard European put option;
- realized variance derivatives.

8.5.1 Put Options Under the Stylized MMM

In this subsection, we motivate how benchmarked Laplace transforms naturally arise when pricing options. For simplicity, we place ourselves in the stylized MMM, see Sect. 3.3, which we now briefly recall, as it is used in this and the next subsection, and Sect. 8.6. The filtered probability space $(\Omega, \mathcal{A}, \underline{\mathcal{A}}, P)$, where the filtration $\underline{\mathcal{A}} = (\mathcal{A}_t)_{t\geq 0}$ is assumed to satisfy the usual conditions, carries one source of uncertainty, a standard Brownian motion $W = \{W_t, t \geq 0\}$. As in Sect. 3.3, we assume a constant short rate and model the savings account using the differential equation

$$dS_t^0 = r S_t^0\, dt,$$

for $t \geq 0$ with $S_0^0 = 1$. We recall that the GOP is modeled using the SDE

$$S_t^{\delta_*} = S_t^0 \, \bar{S}_t^{\delta_*} = S_t^0 \, Y_t \, \alpha_t^{\delta_*}, \tag{8.5.15}$$

where $Y_t = \frac{\bar{S}_t^{\delta_*}}{\alpha_t^{\delta_*}}$ is a square-root process of dimension four, satisfying the SDE

$$dY_t = (1 - \eta \, Y_t) \, dt + \sqrt{Y_t} \, dW_t, \tag{8.5.16}$$

for $t \geq 0$ with initial value $Y_0 > 0$ and net growth rate $\eta > 0$. As before, $\alpha_t^{\delta_*}$ is a deterministic function of time, given by

$$\alpha_t^{\delta_*} = \alpha_0 \exp\{\eta t\},$$

with scaling parameter $\alpha_0 > 0$. The following lemma shows how benchmarked Laplace transforms arise when pricing options.

Lemma 8.5.1 *Let g denote a positive $\mathcal{A}_T$-measurable random variable, and define*

$$h(K) := E\left(\frac{(K - g)^+}{Y_T}\right).$$

We have for $\lambda > 0$,

$$\int_0^\infty \exp\{-\lambda K\} h(K) \, dK = \frac{1}{\lambda^2} E\left(\frac{\exp\{-\lambda g\}}{Y_T}\right).$$

Proof By the Fubini theorem it follows

$$\begin{aligned} \int_0^\infty \exp\{-\lambda K\} h(K) \, dK &= \int_0^\infty \exp\{-\lambda K\} E\left(\frac{(K - g)^+}{Y_T}\right) dK \\ &= E\left(\int_0^\infty \exp\{-\lambda K\} \frac{(K - g)^+}{Y_T} dK\right). \end{aligned}$$

We obtain

$$\begin{aligned} &\int_0^\infty \exp\{-\lambda K\} \frac{(K - g)^+}{Y_T} dK \\ &= \int_g^\infty \exp\{-\lambda K\} \frac{(K - g)}{Y_T} dK \\ &= \frac{1}{Y_T} \int_g^\infty \exp\{-\lambda K\} K \, dK - \frac{g}{Y_T} \int_g^\infty \exp\{-\lambda K\} \, dK \\ &= \frac{1}{Y_T} \left(\frac{g \exp\{-\lambda g\}}{\lambda} + \frac{\exp\{-\lambda g\}}{\lambda^2}\right) - \frac{g}{Y_T} \frac{\exp\{-\lambda g\}}{\lambda} \\ &= \frac{1}{\lambda^2} \frac{\exp\{-\lambda g\}}{Y_T}, \end{aligned}$$

and the result follows. □

Now, for a put option with strike K and maturity date T, we compute

$$p_{T,K}(0) = S_0^{\delta_*} E\left(\frac{(K - S_T^{\delta_*})^+}{S_T^{\delta_*}}\right) = S_0^{\delta_*} E\left(\frac{(\tilde{K} - Y_T)^+}{Y_T}\right),$$

where $\tilde{K} = \frac{K}{S_T^0 \alpha_T^{\delta_*}}$. We are interested in the Laplace transform with respect to the modified strike $\tilde{K}$, and obtain, for

$$h(\tilde{K}) = E\left(\frac{(\tilde{K} - Y_T)^+}{Y_T}\right)$$

the following equality

$$\int_0^\infty \exp\{-\lambda \tilde{K}\} h(\tilde{K})\, d\tilde{K} = \frac{1}{\lambda^2} E\left(\frac{\exp\{-\lambda Y_T\}}{Y_T}\right).$$

We recall from Sect. 3.1, that $Y_t \exp\{\eta t\}/c(t) \sim \chi_4^2(\alpha)$, where $\alpha = \frac{Y_0}{c(t)}$, $c(t) = \frac{\exp\{\eta t\}-1}{4\eta}$, and $\chi_\nu^2(\lambda)$ denotes a non-central χ^2-distributed random variable with ν degrees of freedom and non-centrality parameter λ. Consequently,

$$\begin{aligned} E\left(\frac{\exp\{-\mu Y_T\}}{Y_T}\right) &= E\left(\frac{\exp\{-\tilde{\mu}\chi_4^2(\alpha)\}}{\chi_4^2(\alpha)}\right)\frac{\exp\{\eta T\}}{c(T)} \\ &= \frac{\exp\{-\alpha/2\}(\exp\{\frac{\alpha}{4\tilde{\mu}+2}\} - 1)}{\alpha}\frac{\exp\{\eta T\}}{c(T)}, \end{aligned} \quad (8.5.17)$$

where $\tilde{\mu} = \mu \frac{c(T)}{\exp\{\eta T\}}$. Equality (8.5.17) is easily verified using the probability density function of $\chi_4^2(\alpha)$. This illustrates how benchmarked Laplace transforms arise naturally in the context of option pricing. Finally, we remark that using the techniques from Sect. 13.5, options can now be priced.

8.5.2 Derivatives on Realized Variance Under the Stylized MMM

We remind the reader that in Sect. 3.3, we had already derived the price of a put option under the stylized MMM, without using Laplace transforms. However, we now discuss an example, where the availability of benchmarked Laplace transforms is crucial. In particular, we discuss the pricing of *derivatives on realized variance* of an index. We point out that derivatives on the realized variance of an index, such as the VIX, and options on the VIX, as traded on the Chicago Board Options Exchange, have become important risk management tools.

In this subsection, we show how to price call and put options on realized variance, variance swaps, and volatility swaps. The formulas derived in this subsection are in the spirit of the pricing formulas presented in Sect. 3.3. However, the results needed to price derivatives on realized variance, rely on the benchmarked Laplace transform, see Proposition 7.3.8. Hence we discuss realized variance derivatives in this subsection. Furthermore, we remark that the results presented here have appeared in Baldeaux et al. (2011a), Chan and Platen (2011), and Lennox (2011).

We place ourselves in the stylized MMM and model realized variance as the quadratic variation of the logarithm of the index,

$$\left[\ln\left(S^{\delta_*}\right)\right]_T,$$

which admits the following representation.

Lemma 8.5.2 *The realized variance of the index is given by the integral*

$$\left[\ln\left(S^{\delta_*}\right)\right]_T = \int_0^T \frac{dt}{Y_t}. \tag{8.5.18}$$

Proof Clearly,

$$\left[\ln\left(S^{\delta_*}\right)\right]_T = \left[\ln(Y)\right]_T,$$

as $S^0_\cdot$ and $\alpha^{\delta_*}_\cdot$ are deterministic functions of time. Now one has by the Itô formula

$$\begin{aligned} d\ln(Y_t) &= \frac{dY_t}{Y_t} - \frac{1}{2}\frac{d[Y]_t}{Y_t^2} \\ &= \frac{1}{Y_t}(1-\eta Y_t)\,dt + \frac{dW_t}{\sqrt{Y_t}} - \frac{1}{2}\frac{dt}{Y_t} \\ &= \frac{1}{Y_t}\left(\frac{1}{2} - \eta Y_t\right)dt + \frac{dW_t}{\sqrt{Y_t}}, \end{aligned}$$

which completes the proof. □

We now study call and put options on realized variance. In particular, we present Laplace transforms of prices of options on realized variance and show how benchmarked Laplace transforms naturally arise in this context. We will focus on put options, as prices of call options can be recovered from the following put-call parity.

Lemma 8.5.3 *The following put-call parity relation holds for payoffs of options on realized variance*

$$\begin{aligned} &E\left(\frac{(\frac{1}{T}\int_0^T \frac{dt}{Y_t} - K)^+}{S_T^{\delta_*}}\right) \\ &\quad = E\left(\frac{\frac{1}{T}\int_0^T \frac{dt}{Y_t} - K}{S_T^{\delta_*}}\right) + E\left(\frac{(K - \frac{1}{T}\int_0^T \frac{dt}{Y_t})^+}{S_T^{\delta_*}}\right). \end{aligned}$$

Note that the put-call parity involves the fair zero coupon bond and not the savings bond even when the short rate is constant.

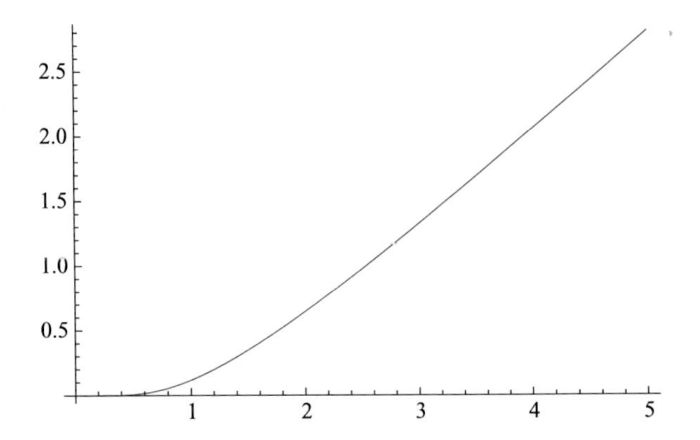

Fig. 8.5.1 Prices of put options on realized variance versus strike prices

We address the problem of pricing put options on realized variance, which by Lemma 8.5.3 also covers the case of call options. For notational convenience, we focus on the case $t = 0$, and are, therefore, interested in computing the expectation

$$h(K) := E\left(\frac{(K - \frac{1}{T}\int_0^T \frac{dt}{Y_t})^+}{Y_T}\right). \tag{8.5.19}$$

Inspired by Carr et al. (2005), we first compute the Laplace transform of $h(K)$ with respect to the strike K, which we obtain from Lemma 8.5.1. Setting $g = \frac{1}{T}\int_0^T \frac{dt}{Y_t}$, we compute

$$\int_0^\infty \exp\{-\lambda K\} h(K)\, dK = \frac{1}{\lambda^2} E\left(\frac{\exp\{-\frac{\lambda}{T}\int_0^T \frac{dt}{Y_t}\}}{Y_T}\right). \tag{8.5.20}$$

The quantity

$$E\left(\frac{\exp\{-\frac{\lambda}{T}\int_0^T \frac{dt}{Y_t}\}}{Y_T}\right)$$

is easily computed using Proposition 7.3.8. We can hence price put options on realized variance by inverting the Laplace transform given in Eq. (8.5.20) and invoking Proposition 7.3.8. To demonstrate that this methodology works reliably, Fig. 8.5.1 displays put option prices for different strikes, that have been confirmed to the shown accuracy via numerical methods to be introduced in Sects. 12.2 and 13.5, where we choose

$$Y_0 = 1, \qquad T = 1, \qquad \eta = 0.052, \qquad r = 0.05.$$

In Sect. 13.5, we will discuss how to invert Laplace transforms, and also present examples relevant to the pricing of realized variance derivatives.

We remark that the approach presented in this subsection cannot immediately be extended to the pricing of call and put options on volatility. This is due to the fact that the approach presented in this subsection requires the computation of the expectation

$$E\left(\frac{\exp\{-\lambda\sqrt{\int_0^T \frac{dt}{Y_t}}\}}{Y_T}\right). \tag{8.5.21}$$

However, there seems to exist no explicit formula for (8.5.21). This motivates us to apply numerical methods to the problem, which we will develop in Sects. 12.2 and 13.5. In particular, we demonstrate how to recover the joint distribution of $(\int_0^t \frac{ds}{Y_s}, Y_t)$ by inverting the one-dimensional Laplace transform given in Eq. (5.4.16). Subsequently, we can apply quadrature methods to compute prices, see Sect. 12.2.

We now discuss variance and volatility swaps. Again, the benchmarked Laplace transforms are useful in this context. The payoff of a *variance swap* maturing at $T > 0$ is given by

$$\left[\ln\left(S^{\delta_*}\right)\right]_T - K,$$

where K is a fixed swap rate, chosen in such a way that the time $t = 0$ value of the variance swap is zero. Hence from the real world pricing formula (1.3.19), we need to solve the following equation for K,

$$S_0^{\delta_*} E\left(\frac{[\ln(S^{\delta_*})]_T - K}{S_T^{\delta_*}}\right) = 0,$$

which by Eq. (8.5.15) and Lemma 8.5.2 is equivalent to

$$\frac{S_0^{\delta_*}}{\alpha_T^{\delta_*} S_T^0} E\left(\frac{\int_0^T \frac{ds}{Y_s}}{Y_T}\right) - K P_T(0) = 0,$$

where $P_T(t)$ denotes the time t price of a zero coupon bond maturing at T. Regarding the computation of

$$E\left(\frac{\int_0^T \frac{ds}{Y_s}}{Y_T}\right),$$

we use the following proposition, see Lennox (2011), Proposition 2.0.41, and also Chan and Platen (2011), Proposition 8.1. We present the result in generality. We consider the square-root process

$$dX_t = (a - bX_t)\,dt + \sqrt{2\sigma X_t}\,dW_t, \tag{8.5.22}$$

where $X_0 = x > 0$, and remark that this proposition follows immediately from the benchmarked Laplace transform given in Proposition 7.3.8.

Proposition 8.5.4 *Let* $X = \{X_t,\ t \geq 0\}$ *be given by* (8.5.22), *let* $\beta(\mu) = 1 + m - \alpha + \frac{\nu(\mu)}{2}$, $m = \frac{1}{2}(\frac{a}{\sigma} - 1)$, *and* $\nu(\mu) = \frac{1}{\sigma}\sqrt{(a-\sigma)^2 + 4\mu\sigma}$, *and assume that* $\frac{2a}{\sigma} \geq 2$.

Then if $m > \alpha - 1$,

$$E\left(\frac{\int_0^t \frac{ds}{X_s}}{X_t^\alpha}\right)$$

$$= -x^{-m} \exp\left\{-\frac{bx}{\sigma(e^{bt}-1)} + bmt\right\} \frac{d}{d\mu}\left(\left(\frac{be^{bt}}{(e^{bt}-1)\sigma}\right)^{-m+\alpha-\frac{\nu(\mu)}{2}}\right.$$

$$\times \left(\frac{b^2 x}{4\sigma^2 \sinh^2(\frac{bt}{2})}\right)^{\nu(\mu)/2} \frac{\Gamma(1+m-\alpha+\frac{\nu(\mu)}{2})}{(1+\nu(\mu))}$$

$$\left.\times {}_1F_1\left(\beta(\mu), 1+\nu(\mu), \frac{bx}{\sigma(e^{bt}-1)}\right)\right)\Bigg|_{\mu=0},$$

where ${}_1F_1$ *denotes the confluent hypergeometric function, see e.g. Chap.* 13 *in Abramowitz and Stegun* (1972).

To price variance swaps, we simply set $a = 1$, $b = \eta$, and $\sigma = \frac{1}{2}$ in (8.5.22) and note that

$$m = \frac{1}{2} > 0 = \alpha - 1,$$

hence the result applies to the stylized MMM.

We now study *volatility swaps*. A volatility swap pays

$$\sqrt{[\ln(S^{\delta_*})]_T} - K,$$

at maturity $T > 0$, where again K is chosen so that the initial value of the volatility swap is zero. Hence we solve the following equation for K:

$$S_0^{\delta_*} E\left(\frac{\sqrt{[\ln(S^{\delta_*})]_T}}{S_T^{\delta_*}}\right) - E\left(\frac{S_0^{\delta_*}}{S_T^{\delta_*}}\right) K = 0,$$

where again $E(\frac{S_0^{\delta_*}}{S_T^{\delta_*}})$ is the time 0 price of a fair zero coupon bond maturing at T. The following representation is useful, and is, for example, also used in Gatheral (2006), Eq. (11.6):

$$\sqrt{x} = \frac{1}{2\pi} \int_0^\infty \frac{1-\exp\{-ux\}}{u^{3/2}} du, \quad x \geq 0. \tag{8.5.23}$$

Hence by Eq. (8.5.15) and Lemma 8.5.2,

$$E\left(\frac{\sqrt{[\ln(S^{\delta_*})]_T}}{S_T^{\delta_*}}\right) = \frac{1}{\alpha_T^{\delta_*} S_T^0} E\left(\frac{\sqrt{\int_0^T \frac{ds}{Y_s}}}{Y_T}\right),$$

and

$$E\left(\frac{\sqrt{\int_0^T \frac{ds}{Y_s}}}{Y_T}\right) = \frac{1}{2\pi} \int_0^\infty \frac{E(\frac{1}{Y_T}) - E(\frac{\exp\{-u\int_0^T \frac{ds}{Y_s}\}}{Y_T})}{u^{3/2}} du.$$

We recall that $E(\frac{1}{Y_T})$ is easily computed using the transition density of Y, see Eq. (3.1). Finally, we observe that

$$E\left(\frac{\exp\{-u\int_0^T \frac{ds}{Y_s}\}}{Y_T}\right)$$

is again the benchmarked Laplace transform, which was computed in Proposition 7.3.8.

We conclude that when pricing derivatives under the benchmark approach, benchmarked Laplace transforms feature prominently, but are easily computed via Lie symmetry methods, for those tractable models we consider in this book under the benchmark approach.

8.6 Pricing Under the Forward Measure Using the Benchmark Approach

In this section, we illustrate how to combine the results from Sect. 8.3 with the benchmark approach. For simplicity, we begin with the one-dimensional case, but we consequently also discuss a two-dimensional example. Assume that the payoff function f admits the representation

$$f(x) = \int_{\Re} \exp\{(w + \imath\lambda)x\}\tilde{f}(\lambda)\,d\lambda, \quad dx\text{-a.s.}$$

Also recall from Sect. 8.3.1, that $f(\cdot)$ is typically a function of the log-price $\ln(S_T^{\delta_*})$. Consequently, Proposition 7.3.10 yields the formula

$$\begin{aligned}
&P_T(t)E_{P^T}\big(f\big(S_T^{\delta_*}\big) \mid \mathcal{A}_t\big) \\
&\quad = P_T(t)\int_{\Re} E_{P^T}\big(\exp\big\{(w + \imath\lambda)\ln\big(S_T^{\delta_*}\big)\big\} \mid \mathcal{A}_t\big)\tilde{f}(\lambda)\,d\lambda \\
&\quad = P_T(t)\int_{\Re} E_{P^T}\big(\big(S_T^{\delta_*}\big)^{w+\imath\lambda} \mid \mathcal{A}_t\big)\tilde{f}(\lambda)\,d\lambda.
\end{aligned}$$

We have

$$E_{P^T}\big(\big(S_T^{\delta_*}\big)^{w+\imath\lambda} \mid \mathcal{A}_t\big) = \frac{S_t^{\delta_*}}{P_T(t)}E\big(\big(S_T^{\delta_*}\big)^{w-1+\imath\lambda} \mid \mathcal{A}_t\big).$$

For the stylized MMM, we use Eq. (8.5.15) to compute, for $u \in \mathbf{C}$,

$$\begin{aligned}
E_{P^T}\big(\exp\big\{u\ln\big(S_T^{\delta_*}\big)\big\} \mid \mathcal{A}_t\big) &= \frac{S_t^{\delta_*}}{P_T(t)}E\left(\frac{\exp\{u\ln(S_T^{\delta_*})\}}{S_T^{\delta_*}} \,\middle|\, \mathcal{A}_t\right) \\
&= \frac{E((S_T^{\delta_*})^{u-1} \mid \mathcal{A}_t)}{E((S_T^{\delta_*})^{-1} \mid \mathcal{A}_t)} \\
&= \big(\alpha_T^{\delta_*} S_T^0\big)^u \frac{E(Y_T^{u-1} \mid \mathcal{A}_t)}{E(Y_T^{-1} \mid \mathcal{A}_t)}.
\end{aligned}$$

Recall that we use $\chi^2_\nu(\lambda)$ to denote a non-central χ^2-distributed random variable with ν degrees of freedom and non-centrality parameter λ. In Sect. 3.1 we established that conditional on $\mathcal{A}_t$, $\frac{Y_T \exp\{\eta(T-t)\}}{c(T-t)}$ follows a non-central χ^2-distribution with 4 degrees of freedom and non-centrality parameter $\beta = \frac{Y_t}{c(T-t)}$, where $c(t) = (\exp\{\eta t\} - 1)/(4\eta)$. We hence obtain

$$\begin{aligned} &E_{P^T}\big(\exp\big\{u \ln\big(S_T^{\delta_*}\big)\big\} \mid \mathcal{A}_t\big) \\ &\quad = \big(S_T^0\big(\varphi(T) - \varphi(t)\big)\big)^u \frac{E((\chi_4^2(\beta))^{u-1} \mid \mathcal{A}_t)}{E((\chi_4^2(\beta))^{-1} \mid \mathcal{A}_t)}, \end{aligned} \tag{8.6.24}$$

where

$$\varphi(t) = \frac{1}{4}\int_0^t \alpha_s^{\delta_*}\, ds.$$

Due to the tractability of the stylized MMM, we can compute explicitly

$$E\big(\big(\chi_4^2(\beta)\big)^{u-1}\big) = 2^{u-1}\Gamma(1+u)\ {}_1F_1\bigg(-u+1, 2, -\frac{\beta}{2}\bigg), \tag{8.6.25}$$

for $\Re(u) > -1$, where ${}_1F_1$ denotes the confluent hypergeometric function, see Chap. 13 in Abramowitz and Stegun (1972). For $u = 0$, this evaluates to

$$E\big(\big(\chi_4^2(\beta)\big)^{-1}\big) = \frac{(1 - \exp\{-\beta/2\})}{\beta}. \tag{8.6.26}$$

The forward measure can also be employed in a bivariate context. We consider the GOP denominated in two currencies: S^a denotes the GOP denominated in the domestic currency, and S^b denotes the GOP denominated in the foreign currency. As in Sect. 3.3, we model both discounted GOPs as independent squared Bessel processes of dimension four, i.e. we assume that

$$S_t^k = S_t^{0,k} \alpha_t^k Y_t^k, \quad k \in \{a, b\},$$

where $S_t^{0,k} = \exp\{r_k t\}$ denotes the savings account denominated in currency k,

$$\alpha_t^k = \alpha_0^k \exp\{\eta^k t\},$$

and

$$dY_t^k = \big(1 - \eta^k Y_t^k\big)\, dt + \sqrt{Y_t^k}\, dW_t^k,$$

where we assume that $d\langle W^a, W^b\rangle_t = 0$.

We consider an exchange option, i.e. the payoff is given by

$$\big(S_T^a - S_T^b\big)^+.$$

Using the forward measure which employs the zero coupon bond in the domestic currency, the real world pricing formula yields

$$S_t^a E\bigg(\frac{(S_T^a - S_T^b)^+}{S_T^a}\,\bigg|\, \mathcal{A}_t\bigg) = P_T^a(t) E_{P^T}\big(\big(S_T^a - S_T^b\big)^+ \mid \mathcal{A}_t\big),$$

where $P_T^k(t)$ denotes the time t price of a zero coupon bond in currency $k \in \{a, b\}$, maturing at T. From Corollary 8.3.3, we get

$$P_T^a(t) E_{P^T}\big((S_T^a - S_T^b)^+ \mid \mathcal{A}_t\big) = \frac{P_T^a(t)}{2\pi} \int_{\Re} \frac{E_{P^T}((S_T^a)^{w+\imath\lambda}(S_T^b)^{-(w-1+\imath\lambda)} \mid \mathcal{A}_t)}{(w+\imath\lambda)(w-1+\imath\lambda)} d\lambda,$$

where $w > 1$. From the assumed independence of S^a and S^b,

$$E_{P^T}\big((S_T^a)^{w+\imath\lambda}(S_T^b)^{-(w-1+\imath\lambda)} \mid \mathcal{A}_t\big) = E_{P^T}\big((S_T^a)^{w+\imath\lambda} \mid \mathcal{A}_t\big) E_{P^T}\big((S_T^b)^{-(w-1+\imath\lambda)} \mid \mathcal{A}_t\big),$$

which can be computed as demonstrated above. This leads to the formula

$$S_t^a E\left(\frac{(S_T^a - S_T^b)^+}{S_T^a} \,\bigg|\, \mathcal{A}_t\right) = \frac{P_T^a(t)}{2\pi} \int_{\Re} \frac{E_{P^T}((S_T^a)^{w+\imath\lambda} \mid \mathcal{A}_t) E_{P^T}((S_T^b)^{-(w-1+\imath\lambda)} \mid \mathcal{A}_t)}{(w+\imath\lambda)(w-1+\imath\lambda)} d\lambda,$$

where $w > 1$. Furthermore, we compute using Eqs. (8.6.24), (8.6.25), and (8.6.26),

$$E_{P^T}\big((S_T^a)^{u_a} \mid \mathcal{A}_t\big) = \frac{\beta_a (S_T^{0,a}(\varphi^a(T) - \varphi^a(t)))^{u_a} 2^{u_a - 1} \Gamma(1+u_a)\ {}_1F_1(-u_a+1, 2, -\frac{\beta_a}{2})}{(1 - \exp\{-\beta_a/2\})},$$

where $u_a = w - 1 + \imath\lambda$,

$$\varphi^k(t) = \frac{1}{4}\int_0^t \alpha_s^k\, ds, \quad k \in \{a, b\},$$

$$\beta^k = \frac{Y_t^k}{c^k(T-t)}, \quad k \in \{a, b\},$$

$$c^k(t) = \frac{(\exp\{\eta^k t\} - 1)}{4\eta^k}, \quad k \in \{a, b\}.$$

We now turn to the computation of $E_{P^T}((S_T^b)^{u_b} \mid \mathcal{A}_t)$. Recall that we used the zero coupon bond in the domestic currency to define the forward measure. It follows that

$$\begin{aligned} E_{P^T}\big((S_T^b)^{u_b} \mid \mathcal{A}_t\big) &= \frac{S_t^a}{P_T^a(t)} E\left(\frac{(S_T^b)^{u_b}}{(S_T^a)} \,\bigg|\, \mathcal{A}_t\right) \\ &= \frac{S_t^a}{P_T^a(t)} E\left(\frac{1}{S_T^a} \,\bigg|\, \mathcal{A}_t\right) E\big((S_T^b)^{u_b} \mid \mathcal{A}_t\big) \\ &= E\big((S_T^b)^{u_b} \mid \mathcal{A}_t\big). \end{aligned}$$

As above, we compute

$$E\big((S_T^b)^{u_b} \mid \mathcal{A}_t\big) = \big(S_T^{0,b}\big(\varphi^b(T) - \varphi^b(t)\big)\big)^{u_b} 2^{u_b} \Gamma(u_b+2)\ {}_1F_1\left(-u_b, 2, -\frac{\beta_b}{2}\right).$$

Chapter 9
Solvable Affine Processes on the Euclidean State Space

In this chapter, we focus on obtaining explicit formulas for the affine transform given in Eq. (7.1.1). Hence the focus of this chapter differs from the focus of Chap. 7 in the following way: in Chap. 7, we studied when the affine transform is well-defined, however, in this chapter we want to know when we can compute the affine transform explicitly. We remark that for specific models, such as the CIR model, Lemma 8.2.1 shows how to compute the affine transform. However, given a particular problem, one would have to come up with a new version of Lemma 8.2.1 in order to solve the problem. In some cases, it might not be possible to identify such a lemma, if the corresponding system of Riccati equations cannot be solved explicitly.

An important application of affine processes is interest rate term structure modeling, as already discussed previously. Dealing with affine processes, the bond price is an exponentially affine function of the state variables. We focus on affine processes in this chapter. However, an obvious question is whether the methodology can be extended to processes resulting in bond prices which are not necessarily affine functions of the state variables, but are polynomial functions of the state variables. From a result presented in Filipović (2002), it is known that the bond price is necessarily affine or quadratic in the state variables. Consequently, for bond prices to be allowed to depend exponentially polynomially on the state variables, we can classify which processes are solvable using the methodology presented in this chapter.

The focal point of this chapter is solving the Riccati equations explicitly. This approach is due to Grasselli and Tebaldi (2008). We now give a very brief non-technical summary of the approach. Further illustrations will be given in Chap. 11.

9.1 A Guided Tour to the Grasselli-Tebaldi Approach

The Grasselli-Tebaldi approach, see Grasselli and Tebaldi (2008), can be summarized as follows: one studies regular continuous Markov processes $\boldsymbol{X}$ assuming values in $\mathcal{D} = \mathcal{D}_+ \times \Re^{n-m}$, where $\mathcal{D}_+$ is a symmetric cone, which is the state space of those elements of $\boldsymbol{X}$ corresponding to positive factors: we remind the reader that in

J. Baldeaux, E. Platen, *Functionals of Multidimensional Diffusions with Applications to Finance*, Bocconi & Springer Series 5, DOI 10.1007/978-3-319-00747-2_9,

Chap. 7 this state space was $(\Re^+)^m$. For a general definition of symmetric cones, see e.g. Definition 2.3 in Grasselli and Tebaldi (2008), for purposes of this book, it suffices to focus on the particular symmetric cones mentioned below in Remark 9.1.3. As we will discuss later, allowing for such a general state space provides the modeler with more flexibility, see the discussion in Sects. 9.2, 9.7, and Chap. 11.

Since affine processes have often been applied to term structure models and because Grasselli and Tebaldi (2008) employ this nomenclature, we refer to the stochastic process $\boldsymbol{X} = \{\boldsymbol{X}_t,\ t \geq 0\}$ as an affine term structure model (ATSM). However, as the examples in subsequent sections illustrate, the methodology is by no means restricted to term structure modeling, but is more generally applicable. To fix ideas, we assume that on the filtered probability space $(\Omega, \mathcal{A}, \underline{\mathcal{A}}, P)$ the Markovian process $\boldsymbol{X} = \{\boldsymbol{X}_t,\ t \geq 0\}$ takes values in $\mathcal{D} \subseteq V$, where $\mathcal{D}$ is the state space, and V is a finite-dimensional real vector space of dimension n. The standard scalar product on V is denoted by $\langle \cdot, \cdot \rangle$. As in Grasselli and Tebaldi (2008), we also take the liberty of denoting the complex extension of the real vector space V by V and similarly the scalar product by $\langle \cdot, \cdot \rangle$. Now, we recall the definition of an affine process on a general state space.

Definition 9.1.1 The regular (continuous) Markov process $\boldsymbol{X}$ is said to be affine if for every $\tau = T - t \in \Re^+$ the discounted conditional characteristic function has an exponential affine dependence on the initial condition $\boldsymbol{x}_t$. That is,

$$\begin{aligned}
&\Psi_X(\boldsymbol{u}, \boldsymbol{x}_t, t, \tau) \\
&\quad = E\left(\exp\left\{ -\int_t^{t+\tau} \left(\eta_0 + \langle \boldsymbol{\eta}, \boldsymbol{X}_s \rangle \right) ds \right\} \exp\left\{ \imath \langle \boldsymbol{u}, \boldsymbol{X}_T \rangle \right\} \,\Big|\, \mathcal{A}_t \right) \\
&\quad = \exp\left\{ V^0(\tau, \imath \boldsymbol{u}) - \left\langle \boldsymbol{V}(\tau, \imath \boldsymbol{u}), \boldsymbol{x}_t \right\rangle \right\}, \qquad (9.1.1)
\end{aligned}$$

where $(\tau, \boldsymbol{u}) \in \Re^+ \times V$, $V^0 : \Re^+ \times V \to \mathbf{C}$, $\boldsymbol{V} : \Re^+ \times V \to V$, and $\eta_0 \in \Re^+$, $\boldsymbol{\eta} \in \mathcal{D}$.

As in Grasselli and Tebaldi (2008), we refer to $\boldsymbol{V}$ as the factor sensitivities.

The infinitesimal generator of a regular affine diffusion process has necessarily the following functional form, see Duffie and Kan (1996),

$$\tilde{\mathcal{A}} = \frac{1}{2} \operatorname{Tr}\left[\left(\boldsymbol{\Sigma}(\boldsymbol{x}) + \boldsymbol{\Sigma}_0 \right) \boldsymbol{D}^\top \boldsymbol{D} \right] + \left\langle \boldsymbol{\Omega}^0 + \boldsymbol{\Omega}(\boldsymbol{x}), \boldsymbol{D}^\top \right\rangle - \left(\eta_0 + \langle \boldsymbol{\eta}, \boldsymbol{x} \rangle \right), \qquad (9.1.2)$$

using the notation $\boldsymbol{D}$ for the row vector gradient operator, Tr denotes the trace over $\mathcal{M}_n(V)$, where $\mathcal{M}_n(V)$ denotes the set of $n \times n$ matrices defined on V, $\boldsymbol{D}^\top \boldsymbol{D}$ the Hessian matrix, $\boldsymbol{\Sigma}(\boldsymbol{x}), \boldsymbol{\Sigma}_0 \in \overline{\mathcal{S}_n^+}(V)$, where $\overline{\mathcal{S}_n^+}(V)$ denotes the set of positive semidefinite matrices taking values in V, and $\boldsymbol{\Omega}(\boldsymbol{x}), \boldsymbol{\Omega}^0 \in V$. The functions $\boldsymbol{\Sigma}(\boldsymbol{x})$, $\boldsymbol{\Omega}(\boldsymbol{x})$ have to be linear in $\boldsymbol{x} \in V$, so that we obtain

$$\left[\boldsymbol{\Sigma}(\boldsymbol{x}) \right]_{i,j} = \sum_{k=1}^n C_{i,j}^k x_k, \qquad \left[\boldsymbol{\Omega}(\boldsymbol{x}) \right]_l = \sum_{k=1}^n \Omega_l^k x_k.$$

We use the Feynman-Kac formula to obtain

$$\frac{\partial \Psi_X}{\partial t} + \tilde{\mathcal{A}} \Psi_X = 0,$$

subject to the initial condition

$$\Psi_X(\boldsymbol{u}, \boldsymbol{x}, T, 0) = \exp\{\imath \langle \boldsymbol{u}, \boldsymbol{x} \rangle\},$$

or, employing $\tau = T - t$ and $\boldsymbol{D}^\top \Psi_X = -\boldsymbol{V}$, it follows from Eq. (9.1.1) that

$$\frac{d}{d\tau} V^0(\tau) - \left\langle \frac{d}{d\tau} \boldsymbol{V}(\tau), \boldsymbol{x} \right\rangle$$
$$= \frac{1}{2} \mathrm{Tr}\big((\boldsymbol{\Sigma}(\boldsymbol{x}) + \boldsymbol{\Sigma}_0)\boldsymbol{V}\boldsymbol{V}^\top\big) - \langle \boldsymbol{\Omega}^0 + \boldsymbol{\Omega}(\boldsymbol{x}), \boldsymbol{V} \rangle - \big(\eta_0 + \langle \boldsymbol{\eta}, \boldsymbol{x} \rangle\big) \quad (9.1.3)$$

subject to

$$V^0(0, \imath \boldsymbol{u}) = 0, \qquad \boldsymbol{V}(0, \imath \boldsymbol{u}) = -\imath \boldsymbol{u}.$$

Hence, $\boldsymbol{V}$ solves a quadratic ODE. Given $\boldsymbol{V}$, one can determine V^0 by integration, and thus we focus on computing $\boldsymbol{V}$. However, for those models which admit a matrix representation, we obtain an explicit expression for $V^0(\tau)$, see Chap. 11.

The first question asked by Grasselli and Tebaldi (2008) investigates in the spirit of Chap. 7 whether the ATSM is admissible, in the following sense:

Definition 9.1.2 An ATSM model is admissible in a state space $\mathcal{D} \subseteq V$ if the generator (9.1.2) and the corresponding regular affine Markov semigroup exist and are unique for any initial condition in $\mathcal{D}$, or equivalently if Ψ_X is uniquely defined by (9.1.1) $\forall \boldsymbol{x} \in \mathcal{D}$ and $\forall \tau \in \Re^+$.

Next, following Grasselli and Tebaldi (2008), we assume that the state space $\mathcal{D}$ is a symmetric cone, i.e. we focus on the positive factors. We remark that it was observed in Duffie et al. (2003), see also Chap. 7, that the extension to the case $\mathcal{D} \times \Re^{n-m}$, where $\mathcal{D}$ is the symmetric cone and $\Re^{n-m}$ the state space supporting the evolution of conditionally Gaussian factors follows easily, as those factor sensitivities in $\boldsymbol{V}$ corresponding to the components of $\boldsymbol{X}$ on $\Re^{n-m}$ satisfy an independent system of linear equations. This issue will be addressed in more detail in Sect. 9.2 and Chap. 11.

We now recall the main observations from Grasselli and Tebaldi (2008). Since the approaches we focus on deal with specific choices of the symmetric cone, namely $(\Re^+)^m$ and positive definite matrix cones, we will discuss these examples in detail in Sect. 9.2 and Chap. 11, and only present here the main ideas of the approach. Readers interested in other state spaces are referred to the original paper.

Having introduced symmetric cones, see Definition 2.3 in Grasselli and Tebaldi (2008), connections between the classification of symmetric cones and Euclidean Jordan algebras (EJA) are used to produce sufficient conditions for an ATSM to be admissible. These conditions are initially only sufficient for admissibility and not necessary. However subsequently, the conditions will be shown to be necessary and sufficient for an ATSM to be solvable. We remind the reader that the ability to compute the affine transform, this means solvability, is the aim of this chapter and not its admissibility. This is an important practical point of view. Before defining solvability, we present an important observation from Grasselli and Tebaldi (2008).

Remark 9.1.3 All possible finite dimensional irreducible symmetric cones are:

- the families of positive definite matrix cones of Hermitian matrices with real, complex, or quaternion entries;
- the Lorentz cones $\Lambda_n = \{x \in \Re^n : x_1^2 - \sum_{i=2}^n x_i^2 > 0,\ x_1 > 0\}$;
- the exceptional cone (27 dimensional cone of 3×3 "positive definite" matrices over the Cayley algebra).

Consequently, the results in Grasselli and Tebaldi (2008) actually produce all possible domains on which one can find solvable ATSMs. The definition of solvability also suggests the solution procedure, which we present in an algorithm below. Subsequently, we will tailor this algorithm to specific state spaces.

Definition 9.1.4 An ATSM in a symmetric cone state space $\mathcal{D}$ is solvable if and only if the corresponding Riccati ODE is linearizable.

For a formal definition of linearizability of ODEs on a symmetric cone state space, we refer the reader to Definition 2.5 in Grasselli and Tebaldi (2008). A consequence of linearizability is that we can explicitly compute the conditional characteristic function in an algorithmic fashion, as will be given below. We point out that for convenience, we again consider the state space $\mathcal{D} = \mathcal{D}_+ \times \Re^{n-m}$, where $\Re^{n-m}$ supports possibly negative, conditionally Gaussian factor sensitivities.

Algorithm 9.1 Linearization Algorithm

1: Solve the ODE corresponding to the conditionally Gaussian factors
2: Decompose the remaining ODEs as a direct sum of Riccati ODEs corresponding to factors living on irreducible symmetric cone state spaces
3: Linearize each of these Riccati ODEs
4: Solve the linearized system and map back to the original system.

The difficult step in Algorithm 9.1, as the examples will illustrate, is Step 4. For autonomous ODEs, this step is straightforward. However, should the coefficient be time dependent, the linear system involves a time ordered exponential, which is a symbolic expression that is hard to compute. As remarked in Grasselli and Tebaldi (2008), this is what is to be expected for linear non-autonomous ODEs. Finally, Grasselli and Tebaldi (2008) point out that Walcher (1991) proposed an alternative procedure for non-autonomous ODEs, but this procedure is explicit up to the knowledge of a particular solution, to be determined on a case by case basis.

We will now discuss the state space $\mathcal{D} = (\Re^+)^m \times \Re^{n-m}$, the Duffie and Kan (1996) state space, and subsequently illustrate it with examples.

9.2 Solvable Affine Processes on the Duffie-Kan State Space

We model the factor process X via the following SDE:

$$dX_t = (\boldsymbol{\Omega} X_t + \boldsymbol{\Omega}^0)\,dt + \mathrm{Diag}\big((C_i X_t + C_i^0)^{1/2}\big)\,dW_t, \tag{9.2.3}$$

$t \geq 0$, where

$$X_0 = x \in (\Re^+)^m \times \Re^{n-m},$$

where C_i denotes the ith row of the matrix C, $m = \mathrm{Rank}(C) \leq n$ and $W = \{W_t, t \geq 0\}$ is an n-dimensional Brownian motion. The parameters $\boldsymbol{\eta}$, η^0, $\boldsymbol{\Omega}$, $\boldsymbol{\Omega}^0$, C, C^0 are defined as follows:

- The drift matrix $\boldsymbol{\Omega}$ satisfies

$$\boldsymbol{\Omega} = \begin{pmatrix} \boldsymbol{\Omega}^{DD}_{m\times m} & \mathbf{0}_{m\times(n-m)} \\ \boldsymbol{\Omega}^{BD}_{(n-m)\times m} & \boldsymbol{\Omega}^{BB}_{(n-m)\times(n-m)} \end{pmatrix},$$

  where the off-diagonal elements of $\boldsymbol{\Omega}^{BB}$ are restricted to be nonnegative;
- $\boldsymbol{\Omega}^0 \in (\Re^+)^m \times \Re^{n-m}$;
- C and C^0 are given by

$$C = \begin{pmatrix} \mathbf{1}_{m\times m} & \mathbf{0}_{m\times(n-m)} \\ C^{BD}_{(n-m)\times m} & \mathbf{0}_{(n-m)\times(n-m)} \end{pmatrix},$$

$$C^0 = \begin{pmatrix} \mathbf{0}_{m\times 1} \\ \mathbf{1}_{(n-m)\times 1} \end{pmatrix};$$

- $\eta_0 \geq 0$, $\boldsymbol{\eta}^D \in (\Re^+)^m$, $\boldsymbol{\eta}^B = 0$.

In the above notation, D corresponds to $\Re^+$-valued factors, and B to $\Re$-valued factors. We point out that the state space of positive factors $(\Re^+)^m$ is the closure of a reducible symmetric cone and obtained by considering m copies of $\Re^+$. We now recall Theorem 4.1 from Grasselli and Tebaldi (2008), which states when affine diffusions on the Duffie-Kan spate space are solvable.

Theorem 9.2.1 *The ATSM* (9.2.3) *is solvable on the state space* $(\Re^+)^m \times \Re^{n-m}$ *if and only if the matrix* $\boldsymbol{\Omega}^{DD}$ *is diagonal.*

We alert the reader to the fact that a necessary condition for an ATSM to be solvable on the state space $(\Re^+)^m \times \Re^{n-m}$ is that positive risk factors are uncorrelated. It is, therefore, clear that retaining analytical tractability comes at the expense of modeling freedom, see also Sect. 9.7. However, in Chap. 11 we will discuss a class of processes allowing for more complex dependence structures.

Example 9.2.2 To illustrate the nature of the condition given in Theorem 9.2.1, consider the following double-mean reverting model:

$$\begin{aligned} dv_t &= -\kappa\big(v_t - v'_t\big)\,dt + \sigma_1\sqrt{v_t}\,dW_t^2, \\ dv'_t &= -c\big(v'_t - z_3\big)\,dt + \sigma_2\sqrt{v'_t}\,dW_t^3, \\ \frac{dS_t}{S_t} &= \sqrt{v_t}\,dW_t^1, \end{aligned}$$

where $\kappa, c, \sigma_, \sigma_2 > 0$. Using the notation $\boldsymbol{X}_t := (v_t, v'_t, Y_t)$, where $Y_t = \log(S_t)$ is a potentially negative factor, we have $m = 2$ and $n = 3$ and the drift matrix $\boldsymbol{\Omega}$ is given by Itô's formula

$$\boldsymbol{\Omega} = \begin{pmatrix} -\kappa & \kappa & 0 \\ 0 & -c & 0 \\ -\frac{1}{2} & 0 & 0 \end{pmatrix},$$

which is not diagonal. In fact, it is clear that the presence of v'_t in the equation for v_t destroys the diagonal structure of $\boldsymbol{\Omega}$. On the other hand, Theorem 7.1.4 yields that the ATSM is admissible, as this only requires $\boldsymbol{\Omega}^{DD}$ to have non-negative off-diagonal elements.

The advantage of the Grasselli-Tebaldi approach is that we have an algorithm at hand, showing how to solve the ATSM. In Grasselli and Tebaldi (2008), the general Algorithm 9.1 was tailored to the special case of the Duffie-Kan state space, which we now recall.

On the Duffie-Kan state space, the Riccati ODEs of a solvable ATSM become

$$\begin{aligned} \frac{d}{d\tau} V_i^D(\tau) = \eta_i^D + \Omega_{i,i}^{DD} V_i^D + \sum_{j=m+1}^{n} \big(\boldsymbol{\Omega}^{BD}\big)_{i,j}^\top V_j^B \\ - \frac{1}{2}\sum_{i=1}^{m}\big(V_i^D(\tau)\big)^2 - \frac{1}{2}\sum_{j=m+1}^{n}\big(\boldsymbol{C}^{BD}\big)_{i,j}^\top \big(V_j^B(\tau)\big)^2, \quad i = 1, \ldots, m \end{aligned} \tag{9.2.4}$$

$$\frac{d}{d\tau} V_i^B(\tau) = \sum_{j=m+1}^{n} \big(\boldsymbol{\Omega}^{BB}\big)_{i,j}^\top V_j^B(\tau) + \eta_i^B, \quad i = m+1, \ldots, n, \tag{9.2.5}$$

with boundary condition $\boldsymbol{V}(T) = \imath\boldsymbol{u} \in \mathbf{C}^n$. As stated above, the factor sensitivities, corresponding to real, possibly negative factors, can be determined by solving an independent system of linear equations, see (9.2.5). We now show how to tailor Algorithm 9.1 to the Duffie-Kan state space.

Step 1 As stated above, the factor sensitivities corresponding to the conditionally Gaussian factors satisfy (9.2.5), which is easily solved to yield

$$\boldsymbol{V}^B(\tau) = \int_0^\tau \exp\big\{(\tau - t)\big(\boldsymbol{\Omega}^{BB}\big)^\top\big\}\boldsymbol{\eta}^B\,dt + \exp\big\{\big(\boldsymbol{\Omega}^{BB}\big)^\top\tau\big\}\boldsymbol{V}^B(0). \tag{9.2.6}$$

Step 2 Each of the remaining factors $V_i^D(\tau)$, $i = 1, \ldots, m$, satisfies a one-dimensional time-dependent Riccati equation, (9.2.4).

Step 3 We linearize the Riccati ODE for $V_i^D(\tau)$ by setting

$$V_i^D(\tau) = \lambda^{-1}(\tau)\pi(\tau),$$

which yields

$$\frac{d}{d\tau}\big(\lambda(\tau)V_i^D(\tau)\big) - \frac{d}{d\tau}\lambda(\tau)V_i^D(\tau) = \lambda(\tau)\frac{d}{d\tau}V_i^D(\tau),$$

and hence substituting (9.2.4), we get

$$\frac{d}{d\tau}\pi(\tau) - \frac{d}{d\tau}\lambda(\tau)V_i^D(\tau) = \tilde{\gamma}_i(\tau)\lambda(\tau) + \Omega_{i,i}^{DD}\pi(\tau) - \frac{1}{2}\pi(\tau)V_i^D(\tau),$$

where

$$\tilde{\gamma}_i(\tau) = \eta_i^D + \sum_{j=m+1}^{n} (\boldsymbol{\Omega}^{BD})_{i,j}^{\top} V_j^B - \frac{1}{2}\sum_{j=m+1}^{n} (\boldsymbol{C}^{BD})_{i,j}^{\top} (V_j^B(\tau))^2. \tag{9.2.7}$$

Matching coefficients yields

$$d\begin{pmatrix} \pi(\tau) \\ \lambda(\tau) \end{pmatrix} = \begin{pmatrix} \Omega_{i,i}^{DD} & \tilde{\gamma}_i(\tau) \\ \frac{1}{2} & 0 \end{pmatrix}\begin{pmatrix} \pi(\tau) \\ \lambda(\tau) \end{pmatrix}. \tag{9.2.8}$$

Step 4 Finally, we exponentiate (9.2.8) to obtain

$$\begin{pmatrix} \pi(\tau) \\ \lambda(\tau) \end{pmatrix} = T\exp\begin{Bmatrix} \tau\Omega_{ii}^{DD} & \int_0^\tau \tilde{\gamma}_i(t)\,dt \\ \frac{\tau}{2} & 0 \end{Bmatrix}\begin{pmatrix} \pi(0) \\ \lambda(0) \end{pmatrix},$$

where $T\exp$ denotes the time ordered exponential. Introducing the notation

$$\begin{pmatrix} M_1^i(\tau) & M_2^i(\tau) \\ M_3^i(\tau) & M_4^i(\tau) \end{pmatrix} := T\exp\begin{Bmatrix} \tau\Omega_{i,i}^{DD} & \int_0^\tau \tilde{\gamma}_i(t)\,dt \\ \frac{\tau}{2} & 0 \end{Bmatrix},$$

we can represent the solution $V_i^D(\tau)$ as follows:

$$V_i^D(\tau) = \frac{V_i^D(0)M_1^i(\tau) + M_2^i(\tau)}{V_i^D(0)M_3^i(\tau) + M_4^i(\tau)}.$$

9.3 Reducing Admissible Affine Processes to the Normal Form

In this section, we show how to reduce an affine process to the normal form, as shown in (9.2.3). Such a transformation is crucial, as it allows us to check if a process is solvable. We start with the following definition of an affine process, which is due to Dai and Singleton (2000), and is essentially the same as in Duffie et al. (2000). We point out that the material presented in this section follows closely Grasselli and Tebaldi (2004b). The following formulation, taken from Dai and Singleton (2000), is convenient to work with.

Definition 9.3.1 A term structure model is an affine ATSM, if the short rate r_t is an affine combination of factors

$$r_t = \delta^0 + \boldsymbol{\delta}' \boldsymbol{Y}_t, \quad t \geq 0,$$

for some $\delta^0 \in \Re$ and $\boldsymbol{\delta} \in \Re^n$, and the dynamics of the factors satisfy the following SDE:

$$\begin{aligned} d\boldsymbol{Y}_t &= \boldsymbol{\kappa}(\boldsymbol{\theta} - \boldsymbol{Y}_t)\,dt + \boldsymbol{\Sigma}\,\mathrm{Diag}\big((\alpha_i + \boldsymbol{\beta}_i^\top \boldsymbol{Y}_t)^{1/2}\big)\,d\boldsymbol{W}_t, \quad t \geq 0, \\ \boldsymbol{Y}_0 &= \boldsymbol{y} \in \Re^n, \end{aligned} \tag{9.3.10}$$

where $\boldsymbol{W} = \{\boldsymbol{W}_t,\, t \geq 0\}$ is an n-dimensional standard Brownian motion and

$$\boldsymbol{\theta}, \boldsymbol{\alpha} \in \Re^n; \qquad \boldsymbol{\kappa}, \boldsymbol{\beta} \in \mathcal{M}_n; \qquad \boldsymbol{\Sigma} \in GL(n).$$

Here α_i indicates the i-th element of the vector $\boldsymbol{\alpha}$. The notation $\boldsymbol{\beta}_i$ refers to the i-th column of matrix $\boldsymbol{\beta}$. For a given vector $z \in \Re^n$, $\mathrm{Diag}(z_i) \in \mathcal{M}_n$ is the diagonal matrix with the elements of the vector z along the diagonal. We require $\boldsymbol{\Sigma}$ to be an element of $GL(n)$, the set of $n \times n$ invertible matrices.

The rank of $\boldsymbol{\beta}$ is defined to be $m \leq n$: in the notation of Dai and Singleton (2000), m classifies the families $\mathbf{A}_m(n)$ of admissible models parameterized by m, n. Without loss of generality we assume that the upper left minor of order m is non-singular. Given an ATSM parameterized by $(\delta^0, \boldsymbol{\delta}, \boldsymbol{\kappa}, \boldsymbol{\theta}, \boldsymbol{\Sigma}, \{\alpha_i, \boldsymbol{\beta}_i\}_{1 \leq i \leq n})$, an affine change of variables

$$\boldsymbol{Y} \to \boldsymbol{X} = \boldsymbol{L}\boldsymbol{Y} + \boldsymbol{\vartheta} \tag{9.3.11}$$

leaves prices unaffected, while the parameters are changed according to

$$\begin{aligned} &\big(\delta^0, \boldsymbol{\delta}, \boldsymbol{\kappa}, \boldsymbol{\theta}, \boldsymbol{\Sigma}, \{\alpha_i, \boldsymbol{\beta}_i\}_{1 \leq i \leq n}\big) \to \\ &\big(\delta^0 - \boldsymbol{\delta}' \boldsymbol{L}^{-1} \boldsymbol{\vartheta}, (\boldsymbol{L}^{-1})^\top \boldsymbol{\delta}, \boldsymbol{L}\boldsymbol{\kappa}\boldsymbol{L}^{-1}, \boldsymbol{L}\boldsymbol{\theta} + \boldsymbol{\vartheta}, \boldsymbol{L}^{-1}\boldsymbol{\Sigma}, \\ &\quad \{(\boldsymbol{\alpha} - \boldsymbol{\beta}^\top \boldsymbol{L}^{-1}\boldsymbol{\vartheta})_i, (\boldsymbol{\beta}^\top \boldsymbol{L}^{-1})_i\}_{1 \leq i \leq n}\big). \end{aligned}$$

Consequently, we can study models in their normal form, see (9.2.3).

Definition 9.3.2 Consider a symmetry transformation $(\boldsymbol{L}, \boldsymbol{\vartheta})$ of the type given in Eq. (9.3.11). Let us fix $\boldsymbol{L} = \boldsymbol{\Sigma}^{-1}$ and let $\boldsymbol{\vartheta} \in \Re^n$ denote any solution to the system of equations

$$\begin{aligned} \boldsymbol{\beta}_i^\top \boldsymbol{\Sigma} \boldsymbol{\vartheta} &= \alpha_i, \quad i = 1, \ldots, m \\ \big\{\boldsymbol{\Sigma}^{-1} \boldsymbol{\kappa}(\boldsymbol{\theta} - \boldsymbol{\vartheta})\big\}_i &= 0, \quad i = m+1, \ldots, n. \end{aligned}$$

Such a transformation maps the original factor dynamics (9.3.10) into the *normal form ATSM*, whose factor dynamics become

$$\begin{aligned} d\boldsymbol{X}_t &= \big(\boldsymbol{A}\boldsymbol{X}_t + \boldsymbol{A}^0\big)\,dt + \mathrm{Diag}\big(S_{i,i}^{1/2}\big)\,d\boldsymbol{W}_t, \quad t \geq 0 \\ S_{i,i} &= \big(\boldsymbol{C}_i \boldsymbol{X}_t + C_i^0\big), \\ \boldsymbol{X}_0 &= \boldsymbol{x}. \end{aligned} \tag{9.3.12}$$

Here $\boldsymbol{C}_i$ denotes the i-th row of the matrix $\boldsymbol{C}$, $m = \text{Rank}(\boldsymbol{C})$, while the short rate is given by

$$r_t = \gamma^0 + \boldsymbol{\gamma}' \boldsymbol{X}_t.$$

The parameters $\boldsymbol{\phi} = (\boldsymbol{\gamma}, \gamma^0, \boldsymbol{A}, \boldsymbol{A}^0, \boldsymbol{C}, \boldsymbol{C}^0)$ are defined as follows:

$$\begin{aligned}
\boldsymbol{X}_t &= \boldsymbol{\Sigma}^{-1} \boldsymbol{Y}_t, \quad t \geq 0, \\
\gamma^0 &= \delta^0 - \boldsymbol{\delta}^\top \boldsymbol{\Sigma} \boldsymbol{\vartheta}, \\
\boldsymbol{\gamma} &= \boldsymbol{\Sigma}^\top \boldsymbol{\delta}, \\
\boldsymbol{A} &= -\boldsymbol{\Sigma}^{-1} \boldsymbol{\kappa} (\boldsymbol{\theta} - \boldsymbol{\vartheta}) \in \mathcal{M}_n, \\
\boldsymbol{A}^0 &= \boldsymbol{\Sigma}^{-1} \boldsymbol{\kappa} (\boldsymbol{\theta} - \boldsymbol{\vartheta}) \in \Re^n, \\
\boldsymbol{C} &= \boldsymbol{\beta}^\top \boldsymbol{\Sigma} \in \mathcal{M}_n, \\
\boldsymbol{C}^0 &= \boldsymbol{\alpha} - \boldsymbol{\beta}^\top \boldsymbol{\Sigma} \boldsymbol{\vartheta},
\end{aligned}$$

where $C_i^0 = 0$, $i = 1, \ldots, m$ and $A_i^0 = 0$, $i = m+1, \ldots, n$.

Recall that in Theorem 7.1.4, we characterized the conditions for admissibility on the state space $\mathcal{D} = (\Re^+)^m \times \Re^{n-m}$. The reason for introducing the normal form is the ability to reduce any model to one whose natural domain is the state space $\mathcal{D}$. Consequently, from Theorem 7.1.4, we can identify parameter restrictions imposed by the admissibility conditions on the normal form. This observation motivates us to introduce the following definition.

Definition 9.3.3 An ATSM is called *admissible in its natural domain* if the corresponding normal form ATSM is admissible within the canonical domain $\mathcal{D}$.

We have the following proposition, which yields parameter restrictions on the normal form.

Proposition 9.3.4 *The normal form corresponding to an admissible ATSM in its natural domain is specified by the parameter set* $\boldsymbol{\phi} = (\boldsymbol{\gamma}, \gamma^0, \boldsymbol{A}, \boldsymbol{A}^0, \boldsymbol{C}, \boldsymbol{C}^0)$ *with:*

- *drift matrix* $\boldsymbol{A}$ *given by*

$$\boldsymbol{A} = \begin{pmatrix} \boldsymbol{A}_{m \times m}^{BB} & \boldsymbol{0}_{m \times (n-m)} \\ \boldsymbol{A}_{(n-m) \times m}^{DB} & \boldsymbol{A}_{(n-m) \times (n-m)}^{DD} \end{pmatrix}$$

  where the off-diagonal elements of $\boldsymbol{A}^{BB}$ *are restricted to be nonnegative;*
- $\boldsymbol{A}^0 \in (\Re^+)^m \times \Re^n$;
- $\boldsymbol{C}$ *and* $\boldsymbol{C}^0$ *are given by*

$$\boldsymbol{C} = \begin{pmatrix} \boldsymbol{I}_{m \times m} & \boldsymbol{0}_{m \times (n-m)} \\ \boldsymbol{C}_{(n-m) \times m}^{DB} & \boldsymbol{0}_{(n-m) \times (n-m)} \end{pmatrix},$$

$$\boldsymbol{C}^0 = \begin{pmatrix} \boldsymbol{0}_{m \times 1} \\ \boldsymbol{C}_{(n-m) \times 1}^{D} \end{pmatrix},$$

  where the elements of $\boldsymbol{C}_{(n-m) \times m}^{DB}$, $\boldsymbol{C}_{(n-m) \times 1}^{D}$ *are nonnegative;*
- $\gamma^0 \in \Re$, $\boldsymbol{\gamma} \in \Re^n$.

Proof We note that in the normal form framework,

$$d\left[X_{\cdot}^{k}, X_{\cdot}^{l}\right]_{t} = \left(\left(\sum_{i=1}^{m} C_{k,i} X_{t}^{i} + C_{k}^{0}\right) \delta_{k,l}\right) dt, \tag{9.3.13}$$

but the admissibility condition in Theorem 7.1.4 implies

$$d\left[X_{\cdot}^{k}, X_{\cdot}^{l}\right]_{t} = \left(\sum_{i=1}^{m} \alpha_{i,kl} X_{t}^{i} + a_{k,l}\right) dt. \tag{9.3.14}$$

We equate the expressions in Eqs. (9.3.13) and (9.3.14), and since the rank of matrix $\boldsymbol{C}$ is m, we can without loss of generality assume that $\alpha_{i,ii} = 1$ and that the non-zero eigenvalues of the matrix $\boldsymbol{A}$ are equal to 1. Now

$$C_{k,i} \delta_{k,l} = \alpha_{i,kl}, \quad k, l, i = 1, \ldots, m,$$

yields that $[C_{k,i}]$, $k, i = 1, \ldots, m$, is the identity matrix. Also, we have

$$C_{k,i} \delta_{k,l} = \alpha_{i,kl}, \quad k = m+1, \ldots, n, \ i = 1, \ldots, m, \ l = 1, \ldots, n,$$

and since $\boldsymbol{\alpha}_i$ is positive semi-definite, setting $k = l$, we obtain that $C_{k,i} \geq 0$. Lastly, regarding the constants

$$C_{k}^{0} \delta_{k,l} = a_{k,l},$$

only the lower diagonal square block of order $n - m$ is non-zero, hence we focus on

$$C_{m+l}^{0} \delta_{k+m,l+m} = a_{k+m,l+m},$$

$l, k = 1, \ldots, n - m$. The condition $C_{m+l}^{0} \geq 0$, $l = 1, \ldots, n - m$, follows since $\boldsymbol{A}$ is positive semidefinite. □

For illustration, we apply below this methodology.

9.4 A First Example: The Balduzzi, Das, Foresi and Sundaram Model

In this section, we wish to apply the methodology developed in this chapter to a particular short rate model, the Balduzzi, Das, Foresi, and Sundaram model, see Balduzzi et al. (1996), where we follow the presentation in Grasselli and Tebaldi (2004a). It is a three factor model, the factors being the short rate, $r = \{r_t, \, t \geq 0\}$, its central tendency $\theta = \{\theta_t, \, t \geq 0\}$, and its variance $v = \{v_t, \, t \geq 0\}$. We immediately present the model under the assumed risk-neutral probability measure, which we employ for pricing:

$$\begin{aligned}
dv_t &= \mu(\bar{v} - v_t)\, dt + \eta \sqrt{v_t}\, dW_t^{v}, \\
d\theta_t &= \alpha(\bar{\theta} - \theta_t)\, dt + \xi\, dW_t^{\theta}, \\
dr_t &= \kappa(\theta_t - r_t)\, dt + \sqrt{v_t}\, dW_t^{r},
\end{aligned}$$

where $\mu, \bar{v}, \eta > 0$, $\alpha, \theta, \xi, \kappa \in \Re$, and $\boldsymbol{W} = (W^v, W^\theta, W^v)$ is a standard Brownian motion under the risk-neutral measure. We can see that the diffusion coefficient of the short rate $r = \{r_t,\, t \geq 0\}$ is stochastic, it is governed by $v = \{v_t,\, t \geq 0\}$, and so is its central tendency $\theta = \{\theta_t,\, t \geq 0\}$. Note that this model nests *Gaussian central tendency models*, see e.g Beaglehole and Tenney (1991), Jegadesh and Pennachi (1996), and *short rate stochastic volatility models*, see e.g. Fong and Vasiček (1991), as special cases. By re-scaling the process v_t, i.e. introducing $\tilde{v} = \{\tilde{v}_t,\, t \geq 0\}$ given by $\tilde{v}_t = \frac{v_t}{\eta^2}$, we obtain the model in normal form:

$$\begin{aligned}
d\tilde{v}_t &= \mu\left(\frac{\bar{v}}{\eta^2} - \tilde{v}_t\right) dt + \sqrt{\tilde{v}_t}\, dW_t^v, \\
d\theta_t &= \alpha(\bar{\theta} - \theta_t)\, dt + \xi\, dW_t^\theta, \\
dr_t &= \kappa(\theta_t - r_t)\, dt + \eta\sqrt{\tilde{v}_t}\, dW_t^r.
\end{aligned}$$

Rewriting this system as in (9.2.3), we obtain for

$$\boldsymbol{X}_t = \begin{pmatrix} \tilde{v}_t \\ \theta_t \\ r_t \end{pmatrix},$$

the vector SDE

$$d\boldsymbol{X}_t = (\boldsymbol{\Omega}\boldsymbol{X}_t + \boldsymbol{\Omega}^0)\, dt + \mathrm{Diag}\big((\boldsymbol{C}_i\boldsymbol{X}_t + C_i^0)^{1/2}\big)\, d\boldsymbol{W}_t^\top,$$

where

$$\boldsymbol{W}_t^\top = \begin{pmatrix} W_t^v \\ W_t^\theta \\ W_t^r \end{pmatrix},$$

and

$$\boldsymbol{\Omega} = \begin{pmatrix} -\mu & 0 & 0 \\ 0 & -\alpha & 0 \\ 0 & \kappa & -\kappa \end{pmatrix}, \quad \boldsymbol{\Omega}^0 = \begin{pmatrix} \frac{\mu\bar{v}}{\eta^2} \\ \alpha\bar{\theta} \\ 0 \end{pmatrix},$$

$$\boldsymbol{C}_1 = \begin{pmatrix} 1 & 0 & 0 \end{pmatrix}, \qquad \boldsymbol{C}_2 = \begin{pmatrix} 0 & 0 & 0 \end{pmatrix}, \qquad \boldsymbol{C}_3 = \begin{pmatrix} \eta^2 & 0 & 0 \end{pmatrix},$$

$$\boldsymbol{C}^0 = \begin{pmatrix} 0 \\ \xi^2 \\ 0 \end{pmatrix}.$$

We point out that the state space is $\Re^+ \times \Re^2$, i.e. $m = 1$ and $n = 3$. Next we focus on computing the bond price, which by (9.1.1) yields the expression

$$\begin{aligned}
P_{t+\tau}(t) &= \Psi_X(\boldsymbol{0}, \boldsymbol{x}_t, t, \tau) \\
&= E\left(\exp\left\{-\int_t^{t+\tau} (\eta_0 + \langle \boldsymbol{\eta}, \boldsymbol{X}_s\rangle)\, ds\right\} \,\middle|\, \mathcal{A}_t\right)
\end{aligned}$$

$$
\begin{aligned}
&= \exp\{V^0(\tau, \mathbf{0}) - \langle \boldsymbol{V}(\tau, \mathbf{0}), \boldsymbol{x}_t \rangle\} \\
&= \exp\left\{ V^0(\tau, \mathbf{0}) - V^v(\tau, 0)\frac{v_t}{\eta^2} - V^\theta(\tau, 0)\theta_t - V^r(\tau, 0)r_t \right\},
\end{aligned}
\tag{9.4.15}
$$

where $\boldsymbol{V}(\tau, \boldsymbol{u}) = (V^v(\tau, u_1), V^\theta(\tau, u_2), V^r(\tau, u_3))$, subject to the initial conditions

$$
V^v(0, u_1) = 0, \qquad V^\theta(0, u_2) = 0, \qquad V^r(0, u_3) = 0, \qquad V^0(0, \boldsymbol{u}) = 0.
$$

First, we compute the functions

$$
\boldsymbol{V}^B(\tau, \mathbf{0}) = \left(V_1^B(\tau, 0), V_2^B(\tau, 0)\right) = \left(V^\theta(\tau, 0), V^r(\tau, 0)\right),
$$

which correspond to the conditionally Gaussian factors. By setting

$$
\boldsymbol{V}^B(\tau, \mathbf{0}) = \boldsymbol{V}^B(\tau) = \left(V_1^B(\tau), V_2^B(\tau)\right) = \left(V^\theta(\tau), V^r(\tau)\right),
$$

we recall (9.2.6), which states that

$$
\boldsymbol{V}^B(\tau) = \int_0^\tau \exp\{(\tau - t)(\boldsymbol{\Omega}^{BB})^\top\}\boldsymbol{\eta}^B\, dt + \exp\{(\boldsymbol{\Omega}^{BB})^\top\}\boldsymbol{V}^B(0).
$$

This results in the relation

$$
V^r(\tau) = \int_0^\tau \exp\{-\kappa(\tau - t)\}\, dt = \frac{1 - \exp\{-\kappa\tau\}}{\kappa}
$$

and

$$
\begin{aligned}
V^\theta(\tau) &= \int_0^\tau \frac{\kappa}{\alpha - \kappa}\left(\exp\{-\kappa(\tau - t)\} - \exp\{-\alpha(\tau - t)\}\right) dt \\
&= -\frac{\exp\{-\kappa\tau\} - \exp\{-\alpha\tau\}}{\alpha - \kappa} + \frac{1 - \exp\{-\alpha\tau\}}{\alpha}.
\end{aligned}
$$

Finally, regarding $V^D(\tau) = V^v(\tau) = V^v(\tau, 0)$, we have

$$
V^v(\tau) = \frac{M_2(\tau)}{M_4(\tau)},
$$

with the time ordered exponential

$$
\begin{pmatrix} M_1(\tau) & M_2(\tau) \\ M_3(\tau) & M_4(\tau) \end{pmatrix} := T \exp\begin{Bmatrix} -\tau\mu & \int_0^\tau \tilde{\gamma}(t)\, dt \\ \tau/2 & 0 \end{Bmatrix}
\tag{9.4.16}
$$

and

$$
\tilde{\gamma}(t) = -\frac{1}{2}\eta^2\left(V^v(t)\right)^2.
$$

We can simplify (9.4.16) using the integral

$$
\int_0^\tau \tilde{\gamma}(t)\, dt = -\frac{1}{2}\frac{\eta^2}{\kappa^2}\left(\tau - \frac{2(1 - \exp\{-\kappa\tau\})}{\kappa} + \frac{1 - \exp\{-2\kappa\tau\}}{2\kappa}\right).
$$

Lastly, from Eq. (9.1.3) it follows that $V^0(\tau)$ can be obtained from

$$
\frac{d}{d\tau}V^0(\tau) = -\frac{\mu\bar{v}}{\eta^2}V^v(\tau) - \alpha\bar{\theta}V^\theta(\tau) + \frac{1}{2}\xi^2\left(V^\theta(\tau)\right)^2,
$$

subject to $V^0(0) = 0$. This provides an explicit description of the zero coupon bond price given in (9.4.15).

9.5 A Second Example: The Heston Model

In this section, we apply the methodology developed in this chapter to the Heston model, as introduced in Heston (1993). We assume that the dynamics of the logarithm of the stock price $Y = \{Y_t,\ t \geq 0\}$ and the variance process $v = \{v_t,\ t \geq 0\}$ are given by

$$dv_t = \kappa(\eta - v_t)\,dt + \lambda\sqrt{v_t}\,dW_t^1,$$
$$dY_t = \left(r - \frac{1}{2}v_t\right)dt + \rho\sqrt{v_t}\,dW_t^1 + \sqrt{1-\rho^2}\sqrt{v_t}\,dW_t^2,$$

where $\kappa, \eta, \lambda > 0$, $r \in \Re$, $\rho \in (-1, 1)$, and $\boldsymbol{W} = \{\boldsymbol{W}_t = (W_t^1,\ W_t^2),\ t \geq 0\}$ denotes a standard Brownian motion. The Heston model does not fit the ATSM model formulation in Definition 9.3.1. Nevertheless, we find the discussion in Sect. 9.3 useful in rewriting the Heston model in normal form. We define a matrix $\boldsymbol{\Sigma}$, which is given by

$$\boldsymbol{\Sigma} = \begin{pmatrix} \lambda & 0 \\ \rho & \sqrt{1-\rho^2} \end{pmatrix},$$

and hence

$$\boldsymbol{\Sigma}^{-1} = \frac{1}{\lambda\sqrt{1-\rho^2}}\begin{pmatrix} \sqrt{1-\rho^2} & 0 \\ -\rho & \lambda \end{pmatrix}.$$

The matrix $\boldsymbol{\Sigma}$ plays a comparable role to the one appearing in Definition 9.3.2. Consequently, we introduce the new processes

$$\begin{pmatrix} \hat{v}_t \\ \hat{Y}_t \end{pmatrix} = \boldsymbol{\Sigma}^{-1}\begin{pmatrix} v_t \\ Y_t \end{pmatrix},$$

and hence $\hat{v}_t$ satisfies the SDE

$$d\hat{v}_t = \kappa\left(\frac{\eta}{\lambda} - \hat{v}_t\right)dt + \sqrt{\hat{v}_t\lambda}\,dW_t^1,$$

and

$$d\hat{Y}_t = \left(\frac{r}{\sqrt{1-\rho^2}} - \frac{v_t}{2\sqrt{1-\rho^2}}\right)dt + \sqrt{v_t}\,dW_t^2 - \frac{\rho\kappa}{\sqrt{1-\rho^2}}\left(\frac{\eta}{\lambda} - \frac{v_t}{\lambda}\right)dt.$$

Finally, we set

$$\tilde{v}_t = \frac{\hat{v}_t}{\lambda} \quad \text{and} \quad \tilde{Y}_t = \hat{Y}_t,$$

to obtain the SDE

$$d\boldsymbol{X}_t = \begin{pmatrix} d\tilde{v}_t \\ d\tilde{Y}_t \end{pmatrix} = \begin{pmatrix} \frac{\kappa\eta}{\lambda^2} + \tilde{v}_t(-\kappa) \\ \frac{r - \frac{\rho\kappa\eta}{\lambda}}{\sqrt{1-\rho^2}} + \tilde{v}_t \frac{\rho\kappa\lambda - \frac{\lambda^2}{2}}{\sqrt{1-\rho^2}} \end{pmatrix} dt + \text{Diag}\begin{pmatrix} \sqrt{\boldsymbol{C}_1 \boldsymbol{X}_t} \\ \sqrt{\boldsymbol{C}_2 \boldsymbol{X}_t} \end{pmatrix} d\boldsymbol{W}_t,$$

where

$$\boldsymbol{C}_1 = (1 \quad 0), \qquad \boldsymbol{C}_2 = (1 \quad 0).$$

This means in terms of (9.2.3) one has

$$\boldsymbol{\Omega} = \begin{pmatrix} -\kappa & 0 \\ \frac{\rho\kappa\lambda - \frac{\lambda^2}{2}}{\sqrt{1-\rho^2}} & 0 \end{pmatrix},$$

$$\boldsymbol{\Omega}_0 = \begin{pmatrix} \frac{\kappa\eta}{\lambda^2} \\ \frac{r - \frac{\rho\kappa\eta}{\lambda}}{\sqrt{1-\rho^2}} \end{pmatrix},$$

$$\boldsymbol{C}^0 = \begin{pmatrix} 0 \\ 0 \end{pmatrix}.$$

We are now ready to compute the characteristic function of the logarithm of the stock price under the Heston model, and obtain

$$\begin{aligned} \Psi_X(u, \boldsymbol{x}_t, t, \tau) &= E\big(\exp\{\imath u Y_T\} \,\big|\, \mathcal{A}_t\big) \\ &= E\big(\exp\big\{\imath u \sqrt{1-\rho^2}\tilde{Y}_T + \imath u \rho\lambda \tilde{v}_T\big\} \,\big|\, \mathcal{A}_t\big) \\ &= \exp\big\{V^0(\tau, \imath u) - \tilde{Y}_t V^B(\tau, \imath u) - \tilde{v}_t V^D(\tau, \imath u)\big\}, \end{aligned}$$

subject to the initial conditions

$$V^0(0, \imath u) = 0, \qquad V^B(0, \imath u) = -\imath u\sqrt{1-\rho^2}, \qquad V^D(0, \imath u) = -\imath u \rho\lambda.$$

We now follow the approach described in Sect. 9.2, and solve

$$V^B(\tau, \imath u) = \exp\{0\} V^D(0, \imath u) = -\imath u \sqrt{1-\rho^2}.$$

Consequently, we follow Steps 2 to 4, yielding

$$V^D(\tau, \imath u) = \frac{M_1(\tau) V^D(0, \imath u) + M_2(\tau)}{M_3(\tau) V^D(0, \imath u) + M_4(\tau)},$$

where

$$\begin{pmatrix} M_1(\tau) & M_2(\tau) \\ M_3(\tau) & M_4(\tau) \end{pmatrix} := \exp\left\{\tau \begin{pmatrix} \Omega^{DD} & \tilde{\gamma} \\ \frac{1}{2} & 0 \end{pmatrix}\right\},$$

and

$$\tilde{\gamma} = \frac{\rho\kappa\lambda - \frac{\lambda^2}{2}}{\sqrt{1-\rho^2}}\big(-\imath u \sqrt{1-\rho^2}\big) - \frac{\lambda^2}{2}(\imath u)^2\big(1-\rho^2\big), \qquad \Omega^{DD} = -\kappa.$$

Using the notation

$$d = \sqrt{(\kappa - \lambda\rho\iota u)^2 + \lambda^2(\iota u + u^2)},$$

we obtain

$$M_1(\tau) = \frac{\exp\{-\kappa\tau/2\}}{d}\left(d\cosh\left(\frac{d\tau}{2}\right) - \kappa\sinh\left(\frac{d\tau}{2}\right)\right),$$

$$M_2(\tau) = \frac{-2}{d}\exp\{-\kappa\tau/2\}\left(\left(\rho\kappa\lambda - \frac{\lambda^2}{2}\right)\iota u + \frac{1}{2}\lambda^2(1-\rho^2)(\iota u)^2\right)\sinh\left(\frac{d\tau}{2}\right),$$

$$M_3(\tau) = \frac{1}{d}\exp\{-\kappa\tau/2\}\sinh\left(\frac{d\tau}{2}\right),$$

$$M_4(\tau) = \frac{\exp\{-\kappa\tau/2\}}{d}\left(d\cosh\left(\frac{d\tau}{2}\right) + \kappa\sinh\left(\frac{d\tau}{2}\right)\right).$$

We hence have

$$\begin{aligned} V^D(\tau, \iota u) = &\big(d(\exp\{d\tau\}+1) - \kappa(\exp\{d\tau\}-1)\big)(-\rho\lambda\iota u) \\ &- 2\left(\left(\rho\kappa\lambda - \frac{\lambda^2}{2}\right)\iota u + \frac{\lambda^2}{2}(1-\rho^2)(\iota u)^2\right)(\exp\{d\tau\}-1) \\ &/\big((\exp\{d\tau\}-1)(-\rho\lambda\iota u) + d(\exp\{d\tau\}+1) + \kappa(\exp\{d\tau\}-1)\big). \end{aligned}$$

The above denominator becomes

$$\begin{aligned} &(\exp\{d\tau\}-1)(-\rho\lambda\iota u) + d\exp\{d\tau\} + d + \kappa\exp\{d\tau\} - \kappa \\ &\quad = (d + \rho\lambda\iota u - \kappa)(1 - g\exp\{d\tau\}), \end{aligned}$$

where

$$g = (\kappa - \rho\lambda\iota u + d)/(\kappa - \rho\lambda\iota u - d).$$

The numerator satisfies the relation

$$\begin{aligned} &\big(d(\exp\{d\tau\}+1) - \kappa(\exp\{d\tau\}-1)\big)(-\rho\lambda\iota u) \\ &\quad - 2\left(\rho\kappa\lambda - \frac{\lambda^2}{2}\right)\iota u(\exp\{d\tau\}-1) \\ &\quad - \lambda^2(1-\rho^2)(\iota u)^2(\exp\{d\tau\}-1) \\ &= \exp\{d\tau\}\Bigg(-d\rho\lambda\iota u + \kappa\rho\lambda\iota u \\ &\quad - 2\left(\rho\kappa\lambda - \frac{\lambda^2}{2}\right)\iota u \\ &\quad - \lambda^2(1-\rho^2)(\iota u)^2\Bigg) \\ &\quad + \Bigg(-d\rho\lambda\iota u - \kappa\rho\lambda\iota u + 2\left(\rho\kappa\lambda - \frac{\lambda^2}{2}\right)\iota u \\ &\quad + \lambda^2(1-\rho^2)(\iota u)^2\Bigg). \end{aligned}$$

Hence one has

$$\begin{aligned}V^D(\tau,\imath u) = \big(&\exp\{d\tau\}\big(-d\rho\lambda\imath u - \kappa\rho\lambda\imath u + \lambda^2\imath u - \lambda^2(\imath u)^2 + \lambda^2\rho^2(\imath u)^2\big)\\ &+ \big(-d\rho\lambda\imath u + \kappa\rho\lambda\imath u - \lambda^2\imath u + \lambda^2(\imath u)^2 - \lambda^2\rho^2(\imath u)^2\big)\big)\\ &/\big((d+\rho\lambda\imath u - \kappa)\big(1 - g\exp\{d\tau\}\big)\big).\end{aligned}$$

Recall that

$$\begin{aligned}&E\big(\exp\{\imath u Y_T\}\,\big|\,\mathcal{A}_t\big)\\ &\quad= \exp\big(V^0(\tau,\imath u) - V^D(\tau,\imath u)\tilde{v}_t - V^B(\tau,\imath u)\tilde{Y}_t\big)\\ &\quad= \exp\left(V^0(\tau,\imath u) - \frac{V^D(\tau,\imath u)v_t}{\lambda^2} - (-\imath u)\left(Y_t - \frac{\rho\lambda v_t}{\lambda^2}\right)\right)\\ &\quad= \exp\left(V^0(\tau,\imath u) + Y_t\imath u - \left(\frac{V^D(\tau,\imath u)}{\lambda^2} + \frac{\rho\lambda\imath u}{\lambda^2}\right)v_t\right),\end{aligned}$$

where we used that

$$\tilde{v}_t = \frac{v_t}{\lambda^2}, \qquad \tilde{Y}_t = \frac{Y_t}{\sqrt{1-\rho^2}} - \frac{\rho v_t}{\lambda\sqrt{1-\rho^2}}, \qquad \text{and} \quad V^B(\tau,\imath u) = -\imath u\sqrt{1-\rho^2}.$$

Hence the following calculation is relevant

$$\begin{aligned}&V^D(\tau,\imath u) + \rho\lambda\imath u\\ &\quad= \big(\exp\{d\tau\}\big(-d\rho\lambda\imath u - \kappa\rho\lambda\imath u + \lambda^2\imath u - \lambda^2(\imath u)^2 + \lambda^2\rho^2(\imath u)^2\big)\\ &\qquad+ \big(-d\rho\lambda\imath u + \kappa\rho\lambda\imath u - \lambda^2\imath u + \lambda^2(\imath u)^2 - \lambda^2\rho^2(\imath u)^2\big)\\ &\qquad+ \rho\lambda\imath u\exp\{d\tau\}(d+\kappa-\rho\lambda\imath u) + \rho\lambda\imath u(d+\rho\lambda\imath u-\kappa)\big)\\ &\qquad/\big((d+\rho\lambda\imath u-\kappa)\big(1-g\exp\{d\tau\}\big)\big)\\ &\quad= \frac{\exp\{d\tau\}(\lambda^2\imath u - \lambda^2(\imath u)^2) + (\lambda^2(\imath u)^2 - \lambda^2\imath u)}{(d+\rho\lambda\imath u-\kappa)(1-g\exp\{d\tau\})}\\ &\quad= \frac{(1-\exp\{d\tau\})\lambda^2(\imath u + u^2)}{(1-g\exp\{d\tau\})(\kappa - d - \rho\lambda\imath u)}.\end{aligned}$$

It is readily checked that

$$\frac{-\lambda^2(\imath u + u^2)}{\kappa - d - \rho\lambda\imath u} = \kappa - \rho\lambda\imath u + d,$$

hence

$$V^D(\tau,\imath u) + \rho\lambda\imath u = -\frac{1-\exp\{d\tau\}}{1-g\exp\{d\tau\}}(\kappa - \rho\lambda\imath u + d). \tag{9.5.17}$$

Lastly, from (9.1.3), it follows that

$$\begin{aligned}\frac{\partial}{\partial\tau}V^0(\tau,\imath u) &= -\frac{V^D(\tau,\imath u)\kappa\eta}{\lambda^2} + r\imath u - \frac{\rho\lambda\eta\imath u\kappa}{\lambda^2}\\ &= -\frac{\kappa\eta}{\lambda^2}\big(V^D(\tau,\imath u) + \rho\imath u\lambda\big) + r\imath u.\end{aligned}$$

Using (9.5.17), we obtain,

$$\frac{\partial}{\partial\tau}V^0(\tau,\imath u)=\frac{\kappa\eta}{\lambda^2}\frac{1-\exp\{d\tau\}}{1-g\exp\{d\tau\}}(\kappa-\rho\lambda\imath u+d)+r\imath u,$$

and hence, using

$$\int_0^\tau\frac{1-\exp\{dt\}}{1-g\exp\{dt\}}dt=\tau+\frac{g-1}{dg}\log\left(\frac{1-g}{1-g\exp\{d\tau\}}\right),$$

we get

$$V^0(\tau,\imath u)=\imath ur\tau+\frac{\kappa\eta}{\lambda^2}(\kappa-\rho\lambda\imath u+d)\tau-2\frac{\kappa\eta}{\lambda^2}\log\left(\frac{1-g\exp\{d\tau\}}{1-g}\right),$$

which yields

$$\begin{aligned}
&E\big(\exp\{\imath uY_T\}\,\big|\,\mathcal{A}_t\big)\\
&=\exp\left\{\imath ur\tau+\imath uY_t+\frac{v_t}{\lambda^2}\frac{(1-\exp\{\tau d\})(\kappa-\rho\lambda\imath u+d)}{1-g\exp\{\tau d\}}\right\}\\
&\quad\times\exp\left\{\frac{\kappa\eta}{\lambda^2}\left((\kappa-\rho\lambda\imath u+d)\tau-2\log\left(\frac{1-g\exp\{\tau d\}}{1-g}\right)\right)\right\},
\end{aligned}$$

which is the characteristic function of the logarithm of the stock price in the Heston model. This function can now be used to calculate option prices and other derivatives.

9.6 A Quadratic Term Structure Model

In the preceding sections, we showed that linearization can be successfully applied to affine processes, or processes which produce bond prices that are exponentially affine in the state variables. The assumption that processes are affine played a crucial role in the derivations.

In this section, we briefly discuss the following questions:

- what processes exist beyond the affine class, which are relevant to interest rate modeling?
- do the techniques developed in previous sections still apply?

The answer to the first question was provided in Filipović (2002): it was proven that the *quadratic class*, which we define below, is the largest class of *polynomial term structure models*, which satisfies the *consistency condition* used in Filipović (2002). The consistency condition used in Filipović (2002) is that discounted bond prices are local martingales under an assumed risk-neutral measure. The answer to the second question is affirmative and will be given in this section. We base our discussion on Grasselli and Tebaldi (2004b). The reader's attention is drawn to the following: in this section, we introduce nonlinearity by relaxing the assumption that the bond price is exponentially affine, whilst retaining the assumption that the process assumes values in the (linear) Euclidean space. In the next chapter, we will allow for processes which take values in a nonlinear domain, involving positive semidefinite matrices, but still produce bond prices which are exponentially affine in the state variables.

The following result from Filipović (2002) allows us to conclude that linearization can essentially be applied to any polynomial term structure model, that is consistent in the sense of Filipović (2002). For corresponding proofs, we refer to the paper Filipović (2002):

Definition 9.6.1 A *polynomial term structure model* is defined by:

(i) the factors $\boldsymbol{Z}_t$ defined in a cone domain $\mathcal{Z} \subseteq \Re^n$, whose dynamics under the risk neutral measure satisfy the following SDE:

$$d\boldsymbol{Z}_t = \boldsymbol{b}(\boldsymbol{Z}_t)\,dt + \boldsymbol{\sigma}(\boldsymbol{Z}_t)\,d\boldsymbol{W}_t, \quad \boldsymbol{Z}_t \in \mathcal{Z}, \tag{9.6.18}$$

where $\boldsymbol{W}_t$ is an n-dimensional Brownian motion, and the drift $\boldsymbol{b}(\cdot)$ and volatility matrix $\boldsymbol{\sigma}(\cdot)$ satisfy the growth constraint:

$$\big\|\boldsymbol{b}(z)\big\| + \big\|\boldsymbol{\sigma}(z)\big\| \leq C\big(1 + \|z\|\big), \quad \forall z \in \mathcal{Z}.$$

(ii) the forward rate curve:

$$r(\boldsymbol{Z}_t, \tau) = \sum_{|\boldsymbol{i}|=0}^{d} g_{\boldsymbol{i}}(\tau)(\boldsymbol{Z}_t)^{\boldsymbol{i}}, \quad \tau \geq 0, \tag{9.6.19}$$

with the notation $\boldsymbol{i} = (i_1, \dots, i_n)$, $|\boldsymbol{i}| = i_1 + \cdots + i_n$ and $z^{\boldsymbol{i}} = z_1^{i_1} \dots z_n^{i_n}$; here d denotes the degree of the polynomial term structure.

Theorem 3.4 from Filipović (2002) establishes that the only relevant cases are $d = 1$ and $d = 2$. This result is very important in the context of linearization: the case $d = 1$ was discussed in Sect. 9.2, we now discuss the case $d = 2$.

Under mild regularity conditions, $\boldsymbol{Z}_t$ is given by a multidimensional OU-process, with constant diffusion coefficient satisfying the SDE:

$$\begin{aligned} d\boldsymbol{Z}_t &= (\boldsymbol{A}_0 + \boldsymbol{A}\boldsymbol{Z}_t)\,dt + d\boldsymbol{W}_t, \\ \boldsymbol{A}_0,\ &\boldsymbol{Z}_t \in \Re^n,\ \boldsymbol{A} \in \mathcal{M}_n. \end{aligned}$$

For the short rate we have the quadratic form

$$\begin{aligned} r(\boldsymbol{Z}_t) &= \boldsymbol{Z}_t^\top \boldsymbol{\Omega}_0 \boldsymbol{Z}_t + \boldsymbol{\Gamma}_0^\top \boldsymbol{Z}_t + \gamma_0, \\ \gamma_0 &\in \Re,\ \boldsymbol{\Gamma}_0 \in \Re^n,\ \boldsymbol{\Omega}_0 \in \overline{\mathcal{S}_n^+}, \end{aligned}$$

where $\overline{\mathcal{S}_n^+}$ denotes the set of positive semidefinite matrices. For the forward rate we set:

$$\begin{aligned} r(\boldsymbol{Z}_t, \tau) &= \boldsymbol{Z}_t^\top \boldsymbol{\Omega}(\tau) \boldsymbol{Z}_t + \boldsymbol{\Gamma}(\tau)^\top \boldsymbol{Z}_t + \gamma(\tau), \\ \gamma(\tau) &\in \Re,\ \boldsymbol{\Gamma}(\tau) \in \Re^n,\ \boldsymbol{\Omega}(\tau) \in \overline{\mathcal{S}_n^+},\ \tau \geq 0. \end{aligned}$$

We remind the reader that the risk neutral zero coupon bond price satisfies the relation

$$E\left(\exp\left\{-\int_t^{t+\tau} r_s\,ds\right\}\,\Big|\,\mathcal{A}_t\right) = \exp\{-r(\boldsymbol{Z}_t,\tau)\}.$$

We obtain the following system of Riccati equations:

$$\frac{d\boldsymbol{\Omega}(\tau)}{d\tau} = \boldsymbol{\Omega}_0 + \boldsymbol{\Omega}(\tau)\boldsymbol{A} + \boldsymbol{A}^\top\boldsymbol{\Omega}(\tau) - 2\boldsymbol{\Omega}(\tau)^2 \tag{9.6.20}$$

$$\frac{d\boldsymbol{\Gamma}(\tau)}{d\tau} = \boldsymbol{\Gamma}_0 + 2\boldsymbol{\Omega}(\tau)\boldsymbol{A}_0 + \boldsymbol{A}^\top\boldsymbol{\Gamma}(\tau) - 2\boldsymbol{\Omega}(\tau)\boldsymbol{\Gamma}(\tau) \tag{9.6.21}$$

$$\frac{d\gamma(\tau)}{d\tau} = \gamma_0 + \boldsymbol{\Gamma}(\tau)^\top\boldsymbol{A}_0 + Tr\big(\boldsymbol{\Omega}(\tau)\big) - \boldsymbol{\Gamma}(\tau)^\top\boldsymbol{\Gamma}(\tau)/2, \tag{9.6.22}$$

subject to the initial conditions

$$\boldsymbol{\Omega}(0) = \boldsymbol{0}_{n\times n}, \qquad \boldsymbol{\Gamma}(0) = \boldsymbol{0}_{n\times 1}, \qquad \gamma(0) = 0. \tag{9.6.23}$$

We now explicitly linearize these equations, where we follow Grasselli and Tebaldi (2004b). This procedure mimics the procedure in Sect. 9.2. Also here, we double the dimensionality of the problem. We firstly solve the equation associated with $\boldsymbol{\Omega}(\tau)$. We introduce

$$\boldsymbol{\Omega}(\tau) = \boldsymbol{F}^{-1}(\tau)\boldsymbol{G}(\tau), \quad \text{for } \boldsymbol{F}(\tau) \in GL(n),\ \boldsymbol{G}(\tau) \in \mathcal{M}_n,$$

then

$$\frac{d}{d\tau}\big(\boldsymbol{F}(\tau)\boldsymbol{\Omega}(\tau)\big) - \frac{d}{d\tau}\big(\boldsymbol{F}(\tau)\big)\boldsymbol{\Omega}(\tau) = \boldsymbol{F}(\tau)\frac{d}{d\tau}\boldsymbol{\Omega}(\tau),$$

and from (9.6.20), we get

$$\frac{d}{d\tau}\boldsymbol{G}(\tau) - \frac{d}{d\tau}\big(\boldsymbol{F}(\tau)\big)\boldsymbol{\Omega}(\tau) = \big(\boldsymbol{F}(\tau)\boldsymbol{\Omega}_0 + \boldsymbol{G}(\tau)\boldsymbol{A}\big) - \big(-\boldsymbol{F}(\tau)\boldsymbol{A}^\top + 2\boldsymbol{G}(\tau)\big)\boldsymbol{\Omega}(\tau).$$

We consequently have the following representation

$$\frac{d}{d\tau}\boldsymbol{G}(\tau) = \boldsymbol{F}(\tau)\boldsymbol{\Omega}_0 + \boldsymbol{G}(\tau)\boldsymbol{A}$$
$$\frac{d}{d\tau}\boldsymbol{F}(\tau) = -\boldsymbol{F}(\tau)\boldsymbol{A}^\top + 2\boldsymbol{G}(\tau),$$

which can be written as follows:

$$\frac{d}{d\tau}\big(\boldsymbol{G}(\tau) \quad \boldsymbol{F}(\tau)\big) = \big(\boldsymbol{G}(\tau) \quad \boldsymbol{F}(\tau)\big)\begin{pmatrix} \boldsymbol{A} & 2\boldsymbol{I}_n \\ \boldsymbol{\Omega}_0 & -\boldsymbol{A}^\top \end{pmatrix}.$$

Exponentiating yields

$$\begin{aligned}\big(\boldsymbol{G}(\tau) \quad \boldsymbol{F}(\tau)\big) &= \big(\boldsymbol{G}(0) \quad \boldsymbol{F}(0)\big)\exp\left\{\tau\begin{pmatrix} \boldsymbol{A} & 2\boldsymbol{I}_n \\ \boldsymbol{\Omega}_0 & -\boldsymbol{A}^\top \end{pmatrix}\right\} \\ &= \big(\boldsymbol{\Omega}(0) \quad \boldsymbol{I}_n\big)\exp\left\{\tau\begin{pmatrix} \boldsymbol{A} & 2\boldsymbol{I}_n \\ \boldsymbol{\Omega}_0 & -\boldsymbol{A}^\top \end{pmatrix}\right\}.\end{aligned} \tag{9.6.24}$$

Analogous to Sect. 9.2, we now introduce the notation

$$\begin{pmatrix} \boldsymbol{A}_1^1(\tau) & \boldsymbol{A}_2^1(\tau) \\ \boldsymbol{A}_1^2(\tau) & \boldsymbol{A}_2^2(\tau) \end{pmatrix} := \exp\left\{\tau \begin{pmatrix} \boldsymbol{A} & 2\boldsymbol{I}_n \\ \boldsymbol{\Omega}_0 & -\boldsymbol{A}^\top \end{pmatrix}\right\},$$

in which case (9.6.24) results in

$$\begin{pmatrix} \boldsymbol{G}(\tau) & \boldsymbol{F}(\tau) \end{pmatrix} = \left(\left(\boldsymbol{A}_1^1(\tau)\boldsymbol{\Omega}(0) + \boldsymbol{A}_1^2(\tau)\right) \quad \left(\boldsymbol{A}_2^1(\tau)\boldsymbol{\Omega}(0) + \boldsymbol{A}_2^2(\tau)\right)\right),$$

and hence

$$\boldsymbol{\Omega}(\tau) = \left(\boldsymbol{A}_2^2(\tau)\right)^{-1}\boldsymbol{A}_1^2(\tau),$$

since $\boldsymbol{\Omega}(0) = \boldsymbol{0}_{n\times n}$. We solve the second equation

$$\frac{d}{d\tau}\boldsymbol{\Gamma}(\tau) = \boldsymbol{\Gamma}_0 + 2\boldsymbol{\Omega}(\tau)\boldsymbol{A}_0 + \boldsymbol{A}^\top\boldsymbol{\Gamma}(\tau) - 2\boldsymbol{\Omega}(\tau)\boldsymbol{\Gamma}(\tau).$$

Again, we set

$$\boldsymbol{\Gamma}(\tau) = \boldsymbol{F}^{-1}(\tau)\tilde{\boldsymbol{\Gamma}}(\tau),$$

and hence

$$\frac{d}{d\tau}\tilde{\boldsymbol{\Gamma}}(\tau) - \frac{d}{d\tau}\boldsymbol{F}(\tau)\boldsymbol{\Gamma}(\tau) = \boldsymbol{F}(\tau)\frac{d}{d\tau}\boldsymbol{\Gamma}(\tau).$$

Consequently, we have

$$\begin{aligned}\frac{d}{d\tau}\boldsymbol{F}(\tau) - \frac{d}{d\tau}\boldsymbol{F}(\tau)\boldsymbol{\Gamma}(\tau) &= \boldsymbol{F}(\tau)\left(\boldsymbol{\Gamma}_0 + 2\boldsymbol{\Omega}(\tau)\boldsymbol{A}_0 + \boldsymbol{A}^\top\boldsymbol{\Gamma}(\tau) - 2\boldsymbol{\Omega}(\tau)\boldsymbol{\Gamma}(\tau)\right) \\ &= \boldsymbol{F}(\tau)\boldsymbol{\Gamma}_0 + 2\boldsymbol{\Gamma}(\tau)\boldsymbol{A}_0 - \left(-\boldsymbol{F}(\tau)\boldsymbol{A}^\top + 2\boldsymbol{G}(\tau)\right)\boldsymbol{\Gamma}(\tau).\end{aligned}$$

We immediately have

$$\frac{d}{d\tau}\tilde{\boldsymbol{\Gamma}}(\tau) = \boldsymbol{F}(\tau)\boldsymbol{\Gamma}_0 + 2\boldsymbol{G}(\tau)\boldsymbol{A}_0.$$

Hence we get the following result for $\boldsymbol{\Gamma}(\tau)$:

$$\begin{aligned}\boldsymbol{\Gamma}(\tau) &= \boldsymbol{F}^{-1}(\tau)\tilde{\boldsymbol{\Gamma}}(\tau) \\ &= \boldsymbol{F}^{-1}(\tau)\int_0^\tau \left(\boldsymbol{F}(\tau')\boldsymbol{\Gamma}_0 + 2\boldsymbol{G}(\tau')\boldsymbol{A}_0\right)d\tau' \\ &= \left(\boldsymbol{A}_2^2(\tau)\right)^{-1}\int_0^\tau \left(\boldsymbol{A}_2^2(\tau')\boldsymbol{\Gamma}_0 + 2\boldsymbol{A}_1^2(\tau')\boldsymbol{A}_0\right)d\tau'.\end{aligned}$$

Finally, we obtain $\gamma(\tau)$ from (9.6.22) by direct integration.

Corollary 9.6.2 *For $n = 1$, we recover the result from Sect.* 4.4 *in Filipović et al.* (2004), *i.e. setting $c = 2\sqrt{A^2 + 2\Omega_0}$ and*

$$\begin{aligned}L_3(t) &= \frac{4\Gamma_0}{c}\left(\exp\{c\tau/2\} - 1\right)\left(\frac{c}{2}\left(\exp\{c\tau/2\} + 1\right) + A\left(1 - \exp\{c\tau/2\}\right)\right) \\ &\quad + \frac{8A_0\Omega_0}{c}\left(\exp\{c\tau/2\} - 1\right)^2, \\ L_5(t) &= c\left(\exp\{c\tau\} + 1\right) - 2A\left(\exp\{c\tau\} - 1\right), \\ L_7(t) &= 2\Omega_0\left(\exp\{c\tau\} - 1\right),\end{aligned}$$

we have

$$\Omega(\tau) = \frac{L_7(\tau)}{L_5(\tau)},$$
$$\Gamma(\tau) = \frac{L_3(\tau)}{L_5(\tau)},$$
$$\gamma(\tau) = \int_0^\tau \left(\Gamma(t)A_0 - \frac{1}{2}\Gamma(t)^2 + \Omega(t) + \gamma_0 \right) dt.$$

9.7 A Multifactor Heston Model

In this section, we discuss a multifactor model, which can be analyzed using the techniques developed in this chapter. The aim is to illustrate that the approach can be used to analyze a powerful, complex model, but at the same time, it also illustrates the limitations of the approach, and motivates the study of the Wishart process in Chap. 11.

The following model is covered by the Dai and Singleton (2000) and Duffie et al. (2000) framework. We introduce two assets and drive their stochastic volatilities using three square-root processes:

$$dS_t^1 = S_t^1 \left(r\,dt + \sqrt{X_t^1}\,dZ_t^1 + \sqrt{X_t^3}\,dZ_t^3 \right), \tag{9.7.25}$$

$$dS_t^2 = S_t^2 \left(r\,dt + \sqrt{X_t^2}\,dZ_t^2 + \sqrt{X_t^3}\,dZ_t^3 \right), \tag{9.7.26}$$

where

$$dX_t^1 = \kappa^1 (\eta^1 - X_t^1)\,dt + \lambda^1 \sqrt{X_t^1}\,dW_t^1,$$
$$dX_t^2 = \kappa^2 (\eta^2 - X_t^2)\,dt + \lambda^2 \sqrt{X_t^2}\,dW_t^2,$$
$$dX_t^3 = \kappa^3 (\eta^3 - X_t^3)\,dt + \lambda^3 \sqrt{X_t^3}\,dW_t^3,$$

and $\kappa^j, \eta^j, \lambda^j > 0$, $j = 1, 2, 3$. To ensure the affinity of the model, we introduce the following correlation structure

$$d[Z^i, W^j] = \rho_j \delta_{i,j}\,dt, \quad i, j = 1, 2, 3,$$

where $\rho_j \in (-1, 1)$, $j = 1, 2, 3$. We now discuss this model, where we follow Da Fonseca et al. (2007). Firstly, we remark that each price process enjoys the dynamics of the *Double Heston model* of Christoffersen et al. (2009). Furthermore, the assets exhibit stochastic volatilities, given by $(X_t^1 + X_t^3)$ and $(X_t^2 + X_t^3)$, respectively, and stochastic covariation, given as

$$d[S^1, S^2]_t = S_t^1 S_t^2 X_t^3\,dt.$$

The assets exhibit stochastic volatility, but the covariation is constrained to remain nonnegative. This issue is addressed in Chap. 11. We encourage the reader to compare the model presented in this section to the two models presented in Sect. 11.5. Finally, we introduce log-asset prices $Y_t^i = \log(S_t^i)$, $i = 1, 2$, and present the joint characteristic function of the log-asset prices, so that the techniques from Chap. 8 are applicable.

Proposition 9.7.1 *Let $Y_t^i = \log(S_t^i)$, $i = 1, 2$, where the dynamics of S_t^1 and S_t^2 are given by* (9.7.25) *and* (9.7.26), *respectively. Then*

$$\begin{aligned} &E\big(\exp\{\iota u_1 Y_T^1 + \iota u_2 Y_T^2\}|\mathcal{A}_t\big) \\ &\quad = \exp\left\{\sum_{j=1}^{3} A_j(\tau)X_t^j + \iota\sum_{k=1}^{2} u_k Y_t^k + c(\tau)\right\}, \end{aligned}$$

where for $j = 1, 2$,

$$\begin{aligned} d_j &= \sqrt{\left(u_j^2 + \iota u_j\right)\lambda_j^2 + (\kappa_j - \iota\lambda_j\rho_j u_j)^2}, \\ g_j &= \frac{\kappa_j - \iota\lambda_j\rho_j u_j + d_j}{\kappa_j - \iota\lambda_j\rho_j u_j - d_j}, \\ A_j(\tau) &= \frac{\kappa_j - \iota\lambda_j\rho_j u_j + d_j}{\lambda_j^2}\left(\frac{1-\exp\{d_j\tau\}}{1-g_j\exp\{d_j\tau\}}\right). \end{aligned}$$

Furthermore,

$$\begin{aligned} d_3 &= \sqrt{\big((u_1+u_2)^2 + \iota(u_1+u_2)\big)\lambda_3^2 + \big(\kappa_3 - \iota\lambda_3\rho_3(u_1+u_2)\big)^2}, \\ g_3 &= \frac{\kappa_3 - \iota\lambda_3\rho_3(u_1+u_2) + d_3}{\kappa_3 - \iota\lambda_3\rho_3(u_1+u_2) - d_3}, \\ A_3(\tau) &= \frac{\kappa_3 - \iota\lambda_3\rho_3(u_1+u_2) + d_3}{\lambda_3^2}\left(\frac{1-\exp\{d_3\tau\}}{1-g_3\exp\{d_3\tau\}}\right), \end{aligned}$$

and

$$\begin{aligned} c(\tau) &= r\iota u_1\tau + r\iota u_2\tau \\ &\quad + \frac{\kappa_1\eta_1}{\lambda_1^2}\left((\kappa_1 - \iota\lambda_1\rho_1 u_1 + d_1)\tau - 2\log\left(\frac{1-g_1\exp\{d_1\tau\}}{1-g_1}\right)\right) \\ &\quad + \frac{\kappa_2\eta_2}{\lambda_2^2}\left((\kappa_2 - \iota\lambda_2\rho_2 u_2 + d_2)\tau - 2\log\left(\frac{1-g_2\exp\{d_2\tau\}}{1-g_2}\right)\right) \\ &\quad + \frac{\kappa_3\eta_3}{\lambda_3^2}\left(\big(\kappa_3 - \iota\lambda_3\rho_3(u_1+u_2) + d_3\big)\tau - 2\log\left(\frac{1-g_3\exp\{d_3\tau\}}{1-g_3}\right)\right). \end{aligned}$$

In conclusion, we remark that on the Euclidean state space, one can introduce powerful multidimensional models. However, it is also clear that the positive factors, in this case the variance processes (X^1, X^2, X^3), have to be orthogonal. This is clear from the discussion in Sect. 9.2. To allow for dependence between positive factors, one needs a more general concept of positivity than $\Re^+$. The Wishart process turns out to be rich enough to allow for dependent positive factors, whilst at the same time the process remains tractable, see Chap. 11.

Chapter 10
An Introduction to Matrix Variate Stochastics

In this chapter, we introduce the reader to matrix variate stochastics. It is intended to set the scene for Wishart processes, which will be covered in the next chapter. We begin by recalling notation and introducing some basic functions used throughout both chapters. This will bring us in a position to discuss matrix valued random variables, matrix valued stochastic processes, and matrix valued stochastic differential equations. To illustrate these concepts, we apply them to the matrix valued version of the Ornstein-Uhlenbeck process and a multidimensional version of the MMM. The main references for this chapter are Gupta and Nagar (2000) and Pfaffel (2008).

10.1 Basic Definitions and Functions

In this section, we fix primarily notation.

Definition 10.1.1 We employ the following notation:

- we denote by $\mathcal{M}_{m,n}(\Re)$ the set of all $m \times n$ matrices with entries in $\Re$. If $m = n$, we write $\mathcal{M}_n(\Re)$ instead;
- we write $GL(p)$ for the group of all invertible matrices of $\mathcal{M}_p(\Re)$;
- let $\mathcal{S}_p$ denote the linear subspace of all symmetric matrices of $\mathcal{M}_p(\Re)$;
- let $\mathcal{S}_p^+$ ($\mathcal{S}_p^-$) denote the set of all symmetric positive (negative) definite matrices of $\mathcal{M}_p(\Re)$;
- denote by $\overline{\mathcal{S}_p^+}$ the closure of $\mathcal{S}_p^+$ in $\mathcal{M}_p(\Re)$, that is the set of all symmetric positive semidefinite matrices of $\mathcal{M}_p(\Re)$.

The next definition provides a one-to-one relationship between vectors and matrices.

J. Baldeaux, E. Platen, *Functionals of Multidimensional Diffusions with Applications to Finance*, Bocconi & Springer Series 5, DOI 10.1007/978-3-319-00747-2_10,

Definition 10.1.2 Let $\boldsymbol{A} \in \mathcal{M}_{m,n}(\Re)$ with columns $\boldsymbol{a}_i \in \Re^m$, $i = 1, \dots, n$. Define the function $vec : \mathcal{M}_{m,n}(\Re) \to \Re^{mn}$ via

$$vec(\boldsymbol{A}) = \begin{pmatrix} \boldsymbol{a}_1 \\ \vdots \\ \boldsymbol{a}_n \end{pmatrix}.$$

Note that $vec(\boldsymbol{A})$ is also an element of $\mathcal{M}_{mn,1}(\Re)$. The next lemma is derived in Gupta and Nagar (2000).

Lemma 10.1.3 *The following properties hold:*

- *for* $\boldsymbol{A}, \boldsymbol{B} \in \mathcal{M}_{m,n}(\Re)$ *it holds that* $tr(\boldsymbol{A}^\top \boldsymbol{B}) = vec(\boldsymbol{A})^\top vec(\boldsymbol{B})$;
- *let* $\boldsymbol{A} \in \mathcal{M}_{p,m}(\Re)$, $\boldsymbol{B} \in \mathcal{M}_{m,n}(\Re)$ *and* $\boldsymbol{C} \in \mathcal{M}_{n,q}(\Re)$. *Then we have*

$$vec(\boldsymbol{A}\boldsymbol{X}\boldsymbol{B}) = \big(\boldsymbol{B}^\top \otimes \boldsymbol{A}\big) vec(\boldsymbol{X}).$$

We now recall from Gupta and Nagar (2000) how a symmetric matrix can be mapped to a vector.

Definition 10.1.4 Let $\boldsymbol{S} \in \mathcal{S}_p$. Define the function $vech : \mathcal{S}_0 \to \Re^{\frac{p(p+1)}{2}}$ via

$$vech(\boldsymbol{S}) = \begin{pmatrix} S_{11} \\ S_{12} \\ S_{22} \\ \vdots \\ S_{1p} \\ \vdots \\ S_{pp} \end{pmatrix},$$

such that $vech(\boldsymbol{S})$ is a vector consisting of the elements of $\boldsymbol{S}$ from above and including the diagonal, taken componentwise.

We point out that the vector *vech* gives access to the $\frac{p(p+1)}{2}$ distinct values of a symmetric $p \times p$ matrix.

10.2 Integrals over Matrix Domains

The aim of this section is to define integrals over matrix domains. These definitions will be employed in the subsequent sections, e.g. when computing characteristic functions and Laplace transforms of matrix valued random variables. Discussing integration, we need a notion of measurability. The following definition is taken from Pfaffel (2008), see also Jacod and Protter (2004).

Definition 10.2.1 Let $(X, \mathcal{T})$ be a topological space. The Borel σ-algebra on X is then given by the smallest σ-algebra that contains $\mathcal{T}$ and is denoted by $\mathcal{B}(X)$.

In this chapter and the following, we focus on $\Re$, $\Re^n$, and $\mathcal{M}_{m,n}(\Re)$, and employ the notation $\mathcal{B}$ for $\mathcal{B}(\Re)$, $\mathcal{B}^n$ for $\mathcal{B}(\Re^n)$ and $\mathcal{B}^{m,n}$ for $\mathcal{B}(\mathcal{M}_{m,n}(\Re))$. We are now in a position to define integrals over matrices, allowing for matrices of size $m \times n$.

Definition 10.2.2 Let $f : \mathcal{M}_{m,n}(\Re) \to \Re$ be a $\mathcal{B}^{m,n} - \mathcal{B}$-measurable function and $M \in \mathcal{B}^{m,n}$ a measurable subset of $\mathcal{M}_{m,n}(\Re)$ and let λ denote the Lebesgue-measure on $(\Re^{mn}, \mathcal{B}^{mn})$. The integral of f over $\boldsymbol{M}$ is then defined by

$$\int_M f(\boldsymbol{X})\, d\boldsymbol{X} := \int_M f(\boldsymbol{X})\, d(\lambda \circ vec)(\boldsymbol{X}) = \int_{vec(M)} f \circ vec^{-1}(\boldsymbol{x})\, d\lambda(\boldsymbol{x}).$$

We call $\lambda \circ vec$ the Lebesgue-measure on $(\mathcal{M}_{m,n}(\Re), \mathcal{B}^{m,n})$.

As pointed out in Pfaffel (2008), $\mathcal{S}_p$ is isomorphic to $\Re^{\frac{p(p+1)}{2}}$, hence for $p \geq 2$, $\mathcal{S}_p$ is a real subspace of $\mathcal{M}_p(\Re)$, and, consequently, of Lebesgue-measure zero. As this means that any integral over subsets of $\mathcal{S}_p$ is zero, we define another Lebesgue measure on the subspace of symmetric matrices $\mathcal{S}_p$:

Definition 10.2.3 Let $f : \mathcal{S}_p \to \Re$ be a $\mathcal{B}(\mathcal{S}_p)$- $\mathcal{B}$-measurable function and $M \in \mathcal{B}(\mathcal{S}_p)$ a Borel-measurable subset of $\mathcal{S}_p$ and let λ denote the Lebesgue-measure on $(\Re^{\frac{p(p+1)}{2}}, \mathcal{B}^{\frac{p(p+1)}{2}})$. The integral of f over M is then defined by

$$\int_M f(\boldsymbol{X})\, d\boldsymbol{X} := \int_M f(\boldsymbol{X})\, d(\lambda \circ vech)(\boldsymbol{X}) = \int_{vech(M)} f \circ vech^{-1}(\boldsymbol{x})\, d\lambda(\boldsymbol{x}).$$

We call $\lambda \circ vech$ the Lebesgue-measure on $(\mathcal{S}_p, \mathcal{B}(\mathcal{S}_p))$.

As in Gupta and Nagar (2000) and Pfaffel (2008), we use the notation

$$etr(\boldsymbol{A}) := \exp\{tr(\boldsymbol{A})\}. \tag{10.2.1}$$

This notation allows the formulation of the next definition.

Definition 10.2.4 The multivariate gamma function is defined as follows:

$$\Gamma_p(a) := \int_{\mathcal{S}_p^+} etr(-\boldsymbol{A}) det(\boldsymbol{A})^{a-\frac{1}{2}(p+1)}\, d\boldsymbol{A} \quad \forall a > \frac{p-1}{2}.$$

The next result from Gupta and Nagar (2000), shows that for $a > \frac{p-1}{2}$, the matrix variate gamma function can be expressed as a finite product of ordinary gamma functions.

Theorem 10.2.5 *For* $a > \frac{1}{2}(p-1)$,

$$\Gamma_p(a) = \pi^{\frac{1}{4}p(p-1)} \prod_{i=1}^{p} \Gamma\left(a - \frac{1}{2}(i-1)\right).$$

Next, we want to introduce hypergeometric functions of matrix arguments. This requires the definition of zonal polynomials, which in turn requires the definition of *symmetric homogeneous polynomials*. Starting with the latter, a symmetric homogeneous polynomial of degree k in $y_1, \dots, y_m$ is a polynomial which is unchanged by a permutation of the subscripts and such that every term in the polynomial has degree k. The following example from Muirhead (1982) illustrates this: set $m = 2$, $k = 3$, then

$$y_1^3 + y_2^3 + 10y_1^2 y_2 + 10y_1 y_2^2$$

is a symmetric homogeneous polynomial of degree 3 in y_1 and y_2. Following Gupta and Nagar (2000), we denote by V_k the vector space of symmetric homogeneous polynomials that are of degree k in the $\frac{1}{2}p(p-1)$ distinct elements of $\boldsymbol{S} \in \mathcal{S}_p^+$. As discussed in Gupta and Nagar (2000), the space V_k can be decomposed into a direct sum of irreducible invariant subspaces V_κ, where κ denotes a partition of k, defined as follows: by a partition of k, we mean the p-tuple $\kappa = (k_1, \dots, k_p)$, where $k_1 \geq \dots \geq k_p \geq 0$, and furthermore $k_1 + \dots + k_p = k$. Then the polynomial $tr(\boldsymbol{S})^k \in V_k$ has the unique decomposition into polynomials $C_\kappa(\boldsymbol{S}) \in V_\kappa$ as

$$tr(\boldsymbol{S})^k = \sum_\kappa C_\kappa(\boldsymbol{S}).$$

We now define *zonal polynomials*.

Definition 10.2.6 The *zonal polynomial* $C_\kappa(\boldsymbol{S})$ is the component of $tr(\boldsymbol{S})^k$ in the subspace V_κ.

The next definition from Gupta and Nagar (2000) introduces *hypergeometric functions of matrix arguments*.

Definition 10.2.7 The *hypergeometric function of matrix argument* is defined by

$${}_mF_n(a_1, \dots, a_m; b_1, \dots, b_n; \boldsymbol{S}) = \sum_{k=0}^{\infty} \sum_\kappa \frac{(a_1)_\kappa \dots (a_m)_\kappa C_\kappa(\boldsymbol{S})}{(b_1)_\kappa \dots (b_n)_\kappa k!}, \tag{10.2.2}$$

where $a_i, b_j \in \Re$, $\boldsymbol{S}$ is a symmetric $p \times p$-matrix and $\sum_\kappa$ the summation over all partitions κ of k and $(a)_\kappa = \prod_{j=1}^p (a - \frac{1}{2}(j-1))_{k_j}$ denotes the generalized hypergeometric coefficient, with $(x)_{k_j} = x(x+1)\dots(x+k_j-1)$.

The following remark provides some properties of hypergeometric functions of matrix arguments.

Remark 10.2.8 Conditions for convergence of the infinite series in Eq. (10.2.2) are of importance, see Gupta and Nagar (2000) for a discussion. The condition $m < n+1$ is sufficient. We also have the special case

$${}_nF_n(a_1, \dots, a_n; a_1, \dots, a_n; \boldsymbol{S}) = \sum_{k=0}^{\infty} \frac{(tr(\boldsymbol{S}))^k}{k!} = etr(\boldsymbol{S}).$$

The next lemma will be subsequently employed when computing expectations of functions of non-central Wishart distributed random variables.

Lemma 10.2.9 *Let $\boldsymbol{Z}, \boldsymbol{T} \in \mathcal{S}_p^+$. Then*

$$\int_{S_p^+} etr(-\boldsymbol{Z}\boldsymbol{S})det(\boldsymbol{S})^{a-\frac{p+1}{2}} {}_mF_n(a_1, \ldots, a_m; b_1, \ldots, b_n; \boldsymbol{S}\boldsymbol{T})\, d\boldsymbol{S}$$

$$= \Gamma_p(a)det(\boldsymbol{Z})^{-a} {}_{m+1}F_n\big(a_1, \ldots, a_m; b_1, \ldots, b_n; \boldsymbol{Z}^{-1}\boldsymbol{T}\big),$$

$\forall a > \frac{p-1}{2}$.

Proof The result is a special case of Theorem 1.6.2 in Gupta and Nagar (2000). □

10.3 Matrix Valued Random Variables

In this section, we discuss matrix valued random variables. First, we need to define what we mean by a matrix valued random variable, and consequently associate with it the concepts well-known from the vector and scalar case, such as probability density functions, characteristic functions, and Laplace transforms. We will discuss two examples, first the normal distribution and second the Wishart distribution. The main reference for this section is Gupta and Nagar (2000), see also Pfaffel (2008). As before, we use $(\Omega, \mathcal{A}, \underline{\mathcal{A}}, P)$ to denote the filtered probability space.

Definition 10.3.1 An $m \times n$ random matrix $\boldsymbol{X}$ is a measurable function

$$\boldsymbol{X} : (\Omega, \mathcal{F}) \to \big(\mathcal{M}_{m,n}(\Re), \mathcal{B}^{m,n}\big).$$

We now discuss probability density functions of random variables.

Definition 10.3.2 A nonnegative measurable function f_X such that

$$P(\boldsymbol{X} \in M) = \int_M f_X(\boldsymbol{A})\, d\boldsymbol{A} \quad \forall M \in \mathcal{B}^{m \times n}$$

defines the probability density function of an $m \times n$ random matrix $\boldsymbol{X}$.

We can now introduce expected values.

Definition 10.3.3 Let X be an $m \times n$-random matrix. For every function $h = (h_{i,j})_{i,j} : \mathcal{M}_{m,n}(\Re) \to \mathcal{M}_{r,s}(\Re)$ with $h_{i,j} : \mathcal{M}_{m,n}(\Re) \to \Re$, $1 \le i \le r$, $1 \le j \le s$, the expected value $E(h(\boldsymbol{X}))$ of $h(\boldsymbol{X})$ is an element of $\mathcal{M}_{r,s}(\Re)$ with elements

$$E\big(h(\boldsymbol{X})\big)_{i,j} = E\big(h_{i,j}(\boldsymbol{X})\big) = \int_{\mathcal{M}_{m,n}(\Re)} h_{i,j}(\boldsymbol{A}) P^{\boldsymbol{X}}(d\boldsymbol{A}).$$

We point out that if $\boldsymbol{X}$ has a probability density function $f_{\boldsymbol{X}}$, then we have

$$E\big(h(\boldsymbol{X})\big)_{i,j} = \int_{\mathcal{M}_{m,n}(\Re)} h_{i,j}(\boldsymbol{A}) f_{\boldsymbol{X}}(\boldsymbol{A})\, d\boldsymbol{A}.$$

The characteristic function or Fourier transform of matrix-valued random variables is now defined.

Definition 10.3.4 The characteristic function of an $m \times n$-random matrix $\boldsymbol{X}$ with probability density function $f_{\boldsymbol{X}}$ is defined as

$$E\big(etr\big(\imath \boldsymbol{X}\boldsymbol{Z}^\top\big)\big) = \int_{\mathcal{M}_{m,n}(\Re)} etr\big(\imath \boldsymbol{A}\boldsymbol{Z}^\top\big) f_{\boldsymbol{X}}(\boldsymbol{A})\, d\boldsymbol{A}, \tag{10.3.3}$$

for every $\boldsymbol{Z} \in \mathcal{M}_p(\Re)$.

Due to the fact that $|\exp(\imath x)| = 1,\ \forall x \in \Re$, the integral in (10.3.3) always exists. Furthermore, (10.3.3) is the Fourier transform of the measure $P^{\boldsymbol{X}}$ at point $\boldsymbol{Z}$.

Definition 10.3.5 The Laplace transform of a $p \times p$-random matrix $\boldsymbol{X} \in \mathcal{S}_p^+$ with probability density function $f_{\boldsymbol{X}}$ is defined as

$$E\big(etr(-\boldsymbol{U}\boldsymbol{X})\big) = \int_{\mathcal{S}_p^+} etr(-\boldsymbol{U}\boldsymbol{A}) f_{\boldsymbol{X}}(\boldsymbol{A})\, d\boldsymbol{A}, \tag{10.3.4}$$

for every $\boldsymbol{U} \in \mathcal{S}_p^+$.

Remark 10.3.6 Recall that the Laplace transform of a positive scalar random variable is always well-defined. For $\boldsymbol{A}, \boldsymbol{U} \in \mathcal{S}_p^+$, we have that $tr(-\boldsymbol{U}\boldsymbol{A}) = -tr(\sqrt{\boldsymbol{U}}\boldsymbol{A}\sqrt{\boldsymbol{U}}) < 0$, since $\sqrt{\boldsymbol{U}}\boldsymbol{A}\sqrt{\boldsymbol{U}}$ is positive definite, hence the integral in Eq. (10.3.4) is well-defined, where $\boldsymbol{X} \in \mathcal{S}_p^+$ is the analogue of a positive random variable.

Next, we introduce covariance matrices for matrix valued random variables.

Definition 10.3.7 Let $\boldsymbol{X}$ be an $m \times n$ random matrix and $\boldsymbol{Y}$ be a $p \times q$ random matrix. Then the $mn \times pq$ covariance matrix is defined as

$$\begin{aligned} cov(\boldsymbol{X}, \boldsymbol{Y}) &= cov\big(vec\big(\boldsymbol{X}^\top\big), vec\big(\boldsymbol{Y}^\top\big)\big) \\ &= E\big(vec\big(\boldsymbol{X}^\top\big) vec\big(\boldsymbol{Y}^\top\big)^\top\big) - E\big(vec\big(\boldsymbol{X}^\top\big)\big) E\big(vec\big(\boldsymbol{Y}^\top\big)\big)^\top, \end{aligned}$$

i.e. $cov(\boldsymbol{X}, \boldsymbol{Y})$ is an $m \times p$ block matrix with blocks $cov(\tilde{\boldsymbol{x}}_i^\top, \tilde{\boldsymbol{y}}_j^\top) \in \mathcal{M}_{n,q}(\Re)$ where $\tilde{\boldsymbol{x}}_i$ (or $\tilde{\boldsymbol{y}}_i$ respectively) denote the rows of $\boldsymbol{X}$ (respectively $\boldsymbol{Y}$).

Having these definitions at hand, we can now discuss some examples. We begin with an example involving the normal distribution.

Definition 10.3.8 A $p \times n$ random matrix is said to have a matrix variate normal distribution with mean $\boldsymbol{M} \in \mathcal{M}_{p,n}(\Re)$ and covariance $\boldsymbol{\Sigma} \otimes \boldsymbol{\Psi}$, where $\boldsymbol{\Sigma} \in \mathcal{S}_p^+$, $\boldsymbol{\Psi} \in \mathcal{S}_n^+$, if $vec(\boldsymbol{X}^\top) \sim \mathcal{N}_{pn}(vec(\boldsymbol{M}^\top), \boldsymbol{\Sigma} \otimes \boldsymbol{\Psi})$, where $\mathcal{N}_{pn}$ denotes the multivariate normal distribution on $\Re^{pn}$ with mean $vec(\boldsymbol{M}^\top)$ and covariance $\boldsymbol{\Sigma} \otimes \boldsymbol{\Psi}$. We will use the notation $\boldsymbol{X} \sim \mathcal{N}_{p,n}(\boldsymbol{M}, \boldsymbol{\Sigma} \otimes \boldsymbol{\Psi})$.

We now recall a result from Gupta and Nagar (2000).

Theorem 10.3.9 *If* $\boldsymbol{X} \sim \mathcal{N}_{p,n}(\boldsymbol{M}, \boldsymbol{\Sigma} \otimes \boldsymbol{\Psi})$, *then* $\boldsymbol{X}^\top \sim \mathcal{N}_{p,n}(\boldsymbol{M}^\top, \boldsymbol{\Psi} \otimes \boldsymbol{\Sigma})$.

Proof The proof is given by the one of Theorem 2.3.1 in Gupta and Nagar (2000). □

The next result gives the characteristic function of the normal distribution.

Theorem 10.3.10 *Let* $\boldsymbol{X} \sim \mathcal{N}_{p,n}(\boldsymbol{M}, \boldsymbol{\Sigma} \otimes \boldsymbol{\Psi})$. *Then the characteristic function of* $\boldsymbol{X}$ *is given by*

$$E\big(etr(\imath \boldsymbol{X}\boldsymbol{Z}^\top)\big) = etr\left(\imath \boldsymbol{Z}^\top \boldsymbol{M} - \frac{1}{2}\boldsymbol{Z}^\top \boldsymbol{\Sigma}\boldsymbol{Z}\boldsymbol{\Psi}\right).$$

By employing Theorem 10.3.10 one proves the matrix analogue of the linear transformation property of normal random variables, see Pfaffel (2008).

Theorem 10.3.11 *Let* $\boldsymbol{X} \sim \mathcal{N}_{p,n}(\boldsymbol{M}, \boldsymbol{\Sigma} \otimes \boldsymbol{\Psi})$, $\boldsymbol{A} \in \mathcal{M}_{m,q}(\Re)$, $\boldsymbol{B} \in \mathcal{M}_{m,p}(\Re)$ *and* $\boldsymbol{C} \in \mathcal{M}_{n,q}(\Re)$. *Then* $\boldsymbol{A} + \boldsymbol{B}\boldsymbol{X}\boldsymbol{C} \sim \mathcal{N}_{m,q}(\boldsymbol{A} + \boldsymbol{B}\boldsymbol{M}\boldsymbol{C}, (\boldsymbol{B}\boldsymbol{\Sigma}\boldsymbol{B}^\top) \otimes (\boldsymbol{C}^\top \boldsymbol{\Psi}\boldsymbol{C}))$.

Next, we discuss an example involving the Wishart distribution.

Definition 10.3.12 A $p \times p$-random matrix $\boldsymbol{X}$ in $\mathcal{S}_p^+$ is said to have a non-central Wishart distribution with parameters $p \in \mathcal{N}$, $n \geq p$, $\boldsymbol{\Sigma} \in \mathcal{S}_p^+$ and $\boldsymbol{\Theta} \in \mathcal{M}_p(\Re)$, if its probability density function is of the form

$$f_{\boldsymbol{X}}(\boldsymbol{S}) = \left(2^{\frac{1}{2}np}\Gamma_p\left(\frac{n}{2}\right)det(\boldsymbol{\Sigma})^{\frac{n}{2}}\right)^{-1} etr\left(-\frac{1}{2}(\boldsymbol{\Theta} + \boldsymbol{\Sigma}^{-1}\boldsymbol{S})\right)$$
$$\times\, det(\boldsymbol{S})^{\frac{1}{2}(n-p-1)}{}_0F_1\left(\frac{n}{2}; \frac{1}{4}\boldsymbol{\Theta}\boldsymbol{\Sigma}^{-1}\boldsymbol{S}\right),$$

where $\boldsymbol{S} \in \mathcal{S}_p^+$ and ${}_0F_1$ is the hypergeometric function. We write

$$\boldsymbol{X} \sim \mathcal{W}_p(n, \boldsymbol{\Sigma}, \boldsymbol{\Theta}).$$

We remark that the requirement $n \geq p$ ensures that the matrix variate gamma function is well-defined. If $\boldsymbol{\Theta} = \boldsymbol{0}$, $\boldsymbol{X}$ is said to follow the central Wishart distribution with parameters p, n and $\boldsymbol{\Sigma} \in \mathcal{S}_p^+$, with probability density function

$$\left(2^{\frac{1}{2}np}\Gamma_p\left(\frac{n}{2}\right)det(\boldsymbol{\Sigma})^{\frac{n}{2}}\right)^{-1} etr\left(-\frac{1}{2}\boldsymbol{\Sigma}^{-1}\boldsymbol{S}\right)det(\boldsymbol{S})^{\frac{1}{2}(n-p-1)},$$

where $S \in \mathcal{S}_p^+$ and $n \geq p$. Next we provide the Laplace transform of the non-central Wishart distribution, see Pfaffel (2008).

Theorem 10.3.13 *Let $S \sim \mathcal{W}_p(n, \Sigma, \Theta)$. Then the Laplace transform of S is given by*

$$E\big(etr(-US)\big) = det(I_p + 2\Sigma U)^{-\frac{n}{2}} etr\big(-\Theta(I_p + 2\Sigma U)^{-1} \Sigma U\big)$$

with $U \in \mathcal{S}_p^+$.

Now, we list the characteristic function of the non-central Wishart distribution according to Gupta and Nagar (2000).

Theorem 10.3.14 *Let $S \sim \mathcal{W}_p(n, \Sigma, \Theta)$. Then the characteristic function of S is given by*

$$E\big(etr(\imath Z S)\big) = det(I_p - 2\imath \Sigma Z)^{-\frac{n}{2}} etr\big(\imath\Theta(I_p - 2\imath \Sigma Z)^{-1} \Sigma Z\big),$$

with $Z \in \mathcal{M}_p(\Re)$.

The next result, which is Theorem 3.5.1 in Gupta and Nagar (2000), shows that the Wishart distribution is the matrix analogue of the non-central χ^2-distribution.

Theorem 10.3.15 *Let $X \sim \mathcal{N}_{p,n}(M, \Sigma \otimes I_n)$, $n \in \{p, p+1, \ldots\}$. Then $XX^\top \sim \mathcal{W}_p(n, \Sigma, \Sigma^{-1} M M^\top)$.*

We remark that Σ can be interpreted as a scale parameter and Θ as a location parameter for S. Consequently, a central Wishart distributed matrix is the square of normally distributed matrix random variables with zero mean.

10.4 Matrix Valued Stochastic Processes

This section closely follows Sect. 3.3 in Pfaffel (2008). First, we define matrix valued stochastic processes. Our first example will be matrix valued Brownian motion. Later, we will introduce matrix valued local martingales and semimartingales, which then allow us to formulate an Itô formula for matrix valued semimartingales. The section concludes with an integration-by-parts formula, which is useful when applying the theory presented in this chapter to examples, such as the Ornstein-Uhlenbeck process. We remind the reader that $\Re^+$ refers to the interval of non-negative real numbers $[0, \infty)$.

Definition 10.4.1 A measurable function $X : \Re^+ \times \Omega \to \mathcal{M}_{m,n}(\Re)$, $(t, \omega) \mapsto X(t, \omega) = X_t(\omega)$ is called a matrix valued stochastic process if $X(t, \omega)$ is a random matrix for all $t \in \Re^+$. Moreover, X is called a stochastic process in $\overline{\mathcal{S}_p^+}$ if $X : \Re_+ \times \Omega \to \overline{\mathcal{S}_p^+}$.

As noticed in Pfaffel (2008), most definitions applicable to scalar processes can be transferred to matrix valued processes by demanding that they apply to every element of the matrix. The first example is Brownian motion.

Definition 10.4.2 A matrix valued Brownian motion $\boldsymbol{W}$ in $\mathcal{M}_{n,p}(\Re)$ is a matrix consisting of independent one-dimensional Brownian motions, i.e. $\boldsymbol{W} = (W_{i,j})_{i,j}$, where $W_{i,j}$ are independent one-dimensional Brownian motions, $1 \leq i \leq n$, $1 \leq j \leq p$. We write $\boldsymbol{W} \sim \mathcal{BM}_{n,p}$ and $\boldsymbol{W} \sim \mathcal{BM}_n$ if $p = n$.

We now show the obvious distributional properties of Brownian motion.

Corollary 10.4.3 *The following distributional properties regarding a matrix valued Brownian motion* $\boldsymbol{W} = \{\boldsymbol{W}_t,\ t \geq 0\}$ *hold*:

- $\boldsymbol{W}_t \sim \mathcal{N}_{n,p}(0, t I_{np})$;
- $\boldsymbol{W} \sim \mathcal{BM}_{n,}$, $\boldsymbol{A} \in \mathcal{M}_{m,q}(\Re)$, $\boldsymbol{B} \in \mathcal{M}_{m,n}(\Re)$ *and* $\boldsymbol{C} \in \mathcal{M}_{p,q}(\Re)$. *Then* $\boldsymbol{A} + \boldsymbol{B}\boldsymbol{W}_t\boldsymbol{C} \sim \mathcal{N}_{m,q}(\boldsymbol{A}, t(\boldsymbol{B}\boldsymbol{B}^\top) \otimes (\boldsymbol{C}^\top \boldsymbol{C}))$.

Proof For the first part, we need to show that $vec(\boldsymbol{W}_t^\top) \sim \mathcal{N}_{np}(0, t\boldsymbol{I}_{np})$, which is easily verified. The second part follows from Theorem 10.3.11 and the observation that $\boldsymbol{I}_{np} = \boldsymbol{I}_n \otimes \boldsymbol{I}_p$. □

We now define a matrix valued local martingale.

Definition 10.4.4 A matrix valued stochastic process $\boldsymbol{X}$ is called a local martingale, if each component of $\boldsymbol{X}$ is a local martingale, i.e. if there exists a sequence of strictly monotonic increasing stopping times $(T_n)_{n \in \mathcal{N}}$, where $T_n \overset{a.s.}{\to} \infty$, such that $\boldsymbol{X}_{\min(t,T_n),ij}$ forms a martingale for all $i, j, t \geq 0$ and $n \in \{1, 2, \ldots\}$.

The next result is the analogue of Lévy's theorem, which allows us to decide if a given matrix valued continuous local martingale is a Brownian motion. This result appeared in Pfaffel (2008).

Theorem 10.4.5 *Let* $\boldsymbol{B}$ *be a* $p \times p$ *dimensional continuous local martingale such that*

$$[B_{i,j}, B_{k,l}]_t = \begin{cases} t & \text{if } i = k \text{ and } j = l \\ 0 & \text{else} \end{cases}$$

for all $i, j, k \in \{1, \ldots, p\}$. *Then* $\boldsymbol{B}$ *is a* $p \times p$*-dimensional Brownian motion,* $\boldsymbol{B} \sim \mathcal{BM}_p$.

Given the definition of a local martingale, as in the scalar case, we can define semimartingales.

Definition 10.4.6 A matrix valued stochastic process $\boldsymbol{X}$ is called a semimartingale if $\boldsymbol{X}$ can be decomposed into $\boldsymbol{X} = \boldsymbol{X}_0 + \boldsymbol{M} + \boldsymbol{A}$, where $\boldsymbol{M}$ is a local martingale and $\boldsymbol{A}$ an adapted process of finite variation.

We can now consider stochastic integrals. As in the scalar and vector case, we focus on continuous semimartingales, and for an $n \times p$-dimensional Brownian motion $\boldsymbol{W} \sim \mathcal{BM}_{n,p}$, stochastic processes $\boldsymbol{X}$ and $\boldsymbol{Y}$ in $\mathcal{M}_{m,n}(\Re)$ and $\mathcal{M}_{p,q}(\Re)$, respectively, and a stopping time T, the matrix variate stochastic integral on $[0, T]$ is a matrix with entries

$$\left(\int_0^T \boldsymbol{X}_t \, d\boldsymbol{W}_t \, \boldsymbol{Y}_t\right)_{i,j} = \sum_{k=1}^{n} \sum_{l=1}^{p} \int_0^T X_{t,ik} Y_{t,lj} \, dW_{t,kl}, \quad \forall 1 \le i \le m, \ 1 \le j \le q.$$

We are now able to state an Itô formula for matrix-variate semimartingales, see Pfaffel (2008).

Theorem 10.4.7 *Let $U \subseteq \mathcal{M}_{m,n}(\Re)$ be open, $\boldsymbol{X}$ a continuous semimartingale with values in U and let $f : U \to \Re$ be a twice continuously differentiable function. Then $f(\boldsymbol{X})$ is a continuous semimartingale and*

$$\begin{aligned} f(\boldsymbol{X}_t) = f(\boldsymbol{X}_0) + tr\left(\int_0^t Df(\boldsymbol{X}_s)^\top d\boldsymbol{X}_s\right) \\ + \frac{1}{2} \int_0^t \sum_{j,l=1}^{n} \sum_{i,k=1}^{n} \frac{\partial^2}{\partial X_{i,j} \partial X_{k,l}} f(\boldsymbol{X}_s) \, d[X_{i,j}, X_{k,l}]_s \end{aligned} \qquad (10.4.5)$$

with $\boldsymbol{D} = (\frac{\partial}{\partial X_{i,j}})_{i,j}$.

The next corollary is given in Pfaffel (2008).

Corollary 10.4.8 *Let $\boldsymbol{X}$ be a continuous semimartingale on a stochastic interval $[0, T]$ with $T = \inf\{t\colon \boldsymbol{X}_t \notin U\} > 0$ for an open set $U \subseteq \mathcal{M}_{m,n}(\Re)$ and let $f : U \to \Re$ be a twice continuously differentiable function. Then $(f(\boldsymbol{X}_t))_{t \in [0,T]}$ is a continuous semimartingale and* (10.4.5) *holds for $t \in [0, T)$.*

In order to state a matrix valued integration by parts formula, we need the following definition of covariation for matrix valued stochastic processes.

Definition 10.4.9 For two semimartingales $\boldsymbol{A} \in \mathcal{M}_{d,m}(\Re)$, $\boldsymbol{B} \in \mathcal{M}_{m,n}(\Re)$ the matrix valued quadratic covariation is defined by

$$[\boldsymbol{A}, \boldsymbol{B}]^M_{t,ij} = \sum_{k=1}^{m} [A_{i,k}, B_{k,j}]_t \in \mathcal{M}_{d,n}(\Re).$$

The following integration-by-parts formula will be useful, see Pfaffel (2008).

Theorem 10.4.10 *Let $\boldsymbol{A} \in \mathcal{M}_{d,m}(\Re)$, $\boldsymbol{B} \in \mathcal{M}_{m,n}(\Re)$ be two semimartingales. Then the matrix product $\boldsymbol{A}_t \boldsymbol{B}_t \in \mathcal{M}_{d,n}(\Re)$ is a semimartingale and*

$$\boldsymbol{A}_t \boldsymbol{B}_t = \boldsymbol{A}_0 \boldsymbol{B}_0 + \int_0^t \boldsymbol{A}_s \, d\boldsymbol{B}_s + \int_0^t d\boldsymbol{A}_s \, \boldsymbol{B}_s + [\boldsymbol{A}, \boldsymbol{B}]^M_t.$$

10.5 Matrix Valued Stochastic Differential Equations

In this section, we briefly discuss matrix valued SDEs. The aim is to be able to make sense of the SDEs presented later describing Wishart processes. We follow Pfaffel (2008) and Stelzer (2007), where additional material on matrix valued Markov processes is presented.

As with scalar valued SDEs, we can distinguish between strong and weak solutions. Recall that a strong solution can roughly be thought of as a function of a given Brownian motion. The next definition can be found in Pfaffel (2008).

Definition 10.5.1 Let $(\Omega, \mathcal{A}, \underline{\mathcal{A}}, P)$ be a filtered probability space satisfying the usual conditions and consider the stochastic differential equation

$$dX_t = b(t, X_t)\,dt + \sigma(t, X_t)\,dW_t, \tag{10.5.1}$$

where $X_0 = x_0$, $b : \Re_+ \times \mathcal{M}_{m,n}(\Re) \to \mathcal{M}_{m,n}(\Re)$ and $\sigma : \Re_+ \times \mathcal{M}_{m,n}(\Re) \to \mathcal{M}_{m,p}(\Re)$ are measurable functions, $x_0 \in \mathcal{M}_{m,n}(\Re)$ and W is a $p \times n$-matrix valued Brownian motion.

(i) A pair (X, W) of $\mathcal{A}_t$-adapted continuous processes defined on $(\Omega, \mathcal{A}, \underline{\mathcal{A}}, P)$ is called a solution of the SDE (10.5.1) on $[0, T)$, $T > 0$, if W is an $\underline{\mathcal{A}}$-Brownian motion and

$$X_t = x_0 + \int_0^t b(s, X_s)\,ds + \int_0^t \sigma(s, X_s)\,dW_s \quad \forall t \in [0, T).$$

(ii) Moreover, the pair (X, W) is said to be a strong solution of (10.5.1), if X is adapted to the filtration $(\mathcal{A}_t^W)_{t\in\Re_+}$, where $\mathcal{G}_t^W = \sigma_c(W_s, s \le t)$ is the σ-algebra generated by W_s, $s \le t$, completed with all P-null sets from $\mathcal{A}$.

(iii) A solution (X, W) of the SDE (10.5.1), which is not a strong solution is termed a weak solution of Eq. (10.5.1).

We now discuss the existence of a solution. As in the scalar case, the local Lipschitz condition turns out to suffice, see Stelzer (2007).

Definition 10.5.2 Let $(U, \|\cdot\|_U)$, $(V, \|\cdot\|_V)$ be two normed spaces and $A \subseteq U$ be open. Then a function $f : A \to V$ is called locally Lipschitz, if for every $x \in A$ there exists an open neighborhood $\mathcal{U}(x) \subset A$ and a constant $C(x) \in \Re^+$ such that

$$\big\| f(z) - f(y) \big\|_V \le C(x)\|z - y\|_U \quad \forall z, y \in \mathcal{U}(x).$$

We term $C(x)$ the local Lipschitz coefficient. If there is a $K \in \Re^+$ such that $C(x) = K$ can be chosen for all $x \in A$, then f is called globally Lipschitz.

The next theorem states that a local Lipschitz condition is a sufficient condition.

Theorem 10.5.3 *Let U be an open subset of $\mathcal{M}_{d,n}(\Re)$ and $(U_n)_{n\in\mathcal{N}}$ a sequence of convex closed sets such that $U_n \subset U$, $U_n \subseteq U_{n+1}$ $\forall n \in \mathcal{N}$ and $\bigcup_{n\in\mathcal{N}} U_n = U$.*

Assume that $f : U \to \mathcal{M}_{d,m}(\Re)$ *is locally Lipschitz and* $\boldsymbol{Z}$ *in* $\mathcal{M}_{m,n}(\Re)$ *is a continuous semimartingale. Then for each* U*-valued* $\mathcal{F}_0$*-measurable initial value* $\boldsymbol{X}_0$ *there exists a stopping time* T *and a unique* U*-valued strong solution* $\boldsymbol{X}$ *to the stochastic differential equation*

$$d\boldsymbol{X}_t = f(\boldsymbol{X}_t)\,d\boldsymbol{Z}_t \tag{10.5.2}$$

up to the time $T > 0$ *a.s., i.e. on the stochastic interval* $[0, T)$*. At* $T < \infty$ *we have that either* $\boldsymbol{X}$ *hits the boundary* ∂U *of* U *at* T*, i.e.* $\boldsymbol{X}_T \in \partial U$*, or explodes, i.e.* $\limsup_{t\to T, t<T} \|\boldsymbol{X}_t\| = \infty$*. If* f *satisfies the linear growth condition*

$$\|f(\boldsymbol{X})\|^2 \leq K\left(1 + \|\boldsymbol{X}\|^2\right)$$

with some constant $K \in \Re_+$*, then no explosion can occur.*

We point out that by unique solution we mean that pathwise uniqueness holds for (10.5.2). Two solutions on the same probability space, started from the same initial value and driven by the same semimartingale are then indistinguishable.

We now present a matrix version of the Girsanov theorem for matrix valued stochastic processes. To do so, we recall the notion of stochastic exponentials.

Definition 10.5.4 Let $\boldsymbol{X}$ be a stochastic process. The unique strong solution $\boldsymbol{Z} = \mathcal{E}(\boldsymbol{X})$ of the SDE

$$d\boldsymbol{Z}_t = \boldsymbol{Z}_t\,d\boldsymbol{X}_t, \quad \boldsymbol{Z}_0 = 1 \tag{10.5.3}$$

is called stochastic exponential of $\boldsymbol{X}$.

Theorem 10.5.3 allows us to conclude that the SDE (10.5.3) has a unique strong solution. We now formulate the Girsanov theorem, which will be employed in the next chapter.

Theorem 10.5.5 *Let* $T > 0$, $\boldsymbol{W} \sim \mathcal{BM}_p$ *and* $\boldsymbol{U}$ *be an adapted, continuous stochastic process with values in* $\mathcal{M}_p(\Re)$ *such that*

$$\left(\mathcal{E}\left(tr\left(-\int_0^t \boldsymbol{U}_s^\top d\boldsymbol{W}_s\right)\right)\right)_{t\in[0,T]} \tag{10.5.4}$$

is a martingale, or, which is a sufficient condition for (10.5.4)*, but not necessary, that the Novikov condition is satisfied*

$$E\left(etr\left(\frac{1}{2}\int_0^T \boldsymbol{U}_t^\top \boldsymbol{U}_t\,dt\right)\right) < \infty.$$

Then

$$\hat{Q} = \int \mathcal{E}\left(tr\left(-\int_0^T \boldsymbol{U}_t^\top d\boldsymbol{W}_t\right)\right) dP$$

is an equivalent probability measure, and

$$\hat{\boldsymbol{W}}_t = \int_0^t \boldsymbol{U}_s\,ds + \boldsymbol{W}_t$$

is a $\hat{Q}$*-Brownian motion on* $[0, T)$.

We point out that the Novikov condition presents a general sufficient condition for

$$\left(\mathcal{E}\left(tr\left(-\int_0^t \boldsymbol{U}_s^\top \, d\boldsymbol{W}_s\right)\right)\right)_{t\in[0,T]}$$

to be a martingale. Clearly, for a given matrix valued process $\boldsymbol{U} = \{\boldsymbol{U}_t, \, t \in [0, T]\}$, it can be possible to improve on this sufficient condition, see e.g. Theorem 4.1 and Remark 4.2. in Mayerhofer (2012) for an example involving Wishart processes.

10.6 Matrix Valued Ornstein-Uhlenbeck Processes

As an example, we discuss the matrix valued OU-process, see Pfaffel (2008), where we direct the reader for additional results.

Definition 10.6.1 Let $\boldsymbol{A}, \boldsymbol{B} \in \mathcal{M}_p(\Re)$, $\boldsymbol{x}_0 \in \mathcal{M}_{n,p}(\Re)$ a.s. and $\boldsymbol{W} \sim \mathcal{BM}_{n,p}$. A solution $\boldsymbol{X}$ of the SDE

$$d\boldsymbol{X}_t = \boldsymbol{X}_t \boldsymbol{B} \, dt + d\boldsymbol{W}_t \, \boldsymbol{A}, \quad \boldsymbol{X}_0 = \boldsymbol{x}_0, \tag{10.6.5}$$

is called an $n \times p$-dimensional Ornstein-Uhlenbeck process. We write $\boldsymbol{X} \sim \mathcal{OUP}_{n,p}(\boldsymbol{A}, \boldsymbol{B}, \boldsymbol{x}_0)$ for its probability law.

Since the coefficients $\boldsymbol{X} \mapsto \boldsymbol{X}\boldsymbol{B}$ and $\boldsymbol{X} \mapsto \boldsymbol{A}$ are globally Lipschitz and satisfy the linear growth condition presented in Theorem 10.5.3, we are assured that the SDE (10.6.5) has a unique strong solution on the interval $[0, \infty)$. We can even go further and solve the SDE explicitly.

Theorem 10.6.2 *For a Brownian motion* $\boldsymbol{W} \sim \mathcal{BM}_{n,p}$, *the unique strong solution of* (10.6.5) *is given by*

$$\boldsymbol{X}_t = \boldsymbol{x}_0 \exp\{\boldsymbol{B}t\} + \left(\int_0^t d\boldsymbol{W}_s \, \boldsymbol{A} \exp\{-\boldsymbol{B}s\}\right) \exp\{\boldsymbol{B}t\}. \tag{10.6.6}$$

Proof The proof is completed by verifying that (10.6.6) solves (10.6.5). In this regard, the integration-by-parts formula, presented in Theorem 10.4.10, is crucial. We compute

$$\begin{aligned}
d\boldsymbol{X}_t &= d\big(\boldsymbol{x}_0 \exp\{\boldsymbol{B}t\}\big) + d\left(\left(\int_0^t d\boldsymbol{W}_s \, \boldsymbol{A} \exp\{-\boldsymbol{B}s\}\right) \exp\{\boldsymbol{B}t\}\right) \\
&= \boldsymbol{x}_0 \exp\{\boldsymbol{B}t\} \boldsymbol{B} \, dt + d\boldsymbol{W}_t \, \boldsymbol{A} \exp\{-\boldsymbol{B}t\} \exp\{\boldsymbol{B}t\} \\
&\quad + \left(\int_0^t d\boldsymbol{W}_s \boldsymbol{A} \exp\{-\boldsymbol{B}s\}\right) \exp\{\boldsymbol{B}t\} \boldsymbol{B} \, dt \\
&\quad + d\left[\int_0^\cdot d\boldsymbol{W}_s \, \boldsymbol{A} \exp\{-\boldsymbol{B}s\}, \exp\{\boldsymbol{B}\cdot\}\right]_t^M
\end{aligned}$$

$$
\begin{aligned}
&= \left(\boldsymbol{x}_0 \exp\{\boldsymbol{B}t\} + \left(\int_0^t d\boldsymbol{W}_s\, \boldsymbol{A} \exp\{-\boldsymbol{B}s\} \right) \exp\{\boldsymbol{B}t\} \right) \boldsymbol{B}\, dt + d\boldsymbol{W}_t\, \boldsymbol{A} \\
&= \boldsymbol{X}_t \boldsymbol{B}\, dt + d\boldsymbol{W}_t\, \boldsymbol{A}.
\end{aligned}
$$

Finally, we note that (10.6.6) is by construction a strong solution. □

We state the following lemma, which is Lemma 3.48 from Pfaffel (2008).

Lemma 10.6.3 *Let* $\boldsymbol{W} \sim \mathcal{BM}_{n,p}$ *and* $\boldsymbol{X} : \Re_+ \to \mathcal{M}_{p,m}(\Re)$, $t \mapsto \boldsymbol{X}_t$ *be a square integrable, deterministic function. Then*

$$
\int_0^t d\boldsymbol{W}_s\, \boldsymbol{X}_s \sim \mathcal{N}_{n,m}\left(\boldsymbol{0}, \boldsymbol{I}_n \otimes \int_0^t \boldsymbol{X}_s^\top \boldsymbol{X}_s\, ds \right).
$$

We conclude this section with a result giving the distribution of the matrix valued Ornstein-Uhlenbeck process, see also Theorem 3.49 in Pfaffel (2008) for an alternative presentation.

Theorem 10.6.4 *Let* $\boldsymbol{X} \sim \mathcal{OUP}_{n,p}(\boldsymbol{A}, \boldsymbol{B}, \boldsymbol{x}_0)$, *then the distribution of* $\boldsymbol{X}$ *is given by*

$$
\boldsymbol{X}_t | \boldsymbol{x}_0 \sim \mathcal{N}_{n,p}\big(\boldsymbol{\mu}, \boldsymbol{\sigma}^2\big),
$$

where

$$
\begin{aligned}
\boldsymbol{\mu} &= \boldsymbol{x}_0 \exp\{\boldsymbol{B}t\}, \\
\boldsymbol{\sigma}^2 &= \boldsymbol{I}_n \otimes \exp\{\boldsymbol{B}^\top t\} \int_0^t \exp\{-\boldsymbol{B}^\top s\} \boldsymbol{A}^\top \boldsymbol{A} \exp\{-\boldsymbol{B}s\}\, ds \exp\{\boldsymbol{B}t\}.
\end{aligned}
$$

Proof The proof follows immediately from Lemma 10.6.3, Theorem 10.6.2, and Theorem 10.3.11. □

Finally, we remark that the stationary distribution of the matrix valued Ornstein-Uhlenbeck process can also be computed, see Pfaffel (2008), which is Gaussian.

10.7 A Two-Dimensional Correlated Minimal Market Model

In this section, we discuss how to extend the model for the GOP when denominated in two currencies, as discussed in Sect. 3.3, to allow for a more complex dependence structure. In particular, we introduce a model which allows us to use our knowledge of the Wishart distribution, see Definition 10.3.12. We denote the GOP denominated in the domestic currency by S^a, and the GOP denominated in the foreign currency by S^b. As discussed e.g. in Heath and Platen (2005), an exchange rate at time t can be expressed in terms of the ratio of the two GOPs, see also Sect. 9.7. Assuming the domestic currency is a, then one would pay, at time t, $\frac{S_t^a}{S_t^b}$ units of currency a to obtain one unit of the foreign currency b. As the domestic currency is a, the price of e.g. a call option on the exchange rate can be expressed as:

$$S_0^a E\left(\frac{(\frac{S_T^a}{S_T^b} - K)^+}{S_T^a}\right). \tag{10.7.7}$$

We now discuss an extension of the model, which is tractable, as we can employ the non-central Wishart distribution to compute (10.7.7). For $k \in \{a, b\}$, we set

$$S_t^k = S_t^{0,k} \bar{S}_t^k,$$

where $S_t^{0,k} = \exp\{r_k t\}$, $S_0^{0,k} = 1$. So $S^{0,k}$ denotes the savings account in currency k, which for simplicity is assumed to be a deterministic function of time. As for the stylized MMM, we model $\bar{S}^k$ as a time-changed squared Bessel process of dimension four. We introduce the 2×4 matrix process $\boldsymbol{X} = \{\boldsymbol{X}_t, t \geq 0\}$ via

$$\boldsymbol{X}_t = \begin{bmatrix} (W^{1,1}_{\varphi^1(t)} + w^{1,1}) & (W^{2,1}_{\varphi^1(t)} + w^{2,1}) & (W^{3,1}_{\varphi^1(t)} + w^{3,1}) & (W^{4,1}_{\varphi^1(t)} + w^{4,1}) \\ (W^{1,2}_{\varphi^2(t)} + w^{1,2}) & (W^{2,2}_{\varphi^2(t)} + w^{2,2}) & (W^{3,2}_{\varphi^2(t)} + w^{3,2}) & (W^{4,2}_{\varphi^2(t)} + w^{4,2}) \end{bmatrix}.$$

The processes $W^{i,1}_{\varphi^1}$, $i = 1, \ldots, 4$, denote independent Brownian motions, subjected to a deterministic time-change

$$\varphi^1(t) = \frac{\alpha_0^1}{4\eta^1}\left(\exp\{\eta^1 t\} - 1\right) = \frac{1}{4}\int_0^t \alpha_s^1\, ds,$$

cf. Sect. 3.3, and $W^{i,2}_{\varphi^2}$, $i = 1, \ldots, 4$, denote independent Brownian motions, subjected to the deterministic time change

$$\varphi^2(t) = \frac{\alpha_0^2}{4\eta^2}\left(\exp\{\eta^2 t\} - 1\right) = \frac{1}{4}\int_0^t \alpha_s^2\, ds.$$

Now consider the process $\boldsymbol{Y} = \{\boldsymbol{Y}_t, t \geq 0\}$, which assumes values in S_2^+, and is given by

$$\boldsymbol{Y}_t := \boldsymbol{X}_t \boldsymbol{X}_t^\top, \quad t \geq 0,$$

which yields

$$\boldsymbol{Y}_t = \begin{bmatrix} \sum_{i=1}^4 (W^{i,1}_{\varphi^1(t)} + w^{i,1})^2 & \sum_{i=1}^4 \sum_{j=1}^2 (W^{i,j}_{\varphi^j(t)} + w^{i,j}) \\ \sum_{i=1}^4 \sum_{j=1}^2 (W^{i,j}_{\varphi^j(t)} + w^{i,j}) & \sum_{i=1}^4 (W^{i,2}_{\varphi^2(t)} + w^{i,2})^2 \end{bmatrix}.$$

We set

$$\bar{S}_t^a = Y_t^{1,1},$$

and

$$\bar{S}_t^b = Y_t^{2,2}.$$

We use the diagonal elements of $\boldsymbol{Y}_t$ to model the GOP in different currency denominations. Next, we introduce the following dependence structure: the Brownian motions $W^{i,1}$ and $W^{i,2}$, $i = 1, \ldots, 4$, covary as follows,

$$\big[W^{i,1}_{\varphi^1(\cdot)}, W^{i,2}_{\varphi^2(\cdot)}\big]_t = \frac{\varrho}{4} \int_0^t \sqrt{\alpha_s^1 \alpha_0^2}\, ds, \quad i = 1, \ldots, 4, \tag{10.7.8}$$

where $-1 < \varrho < 1$. The specification (10.7.8) allows us to employ the non-central Wishart distribution. We work through this example in detail, as it illustrates how to extend the stylized MMM to allow for a non-trivial dependence structure, but still exploit the tractability of the Wishart distribution. As discussed in Sect. 10.3, matrix valued normal random variables are studied by interpreting the matrix as a vector, cf. Definition 10.3.12. We recall that $vec(\boldsymbol{X}_T^\top)$ stacks the two columns of $\boldsymbol{X}_T^\top$, hence

$$vec\big(\boldsymbol{X}_T^\top\big) = \begin{bmatrix} (W^{1,1}_{\varphi^1(T)} + w^{1,1}) \\ \vdots \\ (W^{4,1}_{\varphi^1(T)} + w^{4,1}) \\ (W^{1,2}_{\varphi^2(T)} + w^{1,2}) \\ \vdots \\ (W^{4,2}_{\varphi^2(T)} + w^{4,2}) \end{bmatrix}.$$

It is easily seen that the mean matrix $\boldsymbol{M}$ satisfies

$$vec\big(\boldsymbol{M}^\top\big) = \begin{bmatrix} w^{1,1} \\ \vdots \\ w^{4,1} \\ w^{1,2} \\ \vdots \\ w^{4,2} \end{bmatrix} \tag{10.7.9}$$

and the covariance matrix of $vec(\boldsymbol{X}_T^\top)$ is given by

$$\boldsymbol{\Sigma} \otimes \boldsymbol{I}_4 = \begin{bmatrix} \Sigma^{1,1}\boldsymbol{I}_4 & \Sigma^{1,2}\boldsymbol{I}_4 \\ \Sigma^{2,1}\boldsymbol{I}_4 & \Sigma^{2,2}\boldsymbol{I}_4 \end{bmatrix}, \tag{10.7.10}$$

where $\boldsymbol{\Sigma}$ is a 2×2 matrix with $\Sigma^{1,1} = \varphi^1(T)$, $\Sigma^{2,2} = \varphi^2(T)$, and $\Sigma^{1,2} = \Sigma^{2,1} = \frac{\varrho}{4} \int_0^t \sqrt{\alpha_s^1 \alpha_s^2}\, ds$. We remark that assuming $-1 < \varrho < 1$ results in $\boldsymbol{\Sigma}$ being positive definite. It now immediately follows from Theorem 10.3.15 that

$$\boldsymbol{X}_T \boldsymbol{X}_T^\top \sim W_2\big(4, \boldsymbol{\Sigma}, \boldsymbol{\Sigma}^{-1}\boldsymbol{M}\boldsymbol{M}^\top\big),$$

where $\boldsymbol{M}$ and $\boldsymbol{\Sigma}$ are given in Eqs. (10.7.9) and (10.7.10) respectively.

Recall that we set

$$\begin{aligned} \boldsymbol{Y}_t &= \boldsymbol{X}_t \boldsymbol{X}_t^\top, \\ \bar{S}_t^a &= Y_t^{1,1}, \\ \bar{S}_t^b &= Y_t^{2,2}. \end{aligned}$$

Hence we can compute (10.7.7) using

$$E\big(f(\boldsymbol{Y}_T)\big),$$

where $f : S_2^+ \to \Re$ is given by

$$f(\boldsymbol{M}) = \frac{\left(\frac{\exp\{r_1 T\} M^{1,1}}{\exp\{r_2 T\} M^{2,2}} - K\right)^+}{\exp\{r_1 T\} M^{1,1}},$$

for $\boldsymbol{M} \in S_2^+$, and $M^{i,i}$, $i = 1, 2$, are the diagonal elements of $\boldsymbol{M}$. The probability density function of $\boldsymbol{Y}_T$ is given in Definition 10.3.12.

Chapter 11
Wishart Processes

The aim of this chapter is to introduce Wishart processes as tractable diffusions, which can be used to better capture dependence structures associated with multidimensional stochastic models. The focus of this chapter is on the tractability aspect. We present illustrative examples, which show that we can move beyond the dependence structures possible on the Euclidean state space. As discussed in Chap. 9, we consider a model to be tractable if we have access to its affine transform. As demonstrated in Bru (1991), Grasselli and Tebaldi (2008), Ahdida and Alfonsi (2010), Benabid et al. (2010), Laplace transforms of the Wishart process are available in closed-form and exponentially affine. The Wishart process is, in fact, an affine process. We present results on affine transforms in Sect. 11.4.

Besides computing Laplace transforms, exact simulation schemes play an important role in finance, as they allow the pricing of e.g. path-dependent options, see also Chap. 6. In Sect. 11.3, we will discuss simulation schemes for the Wishart process, where we present the approaches from Benabid et al. (2010) and Ahdida and Alfonsi (2010). The two approaches are different in nature, as they exploit different properties of Wishart processes. We hence present both approaches, as they illustrate interesting properties of Wishart processes.

Subsequently, we present an extension of the model presented in Sect. 9.7 to the case where positive factors are modeled via a Wishart process. This illustrates the additional degrees of freedom given by employing the Wishart process. We begin this chapter with a section which introduces Wishart processes and present existence results. Subsequently, we study some special cases of the Wishart process in detail to gain further insight. One of the special cases motivates immediately one of the simulation schemes to be presented in Sect. 11.3.

11.1 Definition and Existence Results

Wishart processes were introduced in Bru (1991), as a matrix generalization of squared Bessel processes. In her PhD thesis, Bru applied these processes to problems from biology. As we will show below, Wishart processes are S_d^+ or $\overline{S_d^+}$ valued,

J. Baldeaux, E. Platen, *Functionals of Multidimensional Diffusions with Applications to Finance*, Bocconi & Springer Series 5, DOI 10.1007/978-3-319-00747-2_11,

i.e. they assume values as positive definite or positive semidefinite matrices. This makes them natural candidates for the modeling of covariance matrices, as noted in Gouriéroux and Sufana (2004a). Starting with Gouriéroux and Sufana (2004a, 2004b), there is now a substantial body of literature applying Wishart processes to problems in finance, see Gouriéroux et al. (2007), Da Fonseca et al. (2007, 2008a, 2008b, 2008c), and Buraschi et al. (2008, 2010).

All of the above references study Wishart processes in a pure diffusive setting. Recently, matrix valued processes incorporating jumps have been studied, see e.g. Barndorff-Nielsen and Stelzer (2007), Leippold and Trojani (2008). These processes are all contained in the affine framework introduced in Cuchiero et al. (2011), where we direct the interested reader. Furthermore, we mention the recent paper Cuchiero et al. (2011), which extends the results from Cuchiero et al. (2011) to symmetric cones.

We introduce the Wishart process as in the work of Grasselli and collaborators. For $\boldsymbol{x} \in \overline{\mathcal{S}_d^+}$, we introduce the $\overline{\mathcal{S}_d^+}$ valued Wishart process $\boldsymbol{X}^{\boldsymbol{x}} = \boldsymbol{X} = \{\boldsymbol{X}_t, t \geq 0\}$, which satisfies

$$d\boldsymbol{X}_t = \left(\alpha \boldsymbol{a}^\top \boldsymbol{a} + \boldsymbol{b}\boldsymbol{X}_t + \boldsymbol{X}_t \boldsymbol{b}^\top\right) dt + \left(\sqrt{\boldsymbol{X}_t}\, d\boldsymbol{W}_t \boldsymbol{a} + \boldsymbol{a}^\top d\boldsymbol{W}_t^\top \sqrt{\boldsymbol{X}_t}\right), \qquad (11.1.1)$$

where $\alpha \geq 0$, $\boldsymbol{a}, \boldsymbol{b} \in \mathcal{M}_d$ and $\boldsymbol{X}_0 = \boldsymbol{x} \in \mathcal{M}_d$. An obvious question to ask is whether Eq. (11.1.1) admits a solution, and furthermore if such a solution is unique and strong. For results on weak solutions, we refer the reader to Cuchiero et al. (2011), and for results on strong solutions to Mayerhofer et al. (2011b). We now present a summary of their results, see Corollary 3.2 in Mayerhofer et al. (2011b) and also Theorem 1 in Ahdida and Alfonsi (2010).

Theorem 11.1.1 *Assume that* $\boldsymbol{x} \in \overline{\mathcal{S}_d^+}$, *and* $\alpha \geq d - 1$, *then Eq.* (11.1.1) *admits a unique weak solution. If* $\boldsymbol{x} \in \mathcal{S}_d^+$ *and* $\alpha \geq d + 1$, *then this solution is strong.*

In this book, we are primarily interested in explaining the tractability of the processes under consideration, where in this chapter, we focus on Wishart processes. In particular, we present for the Wishart process Laplace transforms and an exact simulation scheme. Weak solutions suffice for our purposes and we assume that $\alpha \geq d-1$, so that the weak solution is also unique. As in Ahdida and Alfonsi (2010), we use $WIS_d(\boldsymbol{x}, \alpha, \boldsymbol{b}, \boldsymbol{a})$ to denote a Wishart process and $WIS_d(\boldsymbol{x}, \alpha, \boldsymbol{b}, \boldsymbol{a}; t)$ for the value of the process at time point t.

11.2 Some Special Cases

In this section, we discuss some particular special cases of Wishart processes. Recall that we defined a Wishart process $WIS_d(\boldsymbol{x}, \alpha, \boldsymbol{b}, \boldsymbol{a})$ to be

$$d\boldsymbol{X}_t = \left(\alpha \boldsymbol{a}^\top \boldsymbol{a} + \boldsymbol{b}\boldsymbol{X}_t + \boldsymbol{X}_t \boldsymbol{b}^\top\right) dt + \sqrt{\boldsymbol{X}_t}\, d\boldsymbol{W}_t \boldsymbol{a} + \boldsymbol{a}^\top d\boldsymbol{W}_t^\top \sqrt{\boldsymbol{X}_t},$$

for $\boldsymbol{a}, \boldsymbol{b} \in \mathcal{M}_d$, and $\alpha \geq d-1$. In Sect. 3.2, we had already introduced Wishart processes. We recover the special case studied in Sect. 3.2 by setting $\boldsymbol{a} = \boldsymbol{I}_d$, $\boldsymbol{b} = \boldsymbol{0}$, and $\alpha = d$, to obtain

$$d\boldsymbol{X}_t = d\boldsymbol{I}_d\, dt + \sqrt{\boldsymbol{X}_t}\, d\boldsymbol{W}_t + d\boldsymbol{W}_t^\top \sqrt{\boldsymbol{X}_t}. \tag{11.2.2}$$

Recall that Eq. (11.2.2) is the analogue of Eq. (3.1.1), which introduces the squared Bessel process as a sum of squared Brownian motions. In Eq. (11.2.2), d is also an integer. Subsequently, in Sect. 3.1 we relaxed the assumption that d is an integer. We now do the same for Wishart processes. However, in Bru (1991), the condition $\alpha \geq d-1$ was used to establish the existence of a unique weak solution, see Theorem 2 in Bru (1991), i.e. she established the existence of a unique weak solution of a $WIS_d(\boldsymbol{x}, d, \boldsymbol{0}, \boldsymbol{I}_d)$ process.

So far, we introduced Wishart processes as squares of matrices of Brownian motions, i.e. the $WIS_d(\boldsymbol{x}, d, \boldsymbol{0}, \boldsymbol{I}_d)$ case. However, as in Bru (1991), we can also establish a connection with squared matrix-valued Ornstein-Uhlenbeck processes. This is an important observation, and will also motivate our first simulation scheme in Sect. 11.3.

Let $\boldsymbol{X} = \{\boldsymbol{X}_t,\ t \geq 0\}$ be an $n \times d$ matrix diffusion solution of

$$d\boldsymbol{X}_t = \gamma\, d\boldsymbol{B}_t + \beta \boldsymbol{X}_t\, dt, \tag{11.2.3}$$

where $\boldsymbol{B} = \{\boldsymbol{B}_t,\ t \geq 0\}$ is an $n \times d$ Brownian motion, and $\boldsymbol{x}$ is an $n \times d$ matrix, $\gamma \in \Re$, and $\beta \in \Re^-$. We set $\boldsymbol{S}_t = \boldsymbol{X}_t^\top \boldsymbol{X}_t$, $\boldsymbol{s} = \boldsymbol{x}^\top \boldsymbol{x}$.

Lemma 11.2.1 *Assume that $\boldsymbol{X}_t$ satisfies Eq. (11.2.3). Then $\boldsymbol{S}_t = \boldsymbol{X}_t^\top \boldsymbol{X}_t$ satisfies the SDE*

$$\begin{aligned} d\boldsymbol{S}_t &= \gamma\big(\sqrt{\boldsymbol{S}_t}\, d\boldsymbol{B}_t + d\boldsymbol{B}_t^\top \sqrt{\boldsymbol{S}_t}\big) + 2\beta\sqrt{\boldsymbol{S}_t}\, dt + n\gamma^2 \boldsymbol{I}_d\, dt, \\ \boldsymbol{S}_0 &= \boldsymbol{s}. \end{aligned} \tag{11.2.4}$$

Proof The technique of the proof follows Theorem 4.19 in Pfaffel (2008), where the result was shown for the more general case that corresponds to Lemma 11.2.2. We define

$$\boldsymbol{S}_t = \boldsymbol{X}_t^\top \boldsymbol{X}_t, \quad t \geq 0,$$

and

$$\boldsymbol{W}_t = \int_0^t \sqrt{\boldsymbol{S}_u^{-1}} \boldsymbol{X}_u^\top\, d\boldsymbol{B}_u \in \mathcal{M}_d,$$

for all $t \geq 0$. We first show that $\boldsymbol{W} = \{\boldsymbol{W}_t,\ t \geq 0\}$ is a Brownian motion. We compute

$$\begin{aligned} &E\left(\int_0^t \left(\sqrt{\boldsymbol{S}_u^{-1}} \boldsymbol{X}_u^\top\right)^\top \left(\sqrt{\boldsymbol{S}_u^{-1}} \boldsymbol{X}_u^\top\right) du\right) \\ &= E\left(\int_0^t \boldsymbol{X}_u \boldsymbol{S}_u^{-1} \boldsymbol{X}_u^\top\, du\right) \\ &= t\boldsymbol{I}_p < \infty, \quad \text{a.s.,} \end{aligned}$$

establishing that $\boldsymbol{W}$ is a local martingale. Also,

$$dW_{t,ij} = \sum_{m=1}^{d} \left[\sqrt{\boldsymbol{S}_u^{-1}} \boldsymbol{X}_u^\top \right]_{i,m} dB_{t,mj}.$$

Hence

$$\begin{aligned} & d[W_{\cdot,ij} W_{\cdot,kl}]_t \\ & = \sum_{m=1}^{d} \left[\sqrt{\boldsymbol{S}_t^{-1}} \boldsymbol{X}_t^\top \right]_{i,m} \left[\sqrt{\boldsymbol{S}_t^{-1}} X_t^\top \right]_{k,m} \mathbf{1}_{j=l}\, dt \\ & = \left[\sqrt{\boldsymbol{S}_t^{-1}} \boldsymbol{X}_t^\top \boldsymbol{X}_t \sqrt{\boldsymbol{S}_t^{-1}} \right]_{i,k} \mathbf{1}_{j=l}\, dt \\ & = \boldsymbol{I}_{ik} \mathbf{1}_{j=l}\, dt \\ & = \mathbf{1}_{i=k} \mathbf{1}_{j=l}\, dt, \end{aligned}$$

where we used that

$$d[W_{\cdot,nj}, W_{\cdot,nl}]_t = dt \iff j = l.$$

By Theorem 10.4.5, $\boldsymbol{W}$ is a Brownian motion. Finally, we compute

$$\begin{aligned} d\boldsymbol{S}_t &= d\big(\boldsymbol{X}_t^\top \boldsymbol{X}_t\big) = (d\boldsymbol{X}_t)^\top \boldsymbol{X}_t + \boldsymbol{X}_t^\top d\boldsymbol{X}_t + d\big[\boldsymbol{X}^\top, \boldsymbol{X}\big]_t^M \\ &= \big(\beta \boldsymbol{X}_t^\top dt + d\boldsymbol{B}_t^\top \gamma\big) \boldsymbol{X}_t + \boldsymbol{X}_t^\top (\beta \boldsymbol{X}_t\, dt + d\boldsymbol{B}_t \gamma) + \gamma d\big[\boldsymbol{B}^\top, \boldsymbol{B}\big]_t^M \gamma \\ &= \beta \boldsymbol{S}_t\, dt + d\boldsymbol{B}_t^\top \gamma \boldsymbol{X}_t + \beta \boldsymbol{S}_t\, dt + \boldsymbol{X}_t^\top d\boldsymbol{B}_t \gamma + \gamma^2 n \boldsymbol{I}_d\, dt \\ &= 2\beta \boldsymbol{S}_t\, dt + \gamma \sqrt{\boldsymbol{S}_t}\, d\boldsymbol{W}_t + \gamma\, d\boldsymbol{W}_t^\top \sqrt{\boldsymbol{S}_t} + \gamma^2 n \boldsymbol{I}_d\, dt, \end{aligned}$$

which yields that $\boldsymbol{S}_t$ solves Eq. (11.2.4). □

The following time-change formula is reminiscent of Proposition 3.1.5, see Eq. (5.3) in Bru (1991). If $\boldsymbol{X} = \{\boldsymbol{X}_t,\ t \geq 0\}$ is a solution of (11.2.3), then there exists a Wishart process $\boldsymbol{\Sigma} = \{\boldsymbol{\Sigma}_t,\ t \geq 0\} \in WIS_d(\boldsymbol{s}, \alpha, \mathbf{0}, \boldsymbol{I}_d)$ such that

$$\boldsymbol{S}_t = \boldsymbol{X}_t^\top \boldsymbol{X}_t = \exp\{2\beta t\} \boldsymbol{\Sigma}_{\gamma^2 \frac{1-\exp\{-2\beta t\}}{2\beta}}.$$

Using this time-change formula, Bru established that the Wishart process $WIS_d(\boldsymbol{x}, \alpha, \boldsymbol{b}, \boldsymbol{a})$, where $\boldsymbol{b} = \beta \boldsymbol{I}_d$, $\boldsymbol{a} = \gamma \boldsymbol{I}_d$, $\beta, \gamma \in \Re$, $\alpha \geq d-1$ and $\boldsymbol{x} \in \overline{\mathcal{S}_d^+}$, with distinct eigenvalues, admits a unique weak solution, see Theorem 2′ in Bru (1991). Finally, she extended this result replacing γ and β by $d \times d$ matrices $\boldsymbol{b}$ and $\boldsymbol{a}$, where $\boldsymbol{a} \in GL(d)$. We consider the following SDE for $\boldsymbol{X} = \{\boldsymbol{X}_t,\ t \geq 0\}$,

$$d\boldsymbol{X}_t = d\boldsymbol{B}_t \boldsymbol{a} + \boldsymbol{X}_t \boldsymbol{b}\, dt, \tag{11.2.5}$$

where $\boldsymbol{X}_0 = \boldsymbol{x}$, $\boldsymbol{B} = \{\boldsymbol{B}_t,\ t \geq 0\}$ is a $n \times d$ matrix valued Brownian motion and $\boldsymbol{X}_t$ is an $n \times d$ matrix. We set $\boldsymbol{S}_t = \boldsymbol{X}_t^\top \boldsymbol{X}_t$, $\boldsymbol{s} = \boldsymbol{x}^\top \boldsymbol{x} \in \mathcal{S}_d^+$, and in the next lemma derive the SDE for $\boldsymbol{S}_t$.

Lemma 11.2.2 *Assuming $\boldsymbol{X} = \{\boldsymbol{X}_t,\ t \geq 0\}$ satisfies Eq.* (11.2.5), *we obtain the following dynamics for $\boldsymbol{S}_t = \boldsymbol{X}_t^\top \boldsymbol{X}_t$,*

$$d\boldsymbol{S}_t = \sqrt{\boldsymbol{S}_t}\, d\boldsymbol{W}_t \sqrt{\boldsymbol{a}^\top \boldsymbol{a}} + \sqrt{\boldsymbol{a}^\top \boldsymbol{a}}\, d\boldsymbol{W}_t^\top \sqrt{\boldsymbol{S}_t} + \left(\boldsymbol{b}^\top \boldsymbol{S}_t + \boldsymbol{S}_t \boldsymbol{b}\right) dt + n\boldsymbol{a}^\top \boldsymbol{a}\, dt, \tag{11.2.6}$$

where $\boldsymbol{S}_0 = \boldsymbol{s}$.

Proof The proof is completed in the same fashion as the proof of Lemma 11.2.1, and is given in Pfaffel (2008). We firstly define

$$\boldsymbol{W}_t = \int_0^t \sqrt{\boldsymbol{S}_u^{-1}}\, \boldsymbol{X}_u^\top\, d\boldsymbol{B}_u \boldsymbol{a} \left(\sqrt{\boldsymbol{a}^\top \boldsymbol{a}}\right)^{-1} du \quad \in \mathcal{M}_d.$$

Note that the matrix square-root is positive definite and hence invertible. We compute

$$\begin{aligned}
&E\left(\int_0^t \left(\sqrt{\boldsymbol{S}_u^{-1}}\, \boldsymbol{X}_u^\top \boldsymbol{a} \left(\sqrt{\boldsymbol{a}^\top \boldsymbol{a}}\right)^{-1}\right)^\top \left(\sqrt{\boldsymbol{S}_u^{-1}}\, \boldsymbol{X}_u^\top \boldsymbol{a} \left(\sqrt{\boldsymbol{a}^\top \boldsymbol{a}}\right)^{-1}\right) du\right) \\
&\quad = E\left(\int_0^t \left(\boldsymbol{a}^\top \boldsymbol{a}\right)^{-\frac{1}{2}} \boldsymbol{a}^\top \boldsymbol{X}_u \boldsymbol{S}_u^{-1} \boldsymbol{X}_u^\top \boldsymbol{a} \left(\boldsymbol{a}^\top \boldsymbol{a}\right)^{-\frac{1}{2}} du\right) \\
&\quad = \int_0^t \left(\boldsymbol{a}^\top \boldsymbol{a}\right)^{-\frac{1}{2}} \boldsymbol{a}^\top \boldsymbol{a} \left(\boldsymbol{a}^\top \boldsymbol{a}\right)^{-\frac{1}{2}} du \\
&\quad = t\boldsymbol{I}_p < \infty \quad \text{a.s.}
\end{aligned}$$

We have that

$$dW_{t,ij} = \sum_{u=1}^{n} \sum_{v=1}^{d} \left[\boldsymbol{a}\left(\sqrt{\boldsymbol{a}^\top \boldsymbol{a}}\right)^{-1}\right]_{v,j} \left[\sqrt{\boldsymbol{S}_t^{-1}}\, \boldsymbol{X}_t^\top\right]_{i,u} d\boldsymbol{B}_{t,uv},$$

and we compute

$$\begin{aligned}
&d[W_{\cdot,ij}, W_{\cdot,kl}]_t \\
&\quad = \sum_{u=1}^{n} \sum_{v=1}^{d} \left[\sqrt{\boldsymbol{S}_t^{-1}}\, \boldsymbol{X}_t^\top\right]_{i,u} \left[\sqrt{\boldsymbol{S}_t^{-1}}\, \boldsymbol{X}_t^\top\right]_{k,u} \left[\boldsymbol{a}\left(\sqrt{\boldsymbol{a}^\top \boldsymbol{a}}\right)^{-1}\right]_{v,j} \left[\boldsymbol{a}\left(\sqrt{\boldsymbol{a}^\top \boldsymbol{a}}\right)^{-1}\right]_{v,l} \\
&\quad = \left[\sqrt{\boldsymbol{S}_t^{-1}}\, \boldsymbol{X}_t^\top \boldsymbol{X}_t \sqrt{\boldsymbol{S}_t^{-1}}\right]_{i,k} \left[\left(\sqrt{\boldsymbol{a}^\top \boldsymbol{a}}\right)^{-1} \boldsymbol{a}^\top \boldsymbol{a} \left(\sqrt{\boldsymbol{a}^\top \boldsymbol{a}}\right)^{-1}\right]_{j,l} dt \\
&\quad = \boldsymbol{I}_{d,ik} \boldsymbol{I}_{d,jl}\, dt \\
&\quad = \mathbf{1}_{i=k} \mathbf{1}_{j=l}\, dt.
\end{aligned}$$

By Theorem 10.4.5, $\boldsymbol{W}$ is a Brownian motion. Finally, we compute

$$\begin{aligned}
d\boldsymbol{S}_t &= d\left(\boldsymbol{X}_t^\top \boldsymbol{X}_t\right) = (d\boldsymbol{X}_t)^\top \boldsymbol{X}_t + \boldsymbol{X}_t^\top d\boldsymbol{X}_t + d\left[\boldsymbol{X}^\top, \boldsymbol{X}\right]_t^M \\
&= \left(\boldsymbol{a}^\top d\boldsymbol{B}_t^\top + \boldsymbol{b}^\top \boldsymbol{X}_t^\top dt\right) \boldsymbol{X}_t + \boldsymbol{X}_t^\top \left(d\boldsymbol{B}_t \boldsymbol{a} + \boldsymbol{X}_t \boldsymbol{b}\right) dt + \boldsymbol{a}^\top d\left[\boldsymbol{B}^\top, \boldsymbol{B}\right]_t^M \boldsymbol{a} \\
&= \boldsymbol{a}^\top d\boldsymbol{B}_t^\top \boldsymbol{X}_t + \boldsymbol{b}^\top \boldsymbol{X}_t^\top \boldsymbol{X}_t\, dt + \boldsymbol{X}_t^\top d\boldsymbol{B}_t \boldsymbol{a} + \boldsymbol{X}_t^\top \boldsymbol{X}_t \boldsymbol{b}\, dt + \boldsymbol{a}^\top \boldsymbol{a} n\, dt \\
&= \boldsymbol{a}^\top d\boldsymbol{B}_t^\top \boldsymbol{X}_t + \boldsymbol{b}^\top \boldsymbol{S}_t\, dt + \boldsymbol{X}_t^\top d\boldsymbol{B}_t \boldsymbol{a} + \boldsymbol{S}_t \boldsymbol{b}\, dt + n\boldsymbol{a}^\top \boldsymbol{a}\, dt \\
&= \left(\boldsymbol{b}^\top \boldsymbol{S}_t + \boldsymbol{S}_t \boldsymbol{b}\right) dt + \sqrt{\boldsymbol{S}_t}\, d\boldsymbol{W}_t \sqrt{\boldsymbol{a}^\top \boldsymbol{a}} + \sqrt{\boldsymbol{a}^\top \boldsymbol{a}}\, d\boldsymbol{W}_t^\top \sqrt{\boldsymbol{S}_t} + n\boldsymbol{a}^\top \boldsymbol{a}\, dt,
\end{aligned}$$

where we used that

$$d\left[\boldsymbol{B}_{\cdot}^{\top}, \boldsymbol{B}_{\cdot}\right]_{t,ij}^{M} = \sum_{k=1}^{n} d\langle B_{\cdot,ki}, B_{\cdot,kj}\rangle_t = n\mathbf{1}_{i=j}\, dt,$$

which establishes that $\boldsymbol{S}_t$ solves Eq. (11.2.6). □

We now state Theorem 2″ from Bru (1991).

Theorem 11.2.3 *Let* $\alpha \in \{1, \ldots, d-1\} \cup (d-1, \infty)$, $\boldsymbol{a} \in GL(d)$, $\boldsymbol{b} \in \mathcal{S}_d^-$, $\boldsymbol{s} \in \mathcal{S}_d^+$ *and all eigenvalues of* $\boldsymbol{s}$ *be distinct, and* $\boldsymbol{B}_t$ *is a* $d \times d$ *matrix valued Brownian motion, then on* $[0, \tau)$, *where* τ *denotes the first time that the eigenvalues of* $\boldsymbol{S}_t$ *collide, the stochastic differential equation*

$$d\boldsymbol{S}_t = \sqrt{\boldsymbol{S}_t}\, d\boldsymbol{W}_t \sqrt{\boldsymbol{a}^{\top}\boldsymbol{a}} + \sqrt{\boldsymbol{a}^{\top}\boldsymbol{a}}\, d\boldsymbol{W}_t^{\top}\sqrt{\boldsymbol{S}_t} + (\boldsymbol{b}\boldsymbol{S}_t + \boldsymbol{S}_t\boldsymbol{b})\, dt + \alpha\sqrt{\boldsymbol{a}^{\top}\boldsymbol{a}}\, dt,$$

where $\boldsymbol{S}_0 = \boldsymbol{s}$ *has a unique weak solution if* $\boldsymbol{b}$ *and* $\sqrt{\boldsymbol{a}^{\top}\boldsymbol{a}}$ *commute.*

We remind the reader that the preceding examples were all studied in the original paper on Wishart processes, Bru (1991).

Now, we turn again to the Wishart process as discussed in Sect. 11.1. We introduce the process $\boldsymbol{X} = \{\boldsymbol{X}_t,\ t \geq 0\} \in WIS_d(\boldsymbol{x}, \alpha, \boldsymbol{b}, \boldsymbol{a})$, given by

$$d\boldsymbol{X}_t = \left(\alpha\boldsymbol{a}^{\top}\boldsymbol{a} + \boldsymbol{b}\boldsymbol{X}_t + \boldsymbol{X}_t\boldsymbol{b}^{\top}\right) dt + \left(\sqrt{\boldsymbol{X}_t}\, d\boldsymbol{W}_t\boldsymbol{a} + \boldsymbol{a}^{\top}\, d\boldsymbol{W}_t^{\top}\sqrt{\boldsymbol{X}_t}\right),$$

and we firstly investigate the following case, which was investigated in Benabid et al. (2010). It shows how to link a Wishart process to a multidimensional square root process, see also Sect. 6.7. We assume that $\boldsymbol{a}$ and $\boldsymbol{b}$ are diagonal matrices and that the elements of $\boldsymbol{a}$ are positive, whereas the elements of $\boldsymbol{b}$ are negative. Then one can show that the diagonal elements of $\boldsymbol{X}_t$ satisfy

$$dX_{t,ii} = \left(\alpha a_{i,i}^2 + 2b_{i,i}X_{t,ii}\right) + 2a_{i,i}\sum_{k=1}^{d}[\sqrt{\boldsymbol{X}_t}]_{i,k}\, dW_{t,ki}.$$

Now we define for $i \in \{1, \ldots, d\}$,

$$B_{t,i} = \int_0^t \sqrt{X_{t,ii}}^{-1} \sum_{k=1}^{d}[\sqrt{\boldsymbol{X}_s}]_{k,i}\, dW_{s,ki}.$$

We have

$$E\left(\int_0^t (\sqrt{X_{s,ii}})^{-1} \sum_{k=1}^{d}[\sqrt{\boldsymbol{X}_s}]_{k,i}[\sqrt{\boldsymbol{X}_s}]_{k,i}[\sqrt{X_{s,ii}}]^{-1}\, ds\right)$$

$$= E\left(\int_0^t (\sqrt{X_{s,ii}})^{-1}[\sqrt{\boldsymbol{X}_s}\sqrt{\boldsymbol{X}_s}]_{i,i}(\sqrt{X_{s,ii}})^{-1}\, ds\right)$$

$$= E\left(\int_0^t (\sqrt{X_{s,ii}})^{-1} X_{s,ii}(\sqrt{X_{s,ii}})^{-1}\, ds\right)$$

$$= t.$$

Hence, $\boldsymbol{B} = (B_1, \ldots, B_d)$ is a vector of d independent Brownian motions, and we obtain

$$dX_{t,ii} = \left(\alpha a_{i,i}^2 + 2b_{i,i} X_{t,ii}\right) dt + 2a_{i,i}\sqrt{X_{t,ii}}\, dB_{t,i}.$$

Consequently, for diagonal matrices $\boldsymbol{a}$ and $\boldsymbol{b}$, the diagonal elements of the Wishart process are square-root processes.

We now discuss how to construct a matrix-valued Wishart process from vector-valued Ornstein-Uhlenbeck processes. This construction will also motivate the first simulation scheme in Sect. 11.3. In particular, we set

$$\boldsymbol{V}_t = \sum_{k=1}^{\beta} \boldsymbol{X}_{t,k} \boldsymbol{X}_{t,k}^\top \in \mathcal{M}_d, \tag{11.2.7}$$

where

$$d\boldsymbol{X}_{t,k} = \boldsymbol{M}\boldsymbol{X}_{t,k}\, dt + \boldsymbol{Q}^\top d\boldsymbol{W}_{t,k}, \quad k = 1, \ldots, \beta, \tag{11.2.8}$$

where $\boldsymbol{M} \in \mathcal{M}_d$, $\boldsymbol{X}_t \in \Re^d$, $\boldsymbol{Q} \in \mathcal{M}_d$, $\boldsymbol{W}_k \in \Re^d$, so that $\boldsymbol{V}_t \in \mathcal{M}_d$, $\boldsymbol{V}_0 = \boldsymbol{v} \in \mathcal{S}_d^+$ and $\beta \geq d+1$. The following lemma gives the dynamics of $\boldsymbol{V} = \{\boldsymbol{V}_t,\, t \geq 0\}$.

Lemma 11.2.4 *Assume that $\boldsymbol{V}_t$ is given by Eq.* (11.2.7), *where $\boldsymbol{X}_t$ satisfies Eq.* (11.2.8). *Then*

$$d\boldsymbol{V}_t = \left(\beta \boldsymbol{Q}^\top \boldsymbol{Q} + \boldsymbol{M}\boldsymbol{V}_t + \boldsymbol{V}_t \boldsymbol{M}^\top\right) dt + \sqrt{\boldsymbol{V}_t}\, d\boldsymbol{W}_t \boldsymbol{Q} + \boldsymbol{Q}^\top d\boldsymbol{W}_t^\top \sqrt{\boldsymbol{V}_t},$$

where $\boldsymbol{W} = \{\boldsymbol{W}_t,\, t \geq 0\}$ is a $d \times d$ matrix valued Brownian motion that is determined by

$$\sqrt{\boldsymbol{V}_t}\, d\boldsymbol{W}_t = \sum_{k=1}^{\beta} \boldsymbol{X}_{t,k}\, d\boldsymbol{W}_{t,k}^\top.$$

Proof We compute

$$\begin{aligned}
d\left(\boldsymbol{X}_{t,k}\boldsymbol{X}_{t,k}^\top\right) &= (d\boldsymbol{X}_{t,k})\boldsymbol{X}_{t,k}^\top + \boldsymbol{X}_{t,k}(d\boldsymbol{X}_{t,k})^\top + d\left[\boldsymbol{X}_k, \boldsymbol{X}_k^\top\right]_t^M \\
&= \left(\boldsymbol{M}\boldsymbol{X}_{t,k} + \boldsymbol{Q}^\top d\boldsymbol{W}_{t,k}\right)\boldsymbol{X}_{t,k}^\top \\
&\quad + \boldsymbol{X}_{t,k}\left(\boldsymbol{X}_{t,k}^\top \boldsymbol{M}^\top dt + d\boldsymbol{W}_{t,k}^\top \boldsymbol{Q}\right) \\
&\quad + \boldsymbol{Q}^\top d\left[\boldsymbol{W}_k, \boldsymbol{W}_k^\top\right]_t^M \boldsymbol{Q} \\
&= \boldsymbol{M}\boldsymbol{X}_{t,k}\boldsymbol{X}_{t,k}^\top dt + \boldsymbol{Q}^\top d\boldsymbol{W}_{t,k}\boldsymbol{X}_{t,k}^\top + \boldsymbol{X}_{t,k}\boldsymbol{X}_{t,k}^\top \boldsymbol{M}^\top dt \\
&\quad + \boldsymbol{X}_{t,k} d\boldsymbol{W}_{t,k}^\top \boldsymbol{Q} + \boldsymbol{Q}^\top \boldsymbol{I}_d \boldsymbol{Q}\, dt,
\end{aligned}$$

where we used that

$$d\left[\boldsymbol{W}_{\cdot,k}, \boldsymbol{W}_{\cdot,k}^\top\right]_t^M = \boldsymbol{I}_d\, dt.$$

Hence

$$\begin{aligned}
d\boldsymbol{V}_t &= \boldsymbol{M}\sum_{k=1}^{\beta} \boldsymbol{X}_{t,k}\boldsymbol{X}_{t,k}^\top dt + \boldsymbol{Q}^\top \sum_{k=1}^{\beta} d\boldsymbol{W}_{t,k}\boldsymbol{X}_{t,k}^\top \\
&\quad + \sum_{k-1}^{\beta} \boldsymbol{X}_{t,k}\boldsymbol{X}_{t,k}^\top \boldsymbol{M}^\top dt + \sum_{k-1}^{\beta} \boldsymbol{X}_{t,k}\, d\boldsymbol{W}_{t,k}^\top \boldsymbol{Q} \\
&\quad + \beta \boldsymbol{Q}^\top \boldsymbol{Q}\, dt \\
&= \boldsymbol{M}\boldsymbol{V}_t\, dt + \boldsymbol{V}_t \boldsymbol{M}^\top dt + \boldsymbol{Q}^\top d\boldsymbol{W}_t^\top \sqrt{\boldsymbol{V}_t} + \sqrt{\boldsymbol{V}_t}\, d\boldsymbol{W}_t \boldsymbol{Q} + \beta \boldsymbol{Q}^\top \boldsymbol{Q}\, dt,
\end{aligned}$$

since

$$\sqrt{\boldsymbol{V}_t} d\boldsymbol{W}_t = \sum_{k=1}^{\beta} \boldsymbol{X}_{t,k}\, d\boldsymbol{W}_{t,k}^\top.$$

To complete the proof, we need to show that $\boldsymbol{W}_t$ is a Brownian motion. As before, we use Theorem 10.4.5. We define

$$\boldsymbol{W}_t = \int_0^t \sqrt{\boldsymbol{V}_u^{-1}} \sum_{k=1}^{\beta} \boldsymbol{X}_{t,k}\, d\boldsymbol{W}_{t,k}^\top$$

and it is easily seen that $\boldsymbol{W}$ is a local martingale. Furthermore,

$$\begin{aligned}
dW_{t,ij} &= \sum_{m=1}^{d} \left[\sqrt{\boldsymbol{V}_t^{-1}}\right]_{i,m} \sum_{k=1}^{\beta} [\boldsymbol{X}_{t,k}]_m \left[d\boldsymbol{W}_{t,k}^\top\right]_j \\
&= \sum_{k=1}^{\beta} \sum_{m=1}^{d} \left[\sqrt{\boldsymbol{V}_t^{-1}}\right]_{i,m} [\boldsymbol{X}_{t,k}]_m \left[d\boldsymbol{W}_{t,k}^\top\right]_j.
\end{aligned}$$

Now we have

$$\begin{aligned}
& d[W_{i,j}, W_{k,l}]_t \\
&= \sum_{k'=1}^{\beta} \sum_{m'=1}^{d} \left[\sqrt{\boldsymbol{V}_t^{-1}}\right]_{i,m'} [\boldsymbol{X}_{t,k'}]_{m'} \sum_{m''=1}^{d} \left[\sqrt{\boldsymbol{V}_t^{-1}}\right]_{k,m''} [\boldsymbol{X}_{t,k'}]_{m''} \mathbf{1}_{j=l}\, dt \\
&= \sum_{k'=1}^{\beta} \left[\sqrt{\boldsymbol{V}_t^{-1}} \boldsymbol{X}_{t,k'}\right]_i \left[\boldsymbol{X}_{t,k'}^\top \sqrt{\boldsymbol{V}_t^{-1}}\right]_k \mathbf{1}_{j=l}\, dt \\
&= \sum_{k'=1}^{\beta} \left[\sqrt{\boldsymbol{V}_t^{-1}} \boldsymbol{X}_{t,k'} \boldsymbol{X}_{t,k'}^\top \sqrt{\boldsymbol{V}_t^{-1}}\right]_{i,k} \mathbf{1}_{j=l}\, dt \\
&= \left[\sqrt{\boldsymbol{V}_t^{-1}} \sum_{k'=1}^{\beta} \boldsymbol{X}_{t,k'} \boldsymbol{X}_{t,k'}^\top \sqrt{\boldsymbol{V}_t^{-1}}\right]_{i,k} \mathbf{1}_{j=l}\, dt \\
&= [\boldsymbol{I}_d]_{i,k} \mathbf{1}_{j=l}\, dt \\
&= \mathbf{1}_{i=k} \mathbf{1}_{j=l}\, dt.
\end{aligned}$$

□

11.3 Exact and Almost Exact Simulation Schemes for Wishart Processes

In this section, we discuss two simulation schemes for Wishart processes. The first is based on Benabid et al. (2010), Sect. 1.3, the second on Ahdida and Alfonsi (2010), Sect. 2. We also alert the reader to Chap. 2 in Platen and Bruti-Liberati (2010), where the simulation of the process $WIS_d(\boldsymbol{x}, d, \boldsymbol{0}, \boldsymbol{I}_d)$ was discussed.

11.3.1 Change of Measure Approach

First we present the approach from Benabid et al. (2010). Recall that in Sect. 11.2, we showed that if α assumes integer values, we can simulate a Wishart process by simulating vectors of Ornstein-Uhlenbeck processes. The simulation of multi-dimensional Ornstein-Uhlenbeck processes was discussed in Sects. 6.7 and 10.6. Intuitively, the approach can be described as follows: starting under a probability measure P, where the Wishart process is given by its general form in Eq. (11.1.1) with $\alpha \geq d+1$, $\alpha \in \Re$, we aim to find a change of probability measure, so that under the new measure the corresponding value of α, say $\tilde{\alpha}$, assumes integer values, i.e. $\tilde{\alpha} \in \mathcal{N}$ and $\tilde{\alpha} \geq d+1$. Consequently, we can simulate the Wishart process under the new measure by using Ornstein-Uhlenbeck processes, as explained in Sect. 11.2. In particular, following Benabid et al. (2010), we represent α as follows,

$$\alpha = K + 2\nu,$$

where $K = \lfloor \alpha \rfloor \geq d + 1$, where $\lfloor a \rfloor$ denotes the largest integer less than or equal to a, and ν is a real number satisfying $0 \leq \nu \leq \frac{1}{2}$. The next result, Theorem 2 in Benabid et al. (2010), shows how to introduce a new measure, say P^*, under which the Wishart process can be simulated via Ornstein-Uhlenbeck processes.

Theorem 11.3.1 *Let $q = K + \nu - d - 1$. If*

$$\Lambda_T = \frac{dP^*}{dP}\bigg|_{\mathcal{A}_T}$$

defines the Radon-Nikodym derivative of dP^* with respect to dP, then

$$\Lambda_T = \left(\frac{det(\boldsymbol{X}_T)}{det(\boldsymbol{X}_0)}\right)^{-\frac{\nu}{2}} \exp\left\{\nu T\left(Tr(\boldsymbol{b})\right)\right\} \exp\left\{\frac{\nu q}{2}\int_0^T Tr\left(\boldsymbol{X}_s^{-1}\boldsymbol{a}^\top\boldsymbol{a}\right) ds\right\}.$$

Proof The proof is given in Benabid et al. (2010), see Theorem 2. We present here only the basic ideas of the proof. As in Definition 10.5.4 and Theorem 10.5.5, we specify the new measure via

$$\frac{dP^*}{dP}\bigg|_{\mathcal{A}_T} = \exp\left\{-\nu\int_0^T Tr\left(\sqrt{\boldsymbol{X}_s^{-1}}\, d\boldsymbol{W}_s\boldsymbol{a}\right) - \frac{\nu^2}{2}\int_0^T Tr\left(\boldsymbol{X}_s^{-1}\boldsymbol{a}^\top\boldsymbol{a}\right) ds\right\},$$

where the Wishart process $\boldsymbol{X} \in WIS_d(\boldsymbol{x}, \alpha, \boldsymbol{b}, \boldsymbol{a})$ satisfies under P

$$d\boldsymbol{X}_t = \left(\alpha \boldsymbol{a}^\top \boldsymbol{a} + \boldsymbol{b}\boldsymbol{X}_t + \boldsymbol{X}_t \boldsymbol{b}^\top\right) dt + \sqrt{\boldsymbol{X}_t}\, d\boldsymbol{W}_t \boldsymbol{a} + \boldsymbol{a}^\top d\boldsymbol{W}^\top \sqrt{\boldsymbol{X}_t}.$$

Under the new measure P^*,

$$\boldsymbol{W}_t^* = \nu \int_0^t \sqrt{\boldsymbol{X}_t^{-1}}\boldsymbol{a}^\top dt + \boldsymbol{W}_t,$$

is a Brownian motion, see Benabid et al. (2010). Consequently, the dynamics of $\boldsymbol{X}_t$ are given by

$$\begin{aligned} d\boldsymbol{X}_t &= \left(\alpha \boldsymbol{a}^\top \boldsymbol{a} + \boldsymbol{b}\boldsymbol{X}_t + \boldsymbol{X}_t \boldsymbol{b}^\top\right) dt + \sqrt{\boldsymbol{X}_t}\, d\boldsymbol{W}_t \boldsymbol{a} + \boldsymbol{a}^\top d\boldsymbol{W}_t^\top \sqrt{\boldsymbol{X}_t} \\ &= \left(K \boldsymbol{a}^\top \boldsymbol{a} + \boldsymbol{b}\boldsymbol{X}_t + \boldsymbol{X}_t \boldsymbol{b}^\top\right) dt + \sqrt{\boldsymbol{X}_t}\, d\boldsymbol{W}_t^* \boldsymbol{a} + \boldsymbol{a}^\top \left(d\boldsymbol{W}_t^*\right)^\top \sqrt{\boldsymbol{X}_t}. \end{aligned}$$

As shown in Benabid et al. (2010), Sect. 1.3.1, the dynamics of the determinant of $\boldsymbol{X}_t$ satisfy

$$\begin{aligned} &\log\left(\frac{det(\boldsymbol{X}_T)}{det(\boldsymbol{X}_0)}\right) \\ &= 2T\left(Tr(\boldsymbol{b})\right) + (K - d - 1) \int_0^T Tr\left(\boldsymbol{X}_t^{-1} \boldsymbol{a}^\top \boldsymbol{a}\right) dt + 2 \int_0^T Tr\left(\sqrt{\boldsymbol{X}_t^{-1}}\, d\boldsymbol{W}_t^* \boldsymbol{a}\right). \end{aligned}$$

Substituting, we get

$$\begin{aligned} &\left.\frac{dP^*}{dP}\right|_{\mathcal{A}_T} \\ &= \exp\left\{-\nu \int_0^T Tr\left(\sqrt{\boldsymbol{X}_t^{-1}} d\boldsymbol{W}_t \boldsymbol{a}\right) - \frac{\nu^2}{2} \int_0^T Tr\left(\boldsymbol{X}_t^{-1} \boldsymbol{a}^\top \boldsymbol{a}\right) dt\right\} \\ &= \exp\left\{-\nu \int_0^T Tr\left(\sqrt{\boldsymbol{X}_t^{-1}} d\boldsymbol{W}_t^* \boldsymbol{a}\right) + \frac{\nu^2}{2} \int_0^T Tr\left(\sqrt{\boldsymbol{X}_t^{-1}} \boldsymbol{a}^\top \boldsymbol{a}\right) dt\right\} \\ &= \exp\left\{-\frac{\nu}{2}\left(\log\left(\frac{det(\boldsymbol{X}_T)}{det(\boldsymbol{X}_0)}\right) - 2T\left(Tr(\boldsymbol{b})\right) - (K - d - 1) \int_0^T Tr\left(\boldsymbol{X}_t^{-1} \boldsymbol{a}^\top \boldsymbol{a}\right) dt\right)\right. \\ &\quad \left. + \frac{\nu^2}{2} \int_0^T Tr\left(\sqrt{\boldsymbol{X}_t^{-1}} \boldsymbol{a}^\top \boldsymbol{a}\right) dt\right\} \\ &= \left(\frac{det(\boldsymbol{X}_T)}{det(\boldsymbol{X}_0)}\right)^{-\frac{\nu}{2}} \exp\left\{T\nu Tr(\boldsymbol{b}) + \frac{\nu}{2}(K - d - 1 + \nu) \int_0^T Tr\left(\sqrt{\boldsymbol{X}_t^{-1}} \boldsymbol{a}^\top \boldsymbol{a}\right) dt\right\}. \end{aligned}$$

□

Consequently, if we are interested in computing

$$E_P\left(f(\boldsymbol{X}_T)\right),$$

for a suitable function $f(\cdot)$, we use Theorem 11.3.1 and obtain

$$
\begin{aligned}
&E\big(f(\boldsymbol{X}_T)\big) \\
&= \exp\big\{-\nu T\big(Tr(\boldsymbol{b})\big)\big\} E_{P^*}\left(\left(\frac{det(\boldsymbol{X}_T)}{det(\boldsymbol{X}_0)}\right)^{\frac{\nu}{2}}\right. \\
&\quad \times \exp\left\{-\frac{\nu}{2}(K+\nu-d-1)\int_0^T Tr\big(\boldsymbol{X}_t^{-1}\boldsymbol{a}^\top\boldsymbol{a}\big)\,dt\right\} f(\boldsymbol{X}_T)\Bigg).
\end{aligned}
$$

The simulation of the integral, which appears in the Radon-Nikodym derivative, can be discretized and approximated as follows:

$$
\int_t^{t+\Delta t} Tr\big(\boldsymbol{X}_s^{-1}\boldsymbol{a}^\top\boldsymbol{a}\big)\,ds \sim \frac{1}{2}\Delta t\, Tr\big(\big(\boldsymbol{X}_t^{-1}+\boldsymbol{X}_{t+\Delta t}^{-1}\big)\boldsymbol{a}^\top\boldsymbol{a}\big).
$$

11.3.2 An Exact Simulation Method

We now discuss an exact simulation scheme for Wishart processes, which is based on Ahdida and Alfonsi (2010). To produce the result, we firstly recall the characteristic function associated with $WIS_d(\boldsymbol{x},\alpha,\boldsymbol{b},\boldsymbol{a})$. The result was presented in Gouriéroux and Sufana (2004a), see also Gouriéroux and Sufana (2004b), and we point out that this result led to the realization that there are affine processes that do not assume values on the Euclidean state space. Furthermore, we point out that additional Laplace transform identities are presented in Sect. 11.4. We now follow the presentation in Ahdida and Alfonsi (2010).

Proposition 11.3.2 *Let $\boldsymbol{X}_t \sim WIS_d(\boldsymbol{x},\alpha,\boldsymbol{b},\boldsymbol{a};t)$,*

$$
\boldsymbol{q}_t = \int_0^t \exp(s\boldsymbol{b})\boldsymbol{a}^\top\boldsymbol{a}\exp\big(s\boldsymbol{b}^\top\big)\,ds
$$

and $\boldsymbol{m}_t = \exp\{t\boldsymbol{b}\}$. The Laplace transform of $\boldsymbol{X}_t$, for $\boldsymbol{v}\in\mathcal{D}_{\boldsymbol{b},\boldsymbol{a};t}$, is given by

$$
E\big(\exp\big\{Tr(\boldsymbol{v}\boldsymbol{X}_t)\big\}\big) = \frac{\exp\{Tr(\boldsymbol{v}(\boldsymbol{I}_d-2\boldsymbol{q}_t\boldsymbol{v})^{-1}\boldsymbol{m}_t\boldsymbol{x}\boldsymbol{m}_t^\top)\}}{det(\boldsymbol{I}_d-2\boldsymbol{q}_t\boldsymbol{v})^{\frac{\alpha}{2}}}, \qquad (11.3.9)
$$

where $\mathcal{D}_{\boldsymbol{b},\boldsymbol{a};t} = \{\boldsymbol{v}\in\mathcal{S}_d, E(\exp\{Tr(\boldsymbol{v}\boldsymbol{X}_t)\}) < \infty\}$ is the set of convergence of the Laplace transform, which is given explicitly by

$$
\mathcal{D}_{\boldsymbol{b},\boldsymbol{a};t} = \big\{\boldsymbol{v}\in\mathcal{S}_d,\ \forall s\in[0,t],\ \boldsymbol{I}_d-2\boldsymbol{q}_s\boldsymbol{v}\in GL(d)\big\}.
$$

We remark that for $\boldsymbol{v}=\boldsymbol{v}_{\Re}+\imath\boldsymbol{v}_I$, $\boldsymbol{v}_{\Re}\in\mathcal{D}_{\boldsymbol{b},\boldsymbol{a};t}$ and $\boldsymbol{v}_I\in\mathcal{S}_d$, the Laplace transform in Eq. (11.3.9) *is well-defined. For $\boldsymbol{X}_t\sim WIS_d(\boldsymbol{x},\alpha,\boldsymbol{0},\boldsymbol{I}_d^n;t)$, we have*

$$
E\big(\exp\big\{Tr(\boldsymbol{v}\boldsymbol{X}_T)\big\}\big) = \frac{\exp\{Tr(\boldsymbol{v}(\boldsymbol{I}_d-2t\boldsymbol{I}_d^n\boldsymbol{v})^{-1}\boldsymbol{x})\}}{det(\boldsymbol{I}_d-2t\boldsymbol{I}_d^n\boldsymbol{v})^{\frac{\alpha}{2}}}.
$$

For a proof of Proposition 11.3.2, we refer the reader to Gouriéroux and Sufana (2004a), and also to Ahdida and Alfonsi (2010). We remark that in Lemma 11.4.3

and Corollary 11.4.5, we will discuss the special cases $\boldsymbol{v}_I = \mathbf{0}, \boldsymbol{b} = \mathbf{0}$, and $\boldsymbol{a} = \boldsymbol{I}_d$ and $\boldsymbol{a} \in GL(d)$, respectively.

Regarding the exact simulation procedure, we need the following result from linear algebra, which shows how to perform an *extended Cholesky decomposition*.

Lemma 11.3.3 *Let $\boldsymbol{q} \in \overline{\mathcal{S}_d^+}$ be a matrix with rank r. Then there is a permutation matrix $\boldsymbol{p}$, an invertible lower triangular matrix $\boldsymbol{c}_r \in GL(r)$ and $\boldsymbol{k}_r \in \mathcal{M}_{d-r\times r}$ such that*

$$\boldsymbol{p}\boldsymbol{q}\boldsymbol{p}^\top = \boldsymbol{c}\boldsymbol{c}^\top, \quad \boldsymbol{c} = \begin{pmatrix} \boldsymbol{c}_r & \mathbf{0} \\ \boldsymbol{k}_r & \mathbf{0} \end{pmatrix}.$$

The triplet $(\boldsymbol{c}_r, \boldsymbol{k}_r, \boldsymbol{p})$ is called extended Cholesky decomposition of $\boldsymbol{q}$. Besides,

$$\tilde{\boldsymbol{c}} = \begin{pmatrix} \boldsymbol{c}_r & \mathbf{0} \\ \boldsymbol{k}_r & \boldsymbol{I}_{d-r} \end{pmatrix} \in GL(d),$$

and we have

$$\boldsymbol{q} = \left(\tilde{\boldsymbol{c}}^\top \boldsymbol{p}\right)^\top \boldsymbol{I}_d^r \tilde{\boldsymbol{c}}^\top \boldsymbol{p},$$

where $\boldsymbol{I}_d^r = [\mathbf{1}_{i=j\leq r}]_{1\leq i,j\leq d}$ and $r \leq d$.

Proof The lemma appeared in this form in Ahdida and Alfonsi (2010), Lemma 33, which refers to Golub and Van Loan (1996), Algorithm 4.2.4. □

We point out that a numerical procedure to obtain such a decomposition can be found in Golub and Van Loan (1996), see Algorithm 4.2.4. When $r = d$, then we can choose $\boldsymbol{p} = \boldsymbol{I}_d$, and $\boldsymbol{c}_r$ is the usual Cholesky decomposition.

The following proposition, which is Proposition 9 in Ahdida and Alfonsi (2010), sits at the heart of the approach. Essentially, it shows that by rescaling, we can represent a general Wishart process $WIS_d(\boldsymbol{x}, \alpha, \boldsymbol{b}, \boldsymbol{a}; t)$ as one which satisfies $\boldsymbol{b} = \mathbf{0}$ and $\boldsymbol{a} = \boldsymbol{I}_d^n$, where $n = Rank(\boldsymbol{q}_t)$. This is crucial, as the law of $WIS_d(\boldsymbol{x}, \alpha, \mathbf{0}, \boldsymbol{I}_d^n; t)$ can be simulated exactly, as we demonstrate below. As in Ahdida and Alfonsi (2010), we remark that one can exactly compute $\boldsymbol{\theta}_t$, which appears in Proposition 11.3.4, using Lemma 11.3.3.

Proposition 11.3.4 *Let $t > 0$, $\boldsymbol{a}, \boldsymbol{b} \in \mathcal{M}_d$, and $\alpha \geq d - 1$. Then $\boldsymbol{m}_t = \exp\{t\boldsymbol{b}\}$, $\boldsymbol{q}_t = \int_0^t \exp\{s\boldsymbol{b}\}\boldsymbol{a}^\top \boldsymbol{a} \exp\{s\boldsymbol{b}^\top\} ds$ and $n = rank(\boldsymbol{q}_t)$, and there is a $\boldsymbol{\theta}_t \in GL(d)$ such that*

$$\boldsymbol{q}_t = t\boldsymbol{\theta}_t \boldsymbol{I}_d^n \boldsymbol{\theta}_t^\top,$$

and we have

$$WIS_d(\boldsymbol{x}, \alpha, \boldsymbol{b}, \boldsymbol{a}; t) \stackrel{d}{=} \boldsymbol{\theta}_t WIS_d\left(\boldsymbol{\theta}_t^{-1}\boldsymbol{m}_t \boldsymbol{x} \boldsymbol{m}_t^\top \left(\boldsymbol{\theta}_t^{-1}\right)^\top, \alpha, \mathbf{0}, \boldsymbol{I}_d^n; t\right)\boldsymbol{\theta}_t^\top.$$

Proof We present the proof from Ahdida and Alfonsi (2010), due to the importance of the result. We apply Lemma 11.3.3 to $\boldsymbol{q}_t/t \in \overline{\mathcal{S}_d^+}$ and obtain the extended Cholesky decomposition $(\boldsymbol{c}_n, \boldsymbol{k}_n, \boldsymbol{p})$. Also, we obtain from Lemma 11.3.3 that

$$\tilde{\boldsymbol{c}} = \begin{pmatrix} \boldsymbol{c}_n & \boldsymbol{0} \\ \boldsymbol{k}_n & \boldsymbol{I}_{d-n} \end{pmatrix}.$$

We define

$$\boldsymbol{\theta}_t = \boldsymbol{p}^{-1}\tilde{\boldsymbol{c}},$$

which by Lemma 11.3.3 is invertible. We now get that

$$\boldsymbol{q}_t = t\boldsymbol{\theta}_t \boldsymbol{I}_d^n \boldsymbol{\theta}_t^\top.$$

Next, we recall that for $\boldsymbol{a}, \boldsymbol{b}, \boldsymbol{c} \in \mathcal{M}_d$, the following equalities hold

$$det(\boldsymbol{ab}) = det(\boldsymbol{ba}), \qquad Tr(\boldsymbol{ab}) = Tr(\boldsymbol{ba}),$$

and also

$$(\boldsymbol{abc})^{-1} = \boldsymbol{c}^{-1}\boldsymbol{b}^{-1}\boldsymbol{a}^{-1},$$

assuming that $\boldsymbol{a}$, $\boldsymbol{b}$, and $\boldsymbol{c}$ are invertible. We hence obtain the following string of equalities,

$$\begin{aligned} det(\boldsymbol{I}_d - 2\imath \boldsymbol{q}_t \boldsymbol{v}) &= det\big(\boldsymbol{I}_d - 2\imath t\boldsymbol{\theta}_t \boldsymbol{I}_d^n \boldsymbol{\theta}_t^\top \boldsymbol{v}\big) \\ &= det\big(\boldsymbol{\theta}_t\big(\boldsymbol{\theta}_t^{-1} - 2\imath t \boldsymbol{I}_d^n \boldsymbol{\theta}_t^\top \boldsymbol{v}\big)\big) \\ &= det\big(\big(\boldsymbol{\theta}_t^{-1} - 2\imath t \boldsymbol{I}_d^n \boldsymbol{\theta}_t^\top \boldsymbol{v}\big)\boldsymbol{\theta}_t\big) \\ &= det\big(\boldsymbol{I}_d - 2\imath t \boldsymbol{I}_d^n \boldsymbol{\theta}_t^\top \boldsymbol{v}\boldsymbol{\theta}_t\big). \end{aligned}$$

Furthermore,

$$\begin{aligned} &Tr\big(\imath \boldsymbol{v}(\boldsymbol{I}_d - 2\imath \boldsymbol{q}_t \boldsymbol{v})^{-1}\boldsymbol{m}_t \boldsymbol{x} \boldsymbol{m}_t^\top\big) \\ &\quad = Tr\big(\imath\big(\boldsymbol{\theta}_t^{-1}\big)^\top \boldsymbol{\theta}_t^\top \boldsymbol{v}\big(\boldsymbol{\theta}_t\boldsymbol{\theta}_t^{-1} - 2\imath t\boldsymbol{\theta}_t \boldsymbol{I}_d^n \boldsymbol{\theta}_t^\top \boldsymbol{v}\boldsymbol{\theta}_t\boldsymbol{\theta}_t^{-1}\big)^{-1}\boldsymbol{m}_t \boldsymbol{x} \boldsymbol{m}_t^\top\big) \\ &\quad = Tr\big(\imath\big(\boldsymbol{\theta}_t^{-1}\big)^\top \boldsymbol{\theta}_t^\top \boldsymbol{v}\big(\boldsymbol{\theta}_t\big(\boldsymbol{I}_d - 2\imath t \boldsymbol{I}_d^n \boldsymbol{\theta}_t^\top \boldsymbol{v}\boldsymbol{\theta}_t\big)\boldsymbol{\theta}_t^{-1}\big)^{-1}\boldsymbol{m}_t \boldsymbol{x} \boldsymbol{m}_t^\top\big) \\ &\quad = Tr\big(\imath\big(\boldsymbol{\theta}_t^{-1}\big)^\top \boldsymbol{\theta}_t^\top \boldsymbol{v}\boldsymbol{\theta}_t\big(\boldsymbol{I}_d - 2\imath t \boldsymbol{I}_d^n \boldsymbol{\theta}_t^\top \boldsymbol{v}\boldsymbol{\theta}_t\big)^{-1}\boldsymbol{\theta}_t^{-1}\boldsymbol{m}_t \boldsymbol{x} \boldsymbol{m}_t^\top\big) \\ &\quad = Tr\big(\imath \boldsymbol{\theta}_t^\top \boldsymbol{v}\boldsymbol{\theta}_t\big(\boldsymbol{I}_d - 2\imath t \boldsymbol{I}_d^n \boldsymbol{\theta}_t^\top \boldsymbol{v}\boldsymbol{\theta}_t\big)^{-1}\boldsymbol{\theta}_t^{-1}\boldsymbol{m}_t \boldsymbol{x} \boldsymbol{m}_t^\top\big(\boldsymbol{\theta}_t^{-1}\big)^\top\big). \end{aligned}$$

We now let $\boldsymbol{X}_t \sim WIS_d(\boldsymbol{x}, \alpha, \boldsymbol{b}, \boldsymbol{a}; t)$ and $\tilde{\boldsymbol{X}}_t \sim WIS_d(\boldsymbol{\theta}_t^{-1}\boldsymbol{m}_t \boldsymbol{x} \boldsymbol{m}_t^\top(\boldsymbol{\theta}_t^{-1})^\top, \alpha, \boldsymbol{0}, \boldsymbol{I}_d^n; t)$ and apply Proposition 11.3.2 to obtain

$$\begin{aligned} &E\big(\exp\big\{\imath Tr(\boldsymbol{v}\boldsymbol{X}_t)\big\}\big) \\ &\quad = \frac{\exp\{Tr(\imath \boldsymbol{v}(\boldsymbol{I}_d - 2\boldsymbol{q}_t \imath \boldsymbol{v})^{-1}\boldsymbol{m}_t \boldsymbol{x} \boldsymbol{m}_t^\top)\}}{det(\boldsymbol{I}_d - 2\boldsymbol{q}_t \imath \boldsymbol{v})^{\frac{\alpha}{2}}} \\ &\quad = \frac{\exp\{Tr(\imath \boldsymbol{\theta}_t^\top \boldsymbol{v}\boldsymbol{\theta}_t(\boldsymbol{I}_d - 2\imath t \boldsymbol{I}_d^n \boldsymbol{\theta}_t^\top \boldsymbol{v}\boldsymbol{\theta}_t)^{-1}\boldsymbol{\theta}_t^{-1}\boldsymbol{m}_t \boldsymbol{x} \boldsymbol{m}_t^\top(\boldsymbol{\theta}_t^{-1})^\top)\}}{det(\boldsymbol{I}_d - 2\imath t \boldsymbol{I}_d^n \boldsymbol{\theta}_t^\top \boldsymbol{v}\boldsymbol{\theta}_t)^{\frac{\alpha}{2}}} \end{aligned}$$

$$= E\big(\exp\big\{Tr\big(\iota \boldsymbol{\theta}_t^\top \boldsymbol{v} \boldsymbol{\theta}_t \tilde{\boldsymbol{X}}_t\big)\big\}\big)$$
$$= E\big(\exp\big\{Tr\big(\iota \boldsymbol{v} \boldsymbol{\theta}_t \tilde{\boldsymbol{X}}_t \boldsymbol{\theta}_t^\top\big)\big\}\big)$$

completing the proof. □

We remark that Lemma 11.3.3 generalizes the well-known one-dimensional link between a square-root and a squared Bessel process. For $d = 1$, Lemma 11.3.3 gives

$$WIS_1(x, \alpha, b, a; t) = \frac{a^2(\exp\{2b\} - 1)}{2bt} WIS_1\left(\frac{2btx}{a^2(1 - \exp\{-2bt\})}, \alpha, 0, 1; t\right). \tag{11.3.10}$$

This identity can easily be obtained from the results in Sect. 3.1. Let $X = \{X_t,\ t \geq 0\}$ be a $WIS_1(x, \alpha, b, a)$, then

$$dX_t = \big(\alpha a^2 + 2bX_t\big)\, dt + 2a\sqrt{X_t}\, dW_t.$$

From Proposition 3.1.5, it follows that

$$X_t \stackrel{d}{=} \exp\{2bt\}\tilde{X}_{c(t)},$$

where

$$c(t) = \frac{a^2(1 - \exp\{-2bt\})}{2b}$$

and $\tilde{X}$ is a squared Bessel process, i.e. a $WIS_1(x, \alpha, 0, 1)$ process. We hence have established that

$$WIS_1(x, \alpha, b, a; t) \stackrel{d}{=} \exp\{2bt\} WIS_1\big(x, \alpha, 0, 1; c(t)\big).$$

Now we apply the linear time-change, see Proposition 3.1.2,

$$WIS_1\left(x, \alpha, 0, 1; \frac{c(t)}{t} t\right) \stackrel{d}{=} \frac{c(t)}{t} WIS_1\left(\frac{xt}{c(t)}, \alpha, 0, 1; t\right).$$

Hence

$$\begin{aligned} WIS_1(x, \alpha, 0, 1; t) &= \exp\{2bt\} WIS_1\big(x, \alpha, 0, 1; c(t)\big) \\ &= \frac{\exp\{2bt\}c(t)}{t} WIS_1\left(\frac{xt}{c(t)}, \alpha, 0, 1; t\right), \end{aligned}$$

which is Eq. (11.3.10).

We now proceed as follows: from Proposition 11.3.4, it is clear that we can focus on the $WIS_d(\boldsymbol{x}, \alpha, \boldsymbol{0}, \boldsymbol{I}_d^n)$ case. For the generator of $WIS_d(\boldsymbol{x}, \alpha, \boldsymbol{0}, \boldsymbol{I}_d^n)$, we recall a remarkable splitting property from Ahdida and Alfonsi (2010). Having split the operator, we show that each of these operators correspond to an SDE which can be solved explicitly.

11.3.3 A Remarkable Splitting Property

The infinitesimal generator of the Wishart process, or rather the splitting property thereof, plays an important role in the development of an exact simulation scheme. We hence recall this generator, which is a special case e.g. of Corollary 4 in Ahdida and Alfonsi (2010).

Lemma 11.3.5 *On $\mathcal{M}_d$, we associate with $WIS_d(\boldsymbol{x}, \alpha, \boldsymbol{b}, \boldsymbol{a})$ the infinitesimal generator*

$$\boldsymbol{L} = Tr\big(\big[\alpha \boldsymbol{a}^\top \boldsymbol{a} + \boldsymbol{b}\boldsymbol{x} + \boldsymbol{x}\boldsymbol{b}^\top\big]\boldsymbol{D}\big) + 2Tr\big(\boldsymbol{x}\boldsymbol{D}\boldsymbol{a}^\top \boldsymbol{a}\boldsymbol{D}\big),$$

where $\boldsymbol{D} = (\frac{\partial}{\partial x_{i,j}})$, $1 \leq i, j \leq d$.

The following result, which is Proposition 10 in Ahdida and Alfonsi (2010), gives the splitting property of the operator $\boldsymbol{L}$, for the special case $WIS_d(\boldsymbol{x}, \alpha, \boldsymbol{0}, \boldsymbol{I}_d^n)$. Recall that by Proposition 11.3.4, the study of the simulation of Wishart processes can be reduced to this case. We use $\boldsymbol{e}_d^i$ to denote the matrix

$$\boldsymbol{e}_d^n = [\boldsymbol{1}_{i=j=n}]_{1 \leq i,j \leq d}.$$

We clearly have, $\boldsymbol{I}_d^n = \sum_{i=1}^n \boldsymbol{e}_d^i$.

Theorem 11.3.6 *Let $\boldsymbol{L}$ be the generator associated with the Wishart process $WIS_d(\boldsymbol{x}, \alpha, \boldsymbol{0}, \boldsymbol{I}_d^n)$ and $\boldsymbol{L}_i$ the generator associated with $WIS_d(\boldsymbol{x}, \alpha, \boldsymbol{0}, \boldsymbol{e}_d^i)$ for $i \in \{1, \dots, d\}$. Then we have*

$$\boldsymbol{L} = \sum_{i=1}^n \boldsymbol{L}_i \quad \text{and } \forall i, j \in \{1, \dots, d\}, \ \boldsymbol{L}_i \boldsymbol{L}_j = \boldsymbol{L}_j \boldsymbol{L}_i. \tag{11.3.11}$$

Proof The first part of the proof follows immediately from Lemma 11.3.5, noting that $\boldsymbol{I}_d^n = \sum_{i=1}^n \boldsymbol{e}_d^i$. The commutativity property is established in Appendix C.1 in Ahdida and Alfonsi (2010). □

As stated in Ahdida and Alfonsi (2010), two features of Eq. (11.3.11) are important:

- the operators $\boldsymbol{L}_i$ and $\boldsymbol{L}_j$ are the same up to the exchange of coordinates i and j;
- the processes $WIS_d(\boldsymbol{x}, \alpha, \boldsymbol{0}, \boldsymbol{e}_d^i)$ and $WIS_d(\boldsymbol{x}, \alpha, \boldsymbol{0}, \boldsymbol{I}_d^n)$ are well defined on $\overline{\mathcal{S}_d^+}$ under the same hypothesis, namely that $\alpha \geq d - 1$ and $\boldsymbol{x} \in \overline{\mathcal{S}_d^+}$.

The latter property motivates the simulation scheme:

$$\begin{aligned}
\boldsymbol{X}_t^{1,\boldsymbol{x}} &\sim WIS_d\big(\boldsymbol{x}, \alpha, \boldsymbol{0}, \boldsymbol{e}_d^1; t\big) \\
\boldsymbol{X}_t^{2,\boldsymbol{X}_t^{1,\boldsymbol{x}}} &\sim WIS_d\big(\boldsymbol{X}_t^{1,\boldsymbol{x}}, \alpha, \boldsymbol{0}, \boldsymbol{e}_d^2; t\big) \\
&\vdots \\
\boldsymbol{X}_t^{n,\dots \boldsymbol{X}_t^{1,\boldsymbol{x}}} &\sim WIS_d\big(\boldsymbol{X}_t^{n-1,\dots \boldsymbol{X}_t^{1,\boldsymbol{x}}}, \alpha, \boldsymbol{0}, \boldsymbol{e}_d^n; t\big).
\end{aligned}$$

Thus, one samples $X_t^{i,\dots X_t^{1,x}}$ according to the distribution at time t of a Wishart process starting from $X_t^{i-1,\dots X_t^{1,x}}$ and with parameters α, $\boldsymbol{a} = \boldsymbol{e}_d^i$ and $\boldsymbol{b} = \boldsymbol{0}$.

Proposition 11.3.7 *Let $X_t^{n,\dots X_t^{1,x}}$ be defined as above. Then*

$$X_t^{n,\dots X_t^{1,x}} \sim WIS_d\big(\boldsymbol{x}, \alpha, \boldsymbol{0}, \boldsymbol{I}_d^n; t\big).$$

Proof For a formal proof, we refer the reader to Ahdida and Alfonsi (2010), here we just present the main ideas of the proof. Consider a smooth function f on $\mathcal{S}_d^+$. Then by iterating Itô's formula, one can establish that

$$E\big(f\big(X_t^{\boldsymbol{x}}\big)\big) = \sum_{k=0}^{\infty} t^k \boldsymbol{L}^k f(\boldsymbol{x})/k!.$$

Next, we employ the tower property, to get

$$\begin{aligned} E\big(f\big(X_t^{n,\dots X_t^{1,x}}\big)\big) &= E\big(E\big(f\big(X_t^{n,\dots X_t^{1,x}}\big)|X_t^{n-1,\dots X_t^{1,x}}\big)\big) \\ &= \sum_{k_n=0}^{\infty} \frac{t^{k_n}}{k_n!} E\big(\boldsymbol{L}_n^{k_n} f\big(X_t^{n-1,\dots X_t^{1,x}}\big)\big). \end{aligned}$$

Repeating this argument, we obtain

$$\begin{aligned} E\big(f\big(X_t^{n,\dots X_t^{1,x}}\big)\big) &= \sum_{k_1,\dots,k_n=0}^{\infty} \frac{t^{\sum_{i=1}^n k_i}}{k_1!\dots k_n!} \boldsymbol{L}_1^{k_1} \dots \boldsymbol{L}_n^{k_n} f(\boldsymbol{x}) \\ &= \sum_{k=0}^{\infty} \frac{t^k}{k!} (\boldsymbol{L}_1 + \dots + \boldsymbol{L}_n)^k = E\big(f\big(X_t^{\boldsymbol{x}}\big)\big). \end{aligned} \tag{11.3.12}$$

Equality (11.3.12) relies on the identification of a Cauchy product and one uses the fact that the operators commute. For example, for $n = 2$,

$$\sum_{k_1,k_2=0}^{\infty} \frac{t^{k_1+k_2}}{k_1!k_2!} \boldsymbol{L}_1^{k_1} \boldsymbol{L}_2^{k_2} f(\boldsymbol{x}) = \sum_{k=0}^{\infty} \boldsymbol{c}_k,$$

where

$$\begin{aligned} \boldsymbol{c}_k &= \sum_{l=0}^{k} \boldsymbol{a}_l \boldsymbol{b}_{k-l} \\ &= \sum_{l=0}^{k} \frac{t^l}{l!} \frac{t^{k-l}}{(k-l)!} \boldsymbol{L}_1^l \boldsymbol{L}_2^{k-l} \\ &= \frac{t^k}{k!} \sum_{l=0}^{k} \frac{k!}{l!(k-l)!} \boldsymbol{L}_1^l \boldsymbol{L}_2^{k-l} \end{aligned}$$

$$= \frac{t^k}{k!} \sum_{l=0}^{k} \binom{k}{l} \boldsymbol{L}_1^l \boldsymbol{L}_2^{k-l}$$

$$= \frac{t^k}{k!} (\boldsymbol{L}_1 + \boldsymbol{L}_2)^k. \qquad \square$$

Proposition 11.3.7 shows that if we can simulate $WIS_d(\boldsymbol{x}, \alpha, \boldsymbol{0}, \boldsymbol{e}_d^k; t)$, for $k \in \{1, \ldots, d\}$, then we can simulate $WIS_d(\boldsymbol{x}, \alpha, \boldsymbol{0}, \boldsymbol{I}_d^n; t)$, which according to Proposition 11.3.4 means that we can simulate $WIS_d(\boldsymbol{x}, \alpha, \boldsymbol{b}, \boldsymbol{a}; t)$. The next lemma shows that we can simulate $WIS_d(\boldsymbol{x}, \alpha, \boldsymbol{0}, \boldsymbol{e}_d^i; t)$ by simulating $WIS_d(\boldsymbol{p}_k \boldsymbol{x} \boldsymbol{p}_k, \alpha, \boldsymbol{0}, \boldsymbol{I}_d^1; t)$, and subsequently changing the first and the kth coordinates, where we use $\boldsymbol{p}_k$ to denote the matrix which changes the first and the kth coordinate. The following lemma formalizes this.

Lemma 11.3.8 *Construct a matrix $\boldsymbol{p}_k \in \mathcal{S}_d$, so that $p_{k,1} = p_{1,k} = p_{i,i} = 1$, for $i \notin \{1, k\}$, and $p_{i,j} = 0$ otherwise. Let the law of $\boldsymbol{X}_t$ be given by $WIS_d(\boldsymbol{p}_k \boldsymbol{x} \boldsymbol{p}_k, \alpha, \boldsymbol{0}, \boldsymbol{I}_d^1; t)$ and the law of $\tilde{\boldsymbol{X}}_t$ by $WIS_d(\boldsymbol{x}, \alpha, \boldsymbol{0}, \boldsymbol{e}_d^k; t)$. Then*

$$WIS_d\big(\boldsymbol{x}, \alpha, \boldsymbol{0}, \boldsymbol{e}_d^k; t\big) \stackrel{d}{=} \boldsymbol{p}_k WIS_d\big(\boldsymbol{p}_k \boldsymbol{x} \boldsymbol{p}_k, \alpha, \boldsymbol{0}, \boldsymbol{I}_d^1; t\big) \boldsymbol{p}_k.$$

Proof This result is proven in the same way as Proposition 11.3.4. In particular, we use the characteristic function given in Proposition 11.3.2 for the case $\boldsymbol{b} = \boldsymbol{0}$ and $\boldsymbol{a} = \boldsymbol{I}_d^n$. The proof is now easily completed by using the facts

$$\boldsymbol{p}_k \boldsymbol{I}_d^1 \boldsymbol{p}_k = \boldsymbol{e}_d^k \quad \text{and} \quad \boldsymbol{p}_k \boldsymbol{p}_k = \boldsymbol{I}_d,$$

which then allows us to establish that

$$E\big(\exp\{\imath Tr(\boldsymbol{v} \boldsymbol{p}_k \boldsymbol{X}_t \boldsymbol{p}_k)\}\big) = E\big(\exp\{\imath Tr(\boldsymbol{v} \tilde{\boldsymbol{X}}_t)\}\big). \qquad \square$$

11.3.4 Exact Simulation for Wishart Processes

In this subsection, we discuss how to simulate a $WIS_d(\boldsymbol{x}, \alpha, \boldsymbol{0}, \boldsymbol{I}_d^1)$ process, with $\alpha \geq d-1$ and $\boldsymbol{x} \in \overline{\mathcal{S}_d^+}$. Due to Proposition 11.3.7 and Lemma 11.3.8, this allows us to sample from the distribution of $WIS_d(\boldsymbol{x}, \alpha, \boldsymbol{0}, \boldsymbol{I}_d^n; t)$. As the presentation is easier, we start with the case $d = 2$.

From Lemma 11.3.5, we obtain the following infinitesimal generator of $WIS_2(\boldsymbol{x}, \alpha, \boldsymbol{0}, \boldsymbol{I}_2^1)$. For $\boldsymbol{x} \in \overline{\mathcal{S}_2^+}$,

$$\boldsymbol{L} f(\boldsymbol{x}) = \alpha \partial_{1,1} f(\boldsymbol{x}) + 2x_{1,1} \partial_{1,1}^2 f(\boldsymbol{x}) + 2x_{1,2} \partial_{1,1} \partial_{1,2} f(\boldsymbol{x}) + \frac{x_{2,2}}{2} \partial_{1,2}^2 f(\boldsymbol{x}). \tag{11.3.13}$$

This generator is associated with an SDE that can be solved explicitly. As in Ahdida and Alfonsi (2010), we denote by $Z^1 = (Z_t^1, t \geq 0)$ and $Z^2 = \{Z_t^2, t \geq 0\}$

two independent standard Brownian motions. We distinguish the two cases when $x_{2,2} = 0$ and $x_{2,2} > 0$. For $x_{2,2} = 0$, we have that $x_{1,2} = 0$, since $\boldsymbol{x} \in \overline{\mathcal{S}_2^+}$. In fact, one has

$$dX_{t,11} = \alpha\, dt + 2\sqrt{X_{t,11}} dZ_t^1, \qquad dX_{t,12} = 0, \qquad dX_{t,22} = 0, \tag{11.3.14}$$

where $\boldsymbol{X}_0 = \boldsymbol{x}$, has the infinitesimal generator given in Eq. (11.3.13). Clearly, $X_{t,11}$ is a squared Bessel process of dimension α, that can be sampled as discussed in Sect. 3.1.

We now turn to the case where $x_{2,2} > 0$. The SDE

$$dX_{t,11} = \alpha\, dt + 2\sqrt{X_{t,11} - \frac{X_{t,12}^2}{X_{t,22}}}\, dZ_t^1 + 2\frac{X_{t,12}}{X_{t,22}}\, dZ_t^2 \tag{11.3.15}$$

$$dX_{t,12} = \sqrt{X_{t,22}}\, dZ_t^2 \tag{11.3.16}$$

$$dX_{t,22} = 0, \tag{11.3.17}$$

started at $\boldsymbol{X}_0 = \boldsymbol{x}$ has an infinitesimal generator as given in Eq. (11.3.13). This system can be solved explicitly. We introduce auxiliary variables

$$U_{t,11} = X_{t,11} - \frac{(X_{t,12})^2}{X_{t,22}}, \qquad U_{t,12} = \frac{X_{t,12}}{\sqrt{x_{2,2}}}, \qquad U_{t,22} = x_{2,2}, \tag{11.3.18}$$

where $\boldsymbol{U}_0 = \boldsymbol{u}$. An application of the Itô formula confirms that

$$dU_{t,11} = (\alpha - 1)\, dt + 2\sqrt{U_{t,11}}\, dZ_t^1, \qquad dU_{t,12} = dZ_t^2, \qquad U_{t,22} = 0.$$

Hence, $U_{t,11}$ is a squared Bessel process of dimension $\alpha - 1$, and $U_{t,12}$ is a Brownian motion. Consequently, we simulate $X_{t,11}$, $X_{t,12}$ and $X_{t,22}$ by inverting Eq. (11.3.18) to yield

$$X_{t,11} = U_{t,11} + (U_{t,12})^2, \qquad X_{t,12} = U_{t,12}\sqrt{U_{t,22}}, \qquad X_{t,22} = U_{t,22}. \tag{11.3.19}$$

The following proposition summarizes the discussion in this subsection.

Proposition 11.3.9 *Let $\boldsymbol{x} \in \overline{\mathcal{S}_2^+}$. Then the process defined by either Eq.* (11.3.14) *or Eq.* (11.3.16) *when $x_{2,2} = 0$ or $x_{2,2} > 0$ respectively, has its infinitesimal generator given by* (11.3.13). *Moreover, the SDE given by Eq.* (11.3.16) *has a unique strong solution that is given by Eq.* (11.3.19) *starting from $u_{1,1} = x_{1,1} - \frac{x_{1,2}^2}{x_{2,2}} \geq 0$, $u_{1,2} = \frac{x_{1,2}}{\sqrt{x_{2,2}}}$, $u_{2,2} = x_{2,2}$.*

Proof This result is a special case of Theorem 13 in Ahdida and Alfonsi (2010). □

As noted in Ahdida and Alfonsi (2010), an interesting property of the result in Proposition 11.3.9 is that the squared Bessel process is well-defined when its dimension $\alpha - 1$ satisfies $\alpha - 1 \geq 0$, which is the same condition under which the Wishart process $WIS_2(\boldsymbol{x}, \alpha, \boldsymbol{0}, \boldsymbol{I}_2^1)$ is well defined, $\alpha \geq d - 1$, since $d = 2$. Lastly, we point out that the process $\boldsymbol{U} = \{\boldsymbol{U}_t,\ t \geq 0\}$ has a squared Bessel process on its diagonal and a Brownian motion on the off-diagonal.

We now discuss how to sample from the distribution $WIS_d(\boldsymbol{x}, \alpha, \boldsymbol{0}, \boldsymbol{I}_d^n; t)$, where $d \geq 2$. It is easy to check that for $WIS_d(\boldsymbol{x}, \alpha, \boldsymbol{0}, \boldsymbol{I}_d^n)$, for $\boldsymbol{x} \in \overline{\mathcal{S}_d^+}$, the infinitesimal generator is given by

$$\begin{aligned} \boldsymbol{L} f(\boldsymbol{x}) &= \alpha \partial_{1,1} f(\boldsymbol{x}) + 2x_{1,1}\partial_{1,1}^2 f(\boldsymbol{x}) \\ &\quad + 2 \sum_{\substack{1 \leq m \leq d \\ m \neq 1}} x_{1,m} \partial_{1,m} \partial_{1,1} f(\boldsymbol{x}) + \frac{1}{2} \sum_{\substack{1 \leq m,l \leq d \\ m \neq 1, l \neq 1}} x_{m,l} \partial_{1,m} \partial_{1,l} f(\boldsymbol{x}). \end{aligned} \tag{11.3.20}$$

The next theorem, which is Theorem 13 in Ahdida and Alfonsi (2010), shows how to construct an SDE with the same infinitesimal generator as Eq. (11.3.20) and that it can be solved explicitly. Recall that for the case $d = 2$, we distinguished two cases depending on whether $x_{2,2} = 0$ or $x_{2,2} > 0$. For the general case, the SDE depends on the rank of the submatrix $[x_{i,j}]_{2 \leq i,j \leq d}$. We set

$$r = Rank\big([x_{i,j}]_{2 \leq i,j \leq d}\big) \in \{0, \ldots, d-1\}.$$

First we consider the case $\exists \boldsymbol{c}_r \in \mathcal{G}_r$ that is lower triangular, $\boldsymbol{k}_r \in \mathcal{M}_{d-1-r \times r}$, so that

$$[x_{i,j}]_{2 \leq i,j \leq d} = \begin{pmatrix} \boldsymbol{c}_r & \boldsymbol{0} \\ \boldsymbol{k}_r & \boldsymbol{0} \end{pmatrix} \begin{pmatrix} \boldsymbol{c}_r^\top & \boldsymbol{k}_r^\top \\ v0 & \boldsymbol{0} \end{pmatrix} =: \boldsymbol{c}\boldsymbol{c}^\top. \tag{11.3.21}$$

The following theorem formally applies to the case where $\boldsymbol{X}_0 = \boldsymbol{x}$ satisfies (11.3.21). However, the subsequent Lemma 11.3.11 shows that such a decomposition can always be obtained by permuting the coordinates $\{2, \ldots, d\}$. As in Ahdida and Alfonsi (2010), we also abuse the notation as follows: when $r = 0$, we still assume that Eq. (11.3.21) holds, in particular with $\boldsymbol{c} = \boldsymbol{0}$. When $r = d - 1$, we recover the usual Cholesky decomposition of $[x_{i,j}]_{2 \leq i,j \leq d}$.

Theorem 11.3.10 *Let us consider $\boldsymbol{x} \in \overline{\mathcal{S}_d^+}$ such that Eq.* (11.3.21) *holds. Let $\boldsymbol{Z} = \{\boldsymbol{Z}_t = (Z_t^1, Z_t^2, \ldots, Z_t^{r+1}),\ t \geq 0\}$ be a vector valued standard Brownian motion. Then the following SDE, where $\sum_{k=1}^r = 0$, for $r = 0$,*

$$\begin{aligned} dX_{t,11} &= \alpha\, dt + 2\sqrt{X_{t,11} - \sum_{k=1}^r \left(\sum_{l=1}^r [\boldsymbol{c}_r^{-1}]_{k,l} X_{t,1(l+1)} \right)^2}\, dZ_t^1 \\ &\quad + 2 \sum_{k=1}^r \sum_{l=1}^r [\boldsymbol{c}_r^{-1}]_{k,l} X_{t,1(l+1)}\, dZ_t^{k+1} \\ dX_{t,1i} &= \sum_{k=1}^r c_{i-1,k} dZ_t^{k+1}, \quad i = 2 \ldots, d \\ dX_{t,lk} &= 0, \quad k, l = 2, \ldots, d, \end{aligned}$$

has a unique strong solution $\boldsymbol{X} = \{\boldsymbol{X}_t,\ t \geq 0\}$ *starting from* $\boldsymbol{x}$. *It assumes values in* $\overline{\mathcal{S}_d^+}$ *and has the infinitesimal generator* $\mathbf{L}$ *given in Eq.* (11.3.20). *Moreover, the explicit solution is given by*

$$\boldsymbol{X}_t = \boldsymbol{C} \begin{pmatrix} U_{t,11} + \sum_{k=1}^r (U_{t,1(k+1)})^2 & [U_{t,1(l+1)}]_{1 \leq l \leq r}^\top & \mathbf{0} \\ [U_{t,1(l+1)}]_{1 \leq l \leq r} & \boldsymbol{I}_r & \mathbf{0} \\ \mathbf{0} & \mathbf{0} & \mathbf{0} \end{pmatrix} \boldsymbol{C}^\top, \qquad (11.3.22)$$

where

$$dU_{t,11} = (\alpha - r)\,dt + 2\sqrt{U_{t,11}}\,dZ_t^1, \qquad u_{1,1} = x_{1,1} - \sum_{k=1}^r u_{1,k+1}^2 \geq 0$$

$$dU_{t,1(l+1)} = dZ_t^{l+1}, \quad 1 \leq l \leq r, \qquad [u_{1,l+1}]_{1 \leq l \leq r} = \boldsymbol{c}_r^{-1}[x_{1,l+1}]_{1 \leq l \leq r},$$

and

$$\boldsymbol{C} = \begin{pmatrix} 1 & \mathbf{0} & \mathbf{0} \\ 0 & \boldsymbol{c}_r & \mathbf{0} \\ 0 & \boldsymbol{k}_r & \boldsymbol{I}_{d-r-1} \end{pmatrix}.$$

When $r = 0$, then Eq. (11.3.22) should simply be read as

$$\boldsymbol{X}_t = \begin{pmatrix} U_{t,11} & 0 & 0 \\ 0 & 0 & 0 \\ 0 & 0 & 0 \end{pmatrix}.$$

Regarding the matrix $\boldsymbol{U}_t$, we point out that the algorithm only accesses the first row and column of this matrix. As in Ahdida and Alfonsi (2010), $\boldsymbol{X}_t$ can be seen as a function of $\boldsymbol{U}_t$ by setting

$$U_{t,ij} = u_{i,j} = x_{i,j}, \quad i, j \geq 2, \qquad U_{t,1i} = u_{1,i} = 0, \quad r + 1 \leq i \leq d.$$

For a proof of Theorem 11.3.10, we refer the reader to Ahdida and Alfonsi (2010). We point out that sampling from the $WIS_d(\boldsymbol{x}, \alpha, \mathbf{0}, \boldsymbol{I}_d^1; t)$ distribution amounts to sampling a non-central chi-squared random variable and a Gaussian random variable. As for the $d = 2$ case, we note that the condition ensuring that the squared Bessel process $U_{1,1}$ is well-defined for all $r \in \{0, \ldots, d-1\}$ is $\alpha - d - 1 \geq 0$, the same as for the Wishart process. We now recall that the procedure in Theorem 11.3.10 assumed that $\boldsymbol{x} \in \overline{\mathcal{S}_d^+}$ satisfied Eq. (11.3.21). This assumption can be relaxed using the extended Cholesky decomposition from Lemma 11.3.3.

Lemma 11.3.11 *Let* $\boldsymbol{X} = \{\boldsymbol{X}_t,\ t \geq 0\}$ *be a* $WIS(\boldsymbol{x}, \alpha, \mathbf{0}, \boldsymbol{I}_d^1)$ *process and* $(\boldsymbol{c}_r, \boldsymbol{k}_r, \boldsymbol{p})$ *be an extended Cholesky decomposition of* $[x_{i,j}]_{2 \leq i,j \leq d}$ *obtained from Lemma* 11.3.3. *Then*

$$\pi = \begin{pmatrix} 1 & \mathbf{0} \\ 0 & \boldsymbol{p} \end{pmatrix}$$

is a permutation matrix, and

$$X = \{X_t, t \geq\} \stackrel{d}{=} \pi^\top WIS_d(\pi x \pi^\top, \alpha, \mathbf{0}, I_d^1)\pi,$$

and

$$[(\pi x \pi^\top)_{i,j}]_{2\leq i,j\leq d} = \begin{pmatrix} c_r & \mathbf{0} \\ k_r & \mathbf{0} \end{pmatrix} \begin{pmatrix} c_r^\top & k_r^\top \\ \mathbf{0} & \mathbf{0} \end{pmatrix}$$

satisfies (11.3.21).

Proof The first part of the proof can be completed in the same way as the proof of Proposition 11.3.4, namely using characteristic functions. The second part of the proof is an immediate consequence of Lemma 11.3.3. □

Hence, using Theorem 11.3.10 and Lemma 11.3.11, we have a simple way of constructing an SDE that has the generator L from (11.3.13) for any initial condition $x \in \overline{\mathcal{S}_d^+}$. It means that we can sample exactly the Wishart distribution $WIS_d(x, \alpha, \mathbf{0}, I_d^1; t)$, which we summarize in Algorithm 11.1.

As discussed in Ahdida and Alfonsi (2010), the computational cost of Algorithm 11.1 is $\mathcal{O}(d^3)$, as this is the computational cost of performing the extended Cholesky decomposition.

We recall that the splitting property established in Theorem 11.3.6 means that if we can sample $WIS_d(x, \alpha, \mathbf{0}, e_d^i; t)$, for $i = 1, \ldots, n$, we can sample $WIS_d(x, \alpha, \mathbf{0}, I_d^n; t)$. However, Lemma 11.3.8 established that sampling $WIS_d(x, \alpha, \mathbf{0}, e_d^i; t)$ amounts to sampling $WIS_d(x, \alpha, \mathbf{0}, I_d^1; t)$, which we discussed in Theorem 11.3.10 and Algorithm 11.1. This is illustrated in Algorithm 11.2.

Algorithm 11.2 performs Algorithm 11.1 n times, resulting in a computational complexity of $\mathcal{O}(nd^3)$, which is bounded by $\mathcal{O}(d^4)$. Concluding this section, we present Algorithm 11.3, which shows how to sample $WIS_d(x, \alpha, b, a; t)$, which uses Proposition 11.3.4 to reformulate the problem into the one solved by Algorithm 11.2. We remind the reader that this algorithm is applicable if $\alpha \geq d - 1$, which is also the requirement for the existence of a unique weak solution of the SDE (11.1.1) describing the Wishart process.

11.4 Affine Transforms of the Wishart Process

In this section, we discuss the explicit computation of affine transforms associated with Wishart processes. These results are crucial, as they make Wishart processes useful for practical applications. We present two approaches to this problem: the first is based on the linearization procedure presented in Chap. 9. As discussed below, it turns out that the Riccati equations associated with affine transforms of the Wishart process can be linearized allowing us to compute the affine transform. Consequently, we present an alternative approach, which generalizes a result from Bru (1991), and, following Bru, we refer to it as *Cameron-Martin formula*. The section concludes with a comparison of the two approaches.

Algorithm 11.1 Exact Simulation for the operator L_1

Require: $\boldsymbol{x} \in \overline{\mathcal{S}_d^+}$, $\alpha \geq d-1$ and $t > 0$

1: Compute the extended Cholesky decomposition $(\boldsymbol{c}_r, \boldsymbol{k}_r, \boldsymbol{p})$ of $[x_{i,j}]_{2\leq i,j\leq d}$ given by Lemma 11.3.3, $r \in \{0, \ldots, d-1\}$

2: Set

$$\boldsymbol{\pi} = \begin{pmatrix} 1 & \boldsymbol{0} \\ 0 & \boldsymbol{p} \end{pmatrix},$$

$$\tilde{\boldsymbol{x}} = \boldsymbol{\pi} \boldsymbol{x} \boldsymbol{\pi}^\top,$$

$$[u_{1,l+1}]_{1\leq l\leq r} = \boldsymbol{c}_r^{-1} [\tilde{\boldsymbol{x}}_{1,l+1}]_{1\leq l\leq r},$$

$$u_{1,1} = \tilde{x}_{1,1} - \sum_{k=1}^{r} (u_{1,k+1})^2 \geq 0.$$

3: Sample independently r normal variates $G_2, \ldots, G_{r+1} \sim N(0,1)$ and a non-central chi-square random variate $\chi^2_{\alpha-r}(\frac{u_{1,1}}{t})$, i.e. a non-central chi-square distributed random variable with $\alpha - r$ degrees of freedom and non-centrality parameter $\frac{u_{1,1}}{t}$.

4: Set $U_{t,1(l+1)} = u_{1,l+1} + \sqrt{t} G_{l+1}$

5: Set $U_{t,11} = t\chi^2_{\alpha-r}(\frac{u_{1,1}}{t})$

6: **return** $X =$

$$\boldsymbol{\pi}^\top \boldsymbol{C} \begin{pmatrix} U_{t,11} + \sum_{k=1}^{r} (U_{t,1(k+1)})^2 & [U_{t,1(l+1)}]_{1\leq l\leq r}^\top & \boldsymbol{0} \\ [U_{t,1(l+1)}]_{1\leq l\leq r} & \boldsymbol{I}_r & \boldsymbol{0} \\ 0 & \boldsymbol{0} & \boldsymbol{0} \end{pmatrix} \boldsymbol{C}^\top \boldsymbol{\pi},$$

where

$$\boldsymbol{C} = \begin{pmatrix} 1 & \boldsymbol{0} & \boldsymbol{0} \\ 0 & \boldsymbol{c}_r & \boldsymbol{0} \\ 0 & \boldsymbol{k}_r & \boldsymbol{I}_{d-r-1} \end{pmatrix}.$$

Algorithm 11.2 Exact Simulation for $WIS_d(\boldsymbol{x}, \alpha, \boldsymbol{0}, \boldsymbol{I}_d^n; t)$

Require: $\boldsymbol{x} \in \overline{\mathcal{S}_d^+}$, $n \leq d$, $\alpha \geq d-1$ and $t > 0$

1: Set $\boldsymbol{y} = \boldsymbol{x}$.

2: **for** $k = 1$ to n **do**

3: Construct the permutation matrix $\boldsymbol{p}$ by setting $p_{k,1} = p_{1,k} = p_{i,i} = 1$ for $i \notin \{1, k\}$, and $p_{i,j} = 0$ otherwise.

4: Set $\boldsymbol{y} = \boldsymbol{p} \boldsymbol{Y} \boldsymbol{p}$, where $\boldsymbol{Y}$ is sampled according to $WIS_d(\boldsymbol{p}\boldsymbol{y}\boldsymbol{p}, \alpha, \boldsymbol{0}, \boldsymbol{I}_d^1; t)$ by using Algorithm 11.1.

5: **end for**

6: **return** $X = \boldsymbol{y}$.

Algorithm 11.3 Exact Simulation for $WIS_d(\boldsymbol{x}, \alpha, \boldsymbol{b}, \boldsymbol{a}; t)$

Require: $\boldsymbol{x} \in \overline{\mathcal{S}_d^+}$, $\alpha \geq d-1$, $\boldsymbol{a}, \boldsymbol{b} \in \mathcal{M}_d$ and $t > 0$

1: Calculate $\boldsymbol{q}_t = \int_0^t \exp\{s\boldsymbol{b}\}\boldsymbol{a}^\top \boldsymbol{a} \exp\{s\boldsymbol{b}^\top\}\, ds$ and $(\boldsymbol{c}_n, \boldsymbol{k}_n, \boldsymbol{p})$ an extended Cholesky decomposition of $\boldsymbol{q}_t/t$.

2: Set

$$\boldsymbol{\theta}_t = \boldsymbol{p}^{-1} \begin{pmatrix} \boldsymbol{c}_n & \boldsymbol{0} \\ \boldsymbol{k}_n & \boldsymbol{I}_{d-n} \end{pmatrix}$$

and $\boldsymbol{m}_t = \exp\{t\boldsymbol{b}\}$.

3: **return** $\boldsymbol{X} = \boldsymbol{\theta}_t \boldsymbol{Y} \boldsymbol{\theta}_t^\top$, where $\boldsymbol{Y} \sim WIS_d(\boldsymbol{\theta}_t^{-1}\boldsymbol{m}_t \boldsymbol{x} \boldsymbol{m}_t^\top (\boldsymbol{\theta}_t^{-1})^\top, \alpha, \boldsymbol{\theta}, \boldsymbol{I}_d^n; t)$ is sampled by Algorithm 11.2.

11.4.1 Linearization Applied to Wishart Processes

We assume the following dynamics for the Wishart process,

$$d\boldsymbol{X}_t = \left(\alpha \boldsymbol{a}^\top \boldsymbol{a} + \boldsymbol{b}\boldsymbol{X}_t + \boldsymbol{X}_t \boldsymbol{b}^\top\right) dt + \left(\sqrt{\boldsymbol{X}_t}\, d\boldsymbol{W}_t \boldsymbol{a} + \boldsymbol{a}^\top\, d\boldsymbol{W}_t \sqrt{\boldsymbol{X}_t}\right), \qquad (11.4.23)$$

and the infinitesimal generator is given by

$$\boldsymbol{L} = Tr\left(\left[\alpha \boldsymbol{a}^\top \boldsymbol{a} + \boldsymbol{b}\boldsymbol{x} + \boldsymbol{x}\boldsymbol{b}^\top\right]\boldsymbol{D}\right) + 2Tr\left(\boldsymbol{x}\boldsymbol{D}\boldsymbol{a}^\top \boldsymbol{a}\boldsymbol{D}\right),$$

see Lemma 11.3.5. As in Chap. 9, we study the discounted conditional characteristic function,

$$\begin{aligned} &\Psi_X(\boldsymbol{u}, \boldsymbol{x}, t, \tau) \\ &= E\left(\exp\left\{-\int_t^T \left(\eta_0 + Tr(\boldsymbol{\eta}\boldsymbol{X}_s)\right) ds\right\} \exp\left\{Tr(\imath \boldsymbol{u}\boldsymbol{X}_T)\right\} | \mathcal{A}_t\right) \\ &= \exp\left\{V^0(\tau, \imath\boldsymbol{u}) - Tr\left(\mathbf{V}(\tau, \imath\boldsymbol{u})\boldsymbol{X}_t\right)\right\}, \end{aligned}$$

where $\tau = T - t$. The Feynman-Kac argument now yields, where we use $\Psi = \Psi_X(\boldsymbol{u}, \boldsymbol{x}, t, \tau)$,

$$\begin{aligned} \frac{\partial \Psi}{\partial \tau} &= \boldsymbol{L}\Psi - \left(\eta_0 + Tr(\boldsymbol{\eta}\boldsymbol{x})\right) \\ &= Tr\left(\left(\alpha \boldsymbol{a}^\top \boldsymbol{a} + \boldsymbol{b}\boldsymbol{x} + \boldsymbol{x}\boldsymbol{b}^\top\right)\boldsymbol{D}\Psi + 2\boldsymbol{x}\boldsymbol{D}\boldsymbol{a}^\top \boldsymbol{a}\boldsymbol{D}\Psi\right) \\ &\quad - \left(\eta_0 + Tr(\boldsymbol{\eta}\boldsymbol{x})\right). \end{aligned}$$

On the other hand,

$$\frac{\partial \Psi}{\partial \tau} = \frac{d}{d\tau} V^0(\tau) - Tr\left(\frac{d}{d\tau}\mathbf{V}(\tau)\boldsymbol{x}\right),$$

which yields

$$\begin{aligned}
&\frac{dV^0(\tau)}{d\tau} - Tr\left(\frac{d}{d\tau}\mathbf{V}(\tau)\boldsymbol{x}\right)\\
&\quad = Tr\big(\big(\alpha \boldsymbol{a}\boldsymbol{a}^\top + \boldsymbol{b}\boldsymbol{x} + \boldsymbol{x}\boldsymbol{b}^\top\big)\boldsymbol{D}\Psi + 2\boldsymbol{x}\boldsymbol{D}\boldsymbol{a}^\top\boldsymbol{a}\boldsymbol{D}\Psi\big)\\
&\qquad - \big(\eta_0 + Tr(\boldsymbol{\eta}\boldsymbol{x})\big)\\
&\quad = Tr\big(\big(\alpha \boldsymbol{a}^\top\boldsymbol{a} + \boldsymbol{b}\boldsymbol{x} + \boldsymbol{x}\boldsymbol{b}^\top\big)(-\mathbf{V}) + 2\boldsymbol{x}\mathcal{V}\boldsymbol{a}^\top\boldsymbol{a}\mathbf{V}\big)\\
&\qquad - \big(\eta_0 + Tr(\boldsymbol{\eta}\boldsymbol{x})\big),
\end{aligned}$$

subject to the initial conditions

$$V^0(0) = 0, \qquad \mathbf{V}(0) = -\imath\boldsymbol{u}.$$

By identifying the coefficients of $\boldsymbol{x}$, we obtain the matrix Riccati ODE satisfied by $\mathbf{V}(\tau)$:

$$-\frac{d}{d\tau}\mathbf{V}(\tau) = -\mathbf{V}(\tau)\boldsymbol{b} - \boldsymbol{b}^\top\mathbf{V}(\tau) + 2\mathbf{V}(\tau)\boldsymbol{a}^\top\boldsymbol{a}\mathbf{V}(\tau) - \boldsymbol{\eta}, \tag{11.4.24}$$

and

$$\frac{dV^0(\tau)}{d\tau} = Tr\big(\alpha\boldsymbol{a}^\top\boldsymbol{a}\big(-\mathbf{V}(\tau)\big)\big) - \eta_0. \tag{11.4.25}$$

From Eq. (11.4.24) we get

$$\frac{d\mathbf{V}(\tau)}{d\tau} = \mathbf{V}(\tau)\boldsymbol{b} + \boldsymbol{b}^\top\mathbf{V}(\tau) - 2\mathbf{V}(\tau)\boldsymbol{a}^\top\boldsymbol{a}\mathbf{V}(\tau) + \boldsymbol{\eta}. \tag{11.4.26}$$

We now employ the linearization idea from Chap. 9, and set

$$\mathbf{V}(\tau) = \mathbf{F}(\tau)^{-1}\mathbf{G}(\tau),$$

where $\mathbf{F}(\tau) \in GL(d)$ and $\mathbf{G}(\tau) \in \mathcal{M}_d$. Now

$$\frac{d}{d\tau}\big(\mathbf{F}(\tau)\mathbf{V}(\tau)\big) - \left(\frac{d}{d\tau}\mathbf{F}(\tau)\right)\mathbf{V}(\tau) = \mathbf{F}(\tau)\frac{d}{d\tau}\mathbf{V}(\tau),$$

and substituting (11.4.26), we get

$$\begin{aligned}
\frac{d}{d\tau}\big(\mathbf{F}(\tau)\mathbf{V}(\tau)\big) - \frac{d}{d\tau}\mathbf{F}(\tau)\mathbf{V}(\tau) &= \mathbf{F}(\tau)\mathbf{V}(\tau)\boldsymbol{b} + \mathbf{F}(\tau)\boldsymbol{b}^\top\mathbf{V}\\
&\quad - 2\mathbf{F}(\tau)\mathbf{V}(\tau)\boldsymbol{a}^\top\boldsymbol{a}\mathbf{V}(\tau) + \mathbf{F}(\tau)\boldsymbol{\eta}.
\end{aligned}$$

Matching coefficients, we obtain

$$\begin{aligned}
\frac{d}{d\tau}\mathbf{G}(\tau) &= \mathbf{G}(\tau)\boldsymbol{b} + \mathbf{F}(\tau)\boldsymbol{\eta}\\
\frac{d}{d\tau}\mathbf{F}(\tau) &= -\mathbf{F}(\tau)\boldsymbol{b}^\top + 2\mathbf{G}(\tau)\boldsymbol{a}^\top\boldsymbol{a},
\end{aligned} \tag{11.4.27}$$

which can be written as

$$\frac{d}{d\tau}\big[\mathbf{G}(\tau) \quad \mathbf{F}(\tau)\big] = \big[\mathbf{G}(\tau) \quad \mathbf{F}(\tau)\big]\begin{bmatrix} \boldsymbol{b} & 2\boldsymbol{a}^\top\boldsymbol{a} \\ \boldsymbol{\eta} & -\boldsymbol{b}^\top \end{bmatrix}.$$

The solution is obtained through exponentiation,

$$\begin{aligned}
\begin{bmatrix}\mathbf{G}(\tau) & \mathbf{F}(\tau)\end{bmatrix} &= \begin{bmatrix}\mathbf{G}(0) & \mathbf{F}(0)\end{bmatrix}\exp\left\{\tau\begin{bmatrix}\boldsymbol{b} & 2\boldsymbol{a}^\top\boldsymbol{a}\\ \boldsymbol{\eta} & -\boldsymbol{b}^\top\end{bmatrix}\right\}\\
&= \begin{bmatrix}\mathbf{V}(0) & \boldsymbol{I}_d\end{bmatrix}\exp\left\{\tau\begin{bmatrix}\boldsymbol{b} & 2\boldsymbol{a}^\top\boldsymbol{a}\\ \boldsymbol{\eta} & -\boldsymbol{b}^\top\end{bmatrix}\right\}\\
&= \big[\big(\mathbf{V}(0)\boldsymbol{A}_{11}(\tau)+\boldsymbol{A}_{21}(\tau)\big) \quad \big(\mathbf{V}(0)\boldsymbol{A}_{12}(\tau)+\boldsymbol{A}_{22}(\tau)\big)\big],
\end{aligned}$$

where we use the notation

$$\begin{bmatrix}\boldsymbol{A}_{11}(\tau) & \boldsymbol{A}_{12}(\tau)\\ \boldsymbol{A}_{21}(\tau) & \boldsymbol{A}_{22}(\tau)\end{bmatrix} := \exp\left\{\tau\begin{pmatrix}\boldsymbol{b} & 2\boldsymbol{a}^\top\boldsymbol{a}\\ \boldsymbol{\eta} & -\boldsymbol{b}^\top\end{pmatrix}\right\}$$

for the matrix exponential. Hence

$$\begin{aligned}
\mathbf{V}(\tau) &= \big[\mathbf{V}(0)\boldsymbol{A}_{12}(\tau)+\boldsymbol{A}_{22}(\tau)\big]^{-1}\big[\mathbf{V}(0)\boldsymbol{A}_{11}(\tau)+\boldsymbol{A}_{21}(\tau)\big]\\
&= \big[-\imath\boldsymbol{u}\boldsymbol{A}_{12}(\tau)+\boldsymbol{A}_{22}(\tau)\big]^{-1}\big[-\imath\boldsymbol{u}\boldsymbol{A}_{11}(\tau)+\boldsymbol{A}_{21}(\tau)\big],
\end{aligned}$$

since $\mathbf{V}(0) = -\imath\boldsymbol{u}$. As usual, a direct integration allows us to compute

$$\frac{d}{d\tau}V^0(\tau) = -Tr\big(\alpha\boldsymbol{a}^\top\boldsymbol{a}\mathbf{V}(\tau)\big) - \eta_0, \tag{11.4.28}$$

which implies that

$$V^0(\tau) = -\int_0^\tau Tr\big(\alpha\boldsymbol{a}^\top\boldsymbol{a}\mathbf{V}(s)\big)\,ds - \eta_0\tau. \tag{11.4.29}$$

Performing the integration in (11.4.29) can be cumbersome, hence we employ the following technique from Da Fonseca et al. (2008c). Equation (11.4.27) can be rewritten as

$$\frac{1}{2}\left(\frac{d}{d\tau}\mathbf{F}(\tau)+\mathbf{F}(\tau)\boldsymbol{b}^\top\right)\big(\boldsymbol{a}^\top\boldsymbol{a}\big)^{-1} = \mathbf{G}(\tau)$$

and from

$$\mathbf{V}(\tau) = \mathbf{F}^{-1}(\tau)\mathbf{G}(\tau),$$

we obtain

$$\mathbf{F}(\tau)\mathbf{V}(\tau) = \frac{1}{2}\left(\frac{d}{d\tau}\mathbf{F}(\tau)+\mathbf{F}(\tau)\boldsymbol{b}^\top\right)\big(\boldsymbol{a}^\top\boldsymbol{a}\big)^{-1},$$

which is equivalent to

$$\mathbf{V}(\tau) = \frac{1}{2}\left(\mathbf{F}^{-1}(\tau)\frac{d}{d\tau}\mathbf{F}(\tau)+\boldsymbol{b}^\top\right)\big(\boldsymbol{a}^\top\boldsymbol{a}\big)^{-1},$$

which we substitute into (11.4.28) to obtain,

$$\frac{d}{d\tau}V^0(\tau) = -Tr\left(\alpha \boldsymbol{a}^\top \boldsymbol{a}\frac{1}{2}\left(\mathbf{F}^{-1}(\tau)\frac{d}{d\tau}\mathbf{F}(\tau) + \boldsymbol{b}^\top\right)\left(\boldsymbol{a}^\top \boldsymbol{a}\right)^{-1}\right) - \eta_0$$
$$= -\frac{\alpha}{2}Tr\left(\mathbf{F}^{-1}(\tau)\frac{d}{d\tau}\mathbf{F}(\tau) + \boldsymbol{b}^\top\right) - \eta_0,$$

which gives

$$V^0(\iota) = -\frac{\alpha}{2}Tr\left(\log(\mathbf{F}(\iota)) + \boldsymbol{b}^\top \iota\right) - \eta_0 \iota.$$

We conclude that the solution can be explicitly represented in terms of blocks of a matrix exponential. Before discussing this solution further, we present a competing method from Gnoatto and Grasselli (2011) and conclude this section with a comparison.

11.4.2 Cameron-Martin Formula

In this subsection, we present an alternative derivation of the Laplace transform. The result is presented in Gnoatto and Grasselli (2011), and it generalizes a result from Bru (1991), namely Eq. (4.7) in Bru (1991). We first state the result and then compare it with the one from the preceding subsection.

Theorem 11.4.1 *Let $X \in WIS_d(X_0, \alpha, \boldsymbol{b}, \boldsymbol{a})$, assume that $\boldsymbol{a} \in GL(d)$,*

$$\boldsymbol{b}^\top \left(\boldsymbol{a}^\top \boldsymbol{a}\right)^{-1} = \left(\boldsymbol{a}^\top \boldsymbol{a}\right)^{-1} \boldsymbol{b},$$

let $\alpha \geq d+1$, and define the set of convergence of the Laplace transform

$$\mathcal{D}_t = \left\{\boldsymbol{w}, \boldsymbol{v} \in \mathcal{S}_d \colon E\left(\exp\left\{-Tr\left(\boldsymbol{w}\boldsymbol{X}_t + \int_0^t \boldsymbol{v}\boldsymbol{X}_u\, du\right)\right\}\right) < \infty\right\}.$$

Then for all $\boldsymbol{w}, \boldsymbol{v} \in \mathcal{D}_t$ the joint moment generating function of the process and its integral is given by:

$$E\left(\exp\left\{-Tr\left(\boldsymbol{w}\boldsymbol{X}_t + \int_0^t \boldsymbol{v}\boldsymbol{X}_u\, du\right)\right\}\right)$$
$$= det\left(\exp\{-\boldsymbol{b}t\}\left(\cosh(\sqrt{\bar{\boldsymbol{v}}}t) + \sinh(\sqrt{\bar{\boldsymbol{v}}}t)\boldsymbol{k}\right)\right)^{\frac{\alpha}{2}}$$
$$\times \exp\left\{Tr\left(\left(\frac{\boldsymbol{a}^{-1}\sqrt{\bar{\boldsymbol{v}}}\boldsymbol{k}(\boldsymbol{a}^\top)^{-1}}{2} - \frac{(\boldsymbol{a}^\top \boldsymbol{a})^{-1}\boldsymbol{b}}{2}\right)X_0\right)\right\},$$

where the matrices $\boldsymbol{k}$, $\bar{\boldsymbol{v}}$, $\bar{\boldsymbol{w}}$ are given by

$$\boldsymbol{k} = -\left(\sqrt{\bar{\boldsymbol{v}}}\cosh(\sqrt{\bar{\boldsymbol{v}}}t) + \bar{\boldsymbol{w}}\sinh(\sqrt{\bar{\boldsymbol{v}}}t)\right)^{-1}\left(\sqrt{\bar{\boldsymbol{v}}}\sinh(\sqrt{\bar{\boldsymbol{v}}}t) + \bar{\boldsymbol{w}}\cosh(\sqrt{\bar{\boldsymbol{v}}}t)\right),$$
$$\bar{\boldsymbol{v}} = \boldsymbol{a}\left(2\boldsymbol{v} + \boldsymbol{b}^\top \boldsymbol{a}^{-1}\left(\boldsymbol{a}^\top\right)^{-1}\boldsymbol{b}\right)\boldsymbol{a}^\top,$$
$$\bar{\boldsymbol{w}} = \boldsymbol{a}\left(2\boldsymbol{w} - \left(\boldsymbol{a}^\top \boldsymbol{a}\right)^{-1}\boldsymbol{b}\right)\boldsymbol{a}^\top.$$

The proof of Theorem 11.4.1 can be found in Gnoatto and Grasselli (2011), it consists of three parts. The first establishes the result for $WIS_d(\boldsymbol{X}_0, \alpha, \boldsymbol{0}, \boldsymbol{I}_d)$, and consequently the conclusion of the first part is extended to $WIS_d(\boldsymbol{X}_0, \alpha, \boldsymbol{0}, \boldsymbol{a})$. The final part establishes the result for $WIS_d(\boldsymbol{X}_0, \alpha, \boldsymbol{b}, \boldsymbol{a})$. Here, we present the first two parts of the proof, for the third part, we refer the reader to Gnoatto and Grasselli (2011). Before proceeding with the proof, we recall two results from Bru (1991). The next result is Proposition 2 in Bru (1991).

Lemma 11.4.2 *If $\boldsymbol{\Phi} : \Re^+ \to \overline{\mathcal{S}_d^+}$ is continuous, constant on $[t, \infty)$ and such that its right derivative (in the distribution sense) $\boldsymbol{\Phi}' : \Re^+ \to \overline{\mathcal{S}_d^-}$ is continuous, with $\boldsymbol{\Phi}(0) = \boldsymbol{I}_d$, and $\boldsymbol{\Phi}'(t) = \boldsymbol{0}$, then for every Wishart process $\boldsymbol{X} \in WIS_d(\boldsymbol{x}, \alpha, \boldsymbol{0}, \boldsymbol{I}_d^n)$, we have*

$$E\left(\exp\left\{-\frac{1}{2}Tr\left(\int_0^t \boldsymbol{\Phi}''(s)\boldsymbol{\Phi}^{-1}(s)\boldsymbol{X}_s\,ds\right)\right\}\right)$$
$$= \left(det\,\boldsymbol{\Phi}(t)\right)^{\frac{\alpha}{2}} \exp\left\{\frac{1}{2}Tr\left(\boldsymbol{X}_0\boldsymbol{\Phi}^+(0)\right)\right\},$$

where

$$\boldsymbol{\Phi}^+(0) = \lim_{t\searrow t} \boldsymbol{\Phi}'(t).$$

Also, we recall Theorem 3 from Bru (1991), see also Proposition 11.3.2.

Lemma 11.4.3 *Let $\boldsymbol{X}$ be a $WIS_d(\boldsymbol{X}_0, \alpha, \boldsymbol{0}, \boldsymbol{I}_d)$ process, where $\alpha \geq d+1$, and $\boldsymbol{u} \in \overline{\mathcal{S}_d^+}$, then*

$$E\left(\exp\left\{-Tr(\boldsymbol{u}\boldsymbol{X}_t)\right\}\right) = \left(det(\boldsymbol{I}_d + 2t\boldsymbol{u})\right)^{\frac{\alpha}{2}} \exp\left\{-Tr\left(\boldsymbol{X}_0(\boldsymbol{I}_d + 2t\boldsymbol{u})^{-1}\boldsymbol{u}\right)\right\}.$$

We now establish the result from Theorem 11.4.1 for a $WIS_d(\boldsymbol{x}, \alpha, \boldsymbol{0}, \boldsymbol{I}_d)$ process, which is Proposition 2 in Gnoatto and Grasselli (2011).

Proposition 11.4.4 *Let $\boldsymbol{\Sigma} \in WIS_d(\boldsymbol{x}, \alpha, \boldsymbol{0}, \boldsymbol{I}_d)$, then*

$$E\left(\exp\left\{-\frac{1}{2}Tr\left(\boldsymbol{w}\boldsymbol{\Sigma}_t + \int_0^t \boldsymbol{v}\boldsymbol{\Sigma}_s\,ds\right)\right\}\right)$$
$$= det\left(\cosh(\sqrt{\boldsymbol{v}}t) + \sinh(\sqrt{\boldsymbol{v}}t)\boldsymbol{k}\right)^{\frac{\alpha}{2}} \exp\left\{\frac{1}{2}Tr(\boldsymbol{\Sigma}_0\sqrt{\boldsymbol{v}}\boldsymbol{k})\right\},$$

where $\boldsymbol{k}$ is given by

$$\boldsymbol{k} = -\left(\sqrt{\boldsymbol{v}}\cosh(\sqrt{\boldsymbol{v}}t) + \boldsymbol{w}\sinh(\sqrt{\boldsymbol{v}}t)\right)^{-1}\left(\sqrt{\boldsymbol{v}}\sinh(\sqrt{\boldsymbol{v}}t) + \boldsymbol{w}\cosh(\sqrt{\boldsymbol{v}}t)\right). \tag{11.4.30}$$

Proof By Lemma 11.4.2, we have to solve the ODE:

$$
\begin{aligned}
\boldsymbol{\Phi}''(s) &= \boldsymbol{v}\boldsymbol{\Phi}(s), \quad s \in (0,t), \\
\boldsymbol{\Phi}'^{-}(t) &= -\boldsymbol{w}\boldsymbol{\Phi}(t), \\
\boldsymbol{\Phi}(0) &= \boldsymbol{I}_d.
\end{aligned}
\tag{11.4.31}
$$

The general solution of (11.4.31) is given by

$$
\boldsymbol{\Phi}(s) = \cosh(\sqrt{\boldsymbol{v}}s)\boldsymbol{k}_1 + \sinh(\sqrt{\boldsymbol{v}}s)\boldsymbol{k}.
$$

From the condition $\boldsymbol{\Phi}(0) = \boldsymbol{I}_d$, we get $\boldsymbol{k}_1 = \boldsymbol{I}_d$. In order to determine $\boldsymbol{k}$, we look at the boundary condition at $\boldsymbol{\Phi}'^{-}(t)$ and hence obtain

$$
\sqrt{\boldsymbol{v}}\sinh(\sqrt{\boldsymbol{v}}t) + \sqrt{\boldsymbol{v}}\cosh(\sqrt{\boldsymbol{v}}t)\boldsymbol{k} = -\boldsymbol{w}\big(\cosh(\sqrt{\boldsymbol{v}}t) + \sinh(\sqrt{\boldsymbol{v}}t)\boldsymbol{k}\big).
$$

This yields the value of $\boldsymbol{k}$ given in Eq. (11.4.30). Next, we compute the derivative of $\boldsymbol{\Phi}$,

$$
\boldsymbol{\Phi}'(s) = \sqrt{\boldsymbol{v}}\sinh(\sqrt{\boldsymbol{v}}s) + \sqrt{\boldsymbol{v}}\cosh(\sqrt{\boldsymbol{v}}s)\boldsymbol{k},
$$

which yields

$$
\lim_{s \searrow 0} \boldsymbol{\Phi}'(s) = \sqrt{\boldsymbol{v}}\boldsymbol{k}.
$$

Since $\boldsymbol{\Phi}$ is constant on $[t,\infty)$, we obtain that $\boldsymbol{\Phi}(\infty) = \boldsymbol{\Phi}(t)$, which completes the proof. □

Now we attend to the second part.

Corollary 11.4.5 *Let $X \in WIS_d(\boldsymbol{x},\alpha,0,\boldsymbol{a})$, where $\alpha \geq d+1$ and $\boldsymbol{a} \in GL(d)$, and let $\boldsymbol{u} \in \overline{\mathcal{S}_d^+}$. Then*

$$
\begin{aligned}
&E\big(\exp\{-Tr(\boldsymbol{u}\boldsymbol{X}_t)\}\big) \\
&\quad = \big(det\big(\boldsymbol{I}_d + 2t\boldsymbol{a}^\top\boldsymbol{a}\boldsymbol{u}\big)\big)^{-\frac{\alpha}{2}} \exp\big\{-Tr\big(\boldsymbol{u}\big(\boldsymbol{I}_d + 2t\boldsymbol{a}^\top\boldsymbol{a}\boldsymbol{u}\big)^{-1}\boldsymbol{x}\big)\big\}.
\end{aligned}
$$

Proof Firstly, we note that since $\boldsymbol{a} \in GL(d)$, $\boldsymbol{a}^\top\boldsymbol{a} \in \mathcal{S}_d^+$, and since $\boldsymbol{u} \in \overline{\mathcal{S}_d^+}$, we have that $\boldsymbol{I}_d + 2t\boldsymbol{a}^\top\boldsymbol{a}\boldsymbol{u} \in \mathcal{S}_d^+$. Furthermore, as demonstrated in Sect. 11.2, for $\boldsymbol{\Sigma} \in WIS_d(\boldsymbol{x},\alpha,\boldsymbol{0},\boldsymbol{I}_d)$, we can set

$$
\boldsymbol{X}_t = \boldsymbol{a}^\top\boldsymbol{\Sigma}_t\boldsymbol{a}, \quad t \geq 0
$$

to obtain

$$
d\boldsymbol{X}_t = \sqrt{\boldsymbol{X}_t}\,d\tilde{\boldsymbol{W}}_t\boldsymbol{a} + \boldsymbol{a}^\top d\tilde{\boldsymbol{W}}_t^\top\sqrt{\boldsymbol{X}_t} + \alpha\boldsymbol{a}^\top\boldsymbol{a}\,dt,
$$

where $d\tilde{\boldsymbol{W}}_t = \sqrt{\boldsymbol{X}_t}^{-1}\boldsymbol{Q}^\top\sqrt{\boldsymbol{\Sigma}_t}\,d\boldsymbol{W}_t$ is a Brownian motion, and $\boldsymbol{W}$ denotes the Brownian motion driving $\boldsymbol{\Sigma}$. We apply Lemma 11.4.3 and use the fact that $\boldsymbol{\Sigma}_0 = (\boldsymbol{a}^\top)^{-1}\boldsymbol{X}_0\boldsymbol{a}^{-1}$ to obtain

$$
\begin{aligned}
E(\exp\{-Tr(\boldsymbol{u}\boldsymbol{X}_t)\}) &= E(\exp\{-Tr(\boldsymbol{u}(\boldsymbol{a}^\top \boldsymbol{\Sigma}_t \boldsymbol{a}))\}) \\
&= (det(\boldsymbol{I}_d + 2t\boldsymbol{aua}^\top))^{-\frac{\alpha}{2}} \\
&\quad \times \exp\{-Tr((\boldsymbol{a}^\top)^{-1}\boldsymbol{X}_0\boldsymbol{a}^{-1}(\boldsymbol{I}_d + 2t\boldsymbol{aua}^\top)^{-1}\boldsymbol{aua}^\top)\} \\
&= (det(\boldsymbol{I}_d + 2t\boldsymbol{aua}^\top))^{-\frac{\alpha}{2}} \\
&\quad \times \exp\{-Tr(\boldsymbol{X}_0\boldsymbol{a}^{-1}(\boldsymbol{I}_d + 2t\boldsymbol{aua}^\top)^{-1}\boldsymbol{au})\}.
\end{aligned}
$$

We now use Sylvester's determinant theorem,

$$
det(\boldsymbol{I}_d + \boldsymbol{AB}) = det(\boldsymbol{I}_d + \boldsymbol{BA}),
$$

to obtain

$$
det(\boldsymbol{I}_d + 2t\boldsymbol{aua}^\top) = det(\boldsymbol{I}_d + 2t\boldsymbol{a}^\top\boldsymbol{au}).
$$

Since

$$
\boldsymbol{a}^{-1}(\boldsymbol{I}_d + 2t\boldsymbol{aua}^\top)^{-1}\boldsymbol{au} = \boldsymbol{u}(\boldsymbol{I}_d + 2t\boldsymbol{a}^\top\boldsymbol{au})^{-1},
$$

we compute

$$
\begin{aligned}
&E(\exp\{-Tr(\boldsymbol{u}\boldsymbol{X}_t)\}) \\
&\quad = (det(\boldsymbol{I}_d + 2t\boldsymbol{a}^\top\boldsymbol{au}))^{-\frac{\alpha}{2}} \exp\{-Tr(\boldsymbol{X}_0\boldsymbol{u}(\boldsymbol{I}_d + 2t\boldsymbol{a}^\top\boldsymbol{au})^{-1})\} \\
&\quad = (det(\boldsymbol{I}_d + 2t\boldsymbol{a}^\top\boldsymbol{au}))^{-\frac{\alpha}{2}} \exp\{-Tr(\boldsymbol{u}(\boldsymbol{I}_d + 2t\boldsymbol{a}^\top\boldsymbol{au})^{-1}\boldsymbol{X}_0)\}.
\end{aligned}
$$

□

We remark that Corollary 11.4.5 is a special case of Proposition 11.3.2. The third step, where one incorporates the drift in Eq. (11.4.23), is completed by employing the Girsanov theorem, we refer the reader to Gnoatto and Grasselli (2011).

11.4.3 A Comparison of the Two Approaches

In this subsection, we recall the discussion in Sect. 3.4 of Gnoatto and Grasselli (2011), which compares the linearization approach to the Cameron-Martin formula. First, in terms of precision and execution speed, the two methods produce identical results. However, the disadvantage of the linearization method is that the functions $\boldsymbol{F}(\tau)$ and $\boldsymbol{G}(\tau)$ are expressed in terms of matrix exponentials, and the matrix exponential depends on the parameters $\boldsymbol{a}$ and $\boldsymbol{b}$ of the Wishart process. Furthermore, to obtain the function $V^0(\tau)$, one multiplies the remaining parameter α by the logarithm of $\boldsymbol{F}(\tau)$, and $\boldsymbol{F}(\tau)$ depends on the matrix exponential. As the matrix exponential is a symbolic expression, it means that the linearization method might be less useful if we want to understand the implications of the various model parameters, which is particularly important in applications. The result in Theorem 11.4.1 is strictly explicit, and furthermore it involves exponentials of $d \times d$ matrices, as opposed to the linearization method, which doubles the dimensionality of the problem, resulting in a $2d \times 2d$ matrix. Also, the Cameron-Martin formula does not require the computation of the matrix logarithm. Finally, with regards to the computation of sensitivities, which play an important role in finance, we can expect the Cameron-Martin formula to be more useful.

11.5 Two Heston Multifactor Volatility Models

In this section, we discuss two Heston multifactor volatility models, firstly a single-asset and secondly a multi-asset model, which were presented in Da Fonseca et al. (2007) and Da Fonseca et al. (2008c), respectively. The aim of this section is to illustrate how to exploit the tractability of the Wishart process. For each of the two models, we firstly discuss how to correlate the Brownian motion driving the asset, or assets respectively, and the Brownian motion driving the Wishart process, to retain the affinity of the model. Finally, we find that once we have an explicit representation of the infinitesimal generator, we can immediately employ the approach from Sect. 11.4 to compute the characteristic function. We remark that we employ linearization, as it follows easily from the presentation. However, instead the Cameron-Martin formula could have been used, see Gnoatto and Grasselli (2011), where the two models were studied using the Cameron-Martin formula.

11.5.1 A Single Asset Heston Multifactor Volatility Model

In this subsection, we present a single-asset model, in which we describe the stochastic volatility via a Wishart process. This model can be considered to be the natural extension of the Heston model, as discussed in Sect. 9.5. Following Da Fonseca et al. (2008c), we model the risky asset under an assumed risk-neutral measure via the SDE,

$$\frac{dS_t}{S_t} = r\,dt + Tr(\sqrt{\boldsymbol{X}_t}\,d\boldsymbol{Z}_t), \tag{11.5.32}$$

where r denotes the risk-free interest rate which, for ease of presentation, is assumed to be constant. The process $\boldsymbol{Z} = \{\boldsymbol{Z}_t,\ t \geq 0\}$ is a matrix-valued Brownian motion, $\boldsymbol{X} = \{\boldsymbol{X}_t,\ t \geq 0\}$ is a $WIS_d(\boldsymbol{x}, \alpha, \boldsymbol{b}, \boldsymbol{a})$ process, given by

$$d\boldsymbol{X}_t = \left(\alpha \boldsymbol{a}^\top \boldsymbol{a} + \boldsymbol{b}\boldsymbol{X}_t + \boldsymbol{b}^\top \boldsymbol{X}_t\right) dt + \left(\sqrt{\boldsymbol{X}_t}\, d\boldsymbol{W}_t \boldsymbol{a} + \boldsymbol{a}^\top d\boldsymbol{W}_t^\top \sqrt{\boldsymbol{X}_t}\right), \tag{11.5.33}$$

where $\alpha \geq d-1$, $\boldsymbol{b} \in \mathcal{M}_d$, and $\boldsymbol{a} \in GL(d)$. Following Da Fonseca et al. (2008c), we assume $\boldsymbol{b} \in \overline{\mathcal{S}_d^-}$, to obtain the mean-reverting behavior of $\boldsymbol{X}$. We now turn to the correlation structure of the Brownian motions $\boldsymbol{Z}$ and $\boldsymbol{W}$. In particular, Da Fonseca et al. (2008c) introduce a correlation matrix $\boldsymbol{R} \in \mathcal{M}_d$ to obtain the following correlation structure,

$$\boldsymbol{Z}_t = \boldsymbol{W}_t \boldsymbol{R}^\top + \boldsymbol{B}_t \sqrt{\boldsymbol{I} - \boldsymbol{R}\boldsymbol{R}^\top}, \quad t \geq 0, \tag{11.5.34}$$

where $\boldsymbol{B} = \{\boldsymbol{B}_t, t \geq 0\}$ is a Brownian motion independent of $\boldsymbol{W}$. The next proposition establishes that $\boldsymbol{Z}$ is a Brownian motion.

Proposition 11.5.1 *The process* $\mathbf{Z} = \{\mathbf{Z}_t, t \geq 0\}$ *defined in Eq.* (11.5.34) *is a Brownian motion.*

Proof We use Theorem 10.4.5 to obtain the proof. Clearly, the process $\boldsymbol{Z}$ is a local martingale. Furthermore,

$$dZ_{t,ij} = \sum_{k=1}^{d} dW_{t,ik} R_{j,k} + \sum_{k=1}^{d} dB_{t,ik} \left(\sqrt{\boldsymbol{I} - \boldsymbol{R}\boldsymbol{R}^\top}\right)_{k,j}.$$

Hence we have

$$\begin{aligned}
&d[Z_{\cdot,ij}, Z_{\cdot,kl}]_t \\
&= \left(\sum_{m=1}^{d} R_{j,m} R_{l,m} + \left(\sqrt{\boldsymbol{I} - \boldsymbol{R}\boldsymbol{R}^\top}\right)_{m,j} \left(\sqrt{\boldsymbol{I} - \boldsymbol{R}\boldsymbol{R}^\top}\right)_{m,l} \right) \mathbf{1}_{i=k}\, dt \\
&= \boldsymbol{I}_{j,l} \mathbf{1}_{i=k}\, dt \\
&= \mathbf{1}_{i=k} \mathbf{1}_{j=l}\, dt,
\end{aligned}$$

which completes the proof. □

The next result discusses the correlation structure of $\boldsymbol{Z}_t$ and $\boldsymbol{W}_t$.

Proposition 11.5.2 *The covariance of* $\mathbf{Z}_t$ *and* $\mathbf{W}_t$ *is given by*

$$Cov(\boldsymbol{Z}_t, \boldsymbol{W}_t) = t \boldsymbol{I}_d \otimes \boldsymbol{R}. \tag{11.5.35}$$

Proof From Definition 10.3.7, we have

$$\begin{aligned}
Cov(\boldsymbol{Z}_t, \boldsymbol{W}_t) &= E\left(vec\left(\boldsymbol{Z}_t^\top\right) vec\left(\boldsymbol{W}_t^\top\right)^\top\right) - E\left(vec\left(\boldsymbol{Z}_t^\top\right)\right) E\left(vec\left(\boldsymbol{W}_t^\top\right)\right)^\top \\
&= E\left(vec\left(\boldsymbol{R}\boldsymbol{W}_t^\top\right) vec\left(\boldsymbol{W}_t^\top\right)^\top\right).
\end{aligned}$$

We find it convenient to denote the i-th row of $\boldsymbol{W}_t$ by $\boldsymbol{w}_i$, and regarding the matrix $\boldsymbol{R}\boldsymbol{W}_t^\top$, we denote its j-th column by $\boldsymbol{r}_j$, so that

$$\boldsymbol{r}_j = \begin{bmatrix} \sum_{k=1}^{d} R_{1,k} W_{j,k} \\ \sum_{k=1}^{d} R_{2,k} W_{j,k} \\ \vdots \\ \sum_{k=1}^{d} R_{d,k} W_{j,k} \end{bmatrix}.$$

Hence

$$
\begin{aligned}
E\left(vec\left(\boldsymbol{R}\boldsymbol{W}_t^\top\right)vec\left(\boldsymbol{W}_t^\top\right)^\top\right) &= E\left(\begin{bmatrix} \boldsymbol{r}_1 \\ \vdots \\ \boldsymbol{r}_d \end{bmatrix} [\boldsymbol{w}_1 \cdots \boldsymbol{w}_d]\right) \\
&= E\left(\begin{bmatrix} \boldsymbol{r}_1\boldsymbol{w}_1 & \cdots & \boldsymbol{r}_1\boldsymbol{w}_d \\ \vdots & \ddots & \vdots \\ \boldsymbol{r}_d\boldsymbol{w}_1 & \cdots & \boldsymbol{r}_d\boldsymbol{w}_d \end{bmatrix}\right) \\
&= \begin{bmatrix} t I_{1,1}\boldsymbol{R} & \cdots & t I_{1,d}\boldsymbol{R} \\ \vdots & \ddots & \vdots \\ t I_{d,1}\boldsymbol{R} & \cdots & t_{d,d}\boldsymbol{R} \end{bmatrix} \qquad (11.5.36) \\
&= t\boldsymbol{I}_d \otimes \boldsymbol{R}.
\end{aligned}
$$

To see equality (11.5.36), we consider an element of the matrix $\boldsymbol{r}_l\boldsymbol{w}_m$, say the element $[\boldsymbol{r}_l\boldsymbol{w}_m]_{i,j}$, where $i, j, l, m \in \{1, \dots, d\}$. This element admits the representation

$$
\sum_{k=1}^{d} R_{i,k} W_{t,lk} W_{t,mj}.
$$

Consequently,

$$
E\left(\sum_{k=1}^{d} R_{i,k} W_{t,lk} W_{t,mj}\right) = \begin{cases} 0 & \text{for } l \neq m \\ t R_{i,j} & \text{for } l = m. \end{cases}
$$

□

Hence, $\boldsymbol{R}$, which is a $d \times d$ matrix, summarizes the covariance structure, which is, in principle, a matrix of size $d^2 \times d^2$. We choose to summarize the covariance structure by $\boldsymbol{R}$, as it preserves the affine structure of the model, which is crucial for analytical tractability.

We now turn to option pricing. It is convenient to work with the logarithm of the stock price, i.e. $Y_t = \log(S_t)$, which satisfies the SDE

$$
dY_t = \left(r - \frac{1}{2}Tr(\boldsymbol{X}_t)\right)dt + Tr\left(\sqrt{\boldsymbol{X}_t}\left(d\boldsymbol{W}_t\,\boldsymbol{R}^\top + d\boldsymbol{B}_t\sqrt{\boldsymbol{I}_d - \boldsymbol{R}\boldsymbol{R}^\top}\right)\right).
$$

As in Da Fonseca et al. (2008c), we work with the infinitesimal generator of the process, which then allows us to employ linearization to compute the Laplace transform. Alternatively, the Cameron-Martin formula could have been employed, we refer the reader to Gnoatto and Grasselli (2011) for this approach. Recall that the Laplace transform is given by

$$
\begin{aligned}
\Psi_{\gamma,t}(\tau) &= E\left(\exp\{\gamma Y_{t+\tau}\}\right) \\
&= \exp\left\{Tr\left(\boldsymbol{A}(\tau)\boldsymbol{X}_t\right) + b(\tau)Y_t + c(\tau)\right\}, \qquad (11.5.37)
\end{aligned}
$$

where $\gamma \in \Re$, $\boldsymbol{A}(\tau) \in \mathcal{M}_d$, $b(\tau) \in \Re$ and $c(\tau) \in \Re$. We use L_X to denote the infinitesimal generator of $\boldsymbol{X}$, and $L_{Y,X}$ to denote the infinitesimal generator of $(Y, \boldsymbol{X})$.

Recall from Lemma 11.3.5 that the infinitesimal generator of $\boldsymbol{X}$ is given by

$$L_X = Tr\big(\big[\alpha \boldsymbol{a}^\top \boldsymbol{a} + \boldsymbol{b}\boldsymbol{x} + \boldsymbol{x}\boldsymbol{b}^\top\big]\boldsymbol{D} + 2\boldsymbol{x}\boldsymbol{D}\boldsymbol{a}^\top \boldsymbol{a}\boldsymbol{D}\big),$$

where $\boldsymbol{D}$ is a matrix differential operator with $D_{i,j} = (\frac{\partial}{\partial x_{i,j}})$, and from Da Fonseca et al. (2008c), Proposition 3.1, we obtain the infinitesimal generator of $(Y_t, \boldsymbol{X}_t)$, which is given by

$$\begin{aligned} L_{Y,X} = {} & \left(r - \frac{1}{2}Tr(\boldsymbol{x})\right)\frac{\partial}{\partial y} + \frac{1}{2}Tr(\boldsymbol{x})\frac{\partial^2}{\partial y^2} \\ & + Tr\big(\big(\alpha \boldsymbol{a}^\top \boldsymbol{a} + \boldsymbol{b}\boldsymbol{x} + \boldsymbol{x}\boldsymbol{b}^\top\big)\boldsymbol{D} + 2\boldsymbol{x}\boldsymbol{D}\boldsymbol{a}^\top \boldsymbol{a}\boldsymbol{D}\big) \\ & + 2Tr(\boldsymbol{x}\boldsymbol{R}\boldsymbol{Q}\boldsymbol{D})\frac{\partial}{\partial y}. \end{aligned} \tag{11.5.38}$$

Using the Feynman-Kac argument, we have

$$\frac{\partial \Psi_{\gamma,t}}{\partial \tau} = L_{Y,X}\Psi_{\gamma,t}$$

and

$$\Psi_{\gamma,t}(0) = \exp\{\gamma Y_t\}.$$

Using Eq. (11.5.38), we obtain that

$$\begin{aligned} \frac{\partial \Psi_{\gamma,t}}{\partial \tau} = {} & \left(r - \frac{1}{2}Tr(\boldsymbol{x})\right)\frac{\partial \Psi_{\gamma,t}}{\partial y} + \frac{1}{2}Tr(\boldsymbol{x})\frac{\partial^2 \Psi_{\gamma,t}}{\partial y^2} \\ & + Tr\big(\big(\alpha \boldsymbol{a}^\top \boldsymbol{a} + \boldsymbol{b}\boldsymbol{x} + \boldsymbol{x}\boldsymbol{b}^\top\big)\boldsymbol{D}\Psi_{\gamma,t} \\ & + 2\big(\boldsymbol{x}\boldsymbol{D}\boldsymbol{a}^\top \boldsymbol{a}\boldsymbol{D}\big)\Psi_{\gamma,t}\big) \\ & + 2Tr(\boldsymbol{x}\boldsymbol{R}\boldsymbol{a}\boldsymbol{D})\frac{\partial \Psi_{\gamma,t}}{\partial \tau}, \end{aligned}$$

subject to $\boldsymbol{A}(0) = \boldsymbol{0}$, $b(0) = \gamma$, and $c(0) = 0$. From Eq. (11.5.37), we obtain that

$$\frac{\partial \Psi_{\gamma,t}}{\partial \tau} = Tr\left(\frac{d}{d\tau}\boldsymbol{A}(\tau)\boldsymbol{x}\right) + \frac{d}{d\tau}b(\tau)y + \frac{d}{d\tau}c(\tau).$$

Identifying the coefficients of y, we obtain

$$\frac{d}{d\tau}b(\tau) = 0,$$

hence $b(\tau) = \gamma$, for $\tau \geq 0$. The remaining part of the argument is identical to the linearization procedure employed in Sect. 11.4. We obtain the following matrix Riccati ODE satisfied by $\boldsymbol{A}(\tau)$,

$$\frac{d}{d\tau}\boldsymbol{A}(\tau) = \boldsymbol{A}(\tau)\boldsymbol{b} + \big(\boldsymbol{b}^\top + 2\gamma \boldsymbol{R}\boldsymbol{a}\big)\boldsymbol{A}(\tau) + 2\boldsymbol{A}(\tau)\boldsymbol{a}^\top \boldsymbol{a}\boldsymbol{A}(\tau) + \frac{\gamma(\gamma-1)}{2}\boldsymbol{I}_d,$$

subject to the condition $\boldsymbol{A}(0) = \boldsymbol{0}$. Again, we compute $c(\tau)$ by direct integration,

$$\frac{d}{d\tau}c(\tau) = Tr\big(\alpha \boldsymbol{a}^\top \boldsymbol{a}\boldsymbol{A}(\tau)\big) + \gamma r,$$

subject to $c(0)=0$. As in Sect. 11.4, we double the dimension of the problem, by setting

$$\mathbf{A}(\tau)=\mathbf{F}^{-1}(\tau)\mathbf{G}(\tau),$$

where $\mathbf{F}(\tau)\in GL(d)$, $\mathbf{G}(\tau)\in\mathcal{M}_d$. Hence we conclude that

$$\begin{bmatrix}\mathbf{G}(\tau) & \mathbf{F}(\tau)\end{bmatrix}=\begin{bmatrix}(\mathbf{A}(0)\mathbf{A}_{11}(\tau)+\mathbf{A}_{21}(\tau)) & (\mathbf{A}(0)\mathbf{A}_{12}(\tau)+\mathbf{A}_{22}(\tau))\end{bmatrix},$$

where

$$\begin{bmatrix}\mathbf{A}_{11}(\tau) & \mathbf{A}_{12}(\tau)\\ \mathbf{A}_{21}(\tau) & \mathbf{A}_{22}(\tau)\end{bmatrix}:=\exp\left\{\tau\begin{pmatrix}\boldsymbol{b} & -2\boldsymbol{a}^\top\boldsymbol{a}\\ \frac{\gamma(\gamma-1)}{2}\boldsymbol{I}_d & -(\boldsymbol{b}^\top+2\gamma\boldsymbol{R}\boldsymbol{a})\end{pmatrix}\right\}.$$

We conclude that

$$\begin{aligned}\mathbf{A}(\tau)&=\left(\mathbf{A}(0)\mathbf{A}_{12}(\tau)+\mathbf{A}_{22}(\tau)\right)^{-1}\left(\mathbf{A}(0)\mathbf{A}_{11}(\tau)+\mathbf{A}_{21}(\tau)\right)\\&=\left(\mathbf{A}_{22}(\tau)\right)^{-1}\mathbf{A}_{21}(\tau),\end{aligned}$$

since $\mathbf{A}(0)=\mathbf{0}$. Lastly, we conclude that

$$c(\tau)=-\frac{\alpha}{2}Tr\left(\log\left(\mathbf{F}(\tau)\right)+\left(\boldsymbol{b}^\top+2\gamma\boldsymbol{R}\boldsymbol{a}\right)\tau\right)+\gamma r\tau,$$

which, as in Sect. 11.4, avoids a numerical integration to compute $c(\tau)$.

11.5.2 A Heston Multi-asset Multifactor Volatility Model

We now discuss Wishart processes in a multi-asset framework. The model presented in this subsection first appeared in Da Fonseca et al. (2007) and extends the models presented in Sect. 6.7. Under an assumed risk-neutral measure, we use the following model for the vector of risky assets,

$$d\boldsymbol{S}_t=Diag(\boldsymbol{S}_t)(r\mathbf{1}\,dt+\sqrt{\boldsymbol{X}_t}\,d\boldsymbol{Z}_t),\tag{11.5.39}$$

where $\mathbf{1}=(1,\ldots,1)^\top$, and $\boldsymbol{Z}=\{\boldsymbol{Z}_t,\ t\geq 0\}\in\Re^d$ is a vector-valued Brownian motion. The process $\boldsymbol{X}=\{\boldsymbol{X}_t,\ t\geq 0\}$ is a $WIS_d(\boldsymbol{x},\alpha,\boldsymbol{b},\boldsymbol{a})$ process with dynamics

$$d\boldsymbol{X}_t=\left(\alpha\boldsymbol{a}^\top\boldsymbol{a}+\boldsymbol{b}\boldsymbol{X}_t+\boldsymbol{X}_t\boldsymbol{b}^\top\right)dt+\sqrt{\boldsymbol{X}_t}\,d\boldsymbol{W}_t\boldsymbol{a}+\boldsymbol{a}^\top d\boldsymbol{W}_t^\top\sqrt{\boldsymbol{X}_t},$$

where $\alpha\geq d-1$, $\boldsymbol{b}\in\mathcal{M}_d$ and $\boldsymbol{a}\in GL(d)$. We now make the following assumptions, cf. Da Fonseca et al. (2007):

Assumption 11.5.3 The following assumptions are in force in this subsection:

1. the continuous-time diffusion model for $\boldsymbol{S}$ is a linear-affine stochastic factor model with respect to the log-returns and variance-covariance factors $X_{\cdot,kl}$;
2. the stochastic covariance matrix is given by the Wishart process $\boldsymbol{X}$;
3. the Brownian motion driving the assets' returns and those driving the instantaneous covariance matrix are linearly correlated.

Now we discuss how the Brownian motions $\boldsymbol{Z}=\{\boldsymbol{Z}_t,\ t\geq 0\}$ and $\boldsymbol{W}=\{\boldsymbol{W}_t,\ t\geq 0\}$ can be correlated in order to satisfy Assumptions 1–3 above.

First we introduce d real-valued matrices $\boldsymbol{R}_k \in \mathcal{M}_d$, $k=1,\dots,d$, so that

$$dZ_t^k = \sqrt{1 - Tr\big(\boldsymbol{R}_k \boldsymbol{R}_k^\top\big)}\, dB_t^k + Tr\big(\boldsymbol{R}_k\, d\boldsymbol{W}_t^\top\big), \quad k = 1,\dots,d,$$

where the vector Brownian motion $\boldsymbol{B} = (B^1,\dots, B^d)$ is independent of $\boldsymbol{W}$. We point out that for a generic choice of $\boldsymbol{R}_k$ the model in Eq. (11.5.39) need not remain affine. Instead, we show the following result from Da Fonseca et al. (2007), which explains how the Brownian motions can be correlated. For a proof, we refer to Da Fonseca et al. (2007).

Proposition 11.5.4 *Assumptions* 1 *and* 2 *imply that for* $k=1,\dots,d$, *the correlation matrix* $\boldsymbol{R}_k$ *is given by*

$$\boldsymbol{R}_k = \begin{pmatrix} 0 & 0 & 0 \\ \rho_1 & \cdots & \rho_d \\ 0 & 0 & 0 \end{pmatrix} \leftarrow k\text{-th row}, \qquad (11.5.40)$$

where $\rho_i \in [-1,1]$, $i=1,\dots,d$ *and* $\boldsymbol{\rho}^\top \boldsymbol{\rho} \le 1$.

Equation (11.5.40) implies that the Brownian motion driving the asset vector has to satisfy

$$d\boldsymbol{Z}_t = \sqrt{1 - \boldsymbol{\rho}^\top \boldsymbol{\rho}}\, d\boldsymbol{B}_t + d\boldsymbol{W}_t\, \boldsymbol{\rho}.$$

In particular, for $d=2$, this means that

$$\begin{aligned} dZ_{t,1} &= \sqrt{1 - \big(\rho_1^2 + \rho_2^2\big)}\, dB_{t,1} + (dW_{t,11}\, \rho_1 + dW_{t,12}\, \rho_2) \\ dZ_{t,2} &= \sqrt{1 - \big(\rho_1^2 + \rho_2^2\big)}\, dB_{t,2} + (dW_{t,21} \rho_1 + dW_{t,22}\, \rho_2). \end{aligned}$$

So all elements of the correlation vector $\boldsymbol{\rho} = (\rho_1, \rho_2)$ feature in both Brownian motions, Z_1 and Z_2.

We now turn to derivative pricing. Recall from Lemma 11.3.5 that the infinitesimal generator of the Wishart process $\boldsymbol{X}$ is given by

$$L_X = Tr\big(\big[\alpha \boldsymbol{a}^\top \boldsymbol{a} + \boldsymbol{b}\boldsymbol{x} + \boldsymbol{x}\boldsymbol{b}^\top\big]\boldsymbol{D} + 2\boldsymbol{x}\boldsymbol{D}\boldsymbol{a}^\top \boldsymbol{a}\boldsymbol{D}\big),$$

and furthermore, the infinitesimal generator of the asset returns, $\boldsymbol{Y}_t = \log(\boldsymbol{S}_t)$, is given by

$$\begin{aligned} L_Y &= \nabla_{\boldsymbol{y}} \left(r\mathbf{1} - \frac{1}{2} Vec(x_{ii}) \right) + \frac{1}{2} \nabla_{\boldsymbol{y}} \boldsymbol{x} \nabla_{\boldsymbol{y}}^\top \\ &= \sum_{i=1}^d \left(r - \frac{1}{2} x_{ii} \right) \frac{\partial}{\partial y_i} + \frac{1}{2} \sum_{i,j=1}^d x_{ij} \frac{\partial^2}{\partial y_i \partial y_j}, \end{aligned}$$

where $\nabla_{\boldsymbol{y}}$ denotes the gradient operator, $\nabla_{\boldsymbol{y}} = (\frac{\partial}{\partial y_1}, \dots, \frac{\partial}{\partial y_d})$. Lastly, from Proposition 4 in Da Fonseca et al. (2007), we have

$$L_{Y,X} = Tr\big((\alpha \boldsymbol{a}^\top \boldsymbol{a} + \boldsymbol{b}\boldsymbol{x} + \boldsymbol{x}\boldsymbol{b}^\top)\boldsymbol{D} + 2\boldsymbol{x}\boldsymbol{D}\boldsymbol{a}^\top \boldsymbol{a}\boldsymbol{D}\big)$$
$$+ \nabla_{\boldsymbol{y}}\left(r\mathbf{1} - \frac{1}{2}Vec(x_{ii})\right) + \frac{1}{2}\nabla_{\boldsymbol{y}}\boldsymbol{x}\nabla_{\boldsymbol{y}}^\top$$
$$+ 2Tr\big(\boldsymbol{D}\boldsymbol{a}^\top \boldsymbol{\rho}\nabla_{\boldsymbol{y}}\boldsymbol{x}\big),$$

where $\boldsymbol{D}$ is a matrix differential operator with elements

$$D_{i,j} = \left(\frac{\partial}{\partial x_{i,j}}\right),$$

and $Vec(x_{ii})$ is the vector comprised of the elements $x_{ii}, i = 1, \ldots, d$. We now attend to the computation of the affine transform of the log-returns under the assumed risk-neutral measure,

$$\Psi_{\gamma,t} = E\big(\exp\{\langle \boldsymbol{\gamma}, \boldsymbol{Y}_{t+\tau}\rangle\}|\mathcal{A}_t\big).$$

As before, we apply the Feynman-Kac argument,

$$\frac{\partial \Psi_{\gamma,t}}{\partial \tau} = L_{Y,X}\Psi_{\gamma,t}. \tag{11.5.41}$$

We guess that $\Psi_{\gamma,t}$ is exponentially affine in $\boldsymbol{X}_t$ and $\boldsymbol{Y}_t$, so we assume that

$$\Psi_{\gamma,t} = \exp\big\{Tr\big(\mathbf{A}(\tau)\boldsymbol{X}_t\big) + \boldsymbol{\beta}^\top(\tau)\boldsymbol{Y}_t + c(\tau)\big\}, \tag{11.5.42}$$

where $\mathbf{A}(\tau) \in \mathcal{M}_d$, $\boldsymbol{\beta}(\tau) \in \Re^d$, and $c(\tau) \in \Re$. From Eq. (11.5.41), we compute

$$\frac{\partial \Psi_{\gamma,t}}{\partial \tau} = Tr\big((\alpha \boldsymbol{a}^\top \boldsymbol{a} + \boldsymbol{b}\boldsymbol{x} + \boldsymbol{x}\boldsymbol{b}^\top)\boldsymbol{D} + 2\boldsymbol{x}\boldsymbol{D}\boldsymbol{a}^\top \boldsymbol{a}\boldsymbol{D}\big)\Psi_{\gamma,t}$$
$$+ \nabla_{\boldsymbol{y}}\left(r\mathbf{1} - \frac{1}{2}Vec\big(Tr(\boldsymbol{e}_{ii}\boldsymbol{x})\big)\right)\Psi_{\gamma,t}$$
$$+ \frac{1}{2}\nabla_{\boldsymbol{y}}\boldsymbol{x}\nabla_{\boldsymbol{y}}^\top \Psi_{\gamma,t}$$
$$+ 2Tr\big(\boldsymbol{D}\boldsymbol{a}^\top \boldsymbol{\rho}\nabla_{\boldsymbol{y}}\boldsymbol{x}\big)\Psi_{\gamma,t},$$

where $\boldsymbol{e}_{ii} = (\delta_{i,j,k})_{j,k=1,\ldots,d}$ denotes the canonical basis of $\mathcal{M}_d$. Replacing $\frac{\partial \Psi_{\gamma,t}}{\partial \tau}$, we get

$$0 = -Tr\left(\frac{d}{d\tau}\mathbf{A}(\tau)\boldsymbol{x}\right) - \frac{d}{d\tau}\boldsymbol{\beta}^\top(\tau)\boldsymbol{y} - \frac{d}{d\tau}c(\tau)$$
$$+ \boldsymbol{\beta}^\top(\tau)\left(r\mathbf{1} - \frac{1}{2}Vec\big(Tr(\boldsymbol{e}_{ii}\boldsymbol{x})\big)\right) + \frac{1}{2}\boldsymbol{\beta}^\top(\tau)\boldsymbol{x}\boldsymbol{\beta}(\tau)$$
$$+ Tr\big((\alpha \boldsymbol{a}^\top \boldsymbol{a} + \boldsymbol{b}\boldsymbol{x} + \boldsymbol{x}\boldsymbol{b}^\top)\mathbf{A}(\tau) + 2\boldsymbol{x}\mathbf{A}(\tau)\boldsymbol{a}^\top \boldsymbol{a}\mathbf{A}(\tau)\big)$$
$$+ 2Tr\big(\mathbf{A}(\tau)\boldsymbol{a}^\top \boldsymbol{\rho}\boldsymbol{\beta}^\top(\tau)\boldsymbol{x}\big),$$

that is

$$0 = -Tr\left(\frac{d}{d\tau}\mathbf{A}(\tau)\boldsymbol{x} + \frac{\partial}{\partial\tau}\boldsymbol{\beta}(\tau)\boldsymbol{y}^\top\right) - \frac{\partial}{\partial\tau}c(\tau)$$
$$+ Tr\left(r\mathbf{1}\boldsymbol{\beta}^\top(\tau) - \frac{1}{2}\sum_{i=1}^{d}\boldsymbol{\beta}_i(\tau)\boldsymbol{e}_{ii}\boldsymbol{x} + \frac{1}{2}\boldsymbol{\beta}(\tau)\boldsymbol{\beta}^\top(\tau)\boldsymbol{x}\right)$$
$$+ Tr\big((\alpha\boldsymbol{a}^\top\boldsymbol{a} + \boldsymbol{b}\boldsymbol{x} + \boldsymbol{x}\boldsymbol{b}^\top)\mathbf{A}(\tau) + 2\boldsymbol{x}\mathbf{A}(\tau)\boldsymbol{a}^\top\boldsymbol{a}\mathbf{A}(\tau) + 2\mathbf{A}(\tau)\boldsymbol{a}^\top\boldsymbol{\rho}\boldsymbol{\beta}^\top(\tau)\boldsymbol{x}\big),$$

subject to the boundary conditions

$$\mathbf{A}(0) = \mathbf{0}, \qquad \boldsymbol{\beta}(0) = \boldsymbol{\gamma}, \qquad c(0) = 0.$$

Identifying the coefficients of $\boldsymbol{y}$ we deduce

$$\frac{d}{d\tau}\boldsymbol{\beta}(\tau) = \mathbf{0},$$

hence $\boldsymbol{\beta}(\tau) = \boldsymbol{\gamma}$, for $\tau \ge 0$. As in Sect. 11.4, by identifying the coefficients of $\boldsymbol{X}$, we obtain the matrix Riccati ODE satisfied by $\boldsymbol{A}(\tau)$,

$$\frac{d}{d\tau}\boldsymbol{A}(\tau) = \boldsymbol{A}(\tau)\boldsymbol{b} + \boldsymbol{b}^\top\boldsymbol{A}(\tau) - \frac{1}{2}\sum_{i=1}^{d}\boldsymbol{\gamma}_i\boldsymbol{e}_{ii} + 2\boldsymbol{A}(\tau)\boldsymbol{a}^\top\boldsymbol{a}\boldsymbol{A}(\tau) + \frac{1}{2}\boldsymbol{\gamma}\boldsymbol{\gamma}^\top$$
$$+ \boldsymbol{A}(\tau)\boldsymbol{a}^\top\boldsymbol{\rho}\boldsymbol{\gamma}^\top + \boldsymbol{\gamma}\boldsymbol{\rho}^\top\boldsymbol{a}\boldsymbol{A}(\tau)$$
$$= \boldsymbol{A}(\tau)\big(\boldsymbol{b} + \boldsymbol{a}^\top\boldsymbol{\rho}\boldsymbol{\gamma}^\top\big) + \big(\boldsymbol{b}^\top + \boldsymbol{\gamma}\boldsymbol{\rho}^\top\boldsymbol{a}\big)\boldsymbol{A}(\tau) + 2\boldsymbol{A}(\tau)\boldsymbol{a}^\top\boldsymbol{a}\boldsymbol{A}(\tau)$$
$$- \frac{1}{2}\sum_{i=1}^{d}\boldsymbol{\gamma}_i\boldsymbol{e}_{ii} + \frac{1}{2}\boldsymbol{\gamma}\boldsymbol{\gamma}^\top,$$

subject to $\boldsymbol{A}(\tau) = \mathbf{0}$. Doubling the dimension of the problem, as in Sect. 11.4, we obtain

$$\boldsymbol{A}(\tau) = \big(\boldsymbol{A}(0)\boldsymbol{A}_{12}(\tau) + \boldsymbol{A}_{22}(\tau)\big)^{-1}\big(\boldsymbol{A}(0)\boldsymbol{A}_{11}(\tau) + \boldsymbol{A}_{21}(\tau)\big),$$

where

$$\begin{bmatrix} \boldsymbol{A}_{11}(\tau) & \boldsymbol{A}_{12}(\tau) \\ \boldsymbol{A}_{21}(\tau) & \boldsymbol{A}_{22}(\tau) \end{bmatrix} := \exp\left\{\tau\begin{pmatrix} \boldsymbol{b} + \boldsymbol{a}^\top\boldsymbol{\rho}\boldsymbol{\gamma}^\top & -2\boldsymbol{a}^\top\boldsymbol{a} \\ \frac{1}{2}(\boldsymbol{\gamma}\boldsymbol{\gamma}^\top - \sum_{i=1}^{d}\boldsymbol{\gamma}_i\boldsymbol{e}_{ii}) & -(\boldsymbol{b}^\top + \boldsymbol{\gamma}\boldsymbol{\rho}^\top\boldsymbol{a}) \end{pmatrix}\right\}.$$

For the function $c(\tau)$, we have

$$\frac{d}{d\tau}c(\tau) = Tr\big(r\mathbf{1}\boldsymbol{\gamma}^\top + \alpha\boldsymbol{a}^\top\boldsymbol{a}\boldsymbol{A}(\tau)\big),$$

subject to the initial condition $c(0) = 0$. We can solve the above equation to yield

$$c(\tau) = -\frac{\alpha}{2}Tr\big(\log(\boldsymbol{A}_{22}(\tau)) + \tau\boldsymbol{b}^\top + \tau\boldsymbol{\gamma}\boldsymbol{\rho}^\top\boldsymbol{a}\big) + \tau r\boldsymbol{\gamma}^\top\mathbf{1}.$$

Consequently, we price derivatives as discussed in Chap. 8.

Chapter 12
Monte Carlo and Quasi-Monte Carlo Methods

In this chapter, we discuss Monte Carlo and Quasi-Monte Carlo methods and show how they can be used to compute functionals of multidimensional diffusions.

12.1 Monte Carlo Methods

Monte Carlo (MC) methods can be employed to compute functionals of multidimensional diffusions, e.g. by using the inverse transform method, see e.g. Sect. 6.1 and Platen and Bruti-Liberati (2010), if the transition density is known explicitly. However, when the transition density is not known explicitly, one can employ discretization schemes, such as the *Euler Scheme*, see Kloeden and Platen (1999), to approximately sample from the transition density. The discretization scheme introduces an error, which can be studied using the techniques in Kloeden and Platen (1999). The aim of this section is to introduce two alternatives to discretization schemes, which allow us to eliminate the discretization error introduced by discretization schemes and to recover the Monte Carlo convergence rate. These alternatives are the *exact simulation methods*, due to Roberts and collaborators, see Beskos et al. (2006, 2008, 2009), Beskos and Roberts (2005), and also Chen and Huang (2012b), and *multilevel methods* due to Giles and coauthors, see Giles (2008a, 2008b).

We firstly provide a very brief introduction to Monte Carlo methods and then briefly illustrate the Euler discretization scheme, which motivates the exact simulation and multilevel methods. For detailed references on Monte Carlo methods applied to finance, we refer the reader to Kloeden and Platen (1999), Glasserman (2004), Jäckel (2002), Platen and Bruti-Liberati (2010), and Korn et al. (2010).

J. Baldeaux, E. Platen, *Functionals of Multidimensional Diffusions with Applications to Finance*, Bocconi & Springer Series 5, DOI 10.1007/978-3-319-00747-2_12,

12.1.1 Monte Carlo Methods

Monte Carlo methods are easily illustrated by considering the problem of estimating the integral

$$a = \int_0^1 f(x)\,dx.$$

This integral can be interpreted as the expected value

$$E\big(f(U)\big),$$

where U is uniformly distributed over the interval [0, 1], assuming that f is integrable. Now consider the i.i.d. random variables $U^1, U^2, \dots, U^N$, uniformly distributed over [0, 1], then

$$\tilde{a}_N = \frac{1}{N}\sum_{i=1}^{N} f(U_i),$$

can be used to approximate a. In fact, since by assumption f is integrable, we have that $\tilde{a}_N$ is unbiased, i.e.

$$E(\tilde{a}_N) = a,$$

and furthermore *strongly consistent*, that is,

$$\tilde{a}_N \to a$$

with probability 1 as $n \to \infty$. Assuming f is square integrable, the variance of $\tilde{a}_N$ is given by

$$\mathrm{Var}(\tilde{a}_N) = \frac{\sigma_f^2}{N}, \tag{12.1.1}$$

where

$$\sigma_f^2 = \int_0^1 \big(f(x) - a\big)^2\,ds.$$

One would expect that it is at least as hard to compute σ_f^2 as it is to compute a, hence for applications it is useful to be able to approximate σ_f^2. For this purpose, we introduce the *sample standard deviation*

$$s_f = \sqrt{\frac{1}{N}\sum_{i=1}^{N}\frac{(f(U_i) - a_N)^2}{N-1}}.$$

The sample standard deviation allows us to compute *confidence intervals*. For example let z_ϵ denote the $1-\epsilon$ quantile of the standard normal distribution, by which we mean that

$$P(Z \le z_\epsilon) = 1 - \epsilon,$$

where $Z \sim N(0, 1)$. Then

$$\tilde{a}_N \pm z_{\frac{\epsilon}{2}} \frac{s_f}{\sqrt{N}} \tag{12.1.2}$$

is a $1 - \epsilon$ confidence interval for a, as $N \to \infty$. This is due to the fact that as $N \to \infty$,

$$\tilde{a}_N - a \sim N\left(0, \frac{\sigma_f}{\sqrt{N}}\right).$$

We remark that at first sight, the convergence rate $N^{-\frac{1}{2}}$ in Eq. (12.1.2) might strike the reader as being slow: for twice differentiable functions, the trapezoidal rule achieves a convergence rate of N^{-2}, and for four times differentiable functions, the Simpson's rule achieves a convergence rate of N^{-4}. However, we remark that for multidimensional integrals, say over the unit cube $[0, 1]^d$, the respective convergence rates become $N^{-\frac{2}{d}}$ and $N^{-\frac{4}{d}}$. One should point out that sparse grid techniques, see Bungartz and Griebel (2004), allow one to obtain convergence rates arbitrarily close to the convergence rates N^{-2} and N^{-4}, respectively. For more details on sparse grids, we refer the reader to Bungartz and Griebel (2004).

The advantage of Monte Carlo methods is that one only requires the function under consideration to be square integrable, and secondly that the convergence rate is independent of the dimension. These two facts make Monte Carlo methods very useful techniques for the computation of functionals of multidimensional diffusions. Lastly, another advantage of Monte Carlo methods is that they allow the user to obtain statistical information on the problem via confidence intervals.

12.1.2 Bias and Computational Complexity

The aim of this subsection is to introduce two concepts, namely *bias* and *computational complexity*. To illustrate these two concepts, we discuss how to price European call options.

Firstly, we recall the Black-Scholes model from Sect. 2.3. Under the real world probability measure P, we model the GOP via the SDE

$$dS_t = S_t\left(r\,dt + \sigma^2\,dt + \sigma\,dW_t\right),$$

where W is a Brownian motion under the real world measure. Alternatively,

$$S_t = S_0 \exp\left(rt + \frac{1}{2}\sigma^2 t + \sigma W_t\right),$$

hence we need to compute

$$c_{T,K}(0) = S_0 E\left(\frac{(S_T - K)^+}{S_T}\right) \tag{12.1.3}$$

to price a European call with maturity T and strike $K \in \Re^+$ at time 0. We need to sample W_T, to obtain i.i.d. samples S_T^i, $i = 1, \dots, N$. Since $W_T \sim N(0, T)$, we can obtain realizations

$$c_{T,K}^i(0) = S_0 \frac{(S_T^i - K)^+}{S_T^i},$$

to obtain the estimator

$$\hat{c}_{T,K}(0) = \frac{1}{N} \sum_{i=1}^{N} c_{T,K}^i(0).$$

As discussed previously, the estimator is unbiased, and we can easily obtain confidence intervals.

However, such a simulation scheme is not always available. Say, we consider the SDE,

$$dS_t = S_t\big(r\,dt + \sigma^2(S_t)\,dt + \sigma(S_t)\,dW_t\big), \tag{12.1.4}$$

for which there is not necessarily an exact simulation scheme as was the case for the Black-Scholes model. One may then consider a discretization scheme, such as the Euler scheme, i.e. consider

$$\tilde{S}_{t+\Delta t} = \tilde{S}_t\big(r\Delta t + \sigma^2(\tilde{S}_t)\Delta t + \sigma(\tilde{S}_t)\Delta W_t\big), \tag{12.1.5}$$

where $\Delta t = \frac{T}{n}$ corresponds to the time step size of an n step Euler scheme, $\Delta W_t = W_{t+\Delta t} - W_t$, and $\tilde{S}_0 = S_0$, where we use the notation $\tilde{S}$ to emphasize that we are approximating S. As the distribution of $\tilde{S}_T$ differs from the distribution of S_T, the resulting call prices can be expected to differ, i.e.

$$E\bigg(\frac{(\tilde{S}_T - K)^+}{\tilde{S}_T}\bigg) \neq E\bigg(\frac{(S_T - K)^+}{S_T}\bigg).$$

It is important to study by how much these prices differ. Useful notions are *strong order of convergence*, which is the exponent α below, if

$$\big(E\big(|\tilde{S}_T - S_T|^2\big)\big)^{\frac{1}{2}} \le c(\Delta)^\alpha,$$

and *weak order of convergence*, which is the exponent β below if for a smooth test function f from a given set one has

$$\big|E\big(f(\tilde{S}_T)\big) - E\big(f(S_T)\big)\big| \le c(\Delta t)^\beta,$$

see Kloeden and Platen (1999) for details. For purposes of this discussion, it suffices to note that the Euler scheme achieves usually strong order $\alpha = 1/2$ and weak order $\beta = 1$, for a sufficiently smooth volatility function $\sigma(\cdot)$ in (12.1.4). Strong convergence is a useful concept when measuring pathwise convergence. Weak convergence is appropriate for Monte Carlo simulation to approximate functionals.

In Monte Carlo simulation we face two errors, one introduced by sampling random variables, the other due to the fact that we may sample from an approximated distribution. Next we present a way of trading off these two errors. This approach will also be used in the subsection discussing multilevel methods. We draw random i.i.d. samples from the distribution of $\tilde{S}_T$ using (12.1.5) to obtain $\tilde{S}_T^i$, $i = 1, \ldots, N$, and form the Monte Carlo estimator

$$\hat{C}_N = \frac{1}{N} \sum_{i=1}^{N} \tilde{C}^i,$$

where

$$\tilde{C}^i = S_0 \frac{(\tilde{S}_T^i - K)^+}{\tilde{S}_T^i},$$

to estimate

$$c_{T,K}(0) = S_0 E\left(\frac{(S_T - K)^+}{S_T} \right),$$

where S_T is given by (12.1.4). We define the *mean square error* (MSE), given by

$$\begin{aligned} MSE &= E\big((\hat{C}_N - c_{T,K}(0))^2 \big) \\ &= \big(E(\hat{C}_N) - c_{T,K}(0) \big)^2 + E\big((\hat{C}_N - E(\hat{C}_N))^2 \big). \end{aligned} \tag{12.1.6}$$

The first term in (12.1.6) is the *bias* squared, the second term is the variance of $\hat{C}_N$. Next we discuss how to trade off bias and variance. The bias depends on the weak order of convergence of a scheme, which we assume is given by β, i.e.

$$\big(E(\hat{C}_N) - c_{T,K}(0) \big)^2 \le c^2 (\Delta t)^{2\beta}.$$

Here $\Delta t = \frac{T}{n}$, where n denotes the number of time steps. The variance is given by

$$E\big((\hat{C}_N - E(\hat{C}_N))^2 \big) = \frac{\mathrm{Var}(\tilde{C}^i)}{N},$$

here N denotes the number of Monte Carlo samples. The computational complexity of the scheme can be assumed to be given by

$$C = Nn,$$

since we use n time steps for each of the N Monte Carlo samples. The *MSE* is given by

$$MSE = c^2 \frac{T^{2\beta}}{n^{2\beta}} + \frac{\mathrm{Var}(\tilde{C}^i)}{N}.$$

The *MSE* is minimized if we balance $n^{2\beta}$ and N, i.e. choose

$$n^{2\beta} \asymp N,$$

i.e. $C \asymp n^{2\beta+1}$. Hence the *MSE* satisfies

$$MSE \asymp C^{-\frac{2\beta}{2\beta+1}}.$$

In particular, if an Euler scheme, which achieves weak order convergence $\beta = 1$, is used, we have

$$MSE \asymp C^{-\frac{2}{3}}.$$

This convergence rate is slower than the convergence rate C^{-1}, which an unbiased Monte Carlo scheme can achieve. The slower convergence is due to the discretization error, i.e. the bias. In the next two subsections, we introduce firstly an exact simulation scheme for diffusions, which is capable of eliminating this bias for a large class of diffusions. In the second subsection, we introduce multilevel methods. The latter methods trade off bias and variance in a manner in which the $C^{-\frac{2}{3}}$ convergence rate can be improved to C^{-1}, i.e. the Monte Carlo convergence rate achieved by unbiased schemes.

12.1.3 Exact Simulation Methods for Diffusions

There has been a growing literature concerned with the topic of exact simulation of diffusions, see e.g. Beskos et al. (2006, 2008, 2009), Beskos and Roberts (2005), and Chen and Huang (2012a, 2012b).

In this subsection, we briefly recall the approach from Chen and Huang (2012a). Assume we are concerned with the following SDE on a filtered probability space $(\Omega, \mathcal{A}, \underline{\mathcal{A}}, P)$:

$$dS_t = \mu(S_t)\,dt + \sigma(S_t)\,dW_t, \tag{12.1.7}$$

$S_0 = s$, and also define $M_T := \max_{0 \le t \le T} S_t$. We point out that instead of considering the maximum of S, we could have also studied the minimum $m_T := \min_{0 \le t \le T} S_t$. For many problems in finance, see Chaps. 2 and 3, one needs to compute expected values of the form

$$E\big(f(M_T, S_T)\big). \tag{12.1.8}$$

In this subsection, we present a Monte Carlo estimator for (12.1.8), which allows for general SDEs of the form specified in (12.1.7). We remark that we could have employed an Euler scheme to compute (12.1.8). However, this introduces a bias, and furthermore, the bias tends to be larger when approximating M_T than when approximating S_T, see Asmussen et al. (1995) for details. We assume that the function μ is continuously differentiable, σ is twice differentiable and $\sigma > 0$ on the state space of S. Now, we introduce the *Lamperti transform*,

$$F(x) = \int_s^x \frac{1}{\sigma(u)}\,du,$$

where s is in the state space of S. We remark that F is strictly increasing as σ is positive. Setting $Y_t = F(S_t)$, we obtain by the Itô formula

$$dY_t = b(Y_t)\,dt + dW_t, \tag{12.1.9}$$

where

$$b(y) = \frac{\mu(F^{-1}(y))}{\sigma(F^{-1}(y))} - \frac{1}{2}\sigma'(F^{-1}(y)).$$

For the remainder of this subsection, we focus on Y given by (12.1.9). We note that due to the monotonicity of F, we simulate Y_T and $\max_{0\le t\le T} Y_t$ and, consequently, set $S_T = F^{-1}(Y_T)$ and $M_T = F^{-1}(\max_{0\le t\le T} Y_t)$. Furthermore, we assume that Y does not explode, see Assumption 2.2. in Chen and Huang (2012a) for a sufficient condition.

The exact simulation scheme in Chen and Huang (2012a) performs importance sampling: the joint density of $(Y_T, \max_{0\le t\le T} Y_t)$ is not known, in general, however, we know the joint density of $(W_T, \max_{0\le t\le T} W_t)$, as we show below. So instead of simulating $(Y_T, \max_{0\le t\le T} Y_t)$, we simulate $(W_T, \max_{0\le t\le T} W_t)$, and adjust our results by multiplying with a likelihood ratio. This likelihood ratio corresponds to a Radon-Nikodym derivative: we change from a measure P under which Y is given by (12.1.9) to another measure $\tilde{P}$ under which the drift is removed from Y, so that Y follows a Brownian motion corresponding to the measure $\tilde{P}$. This is formalized in the next theorem, which is Theorem 2.1 from Chen and Huang (2012a), the proof of which is based on a generalized Girsanov formula. For a proof, we refer the reader to Chen and Huang (2012a).

Theorem 12.1.1 *Suppose that μ is continuously differentiable, σ is twice continuously differentiable, and $\sigma > 0$ on the state space of S. Furthermore, assume that Y does not explode. Then for Borel-measurable $h : \Re^3 \to \Re$,*

$$E_P\Big(h\Big(Y_T, \max_{0\le t\le T} Y_t\Big)\Big) = E_{\tilde{P}}\Big(h\Big(\tilde{W}_T, \max_{0\le t\le T} \tilde{W}_t\Big)L_T\Big),$$

where the likelihood ratio L_T is given by

$$L_T = \exp\bigg\{A(\tilde{W}_T) - \int_0^T \phi(\tilde{W}_s)\,ds\bigg\},$$

where $\tilde{W} = \{\tilde{W}_t,\ t\in[0,T]\}$ is a Brownian motion under $\tilde{P}$ and

$$A(y) = \int_0^y b(u)\,du \quad \text{and} \quad \phi(y) = \frac{b^2(y) + b'(y)}{2}.$$

Theorem 12.1.1 yields the following equality

$$\begin{aligned} E_P\big(f(S_T, M_T)\big) &= E_P\Big(f\Big(F^{-1}(Y_T), F^{-1}\Big(\max_{0\le t\le T} Y_t\Big)\Big)\Big) \\ &= E_{\tilde{P}}\big(f\big(F^{-1}(\tilde{W}_T), F^{-1}(\tilde{K}_T)\big)\exp\big\{A(\tilde{W}_T)\big\}Q\big), \end{aligned}$$

where

$$Q = E_{\tilde{P}}\bigg(\exp\bigg\{-\int_0^T \phi(W_s)\,ds\bigg\}\,\bigg|\,\tilde{\Theta}_T, \tilde{K}_T, \tilde{W}_T\bigg), \tag{12.1.10}$$

with $\tilde{\Theta}_T = \inf\{u \in [0,T]\colon\ \tilde{W}_u = \tilde{K}_T\}$. Computing $E(f(S_T, M_T))$ we hence proceed as follows:

1. simulate $\tilde{\Theta}_T, \tilde{K}_T, \tilde{W}_T$
2. evaluate

$$\exp\{A(\tilde{W}_T)\} E_{\tilde{P}}\left(\exp\left\{-\int_0^T \phi(W_s)\,ds\right\} \,\Big|\, \tilde{\Theta}_T, \tilde{K}_T, \tilde{W}_T\right)$$

3. evaluate $f(F^{-1}(\tilde{W}_T), F^{-1}(\tilde{K}_T))$.

Since the last step is trivial, we focus on the first two steps.

As shown in Chen and Huang (2012a), the joint distribution of $(\tilde{\Theta}_T, \tilde{K}_T, \tilde{W}_T)$ is explicitly known, see Karatzas and Shreve (1991), Problem 2.8.17: we generate three i.i.d. random variables U_1, U_2, U_3, uniformly distributed on [0, 1] and set

$$\begin{aligned}
\tilde{\Theta}_T &= T \sin^2(\pi U/2) \\
\tilde{K}_T &= \sqrt{-2\tilde{\Theta}_T \log(1-V)} \\
\tilde{W}_T &= \tilde{K}_T - \sqrt{2(T-\tilde{\Theta}_T)\left(-\log\left(\frac{\tilde{W}_T}{T-\tilde{\Theta}_T}\right)\right)}.
\end{aligned}$$

We now focus on the second step, where one exploits an interesting connection with a Poisson process, see also Beskos et al. (2006) for a similar observation. The following result is Proposition 3.1 in Chen and Huang (2012a), for which we provide also the proof.

Proposition 12.1.2 *Suppose $\tilde{N}$ is a Poisson random variable with parameter ΛT under $\tilde{P}$, where Λ is a positive constant, and $\{\tau_1, \ldots, \tau_{\tilde{N}}\}$ are $\tilde{N}$ i.i.d. uniform random variables on $[0, T]$. All these random variables are independent of $\tilde{W}$. Then*

$$\begin{aligned}
&E_{\tilde{P}}\left(\exp\left\{-\int_0^T \phi(\tilde{W}_s)\,ds\right\} \,\Big|\, \tilde{\Theta}_T, \tilde{K}_T, \tilde{W}_T\right) \\
&\quad = E_{\tilde{P}}\left(\prod_{i=1}^{\tilde{N}}\left(\frac{\Lambda - \phi(W_{\tau_i})}{\Lambda}\right) \,\Big|\, \tilde{\Theta}_T, \tilde{K}_T, \tilde{W}_T\right).
\end{aligned}$$

Proof We condition on the entire path of $\tilde{W} = \{\tilde{W}_t,\ t \in [0, T]\}$ and have

$$\begin{aligned}
&E_{\tilde{P}}\left(\prod_{i=1}^{\tilde{N}}\left(\frac{\Lambda - \phi(\tilde{W}_{\tau_i})}{\Lambda}\right) \,\Big|\, \sigma(\tilde{W}_t,\ t \in [0, T])\right) \\
&= \sum_{n=0}^{\infty} E\left(\prod_{i=1}^{n}\left(\frac{\Lambda - \phi(W_{\tau_i})}{\Lambda}\right) \,\Big|\, \sigma(\tilde{W}_t,\ t \in [0, T],\ \tilde{N} = n)\right) \frac{\exp\{-\Lambda T\}(\Lambda T)^n}{n!}.
\end{aligned}$$

Since by assumption the random variables τ_i are uniformly distributed on $[0, T]$, we have

Algorithm 12.1 Exact simulation of a diffusion Y given by (12.1.9)

1: Simulate $\tilde{N} \sim$ Poisson (ΛT).
2: Generate independent τ_i, $1 \le i \le \tilde{N}$, each of which is uniform on $[0, T]$.
3: Sort $\{\tau_1, \dots, \tau_{\tilde{N}}\}$ to obtain their order statistics:

$$\tau_{(1)} < \cdots < \tau_{(j-1)} < \tilde{\Theta}_T < \tau_{(j)} < \cdots < \tau_{(N)}.$$

4: Simulate $W_{\tau_{(i)}}$, $1 \le i \le \tilde{N}$, given $\tilde{W}_T, \tilde{\Theta}_T, \tilde{K}_T$.
5: Evaluate

$$\tilde{Q} = \prod_{i=1}^{\tilde{N}} \left(\frac{\Lambda - \phi(\tilde{W}_{\tau_{(i)}})}{\Lambda} \right).$$

$$\begin{aligned} E&\left(\prod_{i=1}^{n} \left(\frac{\Lambda - \phi(W_{\tau_i})}{\Lambda} \right) \,\middle|\, \sigma(W_t,\, 0 \le t \le T),\, \tilde{N} = n \right) \\ &= \left(\frac{1}{T} \int_0^T \left(\frac{\Lambda - \phi(W_t)}{\Lambda} \right) dt \right)^n. \end{aligned}$$

Hence we have

$$\begin{aligned} E&\left(\prod_{i=1}^{\tilde{N}} \left(\frac{\Lambda - \phi(W_{\tau_i})}{\Lambda} \right) \,\middle|\, \sigma(W_t,\, 0 \le t \le T) \right) \\ &= \sum_{n=0}^{\infty} \left(\frac{1}{T} \int_0^T \frac{\Lambda - \phi(W_t)}{\Lambda}\, dt \right)^n \frac{\exp\{-\Lambda T\}(\Lambda T)^n}{n!} \\ &= \exp\left\{ - \int_0^T \phi(W_t)\, dt \right\}. \end{aligned}$$

The proof is completed by taking expectations with respect to $\tilde{W}_T, \tilde{K}_T, \tilde{\Theta}_T$. □

We now state the respective exact simulation algorithm as Algorithm 12.1. Clearly, we have

$$E_{\tilde{P}}\big(f\big(F^{-1}(\tilde{W}_T), F^{-1}(\tilde{K}_T)\big) \exp\big\{A(\tilde{W}_T)\big\} \tilde{Q} \big) = E_P\big(f(S_T, M_T) \big).$$

Lastly, we need to solve the problem of how to simulate $\tilde{W}_{\tau_{(i)}}$ given $\tilde{W}_T, \tilde{K}_T, \tilde{\Theta}_T$.

The answer is given by the Williams path decomposition, see Williams (1974) and Imhof (1984). Denoting $\tilde{\Theta}_T = \theta$, $\tilde{K}_T = k$, and $\tilde{W}_T = w$, we have that

$$\{k - \tilde{W}_{\theta - u},\; 0 \le u \le \theta\} \quad \text{and} \quad \{k - \tilde{W}_{\theta + u},\; 0 \le u \le T - \theta\}$$

are two independent *Brownian meanders*, which are connected at $\tilde{\Theta}_T = \theta$, where the Brownian motion $\tilde{W}$ reaches its maximum over $[0, T]$. Brownian meanders can be represented in terms of *Brownian bridges*, see Imhof (1984). We have that

$$\{\tilde{W}_u,\ 0 \le u \le \theta\} \stackrel{d}{=} k - \sqrt{\left(\frac{k(\theta-u)}{\theta} + B_u^{1,1}\right)^2 + \left(B_u^{1,2}\right)^2 + \left(B_u^{1,3}\right)^2},$$

$$\{\tilde{W}_u,\ \theta \le u \le T\} \stackrel{d}{=} k - \sqrt{\left(\frac{(k-y)(u-\theta)}{T-\theta} + B_u^{2,1}\right)^2 + \left(B_u^{2,2}\right)^2 + \left(B_u^{2,3}\right)^2}.$$

Here $B^{1,j} = \{B_t^{1,j},\ 0 \le t \le \theta\}$, $j = 1, \ldots, 3$, are three independent Brownian motions from 0 to 0 over $[0, \theta]$, and $B^{2,j} = \{B_t^{2,j},\ \theta \le t \le T\}$, $j = 1, \ldots, 3$, are three independent Brownian motions from 0 to 0 over $[\theta, T]$, both under $\tilde{P}$. Hence we see that simulating $\tilde{W}_{\tau(i)}$, $1 \le i \le \tilde{N}$, amounts to simulating Brownian bridges, which is described e.g. in Glasserman (2004).

We remark that it is also shown in Chen and Huang (2012a) how to sample functionals $f(S_T, M_T, m_T)$, recalling that $m_T = \min_{0 \le t \le T} S_t$, see Sect. 4 in Chen and Huang (2012a). Finally, variance reduction techniques tailored to the exact simulation scheme are also discussed in Sect. 5.1 in Chen and Huang (2012a).

We conclude that this importance sampling technique is a useful tool allowing us to obtain unbiased estimators of diffusions and their extremal values, which are important functionals in finance. As these estimators are unbiased, they achieve the Monte Carlo convergence rate.

12.1.4 Multilevel Methods

We note that exact simulation methods eliminate bias, and consequently achieve the Monte Carlo convergence rate. Multilevel methods are markedly different, they show how to trade off bias and variance in a way so that the Monte Carlo convergence rate can be recovered. These methods were introduced by Heinrich in the context of parametric integration, see Heinrich (1998), Heinrich and Sindambiwe (1999), and by Giles in the context of simulating SDEs, see Giles (2008b), and Giles (2008a). The idea behind the multilevel method is to consider discretization schemes, such as the Euler scheme discussed in the preceding subsection, with different step sizes. Recall that above, we used the time step size $\Delta t = \frac{T}{n}$, now we choose

$$\Delta t_l = \frac{T}{n^l},$$

where $l = 0, \ldots, L$. Furthermore, for a stock price given by

$$dS_t = a(S_t)\,dt + b(S_t)\,dW_t,$$

where W denotes a scalar Brownian motion, we assume that we are interested in computing the functional

$$E\big(f(S_T)\big).$$

Denoting $f(S_T)$ by F, we use $\tilde{S}_l$ and $\tilde{F}_l$ to denote the approximations to S_T and F, respectively, obtained by using a discretization scheme with time step size Δt_l. It is useful to introduce for $L \in \{1, 2, \ldots\}$ the following telescoping sum,

$$E(\tilde{F}_L) = E(\tilde{F}_0) + \sum_{l=1}^{L} E(\tilde{F}_l - \tilde{F}_{l-1}). \tag{12.1.11}$$

The multilevel algorithm approximates each expectation on the right hand side of (12.1.11) in a way that minimizes the overall computational complexity. We need to introduce some further notation: we use $\tilde{Y}_0$ as an estimator for $E(\tilde{F}_0)$ using N_0 independent samples and $\tilde{Y}_l$ as an estimator for $E(\tilde{F}_l - \tilde{F}_{l-1})$ based on N_l independent samples. In particular, we set for $l \in \{0, 1, \ldots, L\}$

$$\tilde{Y}_l = N_l^{-1} \sum_{i=1}^{N_l} (\tilde{F}_l^i - \tilde{F}_{l-1}^i), \tag{12.1.12}$$

where $\{\tilde{F}_l^i - \tilde{F}_{l-1}^i\}_{i=1}^{N_l}$ are assumed to be i.i.d and we set $\tilde{F}_{-1}^i = 0$. We now discuss how to obtain realizations of $\tilde{F}_l^i - \tilde{F}_{l-1}^i$: the key point is that for fixed l and i, the same Brownian path is used to construct $\tilde{F}_l^i$ and $\tilde{F}_{l-1}^i$. In Giles (2008b), it is suggested to first construct the Brownian increments to obtain $\tilde{F}_l^i$ and then to sum them in groups of size n to obtain the Brownian increments used to construct $\tilde{F}_{l-1}^i$. For the same fixed l, we repeat this procedure to obtain the i.i.d. samples $\{\tilde{F}_l^i - \tilde{F}_{l-1}^i\}_{i=1}^{N_l}$. For a different level l, we again proceed by first constructing the Brownian increments for the fine path $\tilde{F}_l^i$ and then summing the increments in groups of size n to compute the realization of the coarse path $\tilde{F}_{l-1}^i$. Hence for a fixed level l, the realizations $\{\tilde{F}_l^i - \tilde{F}_{l-1}^i\}_{i=1}^{N_l}$ are independent of each other, but for a fixed i, $\tilde{F}_l^i$ and $\tilde{F}_{l-1}^i$ are obtained using the same Brownian path. The latter property allows us to apply strong convergence results when analyzing the variance of $(\tilde{F}_l^i - \tilde{F}_{l-1}^i)$, as shown below. Finally, we remark that for different levels l_1 and l_2, the estimators Y_{l_1} and Y_{l_2} are independent of each other as independent Brownian paths are used for the construction. By the independence of $\{\tilde{F}_l^i - \tilde{F}_{l-1}^i\}_{i=1}^{N_l}$ for $l \in \{0, \ldots, L\}$, we obtain

$$\mathrm{Var}(\tilde{Y}_l) = N_l^{-1}\,\mathrm{Var}(\tilde{F}_l - \tilde{F}_{l-1}) = N_l^{-1} V_l,$$

i.e. we set $V_l = \mathrm{Var}(\tilde{F}_l - \tilde{F}_{l-1})$. The combined estimator is now given by the formula

$$\tilde{Y} = \sum_{l=0}^{L} \tilde{Y}_l, \tag{12.1.13}$$

and its variance by

$$\mathrm{Var}(\tilde{Y}) = \sum_{l=0}^{L} \mathrm{Var}(\tilde{Y}_l) = \sum_{l=0}^{L} N_l^{-1} V_l, \tag{12.1.14}$$

by the independence of $\tilde{Y}_0, \tilde{Y}_1, \ldots, \tilde{Y}_L$. If we treat the N_l as continuous variables, the variance of $\tilde{Y}$ will be minimized by choosing N_l to be proportional to $\sqrt{V_l \Delta t_l}$.

The idea of the multilevel method can now be summarized as follows: for the coarse level, i.e. those for which Δt_l is large and hence those for which the computational cost is low, one performs a large number of simulations to reduce the overall variance. On the other hand, for the fine levels, i.e. those for which Δt_l is small and hence those for which the computational cost is high, one only performs few simulations. As the bias associated with the estimator $\tilde{Y}$ in (12.1.13) is given by

$$\left|E(\tilde{Y}) - E(F)\right| = \left|E(\tilde{F}_L) - E(F)\right|,$$

one may conclude that on the coarse levels, one reduces the variance, but on the fine levels one reduces the bias. To estimate the bias, as demonstrated in the preceding subsection, one requires a weak convergence result. However, to bound the variance V_l, one can employ a strong convergence result. We note that

$$V_l = \text{Var}(\tilde{F}_l - \tilde{F}_{l-1}) \leq \left(\text{Var}(\tilde{F}_l - F)^{1/2} + \text{Var}(\tilde{F}_{l-1} - F)^{1/2}\right)^2.$$

Now,

$$\text{Var}(\tilde{F}_l - F) \leq E\left((\tilde{F}_l - F)^2\right). \tag{12.1.15}$$

If we have a strong convergence result for the right-hand side in (12.1.15), such as strong order $\frac{1}{2}$ for an Euler scheme,

$$E\left((\tilde{F}_l - F)^2\right) \leq O(\Delta t_l),$$

we have that V_l behaves like $O(\Delta t_l)$. Recall from Eq. (12.1.6) that the MSE is given by bias squared plus variance, where bias depends only on the time step size and variance only on the sample size. Hence to balance the sample size with the time step size, we set N_l proportional to Δt_l. We now fix $\epsilon > 0$, which can be interpreted as the target root mean square error we want to achieve. Then setting $N_l = O(\epsilon^{-2} L \Delta t_l)$, we have from (12.1.14) that $\text{Var}(\tilde{Y}) = O(\epsilon^2)$. Now we choose L so that the bias decreases sufficiently fast. In particular, by setting

$$L = \frac{\log(\epsilon^{-1})}{\log n},$$

we obtain that $\Delta t_L = T n^{-L} = O(\epsilon)$. If we can apply a weak convergence order 1 result, we have

$$\left|E(\tilde{F}_L) - E(F)\right| = O(\epsilon).$$

Hence the MSE of the scheme (12.1.13) is $O(\epsilon^2)$ at a computational cost of

$$\sum_{l=1}^{L} N_l \Delta^{-1} t_l = O\left(\sum_{l=1}^{L} \epsilon^{-2} \Delta t_l \Delta^{-1} t_l L\right) = O\left(\epsilon^{-2} (\log \epsilon)^2\right),$$

which is the Monte Carlo convergence rate of an unbiased scheme, up to the $(\log \epsilon)^2$ factor. This shows that by trading off bias and variance, a scheme based upon biased discretization schemes can recover the Monte Carlo convergence rate achieved by an unbiased scheme.

12.2 Quasi-Monte Carlo Methods

In this section, we briefly review quasi-Monte Carlo (qMC) methods focusing on the application to the computation of functionals of multidimensional diffusions. Quasi-Monte Carlo methods can roughly be divided into integration lattices and nets. For more details on integration lattices, we refer the reader to Niederreiter (1992) and Sloan and Joe (1994). For more information on nets, we refer the reader to Niederreiter (1992) and Dick and Pillichshammer (2010). Finally, for applications of qMC methods to finance, we refer the reader to Glasserman (2004), Chap. 5, Jäckel (2002), Chap. 8, and Korn et al. (2010), Chap. 2.

Quasi-Monte Carlo rules are equal weight integration lattices that can be used to approximate integrals over the unit cube, which are typically of high dimension. In this section, we focus on the construction of qMC point sets, which we are going to apply to finance problems in the next section. We concentrate on nets, in particular *digital nets*, the explicit construction of which we now outline.

12.2.1 The Digital Construction Scheme

We now formally introduce the *digital construction scheme*, which allows us to define digital nets.

Definition 12.2.1 Let b be a prime and m be an integer. Let $\boldsymbol{C}_1, \ldots, \boldsymbol{C}_d$ be $m \times m$ matrices over the finite field $\mathbb{Z}_b$. We construct b^m points in $[0, 1)^d$ as follows: for $0 \leq h < b^m$ let $h = h_0 + h_1 b + \cdots + h_{m-1} b^{m-1}$ be the b-adic expansion of h. Identify h with the vector $\mathbf{h} = (h_0, \ldots, h_{m-1})^\top \in \mathbb{Z}_b^m$. For $1 \leq j \leq d$ we multiply the matrix $\boldsymbol{C}_j$ by $\mathbf{h}$, i.e.

$$\boldsymbol{C}_j \mathbf{h} =: \left(y_{j,1}(h), \ldots, y_{j,m}(h)\right)^\top \in \mathbb{Z}_b^m$$

and set

$$x_{h,j} := \frac{y_{j,1}(h)}{b} + \cdots + \frac{y_{j,m}(h)}{b^m}.$$

The point set $\{\boldsymbol{x}_0, \boldsymbol{x}_1, \ldots, \boldsymbol{x}_{b^m-1}\}$ is called a digital net over $\mathbb{Z}_b$ with generating matrices $\boldsymbol{C}_1, \ldots, \boldsymbol{C}_d$.

Having defined the digital construction scheme, we are now in a position to define digital nets.

Definition 12.2.2 Let b be a prime, t a nonnegative integer, and $m \in \mathcal{N}$, $\mathbb{Z}_b$ the finite field of order b and $\boldsymbol{C}_1, \ldots, \boldsymbol{C}_d \in \mathbb{Z}_b^{m\times m}$ with $\boldsymbol{C}_j = (\boldsymbol{c}_{j,1}, \ldots, \boldsymbol{c}_{j,m})^\top$. If $\forall d_j$, $j = 1, \ldots, d$, $0 \leq d_j \leq m$, such that $\sum_{j=1}^d d_j = m - t$, the vectors $\{\boldsymbol{c}_{j,i}, i = 1, \ldots, d_j, j = 1, \ldots, d\}$ are linearly independent, then the matrices $\boldsymbol{C}_1, \ldots, \boldsymbol{C}_d$ generate a digital (t, m, d)-net over $\mathbb{Z}_b$.

We now briefly comment on the parameters that characterize a digital net:

- m determines the number of points, which is b^m;
- d determines the dimensionality of the point set;
- t is the quality parameter of the digital net, the lower the better the net is.

Having defined digital nets, we note that in order to compute functionals of multidimensional diffusions, we need to have access to the generating matrices $\boldsymbol{C}_1, \ldots, \boldsymbol{C}_d$. Fortunately, many examples of such matrices are known, see e.g. Faure (1982), Niederreiter (1992, 2005, 2008), Niederreiter and Xing (1999), Sobol (1967) and the references therein. Efficient implementations of qMC point sets have been published in the literature, see e.g. Joe and Kuo (2003, 2008), for the Sobol sequence and Pirsic (2002) for an implementation of the Niederreiter-Xing sequence. Implementations of different sequences are also discussed in Hong and Hickernell (2003). We now illustrate the above definitions with some examples.

Example 12.2.3 The Hammersley net, Hammersley (1960), is an example of a digital $(0, m, 2)$-net over $\mathbb{Z}_2$. Its generating matrices are given by

$$\boldsymbol{C}_1 = \begin{pmatrix} 1 & 0 & \cdots & 0 \\ 0 & \ddots & \ddots & \vdots \\ \vdots & \ddots & \ddots & 0 \\ 0 & \cdots & 0 & 1 \end{pmatrix} \quad \text{and} \quad \boldsymbol{C}_2 = \begin{pmatrix} 0 & \cdots & 0 & 1 \\ \vdots & \iddots & \iddots & 0 \\ 0 & \iddots & \iddots & \vdots \\ 1 & 0 & \cdots & 0 \end{pmatrix}.$$

Example 12.2.4 The following matrices generate a strict digital $(1, 3, 4)$-net over $\mathbb{Z}_2$ and stem from a Niederreiter-Xing sequence as implemented in Pirsic (2002):

$$\boldsymbol{C}_1 = \begin{pmatrix} 1 & 1 & 1 \\ 0 & 1 & 0 \\ 0 & 0 & 0 \end{pmatrix}, \qquad \boldsymbol{C}_2 = \begin{pmatrix} 1 & 0 & 0 \\ 0 & 0 & 1 \\ 0 & 1 & 0 \end{pmatrix},$$

$$\boldsymbol{C}_3 = \begin{pmatrix} 1 & 1 & 0 \\ 1 & 0 & 0 \\ 0 & 0 & 1 \end{pmatrix}, \qquad \boldsymbol{C}_4 = \begin{pmatrix} 0 & 1 & 1 \\ 1 & 1 & 0 \\ 1 & 1 & 1 \end{pmatrix}.$$

Example 12.2.5 The following matrices generate a strict digital $(2, 3, 4)$-net over $\mathbb{Z}_2$:

$$\boldsymbol{C}_1 = \begin{pmatrix} 1 & 1 & 0 \\ 1 & 0 & 0 \\ 1 & 1 & 0 \end{pmatrix}, \qquad \boldsymbol{C}_2 = \begin{pmatrix} 1 & 0 & 1 \\ 1 & 0 & 1 \\ 0 & 0 & 0 \end{pmatrix},$$

$$\boldsymbol{C}_3 = \begin{pmatrix} 0 & 0 & 1 \\ 1 & 0 & 0 \\ 0 & 0 & 1 \end{pmatrix}, \qquad \boldsymbol{C}_4 = \begin{pmatrix} 0 & 1 & 0 \\ 0 & 1 & 0 \\ 0 & 0 & 0 \end{pmatrix}.$$

Since Paskov and Traub (1995) it has been known that qMC methods can be successfully applied to finance problems. In theory, see e.g. Niederreiter (1992) and Dick and Pillichshammer (2010), for functions of *bounded variation in the sense of Hardy and Krause*, see Niederreiter (1992), qMC rules can achieve convergence rates arbitrarily close to N^{-1}, which is a significant improvement on the $N^{-1/2}$ convergence rate achieved by MC methods. However, as evident from the discussion in this section, qMC rules are deterministic point sets and provide no practical information on the actual error incurred. Monte Carlo methods on the other hand allow for the computation of confidence intervals, giving statistical information on the error. We now remedy this shortcoming and randomize the qMC point sets. This allows us to supplement the faster convergence rates enjoyed by qMC rules with statistical information. Though different randomization techniques exist, see e.g. Dick and Pillichshammer (2010), we focus on Owen's scrambling algorithm, introduced in Owen (1995), see also Owen (1997), which produces optimal convergence rates, as we explain below.

12.2.2 Owen's Scrambling Algorithm

In this subsection we discuss Owen's scrambling algorithm, which was introduced in Owen (1995). We describe this algorithm using a generic point $\boldsymbol{x} \in [0, 1)^d$, where $\boldsymbol{x} = (x_1, \ldots, x_d)$ and

$$x_j = \frac{\xi_{j,1}}{b} + \frac{\xi_{j,2}}{b^2} + \cdots.$$

Then the scrambled point shall be denoted by $\boldsymbol{y} \in [0, 1)^d$, where $\boldsymbol{y} = (y_1, \ldots, y_s)$,

$$y_j = \frac{\eta_{j,1}}{b} + \frac{\eta_{j,2}}{b^2} + \cdots.$$

The permutation applied to $\xi_{j,l}$, $j = 1, \ldots, d$, depends on $\xi_{j,k}$, for $1 \leq k < l$. In particular, $\eta_{j,1} = \pi_j(\xi_{j,1})$, $\eta_{j,2} = \pi_{j,\xi_{j,1}}(\xi_{j,2})$, $\eta_{j,3} = \pi_{j,\xi_{j,1},\xi_{j,2}}(\xi_{j,3})$ and in general

$$\eta_{j,k} = \pi_{j,\xi_{j,1},\ldots,\xi_{j,k-1}}(\xi_{j,k}), \quad k \geq 2,$$

where π_j and $\pi_{j,\xi_{j,1},\ldots,\xi_{j,k-1}}$, $k \geq 2$, are random permutations of $\{0, \ldots, b-1\}$. We assume that permutations with different indices are mutually independent. It was shown in Owen (1995), Proposition 2, that if we apply Owen's scrambling algorithm to a digital net, each of the resulting points is uniformly distributed in the unit cube. Finally, for an efficient implementation of the scrambling algorithm we refer the reader to Hong and Hickernell (2003) and Matoušek (1998).

12.2.3 Numerical Integration Using Scrambled Digital Nets

We now discuss the effectiveness of scrambled digital nets. The definitive reference on integration using scrambled digital nets is Chap. 13 in Dick and Pillichshammer

(2010), and we briefly discuss one of the main results from this chapter. We introduce the estimator

$$\hat{I}(f) = \frac{1}{N} \sum_{i=0}^{N-1} f(\mathbf{y}_i),$$

where the point set $\{\mathbf{y}_i\}_{i=0}^{N-1}$, for $N = b^m$ and $m \in \mathcal{N}$, is obtained by applying Owen's scrambling algorithm to a digital net. In Dick and Pillichshammer (2010), the authors discuss functions f that enjoy *generalized variation in the sense of Vitali of order* α, where $0 < \alpha \leq 1$, see Chap. 13 in Dick and Pillichshammer (2010). Essentially, α determines the degree of smoothness of f, $\alpha = 0$ corresponds to functions that are only square integrable, and $\alpha = 1$ corresponds to functions with square integrable mixed partial derivatives. One can think of α as a continuity parameter, the larger α, the smoother the function. One of the main results of Chap. 13 in Dick and Pillichshammer (2010) is Theorem 13.25, which states that

$$\mathrm{Var}\big(\hat{I}(f)\big) \leq C_{m,t,d} N^{-(1+2\alpha)},$$

where $C_{m,t,d}$ denotes a constant dependent on m, t, and d. Some comments should be made: for $\alpha = 1$, we have $\mathrm{Var}(\hat{I}(f)) \leq C_{m,t,d} N^{-3}$, which is a significant improvement upon N^{-1}, the Monte Carlo rate. For square integrable functions f, one obtains $\mathrm{Var}(\hat{I}(f)) = o(N^{-1})$, which is still an improvement on the Monte Carlo rate. Finally, we note that scrambled digital nets are adaptive in the following sense: the smoothness α of the integrand under consideration need not be known a priori. Scrambled digital nets will always deliver the optimal convergence rate, $N^{-(1+2\alpha)}$, for $0 < \alpha \leq 1$. These observations suggest that scrambled digital nets are very useful tools when computing functionals of multidimensional diffusions.

12.2.4 Multilevel Quasi-Monte Carlo Methods

We conclude this section by recalling that multilevel methods could be used to combine biased estimators in such a way as to recover the Monte Carlo convergence rate of an unbiased scheme. The same comment applies to *multilevel quasi-Monte Carlo methods*. In a series of papers, see e.g. Gnewuch (2012a, 2012b), Hickernell et al. (2010), Niu et al. (2011), Baldeaux (2012b), Baldeaux and Gnewuch (2012), it was shown how to combine biased qMC rules using a multilevel approach to recover the optimal qMC rate. We refer the reader to these references for details.

12.3 Applications Under the Benchmark Approach

We now apply quasi-Monte Carlo methods to the pricing of realized variance products, see Sect. 8.5.2. In Sect. 13.5 we will discuss how to recover the joint distribution of $(Y_T, \int_0^T \frac{1}{Y_t} dt)$. In this subsection, we discuss how to apply quasi-Monte

Carlo methods to this problem. This approach is applied to the pricing of put options on realized variance and volatility. We compare the results with an almost exact simulation scheme. In the case of the put option on realized variance, we compare the results with the closed-form solution from Sect. 8.5.2.

We firstly discuss how to map the joint density to the unit square $[0,1]^2$. For a detailed discussion, with illustrations, we refer the reader to Baldeaux et al. (2011b), here we only present the outcome of the analysis. Assume that the joint density of $(Y_T, \int_0^T \frac{1}{Y_t}\,dt)$, which we obtain from (5.4.16) by inverting the Laplace transform numerically, is given by $f(y,z)$, where the variable y corresponds to Y_T and z to $\int_0^T \frac{1}{Y_t}\,dt$. We have the following representation for a general functional H of realized variance

$$E\left(\frac{H(\int_0^T \frac{1}{Y_t}\,dt)}{Y_T}\right) = \int_0^\infty \int_0^\infty \frac{H(z)}{y} f(y,z)\,dy\,dz.$$

We alert the reader to the fact that mapping the joint density into the unit square is not a trivial problem, in particular, since we do not have an explicit representation for f. The difficulty in mapping joint densities incorporating dependence structures to the unit cube was also discussed in Kuo et al. (2008). However, the problem studied in Kuo et al. (2008) was slightly different, as the joint density was a multivariate normal density, which is known explicitly. In this section, we map the joint density to the unit square using the transformation

$$\begin{aligned} x_1 - 1 - \exp\{-\lambda_1 y\}, \\ x_2 - 1 - \exp\{-\lambda_2 z\}, \end{aligned} \tag{12.3.1}$$

i.e. we base the transformation on the cumulative distribution function of the exponential distribution. We remark that λ_1 and λ_2 can differ, a feature which turns out to be crucial for the approach. Furthermore, another important feature of the transformation (12.3.1) is that it is easily interpretable, which is important, in particular, since we do not have access to an explicit representation of the joint density. We display in Fig. 12.3.1 the resulting joint density for $\lambda_1 = 0.5$ and $\lambda_2 = 0.18$. This particular transformation is used for numerical integration, for a detailed discussion of this problem, see Baldeaux et al. (2011b).

We now discuss how to employ quasi-Monte Carlo methods where we use the notation

$$\begin{aligned} x_1 = \Psi_1(y) = 1 - \exp\{-\lambda_1 y\}, \\ x_2 = \Psi_2(z) = 1 - \exp\{-\lambda_2 z\}, \end{aligned} \tag{12.3.2}$$

hence

$$\begin{aligned} y = \Psi_1^{-1}(y) = \frac{\log(1-x_1)}{-\lambda_1}, \\ z = \Psi_2^{-1}(z) = \frac{\log(1-x_2)}{-\lambda_2}, \end{aligned} \tag{12.3.3}$$

and

$$\begin{aligned} \psi_1(y) = \lambda_1 \exp\{-\lambda_1 y\}, \\ \psi_2(z) = \lambda_2 \exp\{-\lambda_2 z\}. \end{aligned} \tag{12.3.4}$$

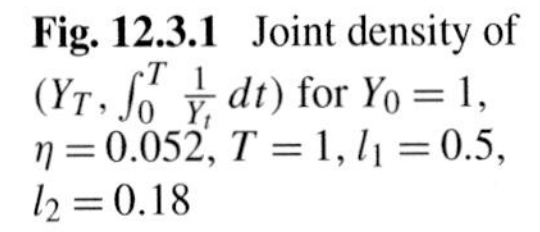

Fig. 12.3.1 Joint density of $(Y_T, \int_0^T \frac{1}{Y_t}\,dt)$ for $Y_0 = 1$, $\eta = 0.052$, $T = 1$, $l_1 = 0.5$, $l_2 = 0.18$

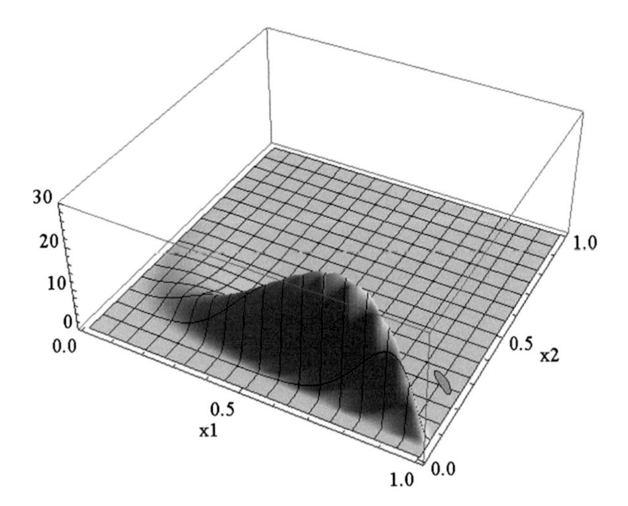

We employ digital nets as discussed in the previous section. Given a digital net $\{(x_{i,1}, x_{i,2})\}_{i=1}^N \in [0,1)^2$, we approximate the integral under consideration as follows:

$$
\begin{aligned}
&\int_0^\infty \int_0^\infty \frac{H(z)}{y} f(y,z)\,dy\,dz \\
&= \int_0^1 \int_0^1 \frac{H(\Psi_2^{-1}(x_2))}{\Psi_1^{-1}(x_1)} \frac{f(\Psi_1^{-1}(x_1), \Psi_2^{-1}(x_2))}{\psi(\Psi_1^{-1}(x_1))\psi(\Psi_2^{-1}(\Psi_2^{-1}(x_2)))}\,dx_1\,dx_2 \\
&= \int_0^1 \int_0^1 \frac{H(\Psi_2^{-1}(x_2))}{\Psi_1^{-1}(x_1)} \frac{f(\Psi_1^{-1}(x_1), \Psi_2^{-1}(x_2))}{\lambda_1(1-x_1)\lambda_2(1-x_2)}\,dx_1\,dx_2 \\
&\approx \frac{1}{N} \sum_{i=1}^N \frac{H(\Psi_2^{-1}(x_{i,2}))}{\Psi_1^{-1}(x_{i,1})} \frac{f(\Psi_1^{-1}(x_{i,1}), \Psi_2^{-1}(x_{i,2}))}{\lambda_1(1-x_{i,1})\lambda_2(1-x_{i,2})}.
\end{aligned}
\tag{12.3.5}
$$

As we deal with a two-dimensional integration problem, we base our quasi-Monte Carlo rule on the Sobol sequence, which, in two dimensions, is well known to have the optimal quality parameter $t = 0$. Furthermore, to be able to estimate standard errors, we use l independent copies of the quadrature rule presented in (12.3.5), each of which is obtained by applying Owen's scrambling algorithm, see Sect. 12.2, to the quasi-Monte Carlo point set $\{(x_{i,1}, x_{i,2})\}_{i=1}^N$. We remark that the scrambling algorithm is implemented according to Hong and Hickernell (2003), and Matoušek (1998). Consequently, we estimate the integral as follows:

$$
I_{RQMC} = \frac{1}{l} \sum_{j=1}^{l} I_j = \frac{1}{l} \sum_{j=1}^{l} \frac{1}{N} \sum_{i=1}^{N} g\big(y_{i,1}^j, y_{i,2}^j\big), \tag{12.3.6}
$$

where $\{(y_{i,1}^j, y_{i,2}^j)\}_{i=1}^N$, $j = 1, \dots, l$, is obtained from the quasi-Monte Carlo point set $\{(x_{i,1}, x_{i,2})\}_{i=1}^N$ by applying Owen's scrambling algorithm. We estimate standard errors via

$$\sigma_{RQMC} = \sqrt{\frac{\sum_{j=1}^l (I_j - I_{RQMC})^2}{l(l-1)}}. \tag{12.3.7}$$

For purposes of comparison, we will also look at Monte Carlo estimators. In this case, we use lN points $\{(u_{i,1}, u_{i,2})\}_{i=1}^{lN}$, independent and identically distributed in $[0,1]^2$, estimate the integral under consideration via

$$I_{MC} = \frac{1}{lN} \sum_{i=1}^{lN} g(u_{i,1}, u_{i,2}), \tag{12.3.8}$$

and compute standard errors using

$$\sigma_{MC} = \sqrt{\frac{\sum_{i=1}^{lN} (g(u_{i,1}, u_{i,2}) - I_{MC})}{lN(lN-1)}}. \tag{12.3.9}$$

Numerical results for puts on realized variance and volatility are presented in Sect. 12.4. Finally, we would like to point out that the numerical scheme presented in this subsection could also be used for variance reduction, e.g. it could serve as a control variate, if models less tractable than the stylized MMM are employed.

12.4 Numerical Results

In this section, we present some numerical results illustrating the method introduced in Sect. 12.2. First, we discuss put options on realized variance, for which analytical solutions are available using the results from Sect. 8.5.2 and, subsequently, we discuss put options on volatility. For the latter, closed form solutions are not available, but we check our results using the almost exact simulation from Sect. 12.4.1.

12.4.1 Almost Exact Simulation for Functionals of Realized Variance

Another method for computing prices of functionals of realized variance is based on a discretization of the integral appearing in the definition of realized variance, i.e.

$$\int_0^T \frac{1}{Y_t} dt \approx \sum_{i=0}^n w_i \frac{1}{Y_{t_i}}, \tag{12.4.10}$$

where we choose $t_0 = 0$, $t_n = T$. Furthermore, in this section, we use the trapezoidal rule, i.e. we set

$$
\begin{aligned}
w_i &= \frac{T}{N} \quad \text{for } i = 1, \ldots, n-1, \\
w_i &= \frac{T}{2N} \quad \text{for } i = 0, n.
\end{aligned}
\tag{12.4.11}
$$

Obviously, Y_{t_i}, $i = 1, \ldots, n$, can be simulated exactly. It is well known that $Y_t \frac{4\eta}{1-\exp\{-\eta t\}}$ follows a $\chi^2(4, \alpha)$ distribution, where $\alpha = \frac{4\eta y}{\exp\{\eta t\}-1}$ and $Y_0 = y$, see e.g. Sect. 3.1, Platen and Rendek (2009) or Jeanblanc et al. (2009). We use a simple Monte Carlo simulation for this approach. A quasi-Monte Carlo point set could have been employed, too. Finally, we remark that the multilevel Monte Carlo method, as discussed in Sect. 12.1, could also be useful in this context.

Since the discretization of the integral introduces a bias, we refer to the scheme as being almost exact. The computational effort needs to be divided up between variance and bias reduction. Following the approach in e.g. Duffie and Glynn (1995), see also Sect. 12.1, we consider the mean square error,

$$
MSE := E\left(\left(\hat{Y} - E\left(f\left(\int_0^T \frac{1}{Y_t} dt\right)\right)\right)^2\right),
$$

where $f(\cdot)$ is the functional of interest and $\hat{Y} = \frac{1}{N}\sum_{j=1}^N (f(\sum_{i=0}^n w_i \frac{1}{Y_{t_i}^j}))$, where $Y_{t_i}^j$, $j = 1, \ldots, N$, are independent copies of Y_{t_i}. Consequently,

$$
MSE = \frac{\mathrm{Var}(f(\sum_{i=0}^n w_i \frac{1}{Y_{t_i}}))}{N} + \left(E\left(f\left(\sum_{i=0}^n w_i \frac{1}{Y_{t_i}}\right)\right) - E\left(f\left(\int_0^T \frac{1}{Y_t} dt\right)\right)\right)^2.
$$

For the functionals under consideration in this paper,

$$
f(z) = \frac{(K-z)^+}{Y_T}
\tag{12.4.12}
$$

and

$$
f(z) = \frac{(K-\sqrt{z})^+}{Y_T},
\tag{12.4.13}
$$

respectively, estimates on the bias

$$
\left| E\left(f\left(\sum_{i=0}^n w_i \frac{1}{Y_{t_i}}\right)\right) - E\left(f\left(\int_0^T \frac{1}{Y_t} dt\right)\right) \right|
$$

do not seem to be known. Consequently, we estimate the bias numerically, using the Laplace transform method from Sect. 8.5.2 and the quasi-Monte Carlo method from Sect. 12.2 to obtain reference values for $E(f(\int_0^T \frac{1}{Y_t} dt))$. This allows us to numerically investigate the bias, which we find to be $O(n^{-1})$ for (12.4.12) and (12.4.13). Consequently, we can now divide the computational effort between variance and bias reduction. Using our numerical estimates on the bias, we find that for both functionals, (12.4.12) and (12.4.13), considered in this paper,

$$MSE = \frac{c_1}{N} + \frac{c_2}{n^2}, \tag{12.4.14}$$

where c_1 and c_2 are positive constants. The computational cost of this scheme is proportional to nN. Consequently, given a computational budget C, (12.4.14) suggests that one should choose $N = C^{2/3}$ and $n = C^{1/3}$, see e.g. Duffie and Glynn (1995) and Sect. 12.1. In this section, we will make use of this choice of n and N, for a given computational budget C.

We remark that this approach paints a very favorable picture for the almost exact simulation approach: the availability of reference solutions allows us to estimate the bias, which in turn allows us to trade-off variance and bias reduction, using the mean-square-error as a criterion. Should reference solutions not be available, one might have to divide the computational effort in a more ad-hoc fashion, resulting in a worse performance of the method.

Finally, we point out that this method could also be applied to *corridor variance swaps*, the floating leg of which is given by

$$\int_0^T \mathbf{1}_{S_t^{\delta_*} \in D} \, d\big[\log\big(S^{\delta_*}\big)\big]_t. \tag{12.4.15}$$

The almost exact simulation scheme presented in this subsection allows us to handle corridor variance swaps.

12.4.2 Numerical Results for Put Options on Realized Variance

In this subsection, we apply the numerical scheme introduced in Sect. 12.2 to compute

$$E\left[\frac{(K - \int_0^T \frac{1}{Y_t} dt)^+}{Y_T}\right], \tag{12.4.16}$$

from which we can compute prices of European puts on realized variance by multiplying by the constant $\frac{S_0^{\delta_*}}{\alpha_T^{\delta_*} S_T^0}$. We use the following set of parameters

$$Y_0 = 1, \qquad K = 5, \qquad \eta = 0.052, \qquad T = 1,$$

and obtain from Sect. 8.5.2 the true value of (12.4.16) as being around 3.11. In Figs. 12.4.2 and 12.4.3, we show estimates of (12.4.16) and standard errors obtained from the quasi-Monte Carlo and the Monte Carlo method, as detailed in Sect. 12.2, and the almost exact simulation scheme from Sect. 12.4.1. For the quasi-Monte Carlo method, estimates of (12.4.16) and standard errors are calculated using (12.3.6) and (12.3.7), where we choose $l = 30$ and $N = 2^m$ and vary m. On the x-axes in Figs. 12.4.2 and 12.4.3, we show the logarithm of the computational complexity of the problem, that is the logarithm of the number of function evaluations performed, which is $\log(l2^m 2)$. As we mentioned before, the computational complexity changes with m.

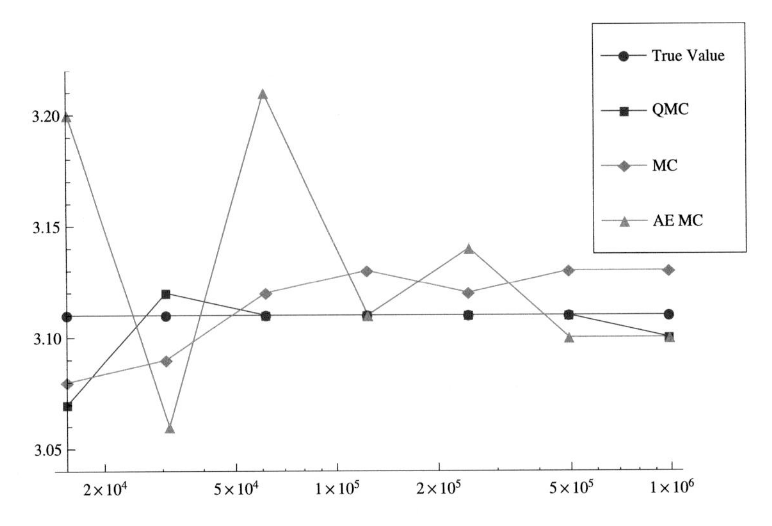

Fig. 12.4.2 Estimates of (12.4.16) versus logarithm of the number of function evaluations

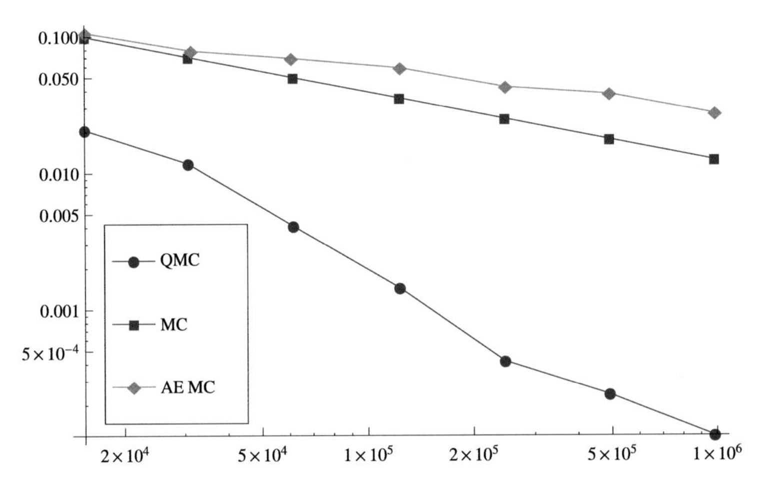

Fig. 12.4.3 Standard errors for the put option on realized variance versus logarithm of the number of function evaluations

We compare the performance of the quasi-Monte Carlo point set to the performance of the standard Monte Carlo method, which uses $l2^m$ two-dimensional points to ensure that the two methods are of the same computational complexity. Estimates of (12.4.16) and standard errors are obtained from (12.3.8) and (12.3.9). We conclude that scrambled quasi-Monte Carlo point sets offer a marked advantage over plain Monte Carlo simulation.

Finally, we use the almost exact simulation scheme (AE MC), discussed in Sect. 12.4.1. As we numerically determined that the bias is $O(n^{-1})$, we choose

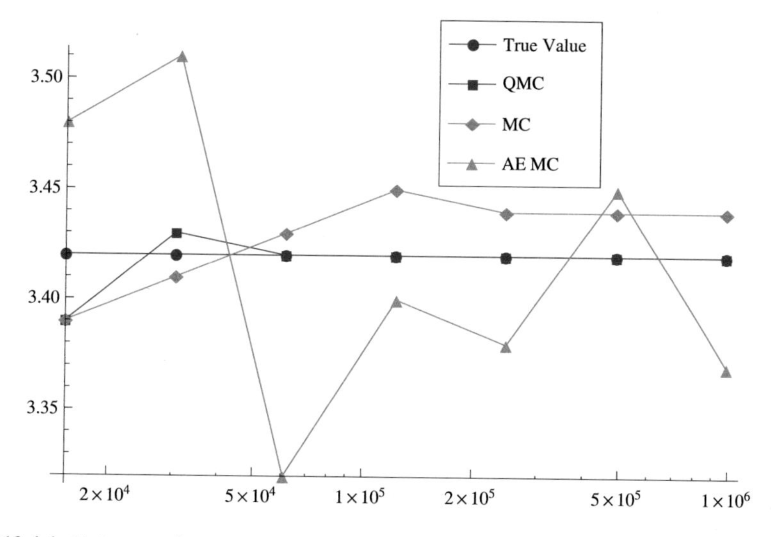

Fig. 12.4.4 Estimates of (12.4.17) versus logarithm of the number of function evaluations

$N = (l2^m 2)^{2/3}$ and $n = (l2^m 2)^{1/3}$, ensuring that the almost exact simulation scheme is of the same computational complexity as the previous two approaches. We note that due to the fact that the computational effort is to be divided up between variance and bias reduction, the two competing methods offer better estimates and standard errors converge at a faster rate, as one would expect. These numerical results highlight the usefulness of the Lie symmetry methods, which allow us to obtain explicit formulae for the Laplace transforms of put option prices and the joint distribution of $(Y_T, \int_0^T \frac{1}{Y_t} dt)$.

12.4.3 Numerical Results for Put Options on Volatility

In this subsection, we present numerical results for put options on volatility. In particular, we compute

$$E\left(\frac{(K - \sqrt{\int_0^T \frac{1}{Y_t} dt})^+}{Y_T}\right), \tag{12.4.17}$$

from which we can obtain prices of European puts on volatility by multiplying by $\frac{S_0^{\delta_*}}{\alpha_T^{\delta_*} S_T^0}$, which is constant. As in Sect. 12.4.2, we choose the set of parameters

$$Y_0 = 1, \qquad K = 5, \qquad \eta = 0.052, \qquad T = 1.$$

As discussed in Sect. 8.5.2, we do not have a closed-form solution for (12.4.17). In Figs. 12.4.4 and 12.4.5, we show estimates of (12.4.17) and standard errors, respectively, obtained from the quasi-Monte Carlo and Monte Carlo method as detailed in Sect. 12.2 and the almost exact simulation scheme from Sect. 12.4.1. As

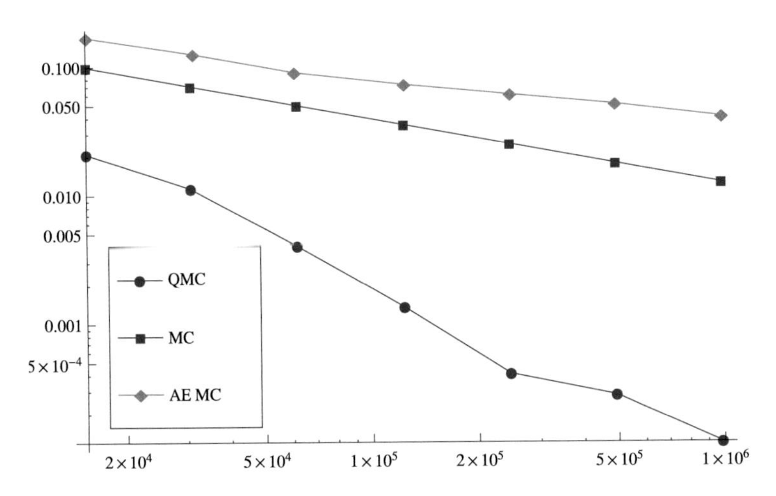

Fig. 12.4.5 Standard errors for the put option on volatility versus logarithm of the number of function evaluations

in Sect. 12.4.2, for the quasi-Monte Carlo method, estimates of (12.4.17) and standard errors are calculated using (12.3.6) and (12.3.7), where we choose $l = 30$ and $N = 2^m$ and again we vary m. Following the discussion in Sect. 12.4.2, on the x-axes in Figs. 12.4.4 and 12.4.5 we show the logarithm of the computational complexity, i.e. the logarithm of the number of function evaluations, which is $\log(l2^m 2)$, where we vary m.

For the Monte Carlo method, we use $l2^m$ two-dimensional points to ensure that the two methods are of the same computational complexity. Again, we obtain estimates of (12.4.17) and standard errors from (12.3.8) and (12.3.9), and conclude that the quasi-Monte Carlo point sets offer a marked advantage over plain Monte Carlo simulation.

Regarding the almost exact simulation scheme, the bias seems to be $O(n^{-1})$, resulting in the same choice for n and N as before, $N = (l2^m 2)^{2/3}$ and $n = (l2^m 2)^{1/3}$. We find again that dividing up the computational effort between variance and bias reduction results in the competing two methods outperforming the almost exact simulation scheme, as expected, highlighting the importance of the results obtained using Lie symmetry methods.

Chapter 13
Computational Tools

It is the aim of this chapter to introduce computational tools, which can be used to implement functionals presented in this book. In particular, we focus on the *non-central chi-squared distribution*, which appeared in the context of the MMM and the TCEV model, and the *non-central beta distribution*, which appeared in the context of pricing exchange options. Lastly, we discuss the inversion of Laplace transforms, which can be used to recover transition densities from the Laplace transforms.

13.1 Some Identities Related to the Non-central Chi-Squared Distribution

The non-central chi-squared distribution featured prominently when pricing European call and put options under the MMM and TCEV model, see Sect. 3.3. In the current section, we recall the distribution, and in Sect. 13.2 we will present an algorithm showing how to implement the distribution, where we follow ideas presented in Hulley (2009).

First, we recall the link between the squared Bessel process and the non-central chi-squared distribution, which is given by

$$\frac{X_t}{t} \stackrel{d}{=} \chi_\delta^2\left(\frac{x}{t}\right),$$

where $X = \{X_t,\ t \geq 0\}$ denotes a squared Bessel process of dimension δ, and $\chi_\delta^2(\lambda)$ denotes a non-central chi-squared random variable with δ degrees of freedom and non-centrality parameter $\lambda > 0$. We recall from Lemma 8.2.2 that the non-central χ^2-distribution with $\delta > 0$ degrees of freedom and non-centrality parameter $\lambda > 0$ has the density function

$$p(x, \delta, \lambda) = \frac{1}{2}\exp\left\{-\frac{x+\lambda}{2}\right\}\left(\frac{x}{\lambda}\right)^{\frac{\delta}{4}-\frac{1}{2}} I_{\frac{\delta}{2}-\frac{1}{2}}(\sqrt{\lambda x}), \quad x \geq 0. \tag{13.1.1}$$

J. Baldeaux, E. Platen, *Functionals of Multidimensional Diffusions with Applications to Finance*, Bocconi & Springer Series 5, DOI 10.1007/978-3-319-00747-2_13,

Here $I_\nu(x) = \sum_{j\geq 0} \frac{1}{j!\Gamma(j+\nu+1)} (\frac{x}{2})^{2j+\nu}$ denotes the modified Bessel function of the first kind of order $\nu > -1$. The following equality is given in Hulley (2009), where

$$\begin{aligned}\frac{\lambda}{x} p(x,4,\lambda) &= \frac{1}{2} e^{-\frac{\lambda+x}{2}} \left(\frac{\lambda}{x}\right)^{\frac{1}{2}} I_1(\sqrt{\lambda x}) \\ &= \frac{1}{2} e^{-\frac{\lambda+x}{2}} \left(\frac{x}{\lambda}\right)^{-\frac{1}{2}} I_{-1}(\sqrt{\lambda x}) = p(x,0,\lambda), \qquad (13.1.2)\end{aligned}$$

for $x \in (0,\infty)$ and $\lambda > 0$, since the modified Bessel function of the first kind satisfies $I_1 = I_{-1}$, see e.g. Abramowitz and Stegun (1972), Eq. (9.6.6). Clearly, this equality entails the probability density function, of a non-central chi-squared random variable of zero degrees of freedom, $p(x,0,\lambda)$. Such a random variable is comprised of a discrete part, as it places positive mass at zero, and a continuous part assuming values in the interval $(0,\infty)$. We return to this issue when discussing this type of probability distributions below. From Eq. (13.1.2), we immediately obtain the following formula, which is employed frequently in the context of the MMM, see Sect. 3.3:

$$\begin{aligned}& E\left(\frac{\lambda(t,S)}{\chi_4^2(\lambda(t,S))} g\left(\chi_4^2(\lambda(t,S))\right)\right) \\ &\quad = E\left(g\left(\chi_4^2(\lambda(t,S))\right)\right) - g(0)\exp\left\{-\frac{\lambda(t,S)}{2}\right\}, \qquad (13.1.3)\end{aligned}$$

for an appropriately integrable function $g(\cdot)$. Next, we introduce the cumulative distribution function of a non-central chi-squared random variable. The following equality, see Eq. (29.3) in Johnson et al. (1995), introduces the non-central chi-squared distribution as a weighted average of central chi-squared distributions, the weights being Poisson weights:

$$P\left(\chi_\delta^2(\lambda) \leq x\right) = \sum_{j=0}^{\infty} \frac{\exp\{-\lambda/2\}(\lambda/2)^j}{j!} P\left(\chi_{\delta+2j}^2 \leq x\right), \qquad (13.1.4)$$

for all $x \in (0,\infty)$, $\delta > 0$ and $\lambda > 0$, where χ_δ^2 denotes the central chi-squared random variable. The distribution of the central chi-squared random variable admits the following presentation in terms of the *regularized incomplete gamma function* $\mathcal{P}(\cdot,\cdot)$, see Johnson et al. (1994), Eq. (18.3):

$$P\left(\chi_\delta^2 \leq x\right) = \mathcal{P}\left(\frac{\delta}{2}, \frac{x}{2}\right), \qquad (13.1.5)$$

for $x \in (0,\infty)$ and $\delta > 0$, where

$$\mathcal{P}(a,z) := \frac{1}{\Gamma(a)} \int_0^z \exp\{-t\} t^{a-1}\, dt, \qquad (13.1.6)$$

for $z \in \Re^+$ and $a > 0$. We can obtain an expression similar to Eq. (13.1.4) for the density of a non-central chi-squared random variable,

$$p(x,\delta,\lambda) = \sum_{j=0}^{\infty} \frac{\exp\{-\lambda/2\}(\lambda/2)^j}{j!} p(x,\delta+2j),$$

for $x \in (0, \infty)$, $\delta > 0$ and $\lambda > 0$, and where $p(x, \delta)$ denotes the probability density function of a chi-squared random variable with $\delta > 0$ degrees of freedom. Finally, we focus on the non-central chi-squared distribution with zero degrees of freedom, which also featured in the context of the MMM in Sect. 3.3. From Eq. (13.1.4), we get

$$P\big(\chi_0^2(\lambda) \leq x\big) = \sum_{j=0}^{\infty} \frac{\exp\{-\lambda/2\}(\frac{\lambda}{2})^j}{j!} P\big(\chi_{2j}^2 \leq x\big), \tag{13.1.7}$$

for $x \geq 0$ and $\lambda > 0$. However, χ_0^2, a central chi-squared random variable of zero degrees of freedom, is simply equal to zero, i.e.,

$$P\big(\chi_0^2(\lambda) \leq x\big) = 1,$$

for all $x \geq 0$, hence

$$P\big(\chi_0^2(\lambda) = 0\big) = \exp\{-\lambda/2\},$$

where $\lambda > 0$. From Eq. (13.1.7) we get

$$\begin{aligned} P\big(\chi_0^2(\lambda) \leq x\big) &= P\big(\chi_0^2(\lambda) = 0\big) + P\big(0 < \chi_0^2(\lambda) \leq x\big) \\ &= \exp\{-\lambda/2\} + \int_0^x p(x, 0, \lambda)\, dx, \end{aligned}$$

for $x \geq 0$, $\lambda > 0$. We remark that a non-central chi-squared random variable of 0 degrees of freedom is not continuous, but places mass at the origin, and hence $p(x, 0, \lambda)$ is not a probability density function. Nevertheless, it is obtained by formally setting $\delta = 0$ in Eq. (13.1.1).

We conclude this section with some useful identities pertaining to the non-central chi-squared distribution. These equalities feature frequently in Sect. 3.3. We recall that $p(\cdot, \delta, \lambda)$ denotes the probability density of a χ^2-distributed random variable, and we use $\Psi(\cdot, \delta, \lambda)$ to denote the distribution of a χ^2-distributed random variable with δ degrees of freedom and non-centrality parameter λ.

Lemma 13.1.1 *The following useful properties hold*:

$$\left(\frac{\lambda}{x}\right)^{\frac{\nu-2}{2}} p(x, \nu, \lambda) = p(\lambda, \nu, x) \tag{13.1.8}$$

$$\int_0^\infty p(x, \nu+2, y)\, dy = \Psi(x, \nu, 0) \tag{13.1.9}$$

$$\int_\lambda^\infty p(x, \nu+2, y)\, dy = \Psi(x, \nu, \lambda) \tag{13.1.10}$$

$$\int_0^\lambda p(x, \nu+2, y)\, dy = \Psi(x, \nu, 0) - \Psi(x, \nu, \lambda). \tag{13.1.11}$$

13.2 Computing the Non-central Chi-Squared Distribution

The aim of this section is to introduce an algorithm allowing us to compute the non-central chi-squared distribution. We recall from Sect. 3.3, that in order to price calls and puts, we need to be able to evaluate this distribution function. Furthermore, we point out that we need to be able to evaluate this distribution function for zero degrees of freedom and for a variety of non-centrality parameters. In particular, for large maturities, the non-centrality parameter is small, whereas for small maturities, the non-centrality parameter is large. This section follows Hulley (2009) closely. As in this reference, we base our approach on an algorithm from Ding (1992), which performs well for small values of the non-centrality parameter, but not for large values. For this reason, we employ an analytic approximation due to Sankaran (1963), for large values. We introduce the *non-central regularized incomplete gamma function*, given by

$$\mathcal{P}(a,b,z) := \sum_{j=0}^{\infty} \frac{\exp\{-b\}b^j}{j!} \mathcal{P}(a+j,z), \tag{13.2.12}$$

for all $z \in \Re^+0$ and $a, b \geq 0$. Formally, we set $\mathcal{P}(0,z) := 1$, as the regularized incomplete gamma function from Eq. (13.1.6) is not well-defined in this case. We can express the distribution function of the non-central chi-squared and the chi-squared random variables in terms of the non-central regularized incomplete gamma function,

$$P(\chi_\delta^2 \leq x) = \mathcal{P}\left(\frac{\delta}{2}, 0, \frac{x}{2}\right),$$

where $x \in (0,\infty)$ and $\delta > 0$, and

$$P(\chi_\delta^2(\lambda) \leq x) = \mathcal{P}\left(\frac{\delta}{2}, \frac{\lambda}{2}, \frac{x}{2}\right),$$

for $x \in (0,\infty)$ (respectively $x \in \Re^+$), $\delta > 0$, (respectively $\delta = 0$) and $\lambda > 0$. We assume for the remainder of this section that one of the following conditions is satisfied:

- $z \in (0,\infty)$ and $a, b > 0$;
- $z \in \Re^+$, $a = 0$ and $b > 0$,

which correspond to the cases $\delta > 0$ and $\delta = 0$, respectively.

In a first step, we rewrite the terms $\mathcal{P}(a+j,z)$ on the right-hand side of Eq. (13.2.12) in terms of an infinite sum. Using integration by parts and the identity $\Gamma(a+j+1) = (a+j)\Gamma(a+j)$, we obtain

$$\mathcal{P}(a+j+1,z) = \mathcal{P}(a+j,z) - \frac{\exp\{-z\}z^{a+j}}{\Gamma(a+j+1)}, \tag{13.2.13}$$

which also holds for $a = j = 0$, as by definition $\mathcal{P}(0, z) = 1$. A recursive application of Eq. (13.2.13) yields

$$\mathcal{P}(a + j, z) = \mathcal{P}(a + j + 1, z) + \frac{\exp\{-z\}z^{a+j}}{\Gamma(a + j + 1)} = \sum_{k=j}^{\infty} \frac{\exp\{-z\}z^{a+k}}{\Gamma(a + k + 1)}, \quad (13.2.14)$$

for $j \in \{0, , 12, \dots\}$. Defining

$$A_k = \sum_{j=0}^{k} \frac{\exp\{-b\}b^j}{j!}$$

and

$$B_k = \frac{\exp\{-z\}z^{a+k}}{\Gamma(a + k + 1)},$$

we have

$$\mathcal{P}(a, b, z) = \sum_{k=0}^{\infty} A_k T_k. \quad (13.2.15)$$

The idea is to truncate the series in Eq. (13.2.15),

$$\begin{aligned}\mathcal{P}(a, b, z) &= \sum_{k=0}^{N-1} A_k T_k + \sum_{k=N}^{\infty} A_k T_k \\ &= \tilde{\mathcal{P}}_N(a, b, z) + \epsilon_N.\end{aligned}$$

We now aim to find an effective bound for $\epsilon_N = \sum_{k=N}^{\infty} A_k T_k$. We have the trivial bound

$$A_k = \sum_{j=0}^{k} \frac{\exp\{-b\}b^j}{j!} < \sum_{j=0}^{\infty} \frac{\exp\{-b\}b^j}{j!} = 1,$$

and hence

$$\epsilon_N = \sum_{k=N}^{\infty} A_k T_k < \sum_{k=N}^{\infty} T_k.$$

We note that the T_k, $k \in \mathcal{N}$, admit the following recursive formula:

$$T_k = \frac{\exp\{-z\}z^{a+k}}{\Gamma(a + k + 1)} = \frac{z}{a + k} \frac{\exp\{-z\}z^{a+k-1}}{\Gamma(a + k)} = \frac{z}{a + k} T_{k-1}, \quad (13.2.16)$$

for $k \in \mathcal{N}$. Hence

$$T_k = \prod_{l=N}^{k} \frac{z}{a + l} T_{N-1} \le \left(\frac{z}{a + N} \right)^{k-N+1} T_{N-1},$$

for $N \in \mathcal{N}$ and $k \in \{N, N+1, N+2, \ldots\}$. This allows us to obtain the following bound on ϵ_N:

$$\epsilon_N < \sum_{k=N}^{\infty} \left(\frac{z}{a+N}\right)^{k-N+1} T_{N-1} = \sum_{k=1}^{\infty} \left(\frac{z}{a+N}\right)^{k} T_{N-1} = \frac{z}{a+N-z} T_{N-1}, \tag{13.2.17}$$

for each $N \in \{N^*, N^*+1, N^*+2, \ldots\}$, where

$$N^* := \min\bigl\{n \in \{0, 1, 2, \ldots\} \bigm| z < a+n\bigr\}.$$

In Algorithm 13.1 below, we present pseudo-code for an algorithm which computes the non-central chi-squared distribution. In words, the algorithm proceeds as follows: we specify a desired level of accuracy, say $\epsilon \in (0, 1)$. Next, we compute N^*. Obtaining N^* is crucial, as our error bound in Eq. (13.2.17) only applies for $N \geq N^*$. We then compute $\tilde{P}_{N^*}(a, b, z)$, and consequently check the truncation error incurred via Eq. (13.2.17). We then proceed to add further terms $A_k T_k$, where $k \in \{N^*, N^*+1, N^*+2, \ldots\}$. As soon as the bound for the truncation error ϵ_N has fallen below ϵ, we truncate the loop and obtain a value $\tilde{\mathcal{P}}(a, b, z) \in (\mathcal{P}(a, b, z) - \epsilon, \mathcal{P}(a, b, z))$, where $N \in \{N^*, N^*+1, N^*+2, \ldots\}$.

Finally, we discuss the implementation of the algorithm. Recall that the T_k can be computed recursively using Eq. (13.2.16), with only one multiplication and division required to compute the next term. Lastly, A_k admits the representation

$$A_k = \sum_{j=0}^{k} \frac{\exp\{-b\} b^j}{j!} = A_{k-1} + \frac{\exp\{-b\} b^k}{k!} = A_{k-1} + B_k,$$

where

$$B_k = \frac{\exp\{-b\} b^{k+1}}{k!} = \frac{b}{k} B_{k-1}.$$

Hence we can also obtain the A_k recursively, with one multiplication, division, and addition required. This means that we can compute $\tilde{P}_N(a, b, z)$ in linear time, i.e. using $O(N)$ operations. In a detailed study, Dyrting (2004) discovered that the algorithm outlined above performs well for small and moderate values of b. For large values of b, the series in Eq. (13.2.12) converges slowly, meaning a large number of terms have to be used to achieve a particular precision ϵ. Furthermore, underflow problems can occur, as the individual terms in the series are small.

To remedy this shortcoming, Hulley (2009) fixed a maximum number of terms to be used in the summation. Once this limit is reached, an analytical approximation to the non-central incomplete gamma function is used. For this there are numerous possibilities, see Johnson et al. (1995), Sect. 29.8. We follow the advice of Schroder (1989), who recommends the analytic approximation due to Sankaran (1963),

$$\mathcal{P}(a, b, z) \approx \Phi(x),$$

where Φ denotes the standard normal cumulative distribution function and

$$x := -\frac{1 - hp\bigl(1 - h + \frac{(2-h)mp}{2}\bigr) - \bigl(\frac{z}{a+b}\bigr)^h}{h\sqrt{2p(1+mp)}}$$

with

$$h = 1 - \frac{2}{3}\frac{(a+b)(a+3b)}{(a+2b)^2}, \qquad p = \frac{1}{2}\frac{a+2b}{(a+b)^2}, \qquad m = (h-1)(1-3h).$$

This is a robust and efficient scheme, see Dyrting (2004). In addition, the approximation improves as the value of b increases. This fact is of particular relevance to us, as the performance of our original scheme performs worse as b increases. We present the pseudo-code of this algorithm in Algorithm 13.1 below.

13.3 The Doubly Non-central Beta Distribution

We firstly introduce the (central) beta random variable, after that the singly non-central beta random variable and finally the doubly non-central beta random variable, all with strictly positive shape parameters. However, in Sect. 3.3, we presented formulas for exchange options in terms of the non-central beta distribution with one shape parameter assuming the value zero, see Eq. (3.3.16). Hence in this section, we follow Hulley (2009) and extend the doubly non-central beta distribution allowing for one shape parameter assuming the value zero. In Sect. 13.4, we show how to compute the doubly non-central beta distribution.

It is well-known that the (central) beta random variable with shape parameters $\delta_1/2 > 0$ and $\delta_2/2 > 0$ admits the following representation in terms of chi-squared random variables,

$$\beta_{\delta_1,\delta_2} := \frac{\chi^2_{\delta_1}}{\chi^2_{\delta_1} + \chi^2_{\delta_2}}, \tag{13.3.18}$$

see Johnson et al. (1995), Chap. 25. As chi-squared random variables are strictly positive, $\beta_{\delta_1,\delta_2}$ assumes values in $(0, 1)$. The distribution of $\beta_{\delta_1,\delta_2}$ can be expressed in terms of the *regularized incomplete beta function*,

$$P(\beta_{\delta_1,\delta_2} \leq x) = I_x\left(\frac{\delta_1}{2}, \frac{\delta_2}{2}\right), \tag{13.3.19}$$

for $x \in (0, 1)$, where

$$I_z(a, b) := \frac{\Gamma(a+b)}{\Gamma(a)\Gamma(b)} \int_0^z t^{a-1}(1-t)^{b-1}\, dt, \tag{13.3.20}$$

for all $z \in [0, 1]$ and $a, b > 0$. We now define the singly non-central beta distribution, with shape parameters $\delta_1/2 > 0$ and $\delta_2/2 > 0$ and non-centrality parameter $\lambda > 0$, which is given by

$$\beta_{\delta_1,\delta_2}(\lambda, 0) := \frac{\chi^2_{\delta_1}(\lambda)}{\chi^2_{\delta_1}(\lambda) + \chi^2_{\delta_2}}. \tag{13.3.21}$$

This distribution was introduced in Tang (1938) and Patnaik (1949), in connection with the power function for the analysis of variance tests. We remark that (13.3.21)

Algorithm 13.1 Non-central regularized incomplete gamma function

Require: $a, b, z \in \Re^+$, $\epsilon \in (0, 1)$ and maxiter $\in \mathcal{N}$

1: errbnd $\leftarrow 1$
2: **if** $z - a \notin \{0, 1, 2, \ldots\}$ **then**
3: $\quad N^* \leftarrow \lceil (z-a)^+ \rceil$
4: **else**
5: $\quad N^* \leftarrow \lceil z - a \rceil + 1$
6: **end if**
7: **if** $N^* - 1 \leq$ maxiter **then**
8: $\quad A \leftarrow \exp\{-b\}$
9: $\quad B \leftarrow A$
10: $\quad T \leftarrow \frac{\exp\{-z\} z^a}{\Gamma(a+1)}$
11: $\quad$ value $\leftarrow A \times T$
12: $\quad k \leftarrow 1$
13: $\quad$ **while** $k \leq N^* - 1$ **do**
14: $\quad\quad B \leftarrow \frac{b}{k} \times B$
15: $\quad\quad A \leftarrow A + B$
16: $\quad\quad T \leftarrow \frac{z}{a+k} \times T$
17: $\quad\quad k \leftarrow k + 1$
18: $\quad$ **end while**
19: $\quad$ errbnd $\leftarrow \frac{z}{a+k-z} \times T$
20: $\quad$ **while** errbnd $\geq \epsilon$ and $k \leq$ maxiter **do**
21: $\quad\quad B \leftarrow \frac{b}{k} \times B$
22: $\quad\quad A \leftarrow A + b$
23: $\quad\quad T \leftarrow \frac{z}{a+k} \times T$
24: $\quad\quad$ value $\leftarrow$ value $+ A \times T$
25: $\quad\quad k \leftarrow k + 1$
26: $\quad\quad$ errbnd $\leftarrow \frac{z}{a+k-z} \times T$
27: $\quad$ **end while**
28: **end if**
29: **if** errbnd $\geq \epsilon$ **then**
30: $\quad h \leftarrow 1 - \frac{2}{3} \frac{(a+b)(a+3b)}{(a+2b)^2}$
31: $\quad p \leftarrow \frac{1}{2} \frac{a+2b}{(a+b)^2}$
32: $\quad m \leftarrow (h-1)(1-3h)$
33: $\quad x \leftarrow -\frac{1 - hp(1-h+\frac{(2-h)mp}{2}) - (\frac{z}{a+b})^h}{h\sqrt{2p(1+mp)}}$
34: $\quad$ value $\leftarrow \Phi(x)$
35: **end if**
36: **return** value

is referred to as Type I non-central beta random variable in Chattamvelli (1995), distinguishing it from a Type II non-central beta random variable, given by

$$\beta_{\delta_1,\delta_2}(0, \lambda) := 1 - \beta_{\delta_1,\delta_2}(\lambda, 0) = \frac{\chi^2_{\delta_2}}{\chi^2_{\delta_1}(\lambda) + \chi^2_{\delta_2}}.$$

The doubly non-central beta distribution, with shape parameters $\delta_1/2 > 0$, $\delta_2 > 0$ and non-centrality parameters $\lambda_1 > 0$ and $\lambda_2 > 0$ is given by

$$\beta_{\delta_1,\delta_2}(\lambda_1,\lambda_2) := \frac{\chi^2_{\delta_1}(\lambda_1)}{\chi^2_{\delta_1}(\lambda_1) + \chi^2_{\delta_2}(\lambda_2)}. \tag{13.3.22}$$

We recall from Eq. (13.1.4) that the distribution of the non-central chi-squared distribution could be expressed as a Poisson weighted mixture of central chi-squared distributions. Analogously, the distribution of the non-central beta distribution can be expressed as a Poisson weighted mixture of central beta distributions

$$P\big(\beta_{\delta_1,\delta_2}(\lambda,0) \leq x\big) = \sum_{j=0}^{\infty} \frac{\exp\{-\lambda/2\}(\lambda/2)^j}{j!} P(\beta_{\delta_1+2j,\delta_2} \leq x), \tag{13.3.23}$$

for all $x \in (0,1)$, $\delta_1, \delta_2 > 0$ and $\lambda > 0$, and

$$P\big(\beta_{\delta_1,\delta_2}(\lambda_1,\lambda_2) \leq x\big)$$
$$= \sum_{j=0}^{\infty} \frac{\exp\{-\lambda_1/2\}(\lambda_1/2)^j}{j!} \sum_{k=0}^{\infty} \frac{\exp\{-\lambda_2\}(\lambda_2/2)^k}{k!} P(\beta_{\delta_1+2j,\delta_2+2k} \leq x),$$

for all $x \in (0,1)$, $\delta_1, \delta_2 > 0$ and $\lambda_1, \lambda_2 > 0$.

Now, we discuss how to extend the singly and doubly non-central beta distributions to the case where one of the shape parameters is zero. We remark that the distributions in (13.3.23) and (13.3.24) do not allow for this, as the gamma function is not defined at zero. We hence follow Hulley (2009), where techniques from Siegel (1979) were used to extend the non-central chi-squared distribution to include the case of zero degrees of freedom. As with the non-central chi-squared distribution, the distribution of the non-central beta distribution with one shape parameter equal to zero is no longer continuous, but comprised of a discrete part placing mass at the end points of the interval $[0,1]$, and a continuous part assuming values in $(0,1)$. Setting $\delta_2 = 0$ in Eq. (13.3.21), results in a random variable identically equal to one. However, setting $\delta_1 = 0$ yields a non-trivial random variable assuming values in $[0,1)$. Similarly, setting $\delta_1 = 0$ in (13.3.22), results in a non-trivial random variable assuming values in $[0,1)$ and setting $\delta_2 = 0$ in Eq. (13.3.22) results in a non-trivial random variable assuming values in $(0,1]$. For the remainder of this section, we set $\delta_1 = 0$ in Eqs. (13.3.23) and (13.3.22) and set $\delta = \delta_2 > 0$ and define

$$\beta_{0,\delta}(\lambda,0) := \frac{\chi^2_0(\lambda)}{\chi^2_0(\lambda) + \chi^2_\delta}, \tag{13.3.24}$$

for all $\delta > 0$ and $\lambda > 0$, and

$$\beta_{0,\delta}(\lambda_1,\lambda_2) := \frac{\chi^2_0(\lambda_1)}{\chi^2_0(\lambda_1) + \chi^2_\delta(\lambda_2)}, \tag{13.3.25}$$

for all $\delta > 0$, and $\lambda_1, \lambda_2 > 0$. The following result from Hulley (2009) shows how to extended the doubly non-central beta distribution to the case where one of the shape parameters assumes the value zero.

Proposition 13.3.1 *Suppose $x \in [0,1)$, $\delta > 0$, and $\lambda_1, \lambda_2 > 0$. Then*

$$P\big(\beta_{0,\delta}(\lambda_1,\lambda_2) \le x\big) = \sum_{j=0}^{\infty} \frac{\exp\{-\lambda_1/2\}(\lambda_1/2)^j}{j!} \sum_{k=0}^{\infty} \frac{\exp\{-\lambda_2\}(\lambda_2/2)^k}{k!} \times P(\beta_{2j,\delta+2k} \le x). \tag{13.3.26}$$

Proof We employ Eqs. (13.3.25) and (13.1.4), (13.1.5), (13.1.6), (13.1.1), to obtain

$$\begin{aligned}
&P\big(\beta_{0,\delta}(\lambda_1,\lambda_2) \le x\big) \\
&= P\left(\chi_0^2(\lambda_1) \le \frac{x}{1-x}\chi_\delta^2(\lambda_2)\right) \\
&= \int_0^\infty P\left(\chi_0^2(\lambda_1) \le \frac{x}{1-x}\xi\right) p(\xi,\delta,\lambda_2)\,d\xi \\
&= \sum_{j=0}^{\infty} \frac{\exp\{-\lambda_1/2\}(\lambda_1/2)^j}{j!} \int_0^\infty P\left(\xi_{2j}^2 \le \frac{x}{1-x}\xi\right) \\
&\quad \times \frac{1}{2}\exp\left\{-\frac{\lambda_2+\xi}{2}\right\}\left(\frac{\xi}{\lambda_2}\right)^{\frac{\delta-2}{4}} \sum_{k=0}^{\infty} \frac{(\sqrt{\lambda_2\xi}/2)^{\frac{\delta-2}{2}+2k}}{k!\,\Gamma(\delta/2+k)}\,d\xi \\
&= \sum_{j=0}^{\infty} \frac{\exp\{-\lambda_1/2\}(\lambda_1/2)^j}{j!} \sum_{k=0}^{\infty} \frac{\exp\{-\lambda/2\}}{k!} \frac{1}{2}\left(\frac{2}{\lambda_2}\right)^{\frac{\delta-2}{2}} \\
&\quad \times \int_0^\infty \frac{(\lambda_2\xi/4)^{\frac{\delta-2}{2}+k}\exp\{-\xi/2\}}{\Gamma(\delta/2+k)} P\left(\chi_{2j}^2 \le \frac{x\xi}{1-x}\right) d\xi \\
&= \exp\{-\lambda_1/2\} + \sum_{j=1}^{\infty} \frac{\exp\{-\lambda_1/2\}(\lambda_1/2)^j}{j!} \sum_{k=0}^{\infty} \frac{\exp\{-\lambda_2\}(\lambda_2/2)^k}{k!} \\
&\quad \times \frac{1}{2}\int_0^\infty \frac{(\xi/2)^{\frac{\delta-2}{2}+k}\exp\{-\xi/2\}}{\Gamma(\delta/2+k)} \mathcal{P}\left(j, \frac{x}{2(1-x)}\xi\right) d\xi \\
&= \exp\{-\lambda_1/2\} + \sum_{j=1}^{\infty} \frac{\exp\{-\lambda_1/2\}(\lambda_1/2)^j}{j!} \sum_{k=0}^{\infty} \frac{\exp\{-\lambda_2/2\}(\lambda_2/2)^k}{k!} \\
&\quad \times \frac{1}{\Gamma(j)\Gamma(\delta/2+k)} \int_0^\infty \zeta^{\delta/2+k-1}\exp\{-\zeta\} \int_0^{\frac{x\zeta}{1-x}} t^{j-1}\exp\{-t\}\,dt\,d\zeta \\
&= \exp\{-\lambda_1/2\} + \sum_{j=1}^{\infty} \frac{\exp\{-\lambda_1/2\}(\lambda_1/2)^j}{j!} \sum_{k=0}^{\infty} \frac{\exp\{-\lambda_2/2\}(\lambda_2/2)^k}{k!} \\
&\quad \times \frac{\Gamma(\delta/2+j+k)}{\Gamma(j)\Gamma(\delta/2+k)} \int_0^{\frac{x}{1-x}} \frac{u^{j-1}}{(1+u)^{\delta/2+j+k}}\,du
\end{aligned}$$

$$
\begin{aligned}
&= \exp\{-\lambda_1/2\} + \sum_{j=1}^{\infty} \frac{\exp\{-\lambda_1/2\}(\lambda_1/2)^j}{j!} \sum_{k=0}^{\infty} \frac{\exp\{-\lambda_2/2\}(\lambda_2/2)^k}{k!} \\
&\quad \times \frac{\Gamma(\delta/2+j+k)}{\Gamma(j)\Gamma(\delta/2+k)} \int_0^x v^{j-1}(1-v)^{\delta/2+k-1}\,dv \\
&= \exp\{-\lambda_1/2\} \\
&\quad + \sum_{j=1}^{\infty} \frac{\exp\{-\lambda_1/2\}(\lambda_1/2)^j}{j!} \sum_{k=0}^{\infty} \frac{\exp\{-\lambda_2\}(\lambda_2/2)^k}{k!} I_x\left(j, \frac{\delta}{2}+k\right) \\
&= \exp\{-\lambda_1/2\} \\
&\quad + \sum_{j=1}^{\infty} \frac{\exp\{-\lambda_1/2\}(\lambda_1/2)^j}{j!} \sum_{k=0}^{\infty} \frac{\exp\{-\lambda_2\}(\lambda_2/2)^k}{k!} P(\beta_{2j,\delta+2k} \leq x),
\end{aligned}
\tag{13.3.27}
$$

where we used the transformations $\xi/2 \mapsto \zeta$, $t/\zeta \mapsto u$, and $u/(1+u) \mapsto v$, together with Eq. (13.3.19). We note that since central chi-squared random variables with zero degrees of freedom are equal to zero, the same applies to $\beta_{0,\delta+2k}$, for all $k \in \{0, 1, 2, \ldots\}$, see Eq. (13.3.18). Hence we have

$$
\exp\{-\lambda_1/2\} = \exp\{-\lambda_1/2\} \sum_{k=0}^{\infty} \frac{\exp\{-\lambda_2\}(\lambda_2/2)^k}{k!} P(\beta_{0,\delta+2k} \leq x),
$$

which completes the proof. □

Inspecting Eq. (13.3.27), we remark that the first term can be interpreted as $P(\beta_{0,\delta}(\lambda_1, \lambda_2) = 0)$ and the double sum as the probability $P(0 < \beta_{0,\delta}(\lambda_1, \lambda_2) \leq x)$ for all $x \in (0, 1)$, $\delta > 0$ and $\lambda_1, \lambda_2 > 0$. Hence we can decompose the distribution of $\beta_{0,\delta}(\lambda_1, \lambda_2)$ into a discrete component placing mass $\exp\{-\lambda_1/2\}$ at zero and a continuous component describing the distribution over $(0, 1)$. Finally, setting $\lambda_2 = 0$ we obtain the distribution of a singly non-central beta random variable

$$
P\big(\beta_{\delta,0}(0, \lambda) \leq x\big) = \sum_{j=0}^{\infty} \frac{\exp\{-\lambda/2\}(\lambda/2)^j}{j!} P(\beta_{\delta,2j} \leq x), \tag{13.3.28}
$$

for all $x \in [0, 1)$, $\delta > 0$ and $\lambda > 0$. Finally, we present the extended versions of the Type II beta random variables,

$$
\beta_{\delta,0}(0, \lambda) := 1 - \beta_{0,\delta}(\lambda, 0) = \frac{\chi_\delta^2}{\chi_0^2(\lambda) + \chi_\delta^2}, \tag{13.3.29}
$$

for all $\delta > 0$ and $\lambda > 0$, and

$$
\beta_{\delta,0}(\lambda_2, \lambda_1) := 1 - \beta_{0,\delta}(\lambda_1, \lambda_2) = \frac{\chi_\delta^2(\lambda_2)}{\chi_0^2(\lambda_1) + \chi_\delta^2(\lambda_2)}, \tag{13.3.30}
$$

where $\delta > 0$ and $\lambda_1, \lambda_2 > 0$, whose values lie in $(0, 1]$. Equations (13.3.30), (13.3.26), and (13.3.18) yield

$$
\begin{aligned}
& P\big(\beta_{\delta,0}(\lambda_2,\lambda_1)\big) \\
&= 1 - P\big(\beta_{0,\delta}(\lambda_1,\lambda_2) < 1-x\big) \\
&= 1 - \sum_{j=0}^{\infty} \frac{\exp\{-\lambda_1/2\}(\lambda_1/2)^j}{j!} \sum_{k=0}^{\infty} \frac{\exp\{-\lambda_2/2\}(\lambda_2/2)^k}{k!} P(\beta_{2j,\delta+2k} < 1-x) \\
&= \sum_{j=0}^{\infty} \frac{\exp\{-\lambda_1/2\}(\lambda_1/2)^j}{j!} \sum_{k=0}^{\infty} \frac{\exp\{-\lambda_2/2\}(\lambda_2/2)^k}{k!} \big(1 - P(\beta_{2j,\delta+2k} < 1-x)\big) \\
&= \sum_{j=0}^{\infty} \frac{\exp\{-\lambda_1/2\}(\lambda_1/2)^j}{j!} \sum_{k=0}^{\infty} \frac{\exp\{-\lambda_2/2\}(\lambda_2/2)^k}{k!} P(\beta_{\delta+2k,2j} \leq x),
\end{aligned}
$$

for all $x \in (0, 1]$, $\delta > 0$ and $\lambda_1, \lambda_2 > 0$. We note that $\beta_{\delta+2k,0}$ is identically equal to one for all $k \in \{0, 1, 2, \ldots\}$. Hence $\beta_{\delta,0}(\lambda_2, \lambda_1)$ can be decomposed into a discrete component that places mass $\exp\{-\lambda_1/2\}$ at one and a continuous component taking values in $(0, 1)$ for all $\delta > 0$, $\lambda_1, \lambda_2 > 0$. Similarly, we obtain

$$
P\big(\beta_{\delta,0}(0,\lambda) \leq x\big) = \sum_{j=0}^{\infty} \frac{\exp\{-\lambda/2\}(\lambda/2)^j}{j!} P(\beta_{\delta,2j} \leq x),
$$

for all $x \in (0, 1]$, $\delta > 0$ and $\lambda > 0$. Again, $\beta_{\delta,0}$ is identically equal to one, hence $\beta_{\delta,0}(0, \lambda)$ can be decomposed into a discrete part placing mass $\exp\{-\lambda\}$ at one and a continuous component assuming values in $(0, 1)$.

13.4 Computing the Doubly Non-central Beta Distribution

In this section, we present an algorithm, which shows how to implement the doubly non-central beta distribution. The algorithm is based on Hulley (2009), where an idea from Posten (1989, 1993) is used to enhance an algorithm presented by Seber (1963) for computing the distribution function of standard singly non-central beta random variables.

We define the doubly non-central regularized incomplete beta function

$$
I_z(a,b,c,d) := \sum_{j=0}^{\infty} \frac{\exp\{-c\}c^j}{j!} \sum_{k=0}^{\infty} \frac{\exp\{-d\}d^k}{k!} I_z(a+j, b+k), \qquad (13.4.31)
$$

for all $z \in [0, 1]$ and $a, b, c, d \geq 0$, such that either $a > 0$ or $b > 0$ and where $I_z(a, b)$ is given by Eq. (13.3.20). We formally set

$$
I_z(0,b) := 1 \quad \text{and} \quad I_z(a,0) := \begin{cases} 0 & \text{if } z < 1; \\ 1 & \text{if } z = 1, \end{cases} \qquad (13.4.32)
$$

for all $z \in [0, 1]$ and $a, b > 0$. This is necessary, as the gamma functions in Eq. (13.3.20) are not well-defined at zero. We note that we can express the distribution functions of both the central and the non-central beta distributions in terms of the doubly non-central regularized incomplete beta function in Eq. (13.4.31). In particular, we have

$$P(\beta_{\delta_1,\delta_2} \le x) = I_x\left(\frac{\delta_1}{2}, \frac{\delta_2}{2}, 0, 0\right),$$

for all $x \in (0, 1)$ and $\delta_1, \delta_2 > 0$. The distribution of the Type I singly non-central beta distribution satisfies

$$P\big(\beta_{\delta_1,\delta_2}(\lambda, 0) \le x\big) = I_x\left(\frac{\delta_1}{2}, \frac{\delta_2}{2}, \frac{\lambda}{2}, 0\right),$$

for all $x \in (0, 1)$ (respectively $x \in [0, 1)$), $\delta_1 > 0$ (respectively $\delta_1 = 0$), $\delta_2 > 0$ and $\lambda > 0$, while for the Type II singly non-central beta distribution we obtain

$$P\big(\beta_{\delta_2,\delta_1}(0, \lambda) \le x\big) = I_x\left(\frac{\delta_2}{2}, \frac{\delta_1}{2}, 0, \frac{\lambda}{2}\right)$$

for all $x \in (0, 1)$ (respectively $x \in (0, 1]$), $\delta_1 > 0$ (respectively $\delta_1 = 0$), $\delta_2 > 0$ and $\lambda > 0$. Lastly, the distribution function of the doubly non-central beta distribution satisfies the equality

$$P\big(\beta_{\delta_1,\delta_2}(\lambda_1, \lambda_2) \le x\big) = I_x\left(\frac{\delta_1}{2}, \frac{\delta_2}{2}, \frac{\lambda_1}{2}, \frac{\lambda_2}{2}\right),$$

for all $x \in (0, 1)$ (respectively $x \in [0, 1)$; $x \in (0, 1]$), $\delta_1, \delta_2 > 0$ (respectively $\delta_1 = 0$, $\delta_2 > 0$; $\delta_1 > 0$, $\delta_2 = 0$) and $\lambda_1, \lambda_2 > 0$. We assume that one of the following parameter combinations is in force:

(i) $z \in (0, 1)$, $a, b \in \mathcal{N}$, $c, d > 0$;
(ii) $z \in [0, 1)$, $a = 0$, $b \in \mathcal{N}$, $c, d > 0$;
(iii) $z \in (0, 1]$, $a \in \mathcal{N}$, $b = 0$, $c, d > 0$.

Assuming condition (i) is in force, we obtain from Seber (1963)

$$\begin{aligned} I_z(a, b, c, d) &= \exp\big\{-c(1-z)\big\} z^a \sum_{k=0}^{\infty} \frac{\exp\{-d\} d^k}{k!} \sum_{n=0}^{b+k-1} (1-z)^n L_n^{(a-1)}(-cz) \\ &= \exp\big\{-c(1-z)\big\} z^a \sum_{k=0}^{\infty} P_k T_k, \end{aligned} \tag{13.4.33}$$

where we have defined the Poisson weights

$$P_k := \frac{\exp\{-d\} d^k}{k!}, \quad k \in \{0, 1, 2, \ldots\},$$

and

$$T_k := \sum_{n=0}^{b+k-1} (1-z)^n L_n^{(a-1)}(-cz), \quad k \in \{0, 1, 2, \ldots\}.$$

When condition (ii) is satisfied with $z = 0$, the problem is trivial, $I_0(0, b, c, d) = \exp\{-c\}$. But for condition (ii) and $z \in (0, 1)$, Eqs. (13.3.20) and (13.4.32) yield $I_z(j, b+k) = 1 - I_{1-z}(b+k, j)$, for each $j, k \in \{0, 1, 2, \ldots\}$, and hence

$$
\begin{aligned}
I_z(0,b,c,d) &= \sum_{j=0}^{\infty} \frac{\exp\{-c\}c^j}{j!} \sum_{k=0}^{\infty} \frac{\exp\{-d\}d^k}{k!} \big(1 - I_{1-z}(b+k,j)\big) \\
&= 1 - \sum_{j=1}^{\infty} \frac{\exp\{-c\}c^j}{j!} \sum_{k=0}^{\infty} \frac{\exp\{-d\}d^k}{k!} I_{1-z}(b+k,j) \\
&= 1 - \exp\{-dz\}(1-z)^b \sum_{j=1}^{\infty} \frac{\exp\{-c\}c^j}{j!} \sum_{n=0}^{j-1} z^n L_n^{(b-1)}\big(-d(1-z)\big) \\
&= 1 - \exp\{-dz\}(1-z)^b \sum_{j=1}^{\infty} P_j \sum_{n=0}^{j-1} T_j,
\end{aligned} \tag{13.4.34}
$$

from (13.4.32) and Seber (1963). Finally, if the arguments satisfy condition (iii) with $z = 1$, then the problem is again trivial since $I_1(a, 0, c, d) = 1$. On the other hand, if the arguments satisfy condition (iii) with $z \in (0, 1)$, then applying (13.4.32) and Seber (1963) yields

$$
\begin{aligned}
I_z(a,0,c,d) &= \sum_{j=0}^{\infty} \frac{\exp\{-c\}c^j}{j!} \sum_{k=1}^{\infty} \frac{\exp\{-d\}d^k}{k!} I_Z(a+j,k) \\
&= \exp\big\{-c(1-z)\big\} z^a \sum_{k=1}^{\infty} \frac{\exp\{-d\}d^k}{k!} \sum_{n=0}^{k-1} (1-z)^n L_n^{(a-1)}(-cz) \\
&= \exp\big\{-c(1-z)\big\} z^a \sum_{k=1}^{\infty} P_k \sum_{n=0}^{k-1} T_k.
\end{aligned} \tag{13.4.35}
$$

We remark that by $L_n^{(\alpha)}$ we denote the Laguerre polynomials, which are defined for $n \in \{0, 1, 2, \ldots\}$, $\alpha \in \Re \setminus \{-1, -2, \ldots\}$. However, for $\alpha \in \{0, 1, 2, \ldots\}$ we have

$$
L_n^{(\alpha)}(\zeta) = \sum_{m=0}^{n} \binom{n+\alpha}{n-m} \frac{\zeta^m}{m!}, \tag{13.4.36}
$$

for all $\zeta \in \Re$ and each $n \in \{0, 1, 2, \ldots\}$. Equation (13.4.36) implies the following recurrence relation, see also Abramowitz and Stegun (1972), Chap. 22,

$$
\begin{aligned}
L_0^{(\alpha)}(\zeta) &= 1, \\
L_1^{(\alpha)}(\zeta) &= \alpha - 1 + \zeta, \\
nL_n^{(\alpha)}(\zeta) &= (2n + \alpha - 1 - \zeta)L_{n-1}^{(\alpha)}(\zeta) - (n + \alpha - 1)L_{n-2}^{(\alpha)}(\zeta),
\end{aligned} \tag{13.4.37}
$$

for all $\zeta \in \Re$, and for all $\alpha \in \{0, 1, 2, \ldots\}$ and $n \in \{2, 3, \ldots\}$. Comparing Eqs. (13.4.33), (13.4.34), and (13.4.35), we note that it suffices to focus on condition (i), as conditions (ii) and (iii) can be covered using the same algorithm. Regarding the outer, infinite sum, we employ an idea from Posten (1989, 1993), and sum the terms in decreasing order of the Poisson weights. The maximal Poisson weight is approximately attained by the index value $k^* = \lceil d \rceil$, hence we truncate the outer sum to the range of index values $(k^* - N_\epsilon)^+, \ldots, k^* + N_\epsilon$, where N_ϵ is given by

$$N_\epsilon := \min\left\{ N \in \{0, 1, 2, \ldots\} \;\middle|\; \sum_{k=(k^*-N)^+}^{k^*+N} P_k > 1 - \epsilon \right\}, \tag{13.4.38}$$

i.e. we approximate $I_z(a, b, c, d)$ using $\tilde{I}_z(a, b, c, d)$, which is given by

$$\tilde{I}_{N_\epsilon,z}(a, b, c, d) := \exp\{-c(1-z)\} z^a \sum_{k=(k^*-N_\epsilon)^+}^{k^*+N_\epsilon} P_k T_k. \tag{13.4.39}$$

We now aim to produce a good bound on the approximation error

$$I_z(a, b, c, d) - \tilde{I}_{N_\epsilon,z}(a, b, c, d). \tag{13.4.40}$$

From Seber (1963) we have

$$\exp\{-c(1-z)\} z^a T_k = I_z(a, b + k, c, 0) \in (0, 1), \tag{13.4.41}$$

hence

$$\begin{aligned} & I_z(a, b, c, d) - \tilde{I}_{N_\epsilon,z}(a, b, c, d) \\ & = \exp\{-c(1-z)\} z^a \left(\sum_{k=0}^{(k^*-N_\epsilon)^+-1} P_k T_k + \sum_{k=k^*+N_\epsilon+1}^{\infty} P_k T_k \right) \\ & \leq \sum_{k=0}^{(k^*-N_\epsilon)^+-1} P_k + \sum_{k=k^*+N_\epsilon+1}^{\infty} P_k = 1 - \sum_{k=(k^*-N_\epsilon)}^{k^*+N_\epsilon} P_k < \epsilon, \end{aligned}$$

where we used the fact that the Poisson weights sum to one and the last inequality follows from the definition of N_ϵ. Hence we have

$$\tilde{I}_{N_\epsilon,z}(a, b, c, d) \in \big(I_z(a, b, c, d) - \epsilon, I_z(a, b, c, d) \big),$$

so the truncation error is bounded by ϵ. Clearly, the value of N_ϵ cannot be determined explicitly in advance, but can only be determined by iteratively adding Poisson weights until their sum exceeds $1 - \epsilon$. The P_k satisfy

$$P_k = P_{k-1} \frac{d}{k},$$

for each $k \in \mathcal{N}$, which allows for a rapid computation of these weights. Finally, we attend to the inner sum in Eq. (13.4.33). In Algorithm 13.2 below, we make use of a list, which stores the values of the Laguerre polynomials, which are used to compute the T_k. Firstly, we calculate the Laguerre polynomials needed to compute T_{k^*}, and store them in a list. Thereafter, we use the following iterative scheme, based on (13.4.37) to compute $T_{k^*+1}, T_{k^*-1}, T_{k^*+2}, T_{k^*-2}, \ldots$:

Algorithm 13.2 Doubly non-central regularized incomplete beta function

```
Require: a, b ∈ 𝒩, c, d ∈ (0, ∞), z ∈ (0, 1) and ε ∈ (0, 1) or
    a = 0, b ∈ 𝒩, c, d ∈ (0, ∞), z ∈ [0, 1) and ε ∈ (0, 1) or
    a ∈ 𝒩, b = 0, c, d ∈ (0, ∞), z ∈ (0, 1] and ε ∈ (0, 1)
 1: switch (z)
 2: case z = 0:
 3:     value ← exp{−c}
 4: case z = 1:
 5:   value ← 1
 6: otherwise:
 7:    if a = 0 then
 8:       a ↔ b
 9:       c ↔ d
10:       z ↔ 1 − z
11:       swapflag ← true
12:    else
13:       swapflag ← false
14:    end if
15:    kmin ← kmax ← ⌊d⌉ ∨ 2
16:    cumPoiss ← exp{−d}d^kmax / kmax!
17:    for n = 0 : b + kmax − 1 do
18:       switch (n)
19:       case n = 0:
20:          Laglist ← ⟨1⟩
21:       case n = 1:
22:          Laglist ← Laglist ⊎ a + cz
23:       otherwise:
24:          Laglist ← Laglist ⊎ ((2n+a−2+cz)×Laglist⟨⟨end⟩⟩−(n+a−2)×Laglist⟨⟨end−1⟩⟩) / n
25:       end switch
26:       Tmax ← Tmax + (1 − z)^n × Laglist⟨⟨end⟩⟩
27:    end for
28:    Tmin ← Tmax
29:    value ← Pmax × Tmax
30:    if b = 0 then
31:       errthrshld ← 1 − exp{−d} − ε
32:    else
33:       errthrshld ← 1 − ε
34:    end if
35:    while cumPoiss ≤ errthrshld do
36:       kmax ← kmax + 1
37:       Pmax ← Pmax × d/kmax
38:       Laglist ← Laglist ⊎ ((a+2b+2kmax−4+cz)×Laglist⟨⟨end⟩⟩−(a+b+kmax−3)×Laglist⟨⟨end−2⟩⟩) / (b+kmax−1)
39:       Tmax ← Tmax + (1 − z)^(b+kmax−1) × Laglist⟨⟨end⟩⟩
40:       value ← value + Pmax × Tmax
41:       cumPoiss ← cumPoiss + Pmax
42:       if kmin ≥ 2 or kmin ≥ 1 and b > 0 then
43:          Pmin ← Pmin × kmin/d
44:          Tmin ← Tmin − (1 − z)^(b+kmin−1) × Laglist⟨⟨b + kmin⟩⟩
45:          value ← value + Pmin × Tmin
46:          cumPoiss ← cumPoiss + Pmin
47:          kmin ← kmin − 1
48:       end if
49:    end while
50:    if swapflag then
51:       value ← 1 − exp{−c(1 − z)}z^a × value
52:    else
53:       value ← exp{−c(1 − z)}z^a × value
54:    end if
55: end switch
56: return value
```

$$T_k = T_{k-1} + L_{b+k-1}^{(a-1)}(-cz)$$
$$= \begin{cases} T_{k-1} + 1 & \text{if } k = 1 - b \\ T_{k-1} + (1-z)(a-2-cz) & \text{if } k = 2 - b \\ T_{k-1} + \frac{(1-z)^{b+k-1}}{b+k-1}(A_k L_{b+k-2}^{(a-1)}(-cz) - B_k L_{b+k-3}^{(a-1)}(-cz)) & \text{if } k \geq 3 - b, \end{cases}$$

for each $k \in \mathcal{N}$, where $A_k := a + 2b + 2k - 4 + cz$ and $B_k := a + b + l - 3$. We present in Algorithm 13.2 below the algorithm, which shows how to compute the doubly non-central regularized incomplete gamma function, which is given in Hulley (2009). The term Laglist denotes the list of Laguerre polynomials, and list $\langle\langle i \rangle\rangle$ references element i of list, and by the symbol list $\uplus x$ we mean that the value x is appended to list.

13.5 Inverting Laplace Transforms

In this section, we discuss how to compute values of a function $f : \Re^+ \to \Re$ from its Laplace transform

$$\hat{f}(s) = \int_0^\infty \exp\{-st\} f(t)\, dt,$$

where s is a complex variable with a nonnegative real part. We present the *Euler method* from Abate and Whitt (1995), which is based on the Bromwich contour inversion integral. We let this contour be any vertical line $s = a$ so that $\hat{f}(s)$ has no singularities on or to the right of it, and hence obtain, as in Abate and Whitt (1995),

$$f(t) = \frac{2\exp\{at\}}{\pi} \int_0^\infty Re\big(\hat{f}(a + \imath u)\big) \cos(ut)\, du.$$

The integral is evaluated numerically using the trapezoidal rule. Specifying the step size as h gives

$$f(t) \approx f_h(t) := \frac{h\exp\{at\}}{\pi} Re\big(\hat{f}(a)\big) + \frac{2h\exp\{at\}}{\pi} \sum_{k=1}^\infty Re\big(\hat{f}(a + \imath kh)\big) \cos(kht).$$

Setting $h = \frac{\pi}{2t}$ and $a = \frac{A}{2t}$, one arrives at the nearly alternating series

$$f_h(t) = \frac{\exp\{A/2\}}{2t} Re\left(\hat{f}\left(\frac{A}{2t}\right)\right) + \frac{\exp\{A/2\}}{t} \sum_{k=1}^\infty (-1)^k Re\left(\hat{f}\left(\frac{A + 2k\pi\imath}{2t}\right)\right). \tag{13.5.42}$$

Regarding the parameters, we need to know how to choose A: In Abate and Whitt (1995) it is shown that to achieve a discretization error $10^{-\gamma}$, we should set $A = \gamma \log 10$. Consequently, truncating the series after n terms, we have

$$s_n(t) = \frac{\exp\{A/2\}}{2t} Re\left(\hat{f}\left(\frac{A}{2t}\right)\right) + \frac{\exp\{A/2\}}{t} \sum_{k=1}^n (-1)^k a_k(t), \tag{13.5.43}$$

where

$$a_k(t) = Re\left(\hat{f}\left(\frac{A + 2k\pi\imath}{2t}\right)\right).$$

Lastly, we apply the Euler summation, which explains the name of the algorithm. In particular, we apply the Euler summation to m terms after the initial n terms, so that the Euler summation, which approximates (13.5.42), is given by

$$E(m, n, t) = \sum_{k=0}^{m} \binom{m}{k} 2^{-m} s_{n+k}(t), \tag{13.5.44}$$

where $s_n(t)$ is given by (13.5.43).We note that $E(m, n, t)$ is the weighted average of the last m partial sums by a binomial probability distribution characterized by parameters m and $p = \frac{1}{2}$. In Abate and Whitt (1995), the parameters $m = 11$ and $n = 15$ are used, and it is suggested to increase n as necessary. In the following subsection, we illustrate how to use this algorithm to recover a bivariate probability density function. Using Lie symmetry methods, the first inversion can be performed analytically, for the second we use the Euler method presented in this section given by (13.5.44).

13.5.1 Recovering the Joint Distribution to Price Realized Variance

In this subsection, we apply the methodology discussed in this section to the pricing of realized variance derivatives, in particular, options on volatility, see Sect. 8.5.2. To price such products, we need to recover the joint distribution of $(Y_T, \int_0^T \frac{1}{Y_t} dt)$. At first sight, obtaining the joint distribution should entail the inversion of a double Laplace transform. However, since Lie symmetry methods provide us with fundamental solutions, we already have the inversion with respect to one of the variables. Consequently, one only needs to invert a one-dimensional Laplace transform numerically, to obtain the joint density over $\Re^+ \times \Re^+$. We subsequently map the joint density into $[0, 1]^2$, following the discussion in Kuo et al. (2008), and hence can employ a randomized quasi-Monte Carlo point set to compute prices. Assuming that the one-dimensional Laplace transform can be inverted at a constant computational complexity, the resulting computational complexity is $O(N)$, where N is the number of two-dimensional quasi-Monte Carlo points employed.

We numerically invert the one-dimensional Laplace transform given in (5.4.16) using the Euler method from Abate and Whitt (1995), which was also employed in Hulley and Platen (2008), see also Craddock et al. (2000). We display the joint density in Fig. 13.5.1.

Inverting the Laplace transform produces the joint density of $(Y_T, \int_0^T \frac{1}{Y_t} dt)$ over $\Re^+ \times \Re^+$. One could now employ a product rule, such as the tensor product of two one-dimensional trapezoidal rules, using N points for each co-ordinate, and perform the numerical integration using N^2 points, at a computational complexity of $O(N^2)$,

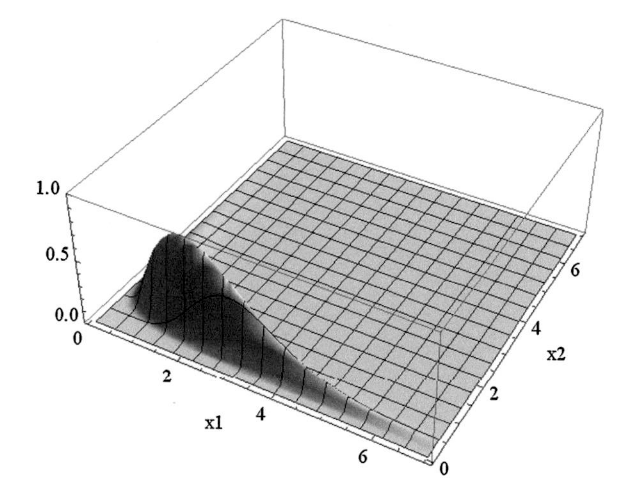

Fig. 13.5.1 Joint density of $(Y_T, \int_0^T \frac{1}{Y_t}\,dt)$ for $Y_0 = 1$, $\eta = 0.052$, $T = 1$

assuming the Laplace inversion can be performed in constant time. However, instead we map the joint distribution into the unit square, and employ an N point quasi-Monte Carlo rule to obtain a quadrature rule whose computational complexity is only $O(N)$, see Sect. 12.2.

Chapter 14
Credit Risk Under the Benchmark Approach

In this chapter, we discuss how to model credit risk under the benchmark approach. We employ the techniques from Sect. 12.3 in Filipović (2009), and we show how under the benchmark approach, the Laplace transforms derived in this book can be incorporated in this framework. The structure of this chapter is as follows: firstly we introduce an affine credit risk model in Sect. 14.1. This model satisfies the key assumptions of Sect. 12.3 in Filipović (2009). Hence the results presented there apply, which we recall for the convenience of the reader. Consequently, in Sect. 14.2, we show how to price *credit default swaps* (CDSs) and introduce *credit valuation adjustment* (CVA) in Sect. 14.3 as an extension of CDSs. In particular, our model can capture *right-way*—and *wrong-way exposure*. This means, we capture the dependence structure of the default event and the value of the transaction under consideration. For simple contracts, we provide closed-form solutions, however, due to the fact that we allow for a dependence between the default event and the value of the transaction, closed-form solutions are difficult to obtain in general. Hence we conclude this chapter with a reduced form model, which is more tractable than the model from Sect. 14.1.

14.1 An Affine Credit Risk Model

In this section, we aim to introduce a realistic, yet tractable model for credit risk. In particular, our model allows for a stochastic interest rate, and a stochastic default intensity, both of which are correlated with the GOP. Mathematically, the model is based on the approach in Sect. 12.3 in Filipović (2009). We point out that our model satisfies the assumptions (**D1**) and (**D2**), see pages 230 and 233 in Filipović (2009), and hence we can employ the results presented in this reference. For further technical background, we refer the reader to this reference.

We fix a probability space $(\Omega, \mathcal{A}, P)$, where P denotes the real world probability measure. Next, we present a model for the evolution of financial information. We remark that in our model, only having access to market information is not sufficient

J. Baldeaux, E. Platen, *Functionals of Multidimensional Diffusions with Applications to Finance*, Bocconi & Springer Series 5, DOI 10.1007/978-3-319-00747-2_14,

to decide whether or not default has occurred or not. We now present this model, which is a doubly stochastic intensity based model. We introduce a filtration $\underline{\mathcal{G}} = (\mathcal{G}_t)_{t\geq 0}$, satisfying the usual conditions and set

$$\mathcal{G}_\infty = \sigma\{\mathcal{G}_t,\, t \geq 0\} \subset \mathcal{A},$$

and a nonnegative $\underline{\mathcal{G}}$-progressively measurable process $\lambda = \{\lambda_t,\, t \geq 0\}$ with the property

$$\int_0^t \lambda_s \, ds < \infty, \qquad P\text{-a.s. for all } t \geq 0.$$

Next, we fix an exponential random variable ϕ with intensity parameter 1, independent of $\mathcal{G}_\infty$, and we define the random time

$$\tau := \inf\left\{t \colon \int_0^t \lambda_s \, ds \geq \phi\right\}$$

assuming values in $(0, \infty]$. From the independence property of ϕ and $\mathcal{G}_\infty$, we have that

$$P(\tau > t \mid \mathcal{G}_\infty) = P\left(\phi > \int_0^t \lambda_s \, ds \,\middle|\, \mathcal{G}_\infty\right) = \exp\left\{-\int_0^t \lambda_s \, ds\right\}. \tag{14.1.1}$$

Lastly, we condition both sides in the preceding equation on $\mathcal{G}_t$ and obtain

$$P(\tau > t \mid \mathcal{G}_t) = \exp\left\{-\int_0^t \lambda_s \, ds\right\}. \tag{14.1.2}$$

Equations (14.1.2) and (14.1.1) are consistent with the assumptions (**D1**) and (**D2**) in Filipović (2009), which are hence satisfied in our model. Next, we set

$$H_t = \mathbf{1}_{\{\tau \leq t\}}$$

and $\mathcal{H}_t = \sigma\{H_s,\, s \leq t\}$ and set $\mathcal{A}_t = \mathcal{G}_t \vee \mathcal{H}_t$, the smallest σ-algebra containing $\mathcal{G}_t$ and $\mathcal{H}_t$. We remark that the inclusion $\mathcal{G}_t \subset \mathcal{A}_t$ is strict, having access to $\mathcal{G}_t$ does not allow us to decide whether default has occurred by t, i.e. the event $\{\tau \leq t\}$ is not included in $\mathcal{G}_t$, so τ is not a $\underline{\mathcal{G}}$-stopping time. We find this realistic, since it means that only by observing financial data such as stock prices and interest rates, one cannot determine whether default has occurred or not, as additional, non-financial factors, can be assumed to be relevant to this decision, too. The following lemma is Lemma 12.1 in Filipović (2009).

Lemma 14.1.1 *Let $t \geq 0$. Then for every $A \in \mathcal{A}_t$, there exists a $B \in \mathcal{G}_t$ such that*

$$A \cap \{\tau > t\} = B \cap \{\tau > t\}.$$

We have the following corollary to Lemma 14.1.1, the proof of which is analogous to the proof of Lemma 12.1 in Filipović (2009).

Corollary 14.1.2 *Let $t \geq 0$. Then for every $A \in \mathcal{A}_t$, there exists a $B \in \mathcal{G}_t$ such that*

$$A \cap \{\tau \leq t\} = B \cap \{\tau \leq t\}. \tag{14.1.3}$$

The first part of the following lemma is Lemma 12.2 in Filipović (2009), the second part of the next lemma forms part of Lemma 12.5 in Filipović (2009).

Lemma 14.1.3 *Let Y be a nonnegative random variable and λ and τ be as defined above. Then*

$$E(\mathbf{1}_{\{\tau>t\}}Y \mid \mathcal{A}_t) = \mathbf{1}_{\{\tau>t\}} \exp\left\{\int_0^t \lambda_s\, ds\right\} E(\mathbf{1}_{\{\tau>t\}}Y \mid \mathcal{G}_t),$$

for all $t \geq 0$. If Y is also $\mathcal{G}_\infty$ measurable, then we have

$$E(\mathbf{1}_{\{\tau\leq t\}}Y \mid \mathcal{A}_t) = \mathbf{1}_{\{\tau\leq t\}} E(Y \mid \mathcal{G}_t).$$

Proof The first part of the lemma is proven in Filipović (2009), see the proof of Lemma 12.2. For the second part, let $A \in \mathcal{A}_t$, and note that by Corollary 14.1.2, there exists a $B \in \mathcal{G}_t$ with property (14.1.3). We now use the definition of conditional expectation, the fact that $\mathbf{1}_{\{\tau\leq t\}}\mathbf{1}_A = \mathbf{1}_{\{\tau\leq t\}}\mathbf{1}_B$, that $Y \in \mathcal{G}_\infty$ and that $P(\tau \leq t \mid \mathcal{G}_\infty) = P(\tau \leq t \mid \mathcal{G}_t)$, which follows from Eqs. (14.1.2) and (14.1.1):

$$\begin{aligned}
\int_A \mathbf{1}_{\{\tau\leq t\}} Y\, dP &= \int_B \mathbf{1}_{\{\tau\leq t\}} Y\, dP \\
&= \int_B E(\mathbf{1}_{\{\tau\leq t\}}Y \mid \mathcal{G}_t)\, dP \\
&= \int_B E\big(E(\mathbf{1}_{\{\tau\leq t\}}Y \mid \mathcal{G}_\infty) \bigm| \mathcal{G}_t\big)\, dP \\
&= \int_B E\big(Y E(\mathbf{1}_{\{\tau\leq t\}} \mid \mathcal{G}_\infty) \bigm| \mathcal{G}_t\big)\, dP \\
&= \int_B E(Y \mid \mathcal{G}_t) E(\mathbf{1}_{\{\tau\leq t\}} \mid \mathcal{G}_t)\, dP \\
&= \int_B E\big(\mathbf{1}_{\{\tau\leq t\}} E(Y \mid \mathcal{G}_t) \bigm| \mathcal{G}_t\big)\, dP \\
&= \int_B \mathbf{1}_{\{\tau\leq t\}} E(Y \mid \mathcal{G}_t)\, dP \\
&= \int_A \mathbf{1}_{\{\tau\leq t\}} E(Y \mid \mathcal{G}_t)\, dP.
\end{aligned}$$

Hence we have

$$E(\mathbf{1}_{\{\tau\leq t\}}Y \mid \mathcal{A}_t) = E\big(\mathbf{1}_{\{\tau\leq t\}} E(Y \mid \mathcal{G}_t) \bigm| \mathcal{A}_t\big) = \mathbf{1}_{\{\tau\leq t\}} E(Y \mid \mathcal{G}_t). \qquad \square$$

The following formula is useful, when considering claims which are independent of default risk. It is an immediate corollary to Lemma 14.1.3.

Corollary 14.1.4 *Let Y be a nonnegative random variable which is $\mathcal{G}_\infty$ measurable. Then*

$$E(Y \mid \mathcal{A}_t) = E(Y \mid \mathcal{G}_t).$$

We now present our specific model, which is based on affine processes. Firstly, we define the square-root process $Y = \{Y_t,\ t \geq 0\}$, given by

$$dY_t = (1 - \eta Y_t)\,dt + \sqrt{Y_t}\,dW_t^1,$$

where W^1 is a $\underline{\mathcal{G}}$-Brownian motion and we define the deterministic time-change

$$\alpha_t = \alpha_0 \exp\{\eta t\},$$

see also Sect. 3.3, and we model the discounted GOP as

$$\bar{S}_t^{\delta_*} = \alpha_t Y_t.$$

We now describe the short-rate using the stochastic process

$$r_t = a_t^r + b^r Z_t^1 + c^r f^r(Y_t), \tag{14.1.4}$$

where $a_\cdot^r$ is a nonnegative deterministic function of time and b^r and c^r are nonnegative constants and $f^r(x) = x$ or $f^r(x) = \frac{1}{x}$. The process $Z^1 = \{Z_t^1,\ t \geq 0\}$ is a square-root process given by

$$dZ_t^1 = \kappa^1\big(\theta^1 - Z_t^1\big)\,dt + \sigma^1\sqrt{Z_t^1}\,dW_t^2, \tag{14.1.5}$$

where $\kappa^1, \theta^1, \sigma^1 > 0$ and $2\kappa^1\theta^1 > (\sigma^1)^2$, where W^2 is an independent $\underline{\mathcal{G}}$-Brownian motion. We now introduce the GOP, which is given by

$$S_t^{\delta^*} = B_t \bar{S}_t^{\delta_*}, \tag{14.1.6}$$

where $B_t = \exp\{\int_0^t r_s\,ds\}$. Furthermore, by setting $f^r(x) = x$ or $f^r(x) = \frac{1}{x}$ respectively, we retain analytical tractability via Propositions 7.3.8 and 7.3.9. Finally, we introduce a model for the stochastic intensity

$$\lambda_t = a_t^\lambda + b^\lambda Z_t^1 + c^\lambda f^r(Y_t) + d^\lambda Z_t^2, \tag{14.1.7}$$

where $\kappa^2, \theta^2, \sigma^2 > 0$, $a_\cdot^\lambda$ is a nonnegative function of time. The constants b^λ, c^λ, and d^λ are nonnegative, and $Z^2 = \{Z_t^2,\ t \geq 0\}$ is a square-root process:

$$dZ_t^2 = \kappa^2\big(\theta^2 - Z_t^2\big)\,dt + \sigma^2\sqrt{Z_t^2}\,dW_t^3,$$

where $2\kappa^2\theta^2 > (\sigma^2)^2$, and W^3 is an independent $\underline{\mathcal{G}}$-Brownian motion. We conclude that λ, r, and S^{δ^*} are dependent, as they share some of their respective stochastic drivers.

We conclude this section with presenting pricing formulas for some standard claims, namely zero coupon bonds and European put options on the GOP. In Sect. 14.3, we will study these products in the presence of CVA. We remark that the affine nature of our model and the Laplace transforms derived using Lie symmetry analysis allow us to obtain these option pricing formulas. Regarding the zero coupon bond with maturity $T > 0$ at time $t \in [0, T]$, we have from the real world pricing formula (1.3.19)

$$\begin{aligned}
P_T(t) &= S_t^{\delta_*} E\left(\frac{1}{S_T^{\delta_*}} \,\Big|\, \mathcal{A}_t\right) \\
&= S_t^{\delta_*} E\left(\frac{1}{S_T^{\delta_*}} \,\Big|\, \mathcal{G}_t\right) \\
&= \frac{\alpha_t}{\alpha_T} Y_t E\left(\frac{1}{Y_T} \exp\left\{-\int_t^T a_s^r\,ds - b^r \int_t^T Z_s^1\,ds - c^r \int_t^T f(Y_s)\,ds\right\} \,\Big|\, \mathcal{G}_t\right) \\
&= \frac{\alpha_t}{\alpha_T} Y_t \exp\left\{-\int_t^T a_s^r\,ds\right\} E\left(\frac{\exp\{-c^r \int_t^T f^r(Y_s)\,ds\}}{Y_T} \,\Big|\, \mathcal{G}_t\right) \\
&\quad \times E\left(\exp\left\{-b^r \int_t^T Z_s^1\,ds\right\} \,\Big|\, \mathcal{G}_t\right).
\end{aligned}$$

We remark that the expectations

$$E\left(\frac{\exp\{-c^r \int_t^T f^r(Y_s)\,ds\}}{Y_T} \,\Big|\, \mathcal{G}_t\right)$$

and

$$E\left(\exp\left\{-b^r \int_t^T Z_s^1\,ds\right\} \,\Big|\, \mathcal{G}_t\right)$$

can be computed using Propositions 7.3.8 and 7.3.9.

Having introduced zero coupon bonds, we now attend to *swaps*, in particular, we consider a *fixed-for-floating forward starting swap settled in arrears*. We fix a finite collection of future dates T_j, $j = 0, \dots, n$, $T_0 \geq 0$, and $T_j - T_{j-1} =: \delta_j > 0$, $j = 1, \dots, n$. The floating rate $L(T_j, T_{j+1})$ received at time T_{j+1} is set at time T_j by reference to a zero coupon bond for the time period $[T_j, T_{j+1})$, in particular,

$$P_{T_{j+1}}^{-1}(T_j) = 1 + \delta_{j+1} L(T_j, T_{j+1}). \tag{14.1.8}$$

The interest rate $L(T_j, T_{j+1})$ is the spot LIBOR that prevails at time T_j for the period of length δ_{j+1}. A long position in a payer swap entitles the investor to receive floating payments in exchange for fixed payments, so the cash flow at time T_j is $L(T_{j-1}, T_j)\delta_j - \kappa\delta_j$. The dates $T_0, \dots, T_{n-1}$ are known as *reset dates*, whereas the dates $T_1, \dots, T_n$ are known as *settlement dates*. The first reset date T_0 is known as the *start date* of the swap. For $t \leq T_0$, the real world pricing formula (1.3.19) gives the following value for a swap:

$$FS_{\kappa,T_0}(t) := E\left(\sum_{j=1}^n \frac{S_t^{\delta_*}}{S_{T_j}^{\delta_*}} \big(L(T_{j-1}, T_j) - \kappa\big)\delta_j \,\Big|\, \mathcal{A}_t\right). \tag{14.1.9}$$

We now show how to rewrite the value of a swap as the difference of a zero coupon bond and a coupon bearing bond. From Eq. (14.1.9), we obtain

$$FS_{\kappa,T_0}(t) = \sum_{j=1}^n E\left(\frac{S_t^{\delta_*}}{S_{T_j}^{\delta_*}} \left(\frac{1}{P_{T_j}(T_{j-1})} - (1 + \kappa\delta_j)\right) \,\Big|\, \mathcal{A}_t\right). \tag{14.1.10}$$

Focusing on the computation of a single term in this sum we obtain

$$\begin{aligned}
&E\left(\frac{S_t^{\delta_*}}{S_{T_j}^{\delta_*}}\left(\frac{1}{P_{T_j}(T_{j-1})}-(1+\kappa\delta_j)\right)\Bigg|\mathcal{A}_t\right)\\
&=E\left(\frac{S_t^{\delta_*}}{S_{T_j}^{\delta_*}P_{T_j}(T_{j-1})}\Bigg|\mathcal{A}_t\right)-(1+\kappa\delta_j)E\left(\frac{S_t^{\delta_*}}{S_{T_j}^{\delta_*}}\Bigg|\mathcal{A}_t\right)\\
&=E\left(\frac{S_t^{\delta_*}}{S_{T_{j-1}}^{\delta_*}}\frac{1}{P_{T_j}(T_{j-1})}E\left(\frac{S_{T_{j-1}}^{\delta_*}}{S_{T_j}^{\delta_*}}\Bigg|\mathcal{A}_{T_{j-1}}\right)\Bigg|\mathcal{A}_t\right)-(1+\kappa\delta_j)P_{T_j}(t)\\
&=P_{T_{j-1}}(t)-(1+\kappa\delta_j)P_{T_j}(t).
\end{aligned}\tag{14.1.11}$$

Substituting Eq. (14.1.11) into (14.1.10), we obtain

$$FS_{\kappa,T_0}(t)=P_{T_0}(t)-\sum_{j=1}^{n}c_jP_{T_j}(t),\tag{14.1.12}$$

where $c_j=\kappa\delta_j$, $j=1,\dots,n-1$ and $c_n=1+\kappa\delta_n$. We remark that Eq. (14.1.12) is analogous to Eq. (13.2) in Musiela and Rutkowski (2005). In Sect. 14.3, we show that in the presence of default risk, even a simple linear product like a swap is in fact treated like an option a swap, or a *swaption*, which we now introduce.

The owner of an option on the above described swap with strike rate κ maturing at $T=T_0$ has the right to enter at time T the underlying fixed-for-floating forward starting swap settled in arrears. The real world pricing formula (1.3.19) yields the following price for such a contract:

$$PS_{\kappa,T_0}:=S_t^{\delta_*}E\left(\frac{(FS_{\kappa,T_0}(T_0))^+}{S_{T_0}^{\delta_*}}\Bigg|\mathcal{A}_t\right).\tag{14.1.13}$$

We remark that, as discussed in Sect. 13.1.2 in Musiela and Rutkowski (2005), it seems difficult to develop closed form solutions for swaptions. However, as we employ a tractable model, we can easily price swaptions via Monte Carlo methods: from Eq. (14.1.13), it is clear that in order to price the swaption, we need to have access to the joint distributions of $(Y_T,\int_t^T f^r(Y_s)\,ds)$ conditional on Y_t, and $(Z_T^1,\int_t^T Z_s^1\,ds)$ conditional on Z_t^1. These were derived in Sects. 6.3 and 6.4, which means that we can price swaptions using an exact Monte Carlo scheme.

For purposes of credit valuation adjustment (CVA), it is convenient to introduce a *forward start swaption*: here the expiry date T of the swaption precedes the initiation date T_0 of the swap, i.e. $T\le T_0$. The real world pricing formula (1.3.19) associates the following value with this contract:

$$PS_{\kappa,T_0,T}(t):=S_t^{\delta_*}E\left(\frac{(FS_{\kappa,T_0}(T))^+}{S_T^{\delta_*}}\Bigg|\mathcal{A}_t\right).$$

We will return to forward start swaptions when discussing CVA.

Finally, we show how to price a European put option, where we employ Lemma 8.3.2 and we explicitly emphasize the dependence on Z_t^1, Y_t and S_t, which will be relevant when discussing CVA. From Corollary 14.1.4, we get

$$\begin{aligned}
p_{T,K}\big(t, Z_t^1, Y_t, S_t\big) &= S_t^{\delta^*} E\bigg(\frac{(K - S_T^{\delta^*})^+}{S_T^{\delta^*}} \,\bigg|\, \mathcal{A}_t\bigg) \\
&= S_t^{\delta^*} E\bigg(\frac{(K - S_T^{\delta^*})^+}{S_T^{\delta^*}} \,\bigg|\, \mathcal{G}_t\bigg) \\
&= K E\bigg(\bigg(\frac{S_t^{\delta_*}}{S_T^{\delta_*}} - \frac{S_t^{\delta_*}}{K}\bigg)^+ \,\bigg|\, \mathcal{G}_t\bigg) \\
&= K E\big(\big(\exp\big\{-\ln\big(Y(t,T)\big)\big\} - \tilde{K}\big)^+ \,\big|\, \mathcal{G}_t\big),
\end{aligned}$$

where $\tilde{K} = \frac{S_t^{\delta_*}}{K}$, $Y(t,T) = \frac{S_T^{\delta_*}}{S_t^{\delta_*}}$. Hence from Lemma 8.3.2, for $w > 1$, it follows

$$\begin{aligned}
&S_t^{\delta^*} E\bigg(\frac{(K - S_T^{\delta^*})^+}{S_T^{\delta^*}} \,\bigg|\, \mathcal{G}_t\bigg) \\
&\quad = \frac{K}{2\pi} \int_{\Re} E\big(\exp\big\{(w + \imath\lambda)\big(-\ln\big(Y(t,T)\big)\big)\big\} \,\big|\, \mathcal{G}_t\big) \frac{\tilde{K}^{-(w-1+\imath\lambda)}}{(w+\imath)(w-1+\imath\lambda)} \, d\lambda.
\end{aligned}$$

We now discuss the computation of the above conditional expectation

$$E\big(\exp\big\{(w + \imath\lambda)\big(-\ln\big(Y(t,T)\big)\big)\big\} \,\big|\, \mathcal{G}_t\big).$$

From Eq. (14.1.4), we have

$$\begin{aligned}
&E\big(\exp\big\{(w + \imath\lambda)\big(-\ln\big(Y(t,T)\big)\big)\big\} \,\big|\, \mathcal{G}_t\big) \\
&\quad = E\bigg(\exp\bigg\{-(w + \imath\lambda)\bigg(\int_t^T r_s \, ds + \ln\bigg(\frac{\alpha_T}{\alpha_t}\bigg) + \ln(Y_T) - \ln(Y_t)\bigg)\bigg\} \,\bigg|\, \mathcal{G}_t\bigg) \\
&\quad = \exp\bigg\{-(w + \imath\lambda)\int_t^T a_s^r \, ds - (w + \imath\lambda)\ln\bigg(\frac{\alpha_T}{\alpha_t}\bigg)\bigg\} Y_t^{(w+\imath)\lambda} \\
&\qquad \times E\bigg(\exp\bigg\{-(w + \imath\lambda)\int_t^T b^r Z_s^1 \, ds\bigg\} \,\bigg|\, \mathcal{G}_t\bigg) \\
&\qquad \times E\bigg(\exp\bigg\{-(w + \imath\lambda)\int_t^T c^r f^1(Y_s) \, ds\bigg\} Y_T^{-(w+\imath\lambda)} \,\bigg|\, \mathcal{G}_t\bigg) \\
&\quad =: f\big(\lambda, Z_t^1, Y_t\big).
\end{aligned}$$

Here

$$E\bigg(\exp\bigg\{-(w + \imath\lambda)\int_t^T b^r Z_s^1 \, ds\bigg\} \,\bigg|\, \mathcal{G}_t\bigg)$$

and

$$E\left(\exp\left\{-(w+\imath\lambda)\int_t^T c^r f^1(Y_s)\,ds\right\}Y_T^{-(w+\imath\lambda)}\;\Big|\;\mathcal{G}_t\right)$$

can be computed using Propositions 7.3.8 and 7.3.9. Hence, we obtain

$$\begin{aligned} &p_{T,K}\left(t, Z_t^1, Y_t, S_t\right) \\ &= \frac{K}{2\pi}\int_{\Re} f\left(\lambda, Z_t^1, Y_t\right) \\ &\quad\times \exp\left\{-(w+\imath\lambda)\left(\int_t^T a_s^r\,ds + \ln\left(\frac{\alpha_T}{\alpha_t}\right)\right)\right\}\frac{(\tilde{K})^{-(w-1+\imath\lambda)}}{(w+\imath\lambda)(w-1+\imath\lambda)}\,d\lambda. \end{aligned}$$

The above formulas will be employed in Sect. 14.3.

14.2 Pricing Credit Default Swaps Under the Benchmark Approach

We now discuss how to price CDSs. In the next section, we show how CVA calculations naturally extend this concept. Firstly, we summarize a CDS transaction. Consider two parties: A, the *protection buyer*, and B, the *protection seller*. If a third party, say C, the *reference company*, defaults at a time τ, where τ is between two fixed times T_a and T_b, B pays A a certain fixed amount, say L. In exchange A pays B coupons at a rate R at time points $T_{a+1}, \dots, T_b$, or until default.

Under the benchmark approach, the techniques from Filipović (2009) can be combined with the Laplace transforms developed in this book. Using the real world pricing formula, the value of this contract to B at a time $t < T_a$ is given by

$$\begin{aligned} CDS_t := &\; S_t^{\delta_*} E\left(\mathbf{1}_{\{T_a<\tau\le T_b\}} R\frac{\tau - T_{\beta(\tau)-1}}{S_\tau^{\delta_*}}\;\Big|\;\mathcal{A}_t\right) \\ &+ S_t^{\delta_*}\sum_{i=a+1}^{b}\alpha_i RE\left(\frac{\mathbf{1}_{\{\tau>T_i\}}}{S_{T_i}^{\delta_*}}\;\Big|\;\mathcal{A}_t\right) \\ &- S_t^{\delta_*} LE\left(\frac{\mathbf{1}_{\{T_a<\tau\le T_b\}}}{S_\tau^{\delta_*}}\;\Big|\;\mathcal{A}_t\right), \end{aligned}$$

where $\delta_i = T_i - T_{i-1}$, and $T_{\beta(\tau)}$ is the first of the T_i's following τ. The interpretation is clear, the first two terms represent payments from party A to party B, where the first term represents the amount accrued between the last payment before default, made at time $T_{\beta(\tau)-1}$, and the default time τ. The last term represents the payment to be made by B in case C defaults. Using the terminology from Filipović (2009), the second term is a zero recovery zero coupon bond, a payment R is only made at T_i if default occurs after T_i. The third term is a partial recovery at default zero coupon bond with payment L, and so is the first term, for which the payment at default is $(\tau - T_{\beta(\tau)-1})R$.

We firstly value the zero recovery zero coupon bond, where we use Lemma 14.1.3:

$$
\begin{aligned}
P_T^0(t) &:= S_t^{\delta_*} E\left(\frac{\mathbf{1}_{\{\tau>T\}}}{S_T^{\delta_*}} \,\middle|\, \mathcal{A}_t\right) \\
&= \mathbf{1}_{\{\tau>t\}} S_t^{\delta_*} \exp\left\{\int_0^t \lambda_s\, ds\right\} E\left(\frac{\mathbf{1}_{\{\tau>T\}}}{S_T^{\delta_*}} \,\middle|\, \mathcal{G}_t\right) \\
&= \mathbf{1}_{\{\tau>t\}} S_t^{\delta_*} \exp\left\{\int_0^t \lambda_s\, ds\right\} E\left(\frac{1}{S_T^{\delta_*}} E(\mathbf{1}_{\{\tau>T\}} \mid \mathcal{G}_T) \,\middle|\, \mathcal{G}_t\right) \\
&= \mathbf{1}_{\{\tau>t\}} S_t^{\delta_*} \exp\left\{\int_0^t \lambda_s\, ds\right\} E\left(\frac{\exp\{-\int_0^T \lambda_s\, ds\}}{S_T^{\delta_*}} \,\middle|\, \mathcal{G}_t\right) \\
&= \mathbf{1}_{\{\tau>t\}} S_t^{\delta_*} E\left(\frac{\exp\{-\int_t^T \lambda_s\, ds\}}{S_T^{\delta_*}} \,\middle|\, \mathcal{G}_t\right) \\
&= \mathbf{1}_{\{\tau>t\}} \frac{\alpha_t}{\alpha_T} E\left(\frac{Y_t}{Y_T} \exp\left\{-\int_t^T (r_s+\lambda_s)\, ds\right\} \,\middle|\, \mathcal{G}_t\right) \qquad (14.2.14) \\
&= \mathbf{1}_{\{\tau>t\}} \frac{\alpha_t}{\alpha_T} Y_t \\
&\quad \times E\left(\frac{\exp\{-\int_t^T a_s\, ds - b\int_t^T Z_s^1\, ds - d\int_t^T Z_s^2\, ds - \int_t^T f(Y_s)\, ds\}}{Y_T} \,\middle|\, \mathcal{G}_t\right) \\
&= \mathbf{1}_{\{\tau>t\}} \frac{\alpha_t}{\alpha_T} Y_t \exp\left\{-\int_t^T a_s\, ds\right\} E\left(\exp\left\{-b\int_t^T Z_s^1\, ds\right\} \,\middle|\, \mathcal{G}_t\right) \qquad (14.2.15) \\
&\quad \times E\left(\exp\left\{-d\int_t^T Z_s^2\, ds\right\} \,\middle|\, \mathcal{G}_t\right) E\left(\frac{\exp\{-\int_t^T f(Y_s)\, ds\}}{Y_T} \,\middle|\, \mathcal{G}_t\right), \\
&\qquad (14.2.16)
\end{aligned}
$$

where $a_t = a_t^r + a_t^\lambda$, $b = b^r + b^\lambda$, $d = d^\lambda$, and $f(x) = c^r f^r(x) + c^\lambda f^\lambda(x)$. We point out that from Eq. (14.2.14), one can confirm the observation from Filipović (2009) that when pricing a zero recovery zero coupon bond, as opposed to a zero coupon bond, one replaces the short rate process by $r_t + \lambda_t$, which results in a lower price. Again, the expected values in Eqs. (14.2.15) and (14.2.16) can be computed using Propositions 7.3.8 and 7.3.9.

We now turn to the remaining two components of the credit default swap pricing formula. We remark that it suffices to focus on

$$
S_t^{\delta_*} E\left(\frac{\tau - T_{\beta(\tau)-1}}{S_\tau^{\delta_*}} \mathbf{1}_{\{T_a<\tau\le T_b\}} \,\middle|\, \mathcal{A}_t\right).
$$

From Sect. 12.3.3.3 in Filipović (2009) we recall that the distribution of τ, conditional on the event $\{\tau > t\}$ for $t \le u$, is given by

$$
\begin{aligned}
&P(t < \tau \le u \mid \mathcal{G}_\infty \vee \mathcal{H}_t) \\
&\quad = \mathbf{1}_{\{\tau>t\}} \exp\left\{\int_0^t \lambda_s\, ds\right\} E(\mathbf{1}_{\{t<\tau\le u\}} \mid \mathcal{G}_\infty)
\end{aligned}
$$

$$= \mathbf{1}_{\{\tau>t\}} \exp\left\{\int_0^t \lambda_s\, ds\right\}\left(\exp\left\{-\int_0^t \lambda_s\, ds\right\} - \exp\left\{-\int_0^u \lambda_s\, ds\right\}\right)$$
$$= \mathbf{1}_{\{\tau>t\}}\left(1 - \exp\left\{-\int_t^u \lambda_s\, ds\right\}\right),$$

which is the regular $\mathcal{G}_\infty \vee \mathcal{H}_t$-conditional distribution of τ given $\{\tau > t\}$. For more details on regular conditional distributions the reader is referred to Sect. 4.1.4 in Filipović (2009). To obtain the density function, we differentiate with respect to u to obtain

$$\mathbf{1}_{\{\tau>t\}} \lambda_u \exp\left\{-\int_t^u \lambda_s\, ds\right\}, \tag{14.2.17}$$

for $u \geq t$. We now price the partial recovery at default bond

$$P_T^p(t) := S_t^{\delta_*} E\left(\frac{(\tau - T_{\beta(\tau)-1})}{S_\tau^{\delta_*}} \mathbf{1}_{\{T_a<\tau\leq T_b\}} \,\Big|\, \mathcal{A}_t\right).$$

Using Eq. (14.2.17) and the Fubini theorem, we compute

$$S_t^{\delta_*} E\left(\frac{\tau - T_{\beta(\tau)-1}}{S_\tau^{\delta_*}} \mathbf{1}_{\{T_a<\tau\leq T_b\}} \,\Big|\, \mathcal{A}_t\right)$$
$$= S_t^{\delta_*} E\left(\frac{f(\tau)\mathbf{1}_{\{T_a<\tau\leq T_b\}}}{S_\tau^{\delta_*}} \,\Big|\, \mathcal{A}_t\right)$$
$$= E\left(E\left(f(\tau)\frac{\alpha_t}{\alpha_\tau} \exp\left\{-\int_t^\tau r_s\, ds\right\} \mathbf{1}_{\{T_a<\tau\leq T_b\}} \frac{Y_t}{Y_\tau} \,\Big|\, \mathcal{G}_\infty \vee \mathcal{H}_t\right) \Big|\, \mathcal{A}_t\right)$$
$$= \mathbf{1}_{\{\tau>t\}} E\left(\int_{T_a}^{T_b} \tilde{f}(u) \exp\left\{-\int_t^u r_s\, ds\right\} \lambda_u \exp\left\{-\int_t^u \lambda_s\, ds\right\} \frac{Y_t}{Y_u}\, du \,\Big|\, \mathcal{A}_t\right)$$
$$= \mathbf{1}_{\{\tau>t\}} \int_{T_a}^{T_b} \tilde{f}(u) E\left(\exp\left\{-\int_t^u r_s\, ds\right\} \lambda_u \exp\left\{-\int_t^u \lambda_s\, ds\right\} \frac{Y_t}{Y_u} \,\Big|\, \mathcal{A}_t\right) du,$$

where $f(x) = (x - T_{\beta(x)-1})$ and $\tilde{f}(x) = \frac{\alpha_t}{\alpha_x} f(x)$. From Corollary 14.1.4, we obtain

$$E\left(\exp\left\{-\int_t^u r_s\, ds\right\} \lambda_u \exp\left\{-\int_t^u \lambda_s\, ds\right\} \frac{Y_t}{Y_u} \,\Big|\, \mathcal{A}_t\right) \tag{14.2.18}$$
$$= E\left(\frac{\lambda_u \exp\{-\int_t^u (r_s + \lambda_s)\, ds\} Y_t}{Y_u} \,\Big|\, \mathcal{G}_t\right). \tag{14.2.19}$$

Hence we conclude that

$$P_T^p(t) = \mathbf{1}_{\{\tau>t\}} Y_t \int_t^T \tilde{f}(u) E\left(\frac{\lambda_u \exp\{-\int_t^u (r_s + \lambda_s)\, ds\}}{Y_u} \,\Big|\, \mathcal{G}_t\right) du.$$

We now discuss how to compute

$$E\left(\frac{\lambda_u \exp\{-\int_t^u (r_s + \lambda_s)\, ds\}}{Y_u} \,\Big|\, \mathcal{G}_t\right).$$

Since we have

$$\begin{aligned}&\exp\left\{-\int_t^u (r_s+\lambda_s)\,ds\right\}\\&\quad=\exp\left(-\int_t^u a_s\,ds - b\int_t^u Z_s^1\,ds - d\int_t^u Z_s^2\,ds - \int_t^u f(Y_s)\,ds\right)\end{aligned}$$

and

$$\lambda_u = a_u^\lambda + b^\lambda Z_u^1 + c^\lambda f^\lambda(Y_u) + d^\lambda Z_u^2,$$

we have

$$\begin{aligned}&E\left(\exp\left\{-\int_t^u a_s\,ds - b\int_t^u Z_s^1\,ds - d\int_t^u Z_s^2 - \int_t^u f(Y_s)\,ds\right\}\frac{\lambda_u}{Y_u}\,\Big|\,\mathcal{G}_t\right)\\&=\exp\left\{-\int_t^u a_s\,ds\right\}\left(a_u^\lambda E\left(\exp\left\{-b\int_t^u Z_s^1\,ds\right\}\Big|\,\mathcal{G}_t\right)\right.\\&\quad\times E\left(\exp\left\{-d\int_t^u Z_s^2\,ds\right\}\Big|\,\mathcal{G}_t\right)E\left(\frac{\exp\{-\int_t^u f(Y_s)\,ds\}}{Y_u}\,\Big|\,\mathcal{G}_t\right)\\&\quad+E\left(b^\lambda Z_u^1\exp\left\{-b\int_t^u Z_s^1\,ds\right\}\Big|\,\mathcal{G}_t\right)E\left(\exp\left\{-d\int_t^u Z_s^2\,ds\right\}\Big|\,\mathcal{G}_t\right)\\&\quad\times E\left(\frac{\exp\{-\int_t^u f(Y_s)\,ds\}}{Y_u}\,\Big|\,\mathcal{G}_t\right)\\&\quad+E\left(\exp\left\{-b\int_t^u Z_s^1\,ds\right\}\Big|\,\mathcal{G}_t\right)E\left(d^\lambda Z_u^2\exp\left\{-d\int_t^u Z_s^2\,ds\right\}\Big|\,\mathcal{G}_t\right)\\&\quad\times E\left(\frac{\exp\{-\int_t^u f(Y_s)\,ds\}}{Y_u}\,\Big|\,\mathcal{G}_t\right)\\&\quad+E\left(\exp\left\{-b\int_t^u Z_s^1\,ds\right\}\Big|\,\mathcal{G}_t\right)E\left(\exp\left\{-d\int_t^u Z_s^2\,ds\right\}\Big|\,\mathcal{G}_t\right)\\&\quad\left.\times E\left(\frac{c^\lambda f^2(Y_u)\exp\{-\int_t^u f(Y_s)\,ds\}}{Y_u}\,\Big|\,\mathcal{G}_t\right)\right),\end{aligned}$$

where all expectations can be computed using Propositions 7.3.8 and 7.3.9. We remark that the third term in the CDS valuation formula can be computed as above, in this case $f(\tau)=1$.

14.3 Credit Valuation Adjustment Under the Benchmark Approach

In this section, we discuss the computation of CVA, in the affine credit risk model introduced in Sect. 14.1. First, we introduce CVA as an extension of a CDS: assume two parties, A and C, have entered into a series of contracts, the aggregate value of which at time t is given by V_t. We take the point of view of party A, and say that $V_t>0$ if the aggregate value of the contracts at time t is profitable to A, and $V_t<0$

if the aggregate value of the contracts is profitable to C. For ease of exposition, we assume that party A cannot default but C can, so we discuss *unilateral CVA*, though of course *bilateral CVA* can also be discussed under the benchmark approach using the techniques introduced in this chapter.

Party A now approaches another party, say B, for protection on its portfolio V with C over the period $[0, T]$: in case C defaults, B pays the value of the part of the portfolio that is not recovered at the time of default, only if the value of the portfolio is positive to A, i.e. only if $V_\tau > 0$, where τ denotes the time of default of C. Hence the payment at default is

$$(1 - R)V_\tau^+,$$

where R is the recovery rate and $V_t^+ := \max(V_t, 0)$. Again, for ease of exposition, we assume that B cannot default. Using the real world pricing formula (1.3.19), we obtain the real world price of this protection as

$$CVA_t := (1 - R)S_t^{\delta_*} E\left(\frac{V_\tau^+}{S_\tau^{\delta_*}} \mathbf{1}_{\{\tau>T\}} \,\bigg|\, \mathcal{A}_t\right), \tag{14.3.20}$$

for $t \geq 0$. It is crucial for CVA computations, that *right-way exposure* and *wrong-way exposure* are taken into account. This requires the modeling of a dependence structure between the portfolio process V and the time of default, τ: under the benchmark approach, the value of V depends on the numéraire, which is the GOP, S^{δ_*}, and hence its stochastic drivers, Y and Z^1. However, τ can in general also be expected to depend on S^{δ_*}: if the GOP drops, which affects the value of V, a default of C can be more likely, or less likely, depending on the nature of company C. In Sect. 14.4, we present an illustrative example including commodities. The exposure is called *right-way* if the value of V is negatively related to the credit quality of the counter party and *wrong-way* is defined analogously, see Cesari et al. (2009). Our specification of λ, which takes into account Z^1 and Y allows us to model this by choosing $f^\lambda(x) = x$ or $f^\lambda(x) = \frac{1}{x}$. We now consider the valuation of some simple contracts.

Firstly, we assume that $V_t = S_t^{\delta_*}$ and that A has bought protection from B for the period $[0, T]$, then

$$\begin{aligned}
CVA_t &= S_t^{\delta_*} E\left(\frac{V_\tau^+}{S_\tau^{\delta_*}} \mathbf{1}_{\{t<\tau\leq T\}} \,\bigg|\, \mathcal{A}_t\right) \\
&= S_t^{\delta_*} P(t < \tau \leq T \mid \mathcal{A}_t) \\
&= S_t^{\delta_*} E(\mathbf{1}_{\{\tau>t\}} - \mathbf{1}_{\{\tau>T\}} \mid \mathcal{A}_t) \\
&= \mathbf{1}_{\{\tau>t\}} S_t^{\delta_*} E(1 - \mathbf{1}_{\{\tau>t\}} \mathbf{1}_{\{\tau>T\}} \mid \mathcal{A}_t).
\end{aligned}$$

Now, we have

$$\begin{aligned}
E(\mathbf{1}_{\{\tau>t\}} \mathbf{1}_{\{\tau>T\}} \mid \mathcal{A}_t) &= \mathbf{1}_{\{\tau>t\}} \exp\left\{\int_0^t \lambda_s\, ds\right\} E(\mathbf{1}_{\{\tau>t\}} \mathbf{1}_{\{\tau>T\}} \mid \mathcal{G}_t) \\
&= \mathbf{1}_{\{\tau>t\}} \exp\left\{\int_0^t \lambda_s\, ds\right\} E\big(E(\mathbf{1}_{\{\tau>T\}} \mid \mathcal{G}_T) \,\big|\, \mathcal{G}_t\big)
\end{aligned}$$

$$= \mathbf{1}_{\{\tau>t\}} \exp\left\{\int_0^t \lambda_s\, ds\right\} E\left(\exp\left\{-\int_0^T \lambda_s\, ds\right\} \,\Big|\, \mathcal{G}_t\right)$$

$$= \mathbf{1}_{\{\tau>t\}} E\left(\exp\left\{-\int_t^T \lambda_s\, ds\right\} \,\Big|\, \mathcal{G}_t\right),$$

where we used Lemma 14.1.3 with $Y = \mathbf{1}_{\{\tau>T\}}$ and Eq. (14.1.2). Finally,

$$CVA_t = \mathbf{1}_{\{\tau>t\}} S_t^{\delta_*} E\left(1 - \exp\left\{-\int_t^T \lambda_s\, ds\right\} \,\Big|\, \mathcal{G}_t\right),$$

and we compute

$$E\left(\exp\left\{-\int_t^T \lambda_s\, ds\right\} \,\Big|\, \mathcal{G}_t\right)$$

as in Sect. 14.2, since λ_t is a function of affine processes, and the relevant Laplace transforms are given in Propositions 7.3.8 and 7.3.9.

Now assume that $V_t = P_T(t)$, a zero coupon bond, which we priced in Sect. 14.1. Again we consider CVA over the period $[0, T]$

$$CVA_t = S_t^{\delta_*} E\left(\frac{V_\tau^+}{S_\tau^{\delta_*}} \mathbf{1}_{\{t<\tau\le T\}} \,\Big|\, \mathcal{A}_t\right)$$

$$= S_t^{\delta_*} E\left(E\left(\frac{1}{S_T^{\delta_*}} \,\Big|\, \mathcal{A}_\tau\right) \mathbf{1}_{\{t<\tau\le T\}} \,\Big|\, \mathcal{A}_t\right)$$

$$= S_t^{\delta_*} E\left(\frac{\mathbf{1}_{\{t<\tau\le T\}}}{S_T^{\delta_*}} \,\Big|\, \mathcal{A}_t\right)$$

$$= \mathbf{1}_{\{\tau>t\}} \left(S_t^{\delta_*} E\left(\frac{1}{S_T^{\delta_*}} \,\Big|\, \mathcal{A}_t\right) - S_t^{\delta_*} E\left(\frac{\mathbf{1}_{\{\tau>T\}}}{S_T^{\delta_*}} \,\Big|\, \mathcal{A}_t\right)\right)$$

$$= \mathbf{1}_{\{\tau>t\}} \left(P_T(t) - P_T^0(t)\right),$$

so, conditional on the event $\{\tau > t\}$, we have represented CVA as the difference between a zero coupon bond and a zero recovery zero coupon bond. We remind the reader that the latter was priced in Sect. 14.2.

Next, we discuss the pricing of a European put option in the presence of counterparty risk. Recall that standard European put options were priced in Sect. 14.1. We use the density function from Eq. (14.2.17), the fact that $\mathbf{1}_{\{\tau>t\}}$ is $\mathcal{A}_t$-measurable, and Corollary 14.1.4 to obtain

$$CVA_t$$

$$= S_t^{\delta_*} E\left(\frac{V_\tau^+}{S_\tau^{\delta_*}} \mathbf{1}_{\{t<\tau\le T\}} \,\Big|\, \mathcal{A}_t\right)$$

$$= S_t^{\delta_*} E\left(\frac{p_{T,K}(\tau, Z_\tau^1, Y_\tau, S_\tau^{\delta_*})}{S_\tau^{\delta_*}} \mathbf{1}_{\{t<\tau\le T\}} \,\Big|\, \mathcal{A}_t\right)$$

$$= S_t^{\delta_*} E\left(\frac{(K - S_T^{\delta_*})^+}{S_T^{\delta_*}} \mathbf{1}_{\{t<\tau\le T\}} \,\Big|\, \mathcal{A}_t\right)$$

$$
\begin{aligned}
&= S_t^{\delta_*} E\left(E\left(\frac{(K - S_T^{\delta_*})^+}{S_T^{\delta_*}} \mathbf{1}_{\{t<\tau\le T\}} \,\Big|\, \mathcal{G}_\infty \vee \mathcal{H}_t \right) \Big|\, \mathcal{A}_t \right) \\
&= S_t^{\delta_*} E\left(\mathbf{1}_{\{\tau>t\}} \int_t^T \frac{(K - S_T^{\delta_*})^+}{S_T^{\delta_*}} \lambda_u \exp\left\{ - \int_t^u \lambda_s \, ds \right\} du \,\Big|\, \mathcal{A}_t \right) \\
&= \mathbf{1}_{\{\tau>t\}} S_t^{\delta_*} \int_t^T E\left(\frac{(K - S_T^{\delta_*})^+}{S_T^{\delta_*}} \lambda_u \exp\left\{ - \int_t^u \lambda_s \, ds \right\} \Big|\, \mathcal{G}_t \right) du \\
&= \mathbf{1}_{\{\tau>t\}} S_t^{\delta_*} \int_t^T E\left(\frac{S_u^{\delta_*}}{S_u^{\delta_*}} E\left(\frac{(K - S_T^{\delta_*})^+}{S_T^{\delta_*}} \,\Big|\, \mathcal{A}_u \right) \lambda_u \exp\left\{ - \int_t^u \lambda_s \, ds \right\} \Big|\, \mathcal{G}_t \right) du \\
&= \mathbf{1}_{\{\tau>t\}} S_t^{\delta_*} \int_t^T E\left(\frac{p_{T,K}(u, Z_u^1, Y_u, S_u^{\delta_*})}{S_u^{\delta_*}} \lambda_u \exp\left\{ - \int_t^u \lambda_s \, ds \right\} \Big|\, \mathcal{G}_t \right) du.
\end{aligned}
\tag{14.3.21}
$$

In general, it seems difficult to simplify the above expression further. Essentially, this is due to the fact that our model incorporates wrong-way and right-way exposure, i.e. we allow for dependence between τ and S^{δ_*}. Hence one would usually employ a Monte Carlo algorithm, see e.g. Cesari et al. (2009). We remark that in Sect. 6.3, we derived the joint law of $(\int_t^u Y_s \, ds, Y_u)$, conditional on Y_t, and in Sect. 6.4, the joint law of $(\int_t^u \frac{ds}{Y_s}, Y_u)$ conditional on Y_t, which are useful in developing a Monte Carlo algorithm.

We now discuss the pricing of swaps in the presence of counterparty risk. In Sect. 14.1, we presented the value of a swap as a linear combination of zero coupon bonds. Hence, if market prices of zero coupon bonds are available, a model would not be required to price swaps in practice. In the presence of counterparty risk, this is different, as we now show. We set $V_t = FS_{\kappa,T_0}(t)$, where $FS_{\kappa,T_0}(t)$ is defined in Eq. (14.1.9), so we consider a swap with start date T_0, and we focus on CVA for the period $[0, T]$ for $T \le T_0$. We have

$$
\begin{aligned}
CVA_t &:= S_t^{\delta_*} E\left(\frac{V_\tau^+}{S_\tau^{\delta_*}} \mathbf{1}_{\{t<\tau\le T\}} \,\Big|\, \mathcal{A}_t \right) \\
&= S_t^{\delta_*} E\left(\frac{(FS_{\kappa,T_0}(\tau))^+}{S_\tau^{\delta_*}} \mathbf{1}_{\{t<\tau\le T\}} \,\Big|\, \mathcal{A}_t \right),
\end{aligned}
\tag{14.3.22}
$$

hence the market price of counterparty risk associated with a swap can be interpreted as a forward start swaption with random expiry date τ. Again, we use the density function from Eq. (14.2.17), the fact that $\mathbf{1}_{\{\tau>t\}}$ is $\mathcal{A}_t$-measurable, and Corollary 14.1.4 to obtain

$$
\begin{aligned}
& S_t^{\delta_*} E\left(\frac{(FS_{\kappa,T_0}(\tau))^+}{S_\tau^{\delta_*}} \mathbf{1}_{\{t<\tau\le T\}} \,\Big|\, \mathcal{A}_t \right) \\
&= S_t^{\delta_*} E\left(E\left(\mathbf{1}_{\{t<\tau\le T\}} \frac{(FS_{\kappa,T_0}(\tau))^+}{S_\tau^{\delta_*}} \,\Big|\, \mathcal{G}_\infty \vee \mathcal{H}_t \right) \Big|\, \mathcal{A}_t \right)
\end{aligned}
\tag{14.3.23}
$$

$$
\begin{aligned}
&= \mathbf{1}_{\{\tau>t\}} S_t^{\delta_*} E\left(\int_t^T \frac{(FS_{K,T_0}(u))^+}{S_u^{\delta_*}} \lambda_u \exp\left\{-\int_t^u \lambda_s\, ds\right\} du \,\bigg|\, \mathcal{A}_t\right) \\
&= \mathbf{1}_{\{\tau>t\}} S_t^{\delta_*} E\left(\int_t^T \frac{(FS_{K,T_0}(u))^+}{S_u^{\delta_*}} \lambda_u \exp\left\{-\int_t^u \lambda_s\, ds\right\} du \,\bigg|\, \mathcal{G}_t\right) \\
&= \mathbf{1}_{\{\tau>t\}} S_t^{\delta_*} \int_t^T E\left(\frac{(FS_{K,T_0}(u))^+}{S_u^{\delta_*}} \lambda_u \exp\left\{-\int_t^u \lambda_s\, ds\right\} \,\bigg|\, \mathcal{G}_t\right) du.
\end{aligned}
\tag{14.3.24}
$$

Hence, as for the European put, we resort to Monte Carlo methods to compute equation (14.3.24). This is due to the fact that accounting for right-way and wrong-way exposure makes it more challenging to compute CVA analytically.

14.4 CVA for Commodities

We now consider counterparty risk for commodities. In particular, we consider the case where the counterparty C is directly affected by the value of the commodity underlying the transaction. For example, say the counterparty C is an airline, in which case it is clear that the company has a large exposure to the price of oil and could be interested in trading forward contracts on oil with company A, which is assumed to be default free. However, in case the price of oil rises, default of company C becomes more likely. Taking into account right-way and wrong-way exposure, it is important to recognize that the value of the commodity impacts both, the value of the transaction V, assumed to be a forward on oil, but also the time of default τ. We hence model this under the benchmark approach following Du and Platen (2012b). In particular, we use S_t^{i,δ_*} to denote the value of the GOP at time t, denominated in units of the i-th security. A general exchange price, which could be a number of units of currency i to be paid for one unit of currency j, or a number of units of currency i to be paid for one unit of commodity j is then given by

$$
X_t^{i,j} = \frac{S_t^{i,\delta_*}}{S_t^{j,\delta_*}}. \tag{14.4.25}
$$

In this section, currency i would be the domestic currency and commodity j the commodity of interest, so j could correspond to oil and i to US dollars. In particular, we note that if S_t^{i,δ_*} appreciates or S_t^{j,δ_*} depreciates, then $X_t^{i,j}$ appreciates, so more units of currency i, say US dollars, have to be paid for one unit of the commodity. We recall the MMM from Du and Platen (2012b). Though parsimonious, the model is tractable and in particular allows us to incorporate right-way and wrong-way exposure. In particular, we set

$$
S_t^{k,\delta_*} = B_t^k Y_t^k A_t^k, \tag{14.4.26}
$$

where $k \in \{i, j\}$ and where

$$
dY_t^k = \xi^k \left(1 - Y_t^k\right) dt + \sqrt{\xi^k Y_t^k}\, dW_t^k,
$$

$k \in \{i, j\}$, where W^i and W^j are independent $\underline{\mathcal{G}}$-Brownian motions. Furthermore,

$$A_t^k = A_0^k \exp\{\xi^k t\},$$

and

$$B_t^k = \exp\left\{\int_0^t r_s^k \, ds\right\}.$$

When considering a currency, say i, $r^i = \{r_t^i, t \geq 0\}$ is interpreted as a short rate process, and for commodities $r^j = \{r_t^j, t \geq 0\}$ can be interpreted as the convenience yield process. Following Du and Platen (2012b), we set

$$r_t^i = a^i + b^{ii} Y_t^i + b^{ij} Y_t^j,$$
$$r_t^j = a^j + b^{ji} Y_t^i + b^{jj} Y_t^j,$$

where a^i, a^j, b^{ii}, b^{ji}, b^{ij}, b^{jj} are nonnegative constants, and b^{ij} corresponds to the sensitivity of the short rate r^i to changes in Y^j. In particular, we note that S^{i,δ_*} and S^{j,δ_*} are dependent, as they share common drivers. We are now in a position to price a standard forward contract, and recall the relevant result from Du and Platen (2012b). Recall that at initiation time t, the forward price $F_t^{i,j,T}$ of one unit of commodity j to be delivered at time T, denominated in currency i, is chosen so that the forward has no value. Using the real world pricing formula (1.3.19), we chose $F_t^{i,j,T}$ so that

$$E\left(\frac{(X_T^{i,j} - F_t^{i,j,T})}{S_T^{i,\delta_*}} \,\middle|\, \mathcal{A}_t\right) = 0. \tag{14.4.27}$$

Solving Eq. (14.4.27) for $F_t^{i,j,T}$ produces Theorem 3.1 from Du and Platen (2012b), which we now present.

Theorem 14.4.1 *The real world price at inception time $t \in [0, T]$ in units of the i-th currency, for one unit of the j-th commodity to be delivered at time $T \in [0, \infty)$ equals*

$$F_t^{i,j,T} = X_t^{i,j} \frac{P_T^j(t)}{P_T^i(t)}.$$

We point out that $P_T^i(t)$ corresponds to a zero coupon bond in currency i, whereas $P_T^j(t)$ is the value of the delivery of one unit of the j-th commodity at maturity T, denominated in units of the commodity j itself. Furthermore, we need to know the value of the forward initiated at time t_0, at an intermediate time, say $t \in [t_0, T]$. The relevant formula is given in Theorem 3.2 in Du and Platen (2012b).

Theorem 14.4.2 *The real world value $U_t^{i,j,t_0,T}$ of a forward contract at time t for one unit of the jth commodity with initiation time t_0 and maturity date T equals*

$$U_t^{i,j,t_0,T} = P_T^i(t)\big(F_t^{i,j,T} - F_{t_0}^{i,j,T}\big),$$

when denominated in units of the i-th currency, $t_0 \in [0, T]$, $t \in [t_0, T]$, $T \in [0, \infty)$.

We remark that for the model introduced in this section, we can derive closed-form solutions for forward prices and the value of forward contracts the way we did in Sect. 14.1.

Now we want to return to our counterparty risk example. As we had discussed before, the airline is more likely to default if the price of the commodity increases. Hence we propose the following model: we introduce an additional square-root process

$$dZ_t = \kappa(\theta - Z_t)\,dt + \sigma\sqrt{Z_t}\,dW_t^k,$$

where W^k is a $\underline{\mathcal{G}}$-Brownian motion independent of W^i and W^j. The default intensity λ is modeled as follows:

$$\lambda_t = a_t^\lambda + b^\lambda Y_t^i + c^\lambda \frac{1}{Y_t^j} + d^\lambda Z_t, \tag{14.4.28}$$

where $b^\lambda, c^\lambda, d^\lambda$ are nonnegative constants and a^λ is a nonnegative function. In particular, we note that if the main driver of S^{i,δ_*}, which is Y^i, increases, then $X_t^{i,j}$ and λ_t increase, i.e. default becomes more likely as the price of the commodity increases. Likewise, as the main driver of S_t^{j,δ_*}, which is Y_t^j, decreases, then $X_t^{i,j}$ and λ_t increase, i.e. default becomes more likely. We now consider *CVA* for $V_t = U_t^{i,j,t_0,T}$ over the period $[0, T]$. We employ the density function from Eq. (14.2.17), the fact that $\mathbf{1}_{\{\tau>t\}}$ is $\mathcal{A}_t$-measurable, and Corollary 14.1.4 to obtain

$$\begin{aligned} CVA_t &= S_t^{i,\delta_*} E\left(\frac{V_\tau^+}{S_\tau^{i,\delta_*}}\mathbf{1}_{\{t<\tau\leq T\}}\,\middle|\,\mathcal{A}_t\right) \\ &= \mathbf{1}_{\{\tau>t\}} S_t^{i,\delta_*} E\left(\int_t^T \frac{(U_u^{i,j,t_0,T})^+}{S_u^{i,\delta_*}}\lambda_u \exp\left\{-\int_t^u \lambda_s\,ds\right\} du\,\middle|\,\mathcal{A}_t\right) \\ &= \mathbf{1}_{\{\tau>t\}} S_t^{i,\delta_*} \int_t^T E\left(\frac{(U_u^{i,j,t_0,T})^+}{S_u^{i,\delta_*}}\lambda_u \exp\left\{-\int_t^u \lambda_s\,ds\right\}\,\middle|\,\mathcal{G}_t\right) du. \end{aligned} \tag{14.4.29}$$

Again, one would resort to Monte Carlo methods to compute (14.4.29), due to the fact that our model takes into account right-way and wrong-way exposure.

14.5 A Reduced-Form Model

The affine credit risk model presented in Sect. 14.1 is able to incorporate right-way and wrong-way exposure, and should hence be useful when performing CVA computations. However, for many products, Monte Carlo algorithms need to be employed when performing computations. Though from e.g. Cesari et al. (2009), this should be expected, we aim to produce a reduced form model in this section. The model assumes independence between default risk and financial risk. Though not necessarily satisfied for all transactions relevant to practice, this is a very tractable model.

We hence modify the model from Sect. 14.1 as follows: firstly, we model S^{δ_*} using the MMM from Sect. 3.3, we set

$$dY_t = (1 - \eta Y_t)\,dt + \sqrt{Y_t}\,dW_t,$$

where W is a $\underline{\mathcal{G}}$-Brownian motion and we set

$$\alpha_t = \alpha_0 \exp\{\eta t\}.$$

A constant interest rate $r \geq 0$ is employed for simplicity, so we have

$$B_t = \exp\{rt\}.$$

Next, we set $\lambda_t = \lambda > 0$, i.e. we employ a constant default intensity. The assumptions of Sect. 12.3 in Filipović (2009) are still satisfied and we have

$$P(\tau > t \mid \mathcal{G}_\infty) = P(\tau > t \mid \mathcal{G}_t) = \exp\{-\lambda t\}$$

and

$$P(t < \tau \leq u \mid \mathcal{G}_\infty \vee \mathcal{H}_t) = \mathbf{1}_{\{\tau > t\}}\big(1 - \exp\{-\lambda(u-t)\}\big),$$

so the conditional density of τ given $\tau > t$ is exponential with parameter λ, i.e.

$$\mathbf{1}_{\{\tau > t\}} \lambda \exp\{-\lambda(u-t)\}. \tag{14.5.30}$$

This facilitates computations greatly, as we now demonstrate.

Assume that $V_t \geq 0$, $\forall t \in [0, T]$, and that the portfolio V does not generate any cash flows on the interval $[0, T]$. Furthermore, we assume that V is fair, see Definition 1.3.4, so $\frac{V}{S^{\delta_*}}$ forms an $(\underline{\mathcal{A}}, P)$-martingale,

$$\begin{aligned}
CVA_t &= S_t^{\delta_*} E\left(\frac{V_\tau^+}{S_\tau} \mathbf{1}_{\{t<\tau\leq T\}} \,\middle|\, \mathcal{A}_t\right)\\
&= S_t^{\delta_*} E\left(E\left(\frac{V_\tau^+}{S_\tau} \mathbf{1}_{\{t<\tau\leq T\}} \,\middle|\, \mathcal{G}_\infty \vee \mathcal{H}_t\right) \,\middle|\, \mathcal{A}_t\right)\\
&= \mathbf{1}_{\{\tau>t\}} S_t^{\delta_*} E\left(\int_t^T \frac{V_u^+}{S_u^{\delta_*}} \lambda \exp\{-\lambda(u-t)\}\,du \,\middle|\, \mathcal{A}_t\right)\\
&= \mathbf{1}_{\{\tau>t\}} S_t^{\delta_*} \int_t^T E\left(\frac{V_u^+}{S_u^{\delta_*}} \,\middle|\, \mathcal{A}_t\right) \lambda \exp\{-\lambda(u-t)\}\,du.
\end{aligned}$$

Since $V_t^+ = V_t$, one can compute

$$\begin{aligned}
S_t^{\delta_*} E\left(\frac{V_u^+}{S_u^{\delta_*}} \,\middle|\, \mathcal{A}_t\right) &= S_t^{\delta_*} E\left(\frac{V_u}{S_u^{\delta_*}} \,\middle|\, \mathcal{A}_t\right)\\
&= S_t^{\delta_*} E\left(E\left(\frac{V_T}{S_T^{\delta_*}} \,\middle|\, \mathcal{A}_u\right) \mathcal{A}_t\right)\\
&= S_t^{\delta_*} E\left(\frac{V_T}{S_T^{\delta_*}} \,\middle|\, \mathcal{A}_t\right)\\
&= V_t.
\end{aligned}$$

Hence we get

$$CVA_t = \mathbf{1}_{\{\tau>t\}} V_t \big(1 - \exp\{-\lambda(T-t)\}\big). \tag{14.5.31}$$

We point out that Eq. (14.5.31) can be used to deal with zero coupon bonds, European call options and swaptions, but not, for example, to deal with swaps, as the latter can assume negative values. We do, however, recall our previous observation that there exists a close link with forward start swaptions, which we now exploit. Set $V_t = FS_{\kappa,T_0}(t)$ and we consider CVA over the period $[0, T]$, where $T \leq T_0$, then using the density in Eq. (14.5.30) we obtain

$$\begin{aligned}
CVA_t &= S_t^{\delta_*} E\left(\frac{V_\tau^+}{S_\tau^{\delta_*}} \mathbf{1}_{\{t<\tau\leq T\}} \,\Big|\, \mathcal{A}_t \right) \\
&= S_t^{\delta_*} E\left(E\left(\frac{V_\tau^+}{S_\tau^{\delta_*}} \mathbf{1}_{\{t<\tau\leq T\}} \,\Big|\, \mathcal{G}_\infty \vee \mathcal{H}_t \right) \Big|\, \mathcal{A}_t \right) \\
&= \mathbf{1}_{\{\tau>t\}} S_t^{\delta_*} E\left(\int_t^T \frac{V_u^+}{S_u^{\delta_*}} \lambda \exp\{-\lambda(u-t)\}\, du \,\Big|\, \mathcal{A}_t \right) \\
&= \mathbf{1}_{\{\tau>t\}} \int_t^T S_t^{\delta_*} E\left(\frac{V_u^+}{S_u^{\delta_*}} \,\Big|\, \mathcal{A}_t \right) \lambda \exp\{-\lambda(u-t)\}\, du \\
&= \mathbf{1}_{\{\tau>t\}} \int_t^T S_t^{\delta_*} E\left(\frac{(FS_{\kappa,T_0}(u))^+}{S_u^{\delta_*}} \,\Big|\, \mathcal{A}_t \right) \lambda \exp\{-\lambda(u-t)\}\, du \\
&= \mathbf{1}_{\{\tau>t\}} \int_t^T PS_{\kappa,T_0,u}(t) \lambda \exp\{-\lambda(u-t)\}\, du.
\end{aligned}$$

We remark that under the MMM, the value of a forward start swaption amounts to the computation of a one-dimensional integral. From Sect. 14.1, we recall

$$\begin{aligned}
PS_{\kappa,T_0,T}(t) &= S_t^{\delta_*} E\left(\frac{(FS_{\kappa,T_0}(T))^+}{S_T^{\delta_*}} \,\Big|\, \mathcal{A}_t \right) \\
&= S_t^{\delta_*} E\left(\frac{(P_{T_0}(T) - \sum_{j=1}^n c_j P_{T_j}(T))^+}{S_T^{\delta_*}} \,\Big|\, \mathcal{A}_t \right).
\end{aligned}$$

For the reduced form model,

$$S_t^{\delta_*} = B_t \bar{S}_t^{\delta_*},$$

and $\bar{S}^{\delta_*}$ is a time-changed squared Bessel process of dimension four, the transition density of which is known in closed-form, see Sect. 3.1:

$$\begin{aligned}
&p_4\big(\varphi(t), x; \varphi(T), y\big) \\
&= \frac{1}{2(\varphi(T)-\varphi(t))} \left(\frac{y}{x}\right)^{\frac{1}{2}} \exp\left\{ -\frac{x+y}{2(\varphi(T)-\varphi(t))} \right\} I_1\left(\frac{\sqrt{xy}}{\varphi(T)-\varphi(t)} \right).
\end{aligned}$$

Furthermore, in Sect. 3.3 we derived

$$P_T(t) = \exp\{-r(T-t)\} \left(1 - \exp\left\{ -\frac{\bar{S}_t^{\delta_*}}{2(\varphi(T)-\varphi(t))} \right\} \right).$$

We define

$$f(t,T,y)=\exp\{-r(T-t)\}\left(1-\exp\left\{-\frac{y}{2(\varphi(T)-\varphi(t))}\right\}\right).$$

Hence we get

$$\begin{aligned}
&PS_{\kappa,T_0,T}(t)\\
&\quad=\bar{S}_t^{\delta_*}\exp\{-r(T-t)\}\\
&\qquad\times\int_0^\infty\frac{(f(T,T_0,y)-\sum_{j=1}^n c_j f(T,T_j,y))}{y}p_4\big(\varphi(t),\bar{S}_t^{\delta_*},\varphi(T),y\big)\,dy,
\end{aligned}$$

and regarding CVA, we obtain

$$\begin{aligned}
CVA_t&=\mathbf{1}_{\{\tau>t\}}\bar{S}_t^{\delta_*}\int_t^T\int_0^\infty\exp\{-(r+\lambda)(u-t)\}\\
&\quad\times\frac{(f(u,T_0,y)-\sum_{j=1}^n c_j f(u,T_j,y))^+}{y}p_4\big(\varphi(t),\bar{S}_t^{\delta_*},\varphi(u),y\big)\lambda\,dy\,du.
\end{aligned}$$

Hence the CVA associated with a swap can be expressed in terms of a two-dimensional integral, which is easily evaluated e.g. using the methods from Chap. 12.

Chapter 15
Continuous Stochastic Processes

It is the aim of this chapter to briefly recall basic definitions and results concerning continuous stochastic processes. We also discuss the Itô formula, the Feynman-Kac formula and existence and uniqueness of solutions of SDEs driven by Wiener processes. This chapter is tailored towards the content presented in this book. For a more detailed discussion, also covering SDEs driven by Poisson processes and Poisson random measures, the reader is referred to e.g. Chap. 1 in Platen and Bruti-Liberati (2010).

15.1 Stochastic Processes

15.1.1 Stochastic Process

If not otherwise stated, throughout the chapter we assume that there exists a common underlying probability space $(\Omega, \mathcal{A}, P)$ consisting of the sample space Ω, the sigma-algebra or collection of events $\mathcal{A}$, and the probability measure P, see for instance Shiryaev (1984). One typically observes a collection of random variables $X_{t_0}, X_{t_1}, \ldots$, which describe the evolution of financial quantities, at the observation times $t_0 < t_1 < \cdots$. The collection of random variables is indexed by the time t, and we call $\mathcal{T}$ the *time set*. The *state space* of X is usually the d-dimensional Euclidean space $\Re^d$, $d \in \mathcal{N} = \{1, 2, \ldots\}$, or a subset of it. However in Chaps. 10 and 11, we also consider matrix-valued diffusions.

Definition 15.1.1 We call a family $X = \{X_t, \, t \in \mathcal{T}\}$ of random variables $X_t \in \Re^d$ a *d-dimensional stochastic process*, where the totality of its finite-dimensional distribution functions

$$F_{X_{t_{i_1}},\ldots,X_{t_{i_j}}}(x_{i_1}, \ldots, x_{i_j}) = P(X_{t_{i_1}} \leq x_{i_1}, \ldots, X_{t_{i_j}} \leq x_{i_j}) \qquad (15.1.1)$$

for $i_j \in \{0, 1, \ldots\}$, $j \in \mathcal{N}$, $x_{i_j} \in \Re^d$ and $t_{i_j} \in \mathcal{T}$ determines its probability law.

J. Baldeaux, E. Platen, *Functionals of Multidimensional Diffusions with Applications to Finance*, Bocconi & Springer Series 5, DOI 10.1007/978-3-319-00747-2_15,

We set the time set to the interval $\mathcal{T} = [0, \infty)$ if not otherwise stated. On some occasions the time set may become the bounded interval $[0, T]$ for $T \in (0, \infty)$ or a set of discrete time points $\{t_0, t_1, t_2, \ldots\}$, where $t_0 < t_1 < t_2 < \cdots$.

15.1.2 Filtration as Information Structure

As already indicated previously, our modeling is based on a given probability space $(\Omega, \mathcal{A}, P)$. On such a probability space we consider dynamics that are based on the observation of a continuous time stochastic vector process $\boldsymbol{X} = \{\boldsymbol{X}_t \in \Re^d, t \geq 0\}$, $d \in \mathcal{N}$. We denote by $\hat{\mathcal{A}}_t$ the time t *information set*, which is the sigma-algebra generated by the events that are known at time $t \geq 0$, see Shiryaev (1984). Our interpretation of $\hat{\mathcal{A}}_t$ is that it represents the information available at time t, which is obtained from the observed values of the vector process $\boldsymbol{X}$ up to time t. More precisely, it is the sigma-algebra

$$\hat{\mathcal{A}}_t = \sigma\{\boldsymbol{X}_s\colon s \in [0, t]\}$$

generated from all observations of $\boldsymbol{X}$ up to time t. Since information is not lost, the increasing family

$$\underline{\hat{\mathcal{A}}} = \{\hat{\mathcal{A}}_t,\, t \geq 0\}$$

of information sets $\hat{\mathcal{A}}_t$ satisfies, for any sequence $0 \leq t_1 < t_2 < \cdots < \infty$ of observation times, the relation $\hat{\mathcal{A}}_{t_1} \subseteq \hat{\mathcal{A}}_{t_2} \subseteq \cdots \subseteq \hat{\mathcal{A}}_\infty = \bigcup_{t \geq 0} \hat{\mathcal{A}}_t$.

For technical reasons one introduces the information set $\mathcal{A}_t$ as the *augmented* sigma-algebra of $\hat{\mathcal{A}}_t$ for each $t \geq 0$. It is augmented by every null set in $\hat{\mathcal{A}}_\infty$ such that it belongs to $\mathcal{A}_0$, and also to each $\hat{\mathcal{A}}_t$, saying that $\mathcal{A}_t$ is *complete*. Define $\mathcal{A}_{t+} = \bigcap_{\varepsilon > 0} \mathcal{A}_{t+\varepsilon}$ as the sigma-algebra of events immediately after $t \in [0, \infty)$. The family $\underline{\mathcal{A}} = \{\mathcal{A}_t, t \geq 0\}$ is called *right continuous* if $\mathcal{A}_t = \mathcal{A}_{t+}$ holds for every $t \geq 0$. Such a right-continuous family $\underline{\mathcal{A}} = \{\mathcal{A}_t, t \geq 0\}$ of information sets one calls a *filtration*. A filtration can model the evolution of information as it becomes available over time. We define $\mathcal{A}$ as the smallest sigma-algebra that contains $\mathcal{A}_\infty = \bigcup_{t \geq 0} \mathcal{A}_t$. From now on, if not stated otherwise, we always assume in this chapter a *filtered probability space* $(\Omega, \mathcal{A}, \underline{\mathcal{A}}, P)$ to be given.

Any right-continuous stochastic process $Y = \{Y_t, t \geq 0\}$ generates its *natural filtration* $\underline{\mathcal{A}}^Y = \{\mathcal{A}_t^Y, t \geq 0\}$, which is the sigma-algebra generated by Y up to time t. For a given model with a vector process $\boldsymbol{X}$ we typically set $\underline{\mathcal{A}} = \underline{\mathcal{A}}^{\boldsymbol{X}}$ with $\mathcal{A}_t = \mathcal{A}_t^{\boldsymbol{X}}$.

If for a process $Z = \{Z_t, t \geq 0\}$ and each time $t \geq 0$ the random variable Z_t is $\mathcal{A}_t^{\boldsymbol{X}}$-measurable, then Z is called *adapted* to $\underline{\mathcal{A}}^{\boldsymbol{X}} = \{\mathcal{A}_t^{\boldsymbol{X}}, t \geq 0\}$. The history of the process Z until time t is then covered by the information set $\mathcal{A}_t^{\boldsymbol{X}}$.

15.1.3 Conditional Expectations

The notion of conditional expectation is central to many of the concepts that arise in applications of stochastic processes. The mean value or expectation $E(X)$ is the coarsest estimate that we have for an *integrable random variable* X, that is, for which $E(|X|) < \infty$, see Shiryaev (1984). If we know that some event A has occurred we may be able to improve on this estimate. For instance, suppose that the event $A = \{\omega \in \Omega\colon X(\omega) \in [a, b]\}$ has occurred. Then in evaluating our estimate of the value of X we need only to consider corresponding values of X in $[a, b]$ and weight them according to their likelihood of occurrence, which thus becomes the conditional probability given this event, see Shiryaev (1984).

The resulting estimate is called the *conditional expectation* of X given the event A and is denoted by $E(X|A)$. For a continuous random variable X with a density function f_X the corresponding *conditional density* is

$$f_X(x \mid A) = \begin{cases} 0 & \text{for } x < a \text{ or } b < x \\ \frac{f_X(x)}{\int_a^b f_X(s)\,ds} & \text{for } x \in [a, b] \end{cases},$$

with the *conditional expectation*

$$E(X \mid A) = \int_{-\infty}^{\infty} x f_X(x \mid A)\,dx = \frac{\int_a^b x f_X(x)\,dx}{\int_a^b f_X(x)\,dx}, \tag{15.1.2}$$

which is conditioned on the event A and is, thus, a number.

More generally let $(\Omega, \mathcal{A}, P)$ be a given probability space with an integrable random variable X. We denote by $\mathcal{S}$ a sub-sigma-algebra of $\mathcal{A}$, thus representing a coarser type of information than is given by $\mathcal{S} \subset \mathcal{A}$. We then define the *conditional expectation* of X with respect to $\mathcal{S}$, which we denote by $E(X \mid \mathcal{S})$, as an $\mathcal{S}$-measurable function satisfying the equation

$$\int_Q E(X \mid \mathcal{S})(\omega)\,dP(\omega) = \int_Q X(\omega)\,dP(\omega), \tag{15.1.3}$$

for all $Q \in \mathcal{S}$. The Radon-Nikodym theorem, see Shiryaev (1984), guarantees the existence and uniqueness of the random variable $E(X \mid \mathcal{S})$. Note that $E(X \mid \mathcal{S})$ is a random variable defined on the coarser probability space $(\Omega, \mathcal{S}, P)$ and thus on $(\Omega, \mathcal{A}, P)$. However, X is usually not a random variable on $(\Omega, \mathcal{S}, P)$, but when it is we have

$$E(X \mid \mathcal{S}) = X, \tag{15.1.4}$$

which is the case when X is $\mathcal{S}$-measurable.

For nested sigma-algebras $\mathcal{S} \subset \mathcal{T} \subset \mathcal{A}$ and an integrable random variable X we have the *iterated conditional expectations*

$$E\big(E(X \mid \mathcal{T}) \bigm| \mathcal{S}\big) = E(X \mid \mathcal{S}) \tag{15.1.5}$$

almost surely. Since most equations and relations formulated in this book hold almost surely, we typically suppress these words. When X is independent of the events in $\mathcal{S}$ we have

$$E(X \mid \mathcal{S}) = E(X). \tag{15.1.6}$$

By setting $\mathcal{S} = \{\emptyset, \Omega\}$ it can be seen that

$$E\big(E(X \mid \mathcal{S})\big) = E(X). \tag{15.1.7}$$

Conditional expectations have similar properties to those of ordinary integrals such as the linearity property

$$E(\alpha X + \beta Y \mid \mathcal{S}) = \alpha E(X \mid \mathcal{S}) + \beta E(Y \mid \mathcal{S}), \tag{15.1.8}$$

where X and Y are integrable random variables and $\alpha, \beta \in \Re$ are deterministic constants. Furthermore, if X is $\mathcal{S}$-measurable, then

$$E(XY \mid \mathcal{S}) = X E(Y \mid \mathcal{S}). \tag{15.1.9}$$

Finally, one has the order preserving property

$$E(X \mid \mathcal{S}) \leq E(Y \mid \mathcal{S}) \tag{15.1.10}$$

if $X \leq Y$ a.s.

The conditional expectation $E(X \mid \mathcal{S})$ is in some sense obtained by smoothing X over the events in $\mathcal{S}$. Thus, the finer the information set $\mathcal{S}$, the more $E(X \mid \mathcal{S})$ resembles the random variable X.

15.1.4 Wiener Process

The Wiener process or Brownian motion is a stochastic process with stationary independent increments. These mathematical properties make it suitable as fundamental building block in stochastic modeling. The random increments $X_{t_{j+1}} - X_{t_j}$, $j \in \{0, 1, \ldots, n-1\}$, of these processes are independent for any sequence of time instants $t_0 < t_1 < \cdots < t_n$ in $[0, \infty)$ for all $n \in \mathcal{N}$. If $t_0 = 0$ is the smallest time instant, then the initial value X_0 and the random increment $X_{t_j} - X_0$ for any other $t_j \in [0, \infty)$ are also required to be independent. Additionally, the increments $X_{t+h} - X_t$ are assumed to be stationary, that is $X_{t+h} - X_t$ has the same distribution as $X_h - X_0$ for all $h > 0$ and $t \geq 0$.

Definition 15.1.2 We define the *standard Wiener process* $W = \{W_t, t \geq 0\}$ as an $\underline{\mathcal{A}}$-adapted process with Gaussian stationary independent increments and continuous sample paths for which

$$W_0 = 0, \qquad \mu(t) = E(W_t) = 0, \qquad \text{Var}(W_t - W_s) = t - s \tag{15.1.11}$$

for all $t \geq 0$ and $s \in [0, t]$.

We now recall the following basic distributional properties of the Wiener process, see e.g. Borodin and Salminen (2002).

Lemma 15.1.3 *The Wiener process* $W = \{W_t, t \geq 0\}$ *enjoys the following distributional properties*:

- spatial homogeneity: *for every* $x \in \Re$, *the process* $x + W$ *is a Brownian motion started at* x;
- symmetry: $-W$ *is a Brownian motion*;

- Scaling: *for every* $c > 0$ *the process* $\{\sqrt{c}W_{t/c},\ t \geq 0\}$ *is a Brownian motion*;
- time inversion: *the process given by*

$$Z_t := \begin{cases} 0 & \text{for } t = 0 \\ tW_{1/t} & \text{for } t > 0 \end{cases} \tag{15.1.12}$$

is a Wiener process;
- time reversibility: *for a given* $t > 0$ *the following equality in law holds*

$$\{W_s\colon 0 \leq s \leq t\} \stackrel{d}{=} \{W_{t-s} - W_t\colon 0 \leq s \leq t\}. \tag{15.1.13}$$

There exists also a multidimensional version of the above Wiener process. We call the vector process $\boldsymbol{W} = \{\boldsymbol{W}_t = (W_t^1, W_t^2, \ldots, W_t^m)^\top,\ t \geq 0\}$ an *m-dimensional standard Wiener process* if each of its components $W^j = \{W_t^j,\ t \geq 0\}$, $j \in \{1, 2, \ldots, m\}$ is a scalar $\underline{\mathcal{A}}$-adapted standard Wiener process and the Wiener processes W^k and W^j are independent for $k \neq j$, $k, j \in \{1, 2, \ldots, m\}$.

This means that according to Definition 15.1.2, each random variable W_t^j is Gaussian and $\mathcal{A}_t$-measurable with

$$E\big(W_t^j \mid \mathcal{A}_0\big) = 0 \tag{15.1.14}$$

and we have independent increments $W_t^j - W_s^j$ such that

$$E\big(W_t^j - W_s^j \mid \mathcal{A}_s\big) = 0 \tag{15.1.15}$$

for $t \geq 0$, $s \in [0, t]$ and $j \in \{1, 2, \ldots, m\}$. Moreover, one has the additional property that

$$E\big(\big(W_t^j - W_s^j\big)\big(W_t^k - W_s^k\big) \mid \mathcal{A}_s\big) = \begin{cases} (t-s) & \text{for } k = j \\ 0 & \text{otherwise} \end{cases} \tag{15.1.16}$$

for $t \geq 0$, $s \in [0, t]$ and $j, k \in \{1, 2, \ldots, m\}$.

15.2 Supermartingales and Martingales

15.2.1 Martingales

We define the quantity F_s for $s \in [0, \infty)$ as the least-squares estimate of a future value X_t at the future time $t \in [s, \infty)$ under the information given by $\mathcal{A}_s$ at time s. This estimate is $\mathcal{A}_s$-measurable and minimizes the error

$$\varepsilon_s = E\big((X_t - F_s)^2\big)$$

over all possible $\mathcal{A}_s$-measurable estimates, assuming that $\varepsilon_s < \infty$. The random variable F_s is simply the least-squares projection of X_t given the information at time $s \in [0, t]$. It is obtained by the conditional expectation

$$F_s = E(X_t \mid \mathcal{A}_s), \tag{15.2.17}$$

for all $s \in [0, t]$. This leads to the following definition:

Definition 15.2.1 A continuous time stochastic process $X = \{X_t,\ t \geq 0\}$ is called an $(\underline{\mathcal{A}},\ P)$-*martingale* or simply a *martingale*, if it satisfies the equation

$$X_s = E(X_t \mid \mathcal{A}_s) \tag{15.2.18}$$

for all $s \in [0, t]$ and the integrability condition

$$E\big(|X_t|\big) < \infty \tag{15.2.19}$$

for all $t \geq 0$.

An example of an $(\underline{\mathcal{A}},\ P)$-martingale is a Wiener process $W = \{W_t, t \geq 0\}$ on a filtered probability space $(\Omega, \mathcal{A}, \underline{\mathcal{A}}, P)$.

There are many other continuous time stochastic processes that form martingales. For example, using again the standard Wiener process W it can be shown that the process

$$X = \big\{X_t = W_t^2 - t,\ t \geq 0\big\} \tag{15.2.20}$$

is an $(\underline{\mathcal{A}},\ P)$-martingale.

15.2.2 Super- and Submartingales

Some systematically upward or downward "trending" stochastic processes can be captured by the following notions:

Definition 15.2.2 An $\underline{\mathcal{A}}$-adapted process $X = \{X_t,\ t \geq 0\}$ is an $(\underline{\mathcal{A}},\ P)$-*supermartingale* (*submartingale*) if

$$E\big(|X_t|\big) < \infty \tag{15.2.21}$$

and

$$X_s \overset{(\leq)}{\geq} E(X_t \mid \mathcal{A}_s) \tag{15.2.22}$$

for all $s \in [0, \infty)$ and $t \in [s, \infty)$.

This means, a supermartingale is "trending" systematically downward or has no trend. We call a supermartingale a *strict supermartingale* (strict submartingale) if the inequality in (15.2.22) is always a strict inequality.

15.2.3 Stopping Times

We define stopping times on the filtered probability space $(\Omega, \mathcal{A}, \underline{\mathcal{A}}, P)$ introduced above.

Definition 15.2.3 A random variable $\tau : \Omega \to [0, \infty)$ is called a *stopping time* with respect to the filtration $\underline{\mathcal{A}}$ if for all $t \geq 0$ one has

$$\{\tau \leq t\} \in \mathcal{A}_t. \tag{15.2.23}$$

The relation (15.2.23) expresses the fact that the event $\{\tau \leq t\}$ is $\mathcal{A}_t$-measurable and, thus, observable at time t. The information set associated with a stopping time τ is defined as

$$\mathcal{A}_\tau = \sigma\big\{A \in \mathcal{A}\colon\ A \cap \{\tau \leq t\} \in \mathcal{A}_t \text{ for } t \geq 0\big\}. \tag{15.2.24}$$

It represents the information available before and at the stopping time τ. For example, the first time

$$\tau(a) = \inf\{t \geq 0\colon\ W_t = a\} \tag{15.2.25}$$

when a Wiener process W reaches a level $a \in \Re$ is a stopping time.

One calls a sigma-algebra *predictable* when it is generated by left-continuous $\underline{\mathcal{A}}$-adapted processes with right hand limits. We exclude in a predictable sigma-algebra, in principle, all information about the time instant when a sudden non predictable event occurs. Immediately after the event a predictable sigma-algebra already contains this information.

A stochastic process $X = \{X_t, t \geq 0\}$, where X_τ is for each stopping time τ measurable with respect to a predictable sigma-algebra, is called *predictable*. For example, all continuous stochastic processes are predictable.

A stopping time is called *predictable*, if $\mathcal{A}_\tau$ is predictable. This means, $\mathcal{A}_\tau$ is generated by left-continuous stochastic processes with right hand limits. The first hitting time $\tau(a)$, given in (15.2.25), of the level a by the Wiener process W is predictable.

For $a, b \in \Re$ we employ the notation $a \wedge b = \min(a, b)$ and $a \vee b = \max(a, b)$. Let us summarize the following useful properties of stopping times τ and τ', see, for instance, Karatzas and Shreve (1991) and Elliott (1982):

(i) τ is $\mathcal{A}_\tau$-measurable;
(ii) for a continuous $\underline{\mathcal{A}}$-adapted process $X = \{X_t, t \geq 0\}$ the random variable X_τ is $\mathcal{A}_\tau$-measurable;
(iii) if $P(\tau \leq \tau') = 1$, then $\mathcal{A}_\tau \subseteq \mathcal{A}_{\tau'}$;
(iv) the random variables $\tau \wedge \tau'$, $\tau \vee \tau'$ and $(\tau + \tau')$ are stopping times;
(v) if for a real valued random variable Y we have $E(|Y|) < \infty$ and $P(\tau \leq \tau') = 1$, then

$$E(Y \mid \mathcal{A}_\tau) = E(Y \mid \mathcal{A}_{\tau \wedge \tau'}) \tag{15.2.26}$$

and

$$E\big(E(Y \mid \mathcal{A}_\tau) \,\big|\, \mathcal{A}_{\tau'}\big) = E(Y \mid \mathcal{A}_\tau). \tag{15.2.27}$$

If $X = \{X_t,\ t \geq 0\}$ is a right continuous $(\underline{\mathcal{A}}, P)$-supermartingale, then the supermartingale property (15.2.18) still holds if the times s and t in (15.2.18) are bounded stopping times. More precisely, Doob's *Optional Sampling Theorem* states the following result, see Doob (1953):

Theorem 15.2.4 (Doob) *If $X = \{X_t, t \geq 0\}$ is a right continuous $(\underline{\mathcal{A}}, P)$-supermartingale on $(\Omega, \mathcal{A}, \underline{\mathcal{A}}, P)$, then it holds for two bounded stopping times τ and τ' with $\tau \leq \tau'$ almost surely that*

$$E(X_{\tau'} \mid \mathcal{A}_\tau) \leq X_\tau. \tag{15.2.28}$$

Furthermore, if X is additionally an $(\underline{\mathcal{A}}, P)$-martingale, then equality holds in (15.2.28).

15.3 Quadratic Variation and Covariation

15.3.1 Quadratic Variation

For simplicity, let us consider an *equidistant time discretization*

$$\{t_k = kh \colon k \in \{0, 1, \ldots\}\}, \tag{15.3.29}$$

with small time steps of length $h > 0$, such that $0 = t_0 < t_1 < t_2 < \cdots$. Thus, we have the discretization times $t_k = k\,h$ for $k \in \{0, 1, \ldots\}$. The specific structure of the time discretization is not essential for the definition below, as long as the maximum time step size vanishes. There is no need to have the time discretization equidistant. However, it makes our presentation simpler.

For a given stochastic process X the *quadratic variation process* $[X] = \{[X]_t, t \geq 0\}$ is defined as the limit in probability as $h \to 0$ of the sums of squared increments of the process X, provided this limit exists and is unique, see Jacod and Shiryaev (2003) and Protter (2005). More precisely, we have at time t the *quadratic variation*

$$[X]_t \overset{P}{=} \lim_{h \to 0} [X]_{h,t}, \tag{15.3.30}$$

where the *approximate quadratic variation* $[X]_{h,t}$ is given by the sum

$$[X]_{h,t} = \sum_{k=1}^{n_t} (X_{t_k} - X_{t_{k-1}})^2. \tag{15.3.31}$$

Here n_t denotes the integer

$$n_t = \max\{k \in \mathcal{N} \colon t_k \leq t\}, \tag{15.3.32}$$

which is the index of the last discretization point before or including $t \geq 0$.

The value of the quadratic variation process $[W] = \{[W]_t, t \geq 0\}$ at time t for a standard Wiener process W is given by the relation

$$[W]_t = t \tag{15.3.33}$$

for $t \geq 0$.

15.3.2 Covariation

For the definition of *covariation* the same equidistant time discretization, as given in (15.3.29), is now used. For two continuous stochastic processes Z_1 and Z_2 the *covariation process* $[Z_1, Z_2] = \{[Z_1, Z_2]_t,\ t \geq 0\}$ is defined as the limit in probability as $h \to 0$ of the values of the *approximate covariation process* $[Z_1, Z_2]_{h,\cdot}$ with

$$[Z_1, Z_2]_{h,t} = \sum_{k=1}^{n_t} \big(Z_1(t_k) - Z_1(t_{k-1})\big)\big(Z_2(t_k) - Z_2(t_{k-1})\big) \qquad (15.3.34)$$

for $t \geq 0$ and $h > 0$, see (15.3.32). More precisely, at time $t \geq 0$ we obtain the *covariation*

$$[Z_1, Z_2]_t \stackrel{P}{=} \lim_{h \to 0} [Z_1, Z_2]_{h,t}, \qquad (15.3.35)$$

where $[Z_1, Z_2]_{h,t}$ is the above approximate covariation.

15.3.3 Local Martingales

As we will see, certain stochastic processes become martingales when properly stopped but are not true martingales.

Definition 15.3.1 A stochastic process $X = \{X_t, t \geq 0\}$ is an $(\underline{\mathcal{A}}, P)$-*local martingale* if there exists an increasing sequence $(\tau_n)_{n \in \mathcal{N}}$ of stopping times, that may depend on X, such that $\lim_{n\to\infty} \tau_n \stackrel{\text{a.s.}}{=} \infty$ and each stopped process

$$X^{\tau_n} = \big\{X_t^{\tau_n} = X_{t \wedge \tau_n},\ t \geq 0\big\} \qquad (15.3.36)$$

is an $(\underline{\mathcal{A}}, P)$-martingale, where $t \wedge \tau_n = \min(t, \tau_n)$.

If X is a local martingale, then the value X_s does, in general, not equal the conditional expectation $E(X_t \mid \mathcal{A}_s)$ for $s \in [0, \infty)$ and $t \in [s, \infty)$. A local martingale that is not a martingale is called a *strict local martingale*.

The following holds, see Protter (2005) and Shiryaev (1984):

Lemma 15.3.2

(i) *An almost surely nonnegative (negative) $(\underline{\mathcal{A}}, P)$-local martingale is an $(\underline{\mathcal{A}}, P)$-supermartingale (submartingale).*

(ii) *An almost surely uniformly bounded $(\underline{\mathcal{A}}, P)$-local martingale is an $(\underline{\mathcal{A}}, P)$-martingale.*

(iii) *A square integrable $(\underline{\mathcal{A}}, P)$-local martingale X is a square integrable $(\underline{\mathcal{A}}, P)$-martingale if and only if*

$$E\big([X]_T\big) < \infty \qquad (15.3.37)$$

for all $T \in [0, \infty)$.

(iv) *A nonnegative $(\underline{\mathcal{A}}, P)$-local martingale $X = \{X_t, t \geq 0\}$ with $E(X_t | \mathcal{A}_s) < \infty$ for all $0 \leq s \leq t < \infty$ is an $(\underline{\mathcal{A}}, P)$-supermartingale.*

15.3.4 Identification of Martingales as Wiener Processes

The Wiener process plays a central role in stochastic calculus and is the basic building block for modeling continuous uncertainty. We saw that the Wiener process is a martingale and from (15.3.33) it follows that its quadratic variation $[W]_t$ equals the time t. Note that the converse can also be shown. *Lévy's Theorem* provides this important result. Its derivation can be found, for instance in Platen and Heath (2010).

Theorem 15.3.3 (Lévy) *For $m \in \mathcal{N}$ let A be a given m-dimensional vector process $\boldsymbol{A} = \{\boldsymbol{A}_t = (A_t^1, A_t^2, \ldots, A_t^m)^\top, t \geq 0\}$ on a filtered probability space $(\Omega, \mathcal{A}, \underline{\mathcal{A}}, P)$. If each of the processes $A^i = \{A_t^i, t \geq 0\}$ is a continuous, square integrable $(\underline{\mathcal{A}}, P)$-martingale that starts at 0 at time $t = 0$ and their covariations are of the form*

$$[A^i, A^k]_t = \begin{cases} t & \text{for } i = k \\ 0 & \text{for } i \neq k \end{cases} \tag{15.3.38}$$

for all $i, k \in \{1, 2, \ldots, m\}$ and $t \geq 0$, then the vector process $\boldsymbol{A}$ is an m-dimensional standard Wiener process on $[0, \infty)$. This means that each process A^i is a one-dimensional Wiener process that is independent of the other Wiener processes A^k for $k \neq i$.

This result implies that a continuous process $X = \{X_t, t \geq 0\}$ is a one-dimensional Wiener process if and only if both the process X and the process $Y = \{Y_t = X_t^2 - t, t \geq 0\}$ are martingales.

15.4 Itô Integral

We now introduce the notion of a stochastic integral. Consider the piecewise constant process $\xi = \{\xi(t), t \in [0, T]\}$ with $\xi(t) = \xi(t_k)$ at time $t \in [t_k, t_{k+1})$, $k \in \{0, 1, \ldots\}$ and $t_k = kh$ for $h > 0$. Then we have, for a suitable stochastic process $X = \{X_t, t \geq 0\}$,

$$\int_0^t \xi(s)\, dX_s = \sum_{k=1}^{n_t} \xi(t_{k-1})\{X_{t_k} - X_{t_{k-1}}\} + \xi(t_{n_t})\{X_t - X_{t_{n_t}}\}, \tag{15.4.1}$$

where

$$n_t = \max\{k \in \mathcal{N}\colon t_k \leq t\} \tag{15.4.2}$$

is the integer index of the latest discretization time before and including t.

For a left continuous, predictable stochastic process $\xi = \{\xi(t), t \geq 0\}$ as integrand with

$$\int_0^T \xi(s)^2\, d[X]_t < \infty \tag{15.4.3}$$

for all $T \in [0, \infty)$ almost surely, the *Itô integral* with respect to X is defined as the left continuous limit in probability

$$\int_0^t \xi(s)\, dX_s \stackrel{P}{=} \lim_{h \to 0} \sum_{k=1}^{n_t} \xi(t_{k-1})\{X_{t_k} - X_{t_{k-1}}\} \tag{15.4.4}$$

of a sequence of corresponding approximating sums for progressively finer time discretizations for $t \geq 0$.

For details on the definition of Itô integrals we refer to Karatzas and Shreve (1991), Kloeden and Platen (1999), Protter (2005) or Platen and Heath (2010). The most important characteristic of the Itô integral is that the evaluation point t_{k-1} for the integrand ξ is always taken at the left hand side of the discretization interval $[t_{k-1}, t_k)$.

To model more general dynamics than those of the Wiener process let $e = \{e_t,\, t \geq 0\}$ and $f = \{f_t, t \geq 0\}$ be predictable stochastic processes. Consider a stochastic process $Y = \{Y_t,\, t \geq 0\}$, where

$$Y_t = y_0 + \int_0^t e_s\, ds + \int_0^t f_s\, dW_s \tag{15.4.5}$$

for $t \geq 0$ and initial value $Y_0 = y_0$. Here $W = \{W_t,\, t \geq 0\}$ is a standard Wiener process. The first integral is a random ordinary Riemann-Stieltjes integral, assuming

$$\int_0^t |e_s|\, ds < \infty \tag{15.4.6}$$

for all $t \geq 0$ almost surely (a.s.). The second integral is an Itô integral with respect to the Wiener process W, see (15.4.4), where we assume that

$$\int_0^t |f_s|^2\, ds < \infty \tag{15.4.7}$$

almost surely for all $t \geq 0$. It is common to express the integral equation (15.4.5) in an equivalent short hand notation, i.e. Itô differential equation in the form

$$dY_t = e_t\, dt + f_t\, dW_t \tag{15.4.8}$$

for $t \geq 0$ with $Y_0 = y_0$. Equation (15.4.8) is simply another way of writing (15.4.5). The processes e and f are called *drift* and *diffusion coefficients*, respectively. The concept of an Itô differential allows the modeling of rather general dynamics.

The above definitions of an Itô integral and Itô differential can be extended to the case of multidimensional integrands ξ and integration with respect to several independent standard Wiener processes.

15.4.1 Some Properties of Itô Integrals

Consider two $\underline{\mathcal{A}}$-adapted independent Wiener processes W^1 and W^2. Recall that $(W_t^i - W_s^i)$ is independent of $\mathcal{A}_s$ for $t \geq 0$, $s \in [0, t]$ and $i \in \{1, 2\}$. It is useful to specify for $T \geq 0$ the class $\mathcal{L}_T^2$ of predictable, square integrable integrands $f = \{f_t, t \in [0, T]\}$ in the form that

$$\int_0^T E(f_t^2)\, dt < \infty. \tag{15.4.9}$$

The Itô integral exhibits the following important properties:

1. *linearity*: for $T \geq 0$, $t \in [0, T]$, $s \in [0, t]$, $Z_1, Z_2 \in \mathcal{L}_T^2$, and $\mathcal{A}_s$-measurable square integrable random variables A and B it holds

$$\int_s^t \big(A\, Z_1(u) + B\, Z_2(u)\big)\, dW_u^1 = A \int_s^t Z_1(u)\, dW_u^1 + B \int_s^t Z_2(u)\, dW_u^1 \tag{15.4.10}$$

2. *local martingale property*: for ξ predictable with

$$\int_0^T \xi(u)^2\, du < \infty \tag{15.4.11}$$

a.s. for all $T \in [0, \infty)$ the Itô integral $I_{\xi,W} = \{I_{\xi,W}(t) = \int_0^t \xi(s)\, dW_s,\ t \geq 0\}$ forms an $(\underline{\mathcal{A}}, P)$-local martingale
3. *martingale property*: assume that $I_{\xi,W}(t)$ forms a square integrable process, then it is a square integrable $(\underline{\mathcal{A}}, P)$-martingale if and only if

$$E\left(\int_0^T \xi(u)^2\, du\right) < \infty \tag{15.4.12}$$

for all $T \geq 0$
4. *correlation property*: for $T \geq 0$, $t \in [0, T]$, independent Wiener processes W^1 and W^2 and $Z_1, Z_2 \in \mathcal{L}_T^2$ it holds that

$$E\left(\int_0^t Z_1(u)\, dW_u^i \int_0^t Z_2(u)\, dW_u^j \,\Big|\, \mathcal{A}_s\right) = \begin{cases} \int_0^t E(Z_1(u) Z_2(u) \mid \mathcal{A}_s)\, du & \text{for } i = j \\ 0 & \text{otherwise} \end{cases} \tag{15.4.13}$$

for $i, j \in \{1, 2\}$
5. *covariation property*: for $t \geq 0$, independent Wiener processes W^1 and W^2 and predictable integrands Z_1 and Z_2 with $\int_0^t |Z_1(u)\, Z_2(u)|\, du < \infty$ almost surely it holds

$$\left[\int_0 Z_1(u)\, dW_u^i, \int_0 Z_2(u)\, dW_u^j\right]_t = \begin{cases} \int_0^t Z_1(u)\, Z_2(u)\, du & \text{for } i = j \\ 0 & \text{otherwise} \end{cases} \tag{15.4.14}$$

for $i, j \in \{1, 2\}$
6. *finite variation property*: for $t \geq 0$ and predictable Z_1 and Z_2 one has

$$\left[\int_0 Z_1(u)\, dW_u^1, \int_0 Z_2(u)\, du\right]_t = 0. \tag{15.4.15}$$

Note that some of the above employed conditions can be weakened, see Protter (2005).

15.5 Itô Formula

15.5.1 One-Dimensional Continuous Itô Formula

To be able to handle functions of solutions to stochastic processes, a chain rule, as it is known in classical calculus, is needed. Let $X = \{X_t,\ t \geq 0\}$ be a continuous stochastic process characterized by the Itô differential

$$dX_t = e_t\,dt + f_t\,dW_t \tag{15.5.16}$$

for $t \geq 0$ with initial value $X_0 = x_0$, see (15.4.8). Here $e = \{e_t, t \geq 0\}$ and $f = \{f_t, t \geq 0\}$ are two predictable stochastic processes.

Consider a function $u : [0, \infty) \times \Re \to \Re$ that is differentiable with respect to time t and twice continuously differentiable with respect to the spatial component x, that is, the partial derivatives $\frac{\partial u}{\partial t}$, $\frac{\partial u}{\partial x}$ and $\frac{\partial^2 u}{\partial x^2}$ exist and are continuous. To quantify the changes in $u(t, X_t)$ caused by changes in X_t one has the *Itô formula*

$$\begin{aligned} du(t, X_t) = {} & \left(\frac{\partial u(t, X_t)}{\partial t} + e_t \frac{\partial u(t, X_t)}{\partial x} + \frac{1}{2} (f_t)^2 \frac{\partial^2 u(t, X_t)}{\partial x^2} \right) dt \\ & + f_t \frac{\partial u(t, X_t)}{\partial x}\, dW_t \end{aligned} \tag{15.5.17}$$

for $t \geq 0$.

By using the notion of quadratic variation, see (15.3.30), we can rewrite the Itô formula (15.5.17) conveniently in the form

$$du(t, X_t) = \frac{\partial u(t, X_t)}{\partial t}\, dt + \frac{\partial u(t, X_t)}{\partial x}\, dX_t + \frac{1}{2} \frac{\partial^2 u(t, X_t)}{\partial x^2}\, d[X]_t \tag{15.5.18}$$

for $t \geq 0$.

15.5.2 Multidimensional Continuous Itô Formula

We now extend the Itô-formula to the multidimensional case. Consider a d-dimensional vector process $\boldsymbol{e} = \{\boldsymbol{e}_t = (e_t^1, \ldots, e_t^d)^T,\ t \geq 0\}$ with predictable components e^k, $k \in \{1, 2, \ldots, d\}$. Assume that

$$\int_0^T \left| e_z^k \right| dz < \infty \tag{15.5.19}$$

almost surely for all $k \in \{1, 2, \ldots, d\}$. The $d \times m$-matrix valued process $\boldsymbol{F} = \{\boldsymbol{F}_t = [F_t^{i,j}]_{i,j=1}^{d,m}, t \geq 0\}$ is assumed to have predictable elements $F^{i,j}$ with

$$\int_0^T \left(F_z^{i,j}\right)^2 dz < \infty \tag{15.5.20}$$

almost surely for $i \in \{1, 2, \ldots, d\}$, $j \in \{1, 2, \ldots, m\}$ and all $T \in (0, \infty)$, see Protter (2005). This allows us to introduce a d-dimensional continuous stochastic vector process $\boldsymbol{X} = \{\boldsymbol{X}_t = (X_t^1, X_t^2, \ldots, X_t^d)^\top, t \geq 0\}$, where the kth component X^k is defined via the Itô differential

$$dX_t^k = e_t^k\, dt + \sum_{j=1}^m F_t^{k,j}\, dW_t^j \tag{15.5.21}$$

for $t \geq 0$ and given $\mathcal{A}_0$-measurable initial value $\boldsymbol{X}_0 = (X_0^1, \ldots, X_0^d)^\top \in \Re^d$.

Consider now a function $u : [0, \infty) \times \Re^d \to \Re$ that has continuous partial derivatives $\frac{\partial u}{\partial t}$, $\frac{\partial u}{\partial x^k}$ and $\frac{\partial^2 u}{\partial x^k \partial x^i}$ for all $k, i \in \{1, 2, \ldots, d\}$, $t \geq 0$ and $\boldsymbol{x} = (x^1, x^2, \ldots, x^d)^\top$. The scalar stochastic process $u = \{u(t, X_t^1, X_t^2, \ldots, X_t^d),\ t \geq 0\}$ satisfies then the Itô differential characterized by the *Itô formula*

$$\begin{aligned} du\big(t, X_t^1, X_t^2, \ldots, X_t^d\big) &= \left\{\frac{\partial u}{\partial t} + \sum_{k=1}^d e_t^k \frac{\partial u}{\partial x^k} + \frac{1}{2} \sum_{j=1}^m \sum_{i,k=1}^d F_t^{i,j} F_t^{k,j} \frac{\partial^2 u}{\partial x^i \partial x^k}\right\} dt \\ &\quad + \sum_{j=1}^m \sum_{i=1}^d F_t^{i,j} \frac{\partial u}{\partial x^i}\, dW_t^j, \end{aligned} \tag{15.5.22}$$

for $t \geq 0$ with initial value $u(0, X_0^1, X_0^2, \ldots, X_0^d)$, where the partial derivatives of the function u are evaluated at $(t, X_t^1, X_t^2, \ldots, X_t^d)$, which we suppressed in our notation.

We can rewrite the multidimensional Itô formula (15.5.22) also in the form

$$du\big(t, X_t^1, X_t^2, \ldots, X_t^d\big) = \frac{\partial u}{\partial t}\, dt + \sum_{i=1}^m \frac{\partial u}{\partial x^i}\, dX_t^i + \frac{1}{2} \sum_{i,k=1}^m \frac{\partial^2 u}{\partial x^i\, \partial x^k}\, d\big[X^i, X^k\big]_t \tag{15.5.23}$$

for all $t \geq 0$ using covariations, see (15.3.35).

15.6 Stochastic Differential Equations

15.6.1 Feedback in Stochastic Dynamics

In the Itô differentials that we have considered we left the specification of the drift coefficient and the diffusion coefficient open. For instance, these coefficients can be made state dependent. A stochastic differential equation (SDE) contains an unknown, which is its solution process. To be useful in modeling, such a solution needs to *exist* in an appropriate sense. Furthermore, some *uniqueness* of the solution of an SDE has to be guaranteed to make sure that one achieves the modeling goal without any ambiguity.

15.6.2 Solution of Continuous SDEs

We consider now a continuous stochastic process $Y = \{Y_t,\ t \geq 0\}$, which is a *solution* of a given SDE

$$dY_t = a(t, Y_t)\,dt + b(t, Y_t)\,dW_t \tag{15.6.24}$$

for $t \geq 0$, with initial value $Y_0 = y_0$. Here W denotes the driving standard Wiener process. If the process Y has for all $t \geq 0$ an Itô differential of the form (15.6.24), then $Y = \{Y_t, t \geq 0\}$ is called a solution of the SDE (15.6.24). More precisely, a solution of the SDE (15.6.24) is a pair (Y, W) of adapted stochastic processes, defined on a given filtered probability space $(\Omega, \mathcal{A}, \underline{\mathcal{A}}, P)$, where the continuous process Y satisfies for each $t \geq 0$ the Itô integral equation

$$Y_t = y_0 + \int_0^t a(s, Y_s)\,ds + \int_0^t b(s, Y_s)\,dW_s. \tag{15.6.25}$$

We need to assume that both integrals on the right hand side of (15.6.25) exist. It is sufficient to request that

$$\int_0^t \big|a(s, Y_s)\big|\,ds + \int_0^t \big|b(s, Y_s)\big|^2\,ds < \infty$$

almost surely for all $t \geq 0$. Note that the SDE (15.6.24) is only a shorthand notation for the integral equation (15.6.25). The existence and uniqueness of a solution of an SDE is not trivially given, and needs to be ensured, as will be discussed later. In addition to the initial value, not only the drift coefficient and the diffusion coefficient need to be given for certain SDEs, additionally also the behavior of its solution at certain boundaries may have to be defined when establishing the uniqueness of the solution of the SDE. For instance, absorption or reflection has to be declared for certain SDEs at the level zero.

15.6.3 Continuous Vector SDEs

We now consider multidimensional solutions

$$\boldsymbol{X} = \big\{\boldsymbol{X}_t = \big(X_t^1, X_t^2, \ldots, X_t^d\big),\ t \geq 0)\big\}$$

of SDEs. We recall that $\boldsymbol{W} = \{\boldsymbol{W}_t = (W_t^1, W_t^2, \ldots, W_t^m)^\top,\ t \geq 0\}$ is an m-dimensional standard Wiener process with components $W^1, W^2, \ldots, W^m$.

Given a d-dimensional vector function $\boldsymbol{a} : [0, \infty) \times \Re^d \to \Re^d$ and a $d \times m$-matrix function $\boldsymbol{b} : [0, \infty) \times \Re^d \to \Re^{d \times m}$, then we can form the d-dimensional *vector stochastic differential equation*

$$d\boldsymbol{X}_t = \boldsymbol{a}(t, \boldsymbol{X}_t)\, dt + \boldsymbol{b}(t, \boldsymbol{X}_t)\, d\boldsymbol{W}_t \tag{15.6.26}$$

for $t \geq 0$ with initial value $\boldsymbol{X}_0 \in \Re^d$. The vector stochastic differential (15.6.26) should be interpreted as an Itô integral equation of the form

$$\boldsymbol{X}_t = \boldsymbol{X}_0 + \int_0^t \boldsymbol{a}(s, \boldsymbol{X}_s)\, ds + \int_0^t \boldsymbol{b}(s, \boldsymbol{X}_s)\, d\boldsymbol{W}_s, \tag{15.6.27}$$

for $t \geq 0$, where the integrals are defined componentwise. Thus, the ith component of (15.6.27) is then given by the SDE

$$X_t^i = X_0^i + \int_0^t a^i(s, \boldsymbol{X}_s)\, ds + \sum_{\ell=1}^m \int_0^t b^{i,\ell}(s, \boldsymbol{X}_s)\, dW_s^\ell, \tag{15.6.28}$$

for $t \geq 0$ and $i \in \{1, 2, \ldots, d\}$. Note that the drift and diffusion coefficients of each component can depend on all other components.

We now address the issue of existence and uniqueness of SDEs, see also Krylov (1980) and Protter (2005).

15.7 Existence and Uniqueness of Solutions of SDEs

15.7.1 Strong Solution

For any model that uses an SDE it is essential that it has a solution. Furthermore, it is important that it has a unique solution according to some appropriate criterion. One such criterion is described below in detail, which is based on a notion of strong uniqueness. Usually one can only formulate sufficient conditions to establish uniqueness. The techniques presented in the literature for proving existence and uniqueness of a solution of an SDE are rather similar. They typically assume Lipschitz continuity of the drift and diffusion coefficients. We briefly discuss here some typical issues that arise when ensuring the existence and uniqueness of a solution of an SDE.

Assume that we have given a filtered probability space $(\Omega, \mathcal{A}, \underline{\mathcal{A}}, P)$. For Eq. (15.6.27) to make sense $\boldsymbol{X}$ needs to be $\underline{\mathcal{A}}$-adapted. This leads to the following definition.

Definition 15.7.1 We call $(\boldsymbol{X}, \boldsymbol{W})$, consisting of a stochastic process $\boldsymbol{X} = \{\boldsymbol{X}_t,\ t \in [0, T]\}$, $T \in (0, \infty)$, and an $\underline{\mathcal{A}}$-adapted standard Wiener process $\boldsymbol{W}$, a *strong solution* of the Itô integral equation (15.6.27) if $\boldsymbol{X}$ is $\underline{\mathcal{A}}$-adapted, the integrals on the right hand side are well-defined and the equality in (15.6.27) holds almost surely.

For fixed coefficient functions $\boldsymbol{a}$ and $\boldsymbol{b}$, any solution $\boldsymbol{X}$ will usually depend on the particular initial value $\boldsymbol{X}_0$ and the sample path of the Wiener process $\boldsymbol{W}$ under consideration. For a specified initial value $\boldsymbol{X}_0$ the *uniqueness* of strong solutions of the SDE (15.6.27) refers to the following notion of *indistinguishability* of the solution processes.

Definition 15.7.2 If any two strong solutions $\boldsymbol{X}$ and $\tilde{\boldsymbol{X}}$ are indistinguishable on $[0, T]$, that is,

$$\boldsymbol{X}_t = \tilde{\boldsymbol{X}}_t \tag{15.7.29}$$

almost surely for all $t \in [0, T]$, then we say that the solution of (15.6.27) on $[0, T]$ is a *unique strong solution*.

15.7.2 Existence and Uniqueness Theorem

Let us now state a standard theorem on the existence and uniqueness of strong solutions of SDEs. It ensures that the objects we model are well defined. For details on the definition of strong solutions of SDEs, we can refer, for instance, to Ikeda and Watanabe (1989) or Protter (2005).

We assume that the coefficient functions of the SDE (15.6.27) satisfy the Lipschitz conditions

$$\left|\boldsymbol{a}(t,\boldsymbol{x}) - \boldsymbol{a}(t,\boldsymbol{y})\right| \le C_1|\boldsymbol{x}-\boldsymbol{y}|, \qquad \left|\boldsymbol{b}(t,\boldsymbol{x}) - \boldsymbol{b}(t,\boldsymbol{y})\right| \le C_2|\boldsymbol{x}-\boldsymbol{y}|, \tag{15.7.30}$$

for every $t \in [0, T]$ and $\boldsymbol{x}, \boldsymbol{y} \in \Re^d$, as well as the linear growth conditions

$$\left|\boldsymbol{a}(t,\boldsymbol{x})\right| \le K_1\big(1+|\boldsymbol{x}|\big), \qquad \left|\boldsymbol{b}(t,\boldsymbol{x})\right| \le K_2\big(1+|\boldsymbol{x}|\big), \tag{15.7.31}$$

for all $t \in [0, T]$ and $\boldsymbol{x} \in \Re^d$. Note that the linear growth conditions can usually be derived from the corresponding Lipschitz conditions.

Moreover, we assume that the initial value $\boldsymbol{X}_0$ is $\mathcal{A}_0$-measurable with

$$E\big(|\boldsymbol{X}_0|^2\big) < \infty. \tag{15.7.32}$$

Theorem 15.7.3 *Suppose that the coefficient functions $\boldsymbol{a}(\cdot)$ and $\boldsymbol{b}(\cdot)$ of the SDE* (15.6.27) *satisfy the Lipschitz conditions* (15.7.30), *the linear growth conditions* (15.7.31) *and the initial condition* (15.7.32). *Then the SDE* (15.6.27) *admits a unique strong solution. Moreover, the solution $\boldsymbol{X}$ of the SDE* (15.6.27) *satisfies the estimate*

$$E\Big(\sup_{0\le s\le T} |\boldsymbol{X}_s|^2\Big) \le C\big(1 + E\big(|\boldsymbol{X}_0|^2\big)\big) \tag{15.7.33}$$

with $T < \infty$, where C is a finite positive constant.

The proof of Theorem 15.7.3 can be found in Ikeda and Watanabe (1989) or Situ (2005). In the mentioned literature one finds also the notion of a weak solution of an SDE. We note that if an SDE has a strong solution, then it has also a weak solution. Within this book we have discussed several classes of tractable diffusions, which do not satisfy the above mentioned Lipschitz and linear growth conditions. Still, there exist unique weak solutions of the respective SDEs. By working with tractable diffusions and functions of these one can access a modeling world with properties that are difficult or impossible to establish under classical Lipschitz and linear growth conditions.

15.8 Functionals of Solutions of SDEs

This book is concerned with the exploration of classes of diffusions for which explicit formulas for important functionals can be computed. These functionals often have the format of conditional expectations. For Markovian state variables these conditional expectations lead to pricing functions that satisfy *partial differential equations* (PDEs). The link between the conditional expectations and respective PDEs can be interpreted as an application of the, so-called, *Feynman-Kac formula.* Below we formulate the Feynman-Kac formula in various ways. Furthermore, this section presents some results on transition probability densities, changes of measures, the Bayes rule, and the Girsanov transformation. For details, see e.g. Platen and Heath (2010).

15.8.1 SDE for Some Factor Process

We consider a fixed time horizon $T \in (0, \infty)$ and a d-dimensional Markovian factor process $\boldsymbol{X}^{t,\boldsymbol{x}} = \{\boldsymbol{X}_s^{t,\boldsymbol{x}}, s \in [t, T]\}$, which satisfies the vector SDE

$$d\boldsymbol{X}_s^{t,\boldsymbol{x}} = \boldsymbol{a}\big(s, \boldsymbol{X}_s^{t,\boldsymbol{x}}\big)\, ds + \sum_{k=1}^{m} \boldsymbol{b}^k\big(s, \boldsymbol{X}_s^{t,\boldsymbol{x}}\big)\, dW_s^k \tag{15.8.34}$$

for $s \in [t, T]$ with initial value $\boldsymbol{X}_t^{t,x} = \boldsymbol{x} \in \Re^d$ at time $t \in [0, T]$. The process $\boldsymbol{W} = \{\boldsymbol{W}_t = (W_t^1, \ldots, W_t^m)^\top, t \in [0, T]\}$ is an m-dimensional standard Wiener process on a filtered probability space $(\Omega, \mathcal{A}, \underline{\mathcal{A}}, P)$. The process $\boldsymbol{X}^{t,\boldsymbol{x}}$ has a drift coefficient $\boldsymbol{a}(\cdot,\cdot)$ and diffusion coefficients $\boldsymbol{b}^k(\cdot,\cdot)$, $k \in \{1, 2, \ldots, m\}$. In general, $\boldsymbol{a} = (a^1, \ldots, a^d)^\top$ and $\boldsymbol{b}^k = (b^{1,k}, \ldots, b^{d,k})^\top$, $k \in \{1, 2, \ldots, m\}$, represent vector valued functions on $[0, T] \times \Re^d$ into $\Re^d$, and we assume that a unique strong solution of the SDE (15.8.34) exists. We motivate various versions of the Feynman-Kac formula by giving them financial interpretations. The values considered will be typically denominated in units of the benchmark.

15.8.2 Terminal Payoff

Let us discuss the case of a European style option, where we have a terminal payoff $H(\boldsymbol{X}_T^{t,\boldsymbol{x}})$ at the maturity date T with some given payoff function $H : \Re^d \to [0, \infty)$ such that

$$E\big(\big|H\big(\boldsymbol{X}_T^{t,\boldsymbol{x}}\big)\big|\big) < \infty. \tag{15.8.35}$$

We can then introduce the pricing function $u : [0, T] \times \Re^d \to [0, \infty)$ as the conditional expectation

$$u(t, \boldsymbol{x}) = E\big(H\big(\boldsymbol{X}_T^{t,\boldsymbol{x}}\big) \mid \mathcal{A}_t\big) \tag{15.8.36}$$

for $(t, \boldsymbol{x}) \in [0, T] \times \Re^d$. The *Feynman-Kac formula* for this payoff refers to the fact that under sufficient regularity on $\boldsymbol{a}, \boldsymbol{b}^{\boldsymbol{1}}, \ldots, \boldsymbol{b}^{\boldsymbol{m}}$ and H the function $u : (0, T) \times \Re^d \to [0, \infty)$ satisfies the PDE

$$\begin{aligned} L^0 u(t, \boldsymbol{x}) &= \frac{\partial u(t, \boldsymbol{x})}{\partial t} + \sum_{i=1}^{d} a^i(t, \boldsymbol{x}) \frac{\partial u(t, \boldsymbol{x})}{\partial x^i} \\ &\quad + \frac{1}{2} \sum_{i,k=1}^{d} \sum_{j=1}^{m} b^{i,j}(t, \boldsymbol{x}) b^{k,j}(t, \boldsymbol{x}) \frac{\partial^2 u(t, \boldsymbol{x})}{\partial x^i \partial x^k} \\ &= 0 \end{aligned} \tag{15.8.37}$$

for $(t, \boldsymbol{x}) \in (0, T) \times \Re^d$ with terminal condition

$$u(T, \boldsymbol{x}) = H(\boldsymbol{x}) \tag{15.8.38}$$

for $\boldsymbol{x} \in \Re^d$. Equation (15.8.37) is also called the *Kolmogorov backward equation.*

Note that, in general, one needs also to specify the behavior of the solution of the PDE at its boundaries. In a benchmark setting when nonnegative value processes become zero they stay afterwards zero and face absorption at the respective boundary.

15.8.3 Discounted Payoff

We now generalize the above payoff function by discounting it, using a given *discount rate process* r, which is obtained as a function of the given vector diffusion process $\boldsymbol{X}^{t,\boldsymbol{x}}$, that is $r : [0, T] \times \Re^d \to \Re$.

Over the period $[t, T]$ we consider for the *discounted payoff*

$$\exp\left\{-\int_t^T r\big(s, \boldsymbol{X}_s^{t,\boldsymbol{x}}\big)\, ds\right\} H\big(\boldsymbol{X}_T^{t,\boldsymbol{x}}\big)$$

the pricing function

$$u(t, \boldsymbol{x}) = E\left(\exp\left\{-\int_t^T r\big(s, \boldsymbol{X}_s^{t,\boldsymbol{x}}\big)\, ds\right\} H\big(\boldsymbol{X}_T^{t,\boldsymbol{x}}\big) \,\middle|\, \mathcal{A}_t\right) \tag{15.8.39}$$

for $(t, \boldsymbol{x}) \in [0, T] \times \Re^d$. It follows rather generally that the pricing function u satisfies the PDE

$$L^0 u(t, \boldsymbol{x}) = r(t, \boldsymbol{x})\, u(t, \boldsymbol{x}) \tag{15.8.40}$$

for $(t, \boldsymbol{x}) \in (0, T) \times \Re^d$ with terminal condition

$$u(T, \boldsymbol{x}) = H(\boldsymbol{x}) \tag{15.8.41}$$

for $\boldsymbol{x} \in \Re^d$. Here the PDE operator L^0 is given as in (15.8.37).

15.8.4 Terminal Payoff and Payoff Rate

Now, we add to the above discounted payoff structure some payoff stream that continuously pays with a *payoff rate* $g : [0, T] \times \Re^d \to [0, \infty)$ some amount per unit of time. The corresponding *discounted payoff with payoff rate* is then of the form

$$\exp\left\{-\int_t^T r\left(s, \boldsymbol{X}_s^{t,\boldsymbol{x}}\right) ds\right\} H\left(\boldsymbol{X}_T^{t,\boldsymbol{x}}\right) + \int_t^T \exp\left\{-\int_t^s r\left(z, \boldsymbol{X}_z^{t,\boldsymbol{x}}\right) dz\right\} g\left(s, \boldsymbol{X}_s^{t,\boldsymbol{x}}\right) ds,$$

which leads to the pricing function

$$\begin{aligned} u(t, \boldsymbol{x}) = E\Bigg(&\exp\left\{-\int_t^T r\left(s, \boldsymbol{X}_s^{t,\boldsymbol{x}}\right) ds\right\} H\left(\boldsymbol{X}_T^{t,\boldsymbol{x}}\right) \\ &+ \int_t^T \exp\left\{-\int_t^s r\left(z, \boldsymbol{X}_z^{t,\boldsymbol{x}}\right) dz\right\} g\left(s, \boldsymbol{X}_s^{t,\boldsymbol{x}}\right) ds \,\Bigg|\, \mathcal{A}_t\Bigg) \end{aligned} \tag{15.8.42}$$

for $(t, \boldsymbol{x}) \in [0, T] \times \Re^d$. This pricing function satisfies the PDE

$$L^0 u(t, \boldsymbol{x}) + g(t, \boldsymbol{x}) = r(t, \boldsymbol{x}) u(t, \boldsymbol{x}) \tag{15.8.43}$$

for $(t, \boldsymbol{x}) \in (0, T) \times \Re^d$ with terminal condition

$$u(T, \boldsymbol{x}) = H(\boldsymbol{x}) \tag{15.8.44}$$

for $\boldsymbol{x} \in \Re^d$. As mentioned earlier, for certain dynamics boundary conditions may have to be added.

15.8.5 Payoff with First Exit Time

Derivatives like barrier options have a, so-called, *continuation region* Φ, which is an open connected subset of $[0, T] \times \Gamma$. The holder of a derivative continues to receive payments as long as the process $\boldsymbol{X}^{t,\boldsymbol{x}}$ stays in the continuation region Φ. For instance, in the case of a, so-called, *knock-out-barrier option* this would mean that $\boldsymbol{X}_s^{t,\boldsymbol{x}}$ has to stay below a given critical barrier to receive the terminal payment. To make this precise, we define the *first exit time* τ_Φ^t from Φ after t as

$$\tau_\Phi^t = \inf\left\{s \in [t, T]\colon \left(s, \boldsymbol{X}_s^{t,\boldsymbol{x}}\right) \notin \Phi\right\}, \tag{15.8.45}$$

which is a stopping time, see (15.2.23).

Consider now a general payoff structure with *terminal payoff function* $H : (0, T] \times \Gamma \to [0, \infty)$ for payments at time τ_Φ^t, a *payoff rate* $g : [0, T] \times \Gamma \to [0, \infty)$ for incremental payments during the time period $[t, \tau_\Phi^t)$ and a *discount rate* $r : [0, T] \times \Gamma \to \Re$. Assume that the process $\boldsymbol{X}^{t,\boldsymbol{x}}$ does not explode or leave Γ before the terminal time T. We then define the *pricing function* $u : \Phi \to [0, \infty)$ by

$$u(t, \boldsymbol{x}) = E\Bigg(H\big(\tau_\Phi^t, \boldsymbol{X}_{\tau_\Phi^t}^{t,\boldsymbol{x}}\big) \exp\bigg\{ -\int_t^{\tau_\Phi^t} r\big(s, \boldsymbol{X}_s^{t,\boldsymbol{x}}\big)\, ds \bigg\} + \int_t^{\tau_\Phi^t} g\big(s, \boldsymbol{X}_s^{t,\boldsymbol{x}}\big) \exp\bigg\{ -\int_t^{s} r\big(z, \boldsymbol{X}_u^{t,\boldsymbol{x}}\big)\, dz \bigg\}\, ds \,\bigg|\, \mathcal{A}_t \Bigg) \tag{15.8.46}$$

for $(t, \boldsymbol{x}) \in \Phi$.

For the formulation of the resulting PDE of the function u we use the operator L^0 given in (15.8.37). Under sufficient regularity of Φ, $\boldsymbol{a}, \boldsymbol{b}^1, \ldots, \boldsymbol{b}^m$, H, g and r one can show by application of the Itô formula (15.5.23) and some martingale argument that the pricing function u satisfies the PDE

$$L^0 u(t, \boldsymbol{x}) + g(t, \boldsymbol{x}) = r(t, \boldsymbol{x}) u(t, \boldsymbol{x}) \tag{15.8.47}$$

for $(t, \boldsymbol{x}) \in \Phi$ with boundary condition

$$u(t, \boldsymbol{x}) = H(t, \boldsymbol{x}) \tag{15.8.48}$$

for $(t, \boldsymbol{x}) \in ((0, T] \times \Gamma) \backslash \Phi$. This result links the functional (15.8.46) to the PDE (15.8.47)–(15.8.48) and is often called a *Feynman-Kac formula*.

15.8.6 Generalized Feynman-Kac Formula

For a rather general situation, where $\Phi = (0, T) \times \Gamma$ and $\tau_\Phi^t = T$, let us now formulate sufficient conditions that ensure that the Feynman-Kac formula holds, see Heath and Schweizer (2000) and Platen and Heath (2010).

(A) The drift coefficient $\boldsymbol{a}$ and diffusion coefficients $\boldsymbol{b}^k$, $k \in \{1, 2, \ldots, m\}$, are assumed to be on $[0, T] \times \Gamma$ locally Lipschitz-continuous in $\boldsymbol{x}$, uniformly in t. That is, for each compact subset Γ^1 of Γ there exists a constant $K_{\Gamma^1} < \infty$ such that

$$\big|\boldsymbol{a}(t, \boldsymbol{x}) - \boldsymbol{a}(t, \boldsymbol{y})\big| + \sum_{k=1}^{m} \big|\boldsymbol{b}^k(t, \boldsymbol{x}) - \boldsymbol{b}^k(t, \boldsymbol{y})\big| \le K_{\Gamma^1} |\boldsymbol{x} - \boldsymbol{y}| \tag{15.8.49}$$

for all $t \in [0, T]$ and $\boldsymbol{x}, \boldsymbol{y} \in \Gamma^1$.

(B) For all $(t, \boldsymbol{x}) \in [0, T) \times \Gamma$ the solution $\boldsymbol{X}^{t,\boldsymbol{x}}$ of (15.8.34) neither explodes nor leaves Γ before T, that is

$$P\Big(\sup_{t \le s \le T} \big|\boldsymbol{X}_s^{t,\boldsymbol{x}}\big| < \infty \Big) = 1 \tag{15.8.50}$$

and

$$P\big(\boldsymbol{X}_s^{t,\boldsymbol{x}} \in \Gamma \text{ for all } s \in [t, T]\big) = 1. \tag{15.8.51}$$

(C) There exists an increasing sequence $(\Gamma_n)_{n\in\mathcal{N}}$ of bounded, open and connected domains of Γ such that $\bigcup_{n=1}^{\infty}\Gamma_n = \Gamma$, and for each $n \in \mathcal{N}$ the PDE

$$L^0 u_n(t,\boldsymbol{x}) + g(t,\boldsymbol{x}) = r(t,\boldsymbol{x})\, u_n(t,\boldsymbol{x}) \tag{15.8.52}$$

has a unique solution u_n in the sense of Friedman (1975) on $(0,T)\times\Gamma_n$ with boundary condition

$$u_n(t,\boldsymbol{x}) = u(t,\boldsymbol{x}) \tag{15.8.53}$$

on $((0,T)\times\partial\Gamma_n)\cup(\{T\}\times\Gamma_n)$, where $\partial\Gamma_n$ denotes the boundary of Γ_n.

(D) The process $b^{i,k}(\cdot,\boldsymbol{X}_\cdot)\frac{\partial u(\cdot,\boldsymbol{X}_\cdot)}{\partial x^i}$ is measurable and square integrable on $[0,T]$ for all $i \in \{1,2,\ldots,d\}$ and $k \in \{1,2,\ldots,m\}$.

The proof of the following theorem is given in Platen and Heath (2010).

Theorem 15.8.1 *Under the conditions* (A), (B), (C) *and* (D), *the function u given by* (15.8.46) *is the unique solution of the PDE* (15.8.47) *with boundary condition* (15.8.48), *where u is differentiable with respect to t and twice differentiable with respect to the components of* $\boldsymbol{x}$.

Condition (A) is satisfied if, for instance, $\boldsymbol{a}$ and $\boldsymbol{b} = (\boldsymbol{b}^1,\ldots,\boldsymbol{b}^m)$ are differentiable in $\boldsymbol{x}$ on the open set $(0,T)\times\Gamma$ with derivatives that are continuous on $[0,T]\times\Gamma$.

To establish condition (B) one needs to exploit specific properties of the process $\boldsymbol{X}^{t,\boldsymbol{x}}$ given by the SDE (15.8.34).

Condition (C) can be shown to be implied by the following assumptions.

(C1) There exists an increasing sequence $(\Gamma_n)_{n\in\mathcal{N}}$ of bounded, open and connected subdomains of Γ with $\Gamma_n\cup\partial\Gamma_n\subset\Gamma$ such that $\cup_{n=1}^{\infty}\Gamma_n=\Gamma$, and each Γ_n has a twice differentiable boundary $\partial\Gamma_n$.

(C2) For each $n\in\mathcal{N}$ the functions $\boldsymbol{a}$ and $\boldsymbol{b}\boldsymbol{b}^\top$ are uniformly Lipschitz-continuous on $[0,T]\times(\Gamma_n\cup\partial\Gamma_n)$.

(C3) For each $n\in\mathcal{N}$ the function $\boldsymbol{b}(t,\boldsymbol{x})\boldsymbol{b}(t,\boldsymbol{x})^\top$ is uniformly elliptic on $\Re^d$ for $(t,\boldsymbol{x})\in[0,T]\times\Gamma_n$, that is, there exists a $\delta_n>0$ such that

$$\boldsymbol{y}^\top\boldsymbol{b}(t,\boldsymbol{x})\,\boldsymbol{b}(t,\boldsymbol{x})^\top\boldsymbol{y} \ge \delta_n\,|\boldsymbol{y}|^2 \tag{15.8.54}$$

for all $\boldsymbol{y}\in\Re^d$.

(C4) For each $n\in\mathcal{N}$ the functions r and g are uniformly *Hölder-continuous* on $[0,T]\times(\Gamma_n\cup\partial\Gamma_n)$, that is, there exists a constant $\bar{K}_n$ and an exponent $q_n>0$ such that

$$\big|r(t,\boldsymbol{x})-r(t,\boldsymbol{y})\big| + \big|g(t,\boldsymbol{x})-g(t,\boldsymbol{y})\big| \le \bar{K}_n|\boldsymbol{x}-\boldsymbol{y}|^{q_n} \tag{15.8.55}$$

for $t\in[0,T]$ and $\boldsymbol{x},\boldsymbol{y}\in(\Gamma_n\cup\partial\Gamma_n)$.

(C5) For each $n\in\mathcal{N}$ the function u is finite and continuous on $([0,T]\times\partial\Gamma_n)\cup(\{T\}\times(\Gamma_n\cup\partial\Gamma_n))$.

Condition (D) is satisfied, for instance, when

$$\int_0^T E\left(\left(b^{i,k}(t,\boldsymbol{X}_t)\frac{\partial u(t,\boldsymbol{X}_t)}{\partial x^i}\right)^2\right)dt<\infty$$

for all $i \in \{1,2,\dots,d\}$ and $k \in \{1,2,\dots,m\}$. This condition ensures that the process $u(\cdot,\boldsymbol{X}_\cdot)$ is a martingale and that the PDE (15.8.47)–(15.8.48) has a unique solution.

Note that in the case when local Lipschitz continuity is not guaranteed, one may have to specify particular boundary conditions to obtain an appropriate description of the pricing function. This is a consequence of the fact that strict local martingales may drive the factor dynamics. These need extra care when defining the behavior of PDE solutions at boundaries. Within this book we have given many explicit formulas for pricing functions, which satisfy a PDE of the form (15.8.47)–(15.8.48) but may not satisfy some of the above mentioned conditions. This reveals another advantage of working with tractable diffusions, where one has not to rely on restrictive Lipschitz conditions or similar constraining assumptions and still obtains a solution for the problem at hand.

15.8.7 Kolmogorov Equations

When the drift coefficient $a(\cdot)$ and diffusion coefficient $b(\cdot)$ of the solution of a scalar SDE are appropriate functions, then the corresponding transition probability density $p(s,x;t,y)$ of the solution of the SDE satisfies a certain PDE. This is the *Kolmogorov forward equation* or *Fokker-Planck equation*

$$\frac{\partial p(s,x;t,y)}{\partial t}+\frac{\partial}{\partial y}\{a(t,y)p(s,x;t,y)\}-\frac{1}{2}\frac{\partial^2}{\partial y^2}\{b^2(t,y)p(s,x;t,y)\}=0, \tag{15.8.56}$$

for (s,x) fixed. However, $p(s,x;t,y)$ satisfies also the *Kolmogorov backward equation*

$$\frac{\partial p(s,x;t,y)}{\partial s}+a(s,x)\frac{\partial p(s,x;t,y)}{\partial x}+\frac{1}{2}b^2(s,x)\frac{\partial^2 p(s,x;t,y)}{\partial x^2}=0, \tag{15.8.57}$$

for (t,y) fixed. Obviously, the *initial condition* for both PDEs is given by the *Dirac delta function*

$$p(s,x;s,y)=\delta(y-x)=\begin{cases}\infty & \text{for } y=x\\ 0 & \text{for } y\neq x,\end{cases} \tag{15.8.58}$$

where

$$\int_{-\infty}^{\infty}\delta(y-x)\,dy=1 \tag{15.8.59}$$

for given x. The Kolmogorov equations have multidimensional counterparts. In Sect. 5.3 we have given various examples of explicit transition probability densities.

15.8.8 Change of Probability Measure

We denote by $\boldsymbol{W} = \{\boldsymbol{W}_t = (W_t^1, \ldots, W_t^m)^\top, t \in [0,T]\}$ an m-dimensional standard Wiener process on a filtered probability space $(\Omega, \mathcal{A}, \underline{\mathcal{A}}, P)$ with $\mathcal{A}_0$ being the trivial σ-algebra, augmented by the sets of zero probability. For an $\underline{\mathcal{A}}$-predictable m-dimensional stochastic process $\boldsymbol{\theta} = \{\boldsymbol{\theta}_t = (\theta_t^1, \ldots, \theta_t^m)^\top, t \in [0,T]\}$ with

$$\int_0^T \sum_{i=1}^m (\theta_t^i)^2 \, dt < \infty \tag{15.8.60}$$

almost surely, let us assume that the strictly positive *Radon-Nikodym derivative process* $\Lambda_{\boldsymbol{\theta}} = \{\Lambda_{\boldsymbol{\theta}}(t), t \in [0,T]\}$, where

$$\Lambda_{\boldsymbol{\theta}}(t) = \exp\left\{-\int_0^t \boldsymbol{\theta}_s^\top \, d\boldsymbol{W}_s - \frac{1}{2}\int_0^t \boldsymbol{\theta}_s^\top \boldsymbol{\theta}_s \, ds\right\} < \infty \tag{15.8.61}$$

almost surely for $t \in [0,T]$, is an *$(\underline{\mathcal{A}}, P)$-martingale*. By the Itô formula (15.5.22) it follows from (15.8.61) that

$$\Lambda_{\boldsymbol{\theta}}(t) = 1 - \sum_{i=1}^m \int_0^t \Lambda_{\boldsymbol{\theta}}(s) \theta_s^i \, dW_s^i \tag{15.8.62}$$

for $t \in [0,T]$. Since $\Lambda_{\boldsymbol{\theta}}$ is assumed here to be an $(\underline{\mathcal{A}}, P)$-martingale we have for $t \in [0,T]$

$$E\big(\Lambda_{\boldsymbol{\theta}}(t) \,\big|\, \mathcal{A}_0\big) = \Lambda_{\boldsymbol{\theta}}(0) = 1. \tag{15.8.63}$$

We can now define a measure $P_{\boldsymbol{\theta}}$ via the Radon-Nikodym derivative

$$\frac{dP_{\boldsymbol{\theta}}}{dP} = \Lambda_{\boldsymbol{\theta}}(T) \tag{15.8.64}$$

by setting

$$P_{\boldsymbol{\theta}}(A) = E\big(\Lambda_{\boldsymbol{\theta}}(T) \mathbf{1}_A\big) = E_{\boldsymbol{\theta}}(\mathbf{1}_A) \tag{15.8.65}$$

for $A \in \mathcal{A}_T$. Here $\mathbf{1}_A$ is the indicator function for A and $E_{\boldsymbol{\theta}}$ means expectation with respect to $P_{\boldsymbol{\theta}}$.

Note that $P_{\boldsymbol{\theta}}$ is not just a measure but also a probability measure because

$$P_{\boldsymbol{\theta}}(\Omega) = E\big(\Lambda_{\boldsymbol{\theta}}(T)\big) = E\big(\Lambda_{\boldsymbol{\theta}}(T) \,\big|\, \mathcal{A}_0\big) = \Lambda_{\boldsymbol{\theta}}(0) = 1 \tag{15.8.66}$$

as a result of the martingale property of $\Lambda_{\boldsymbol{\theta}}$. For several asset price models discussed in this book, the Radon-Nikodym derivative for the putative risk neutral probability measure is only a strict local martingale and a risk neutral probability measure does not exist.

15.8.9 Bayes Rule

It is useful to be able to change the probability measure when taking conditional expectations. The following *Bayes rule* establishes a relationship between conditional expectations with respect to different equivalent probability measures.

Assume for an equivalent probability measure P_{θ} that the corresponding strictly positive Radon-Nikodym derivative process Λ_{θ} is an $(\underline{\mathcal{A}}, P)$-martingale. Then for any given stopping time $\tau \in [0, T]$ and any $\mathcal{A}_{\tau}$-measurable random variable Y, satisfying the integrability condition

$$E_{\theta}\big(|Y|\big) < \infty, \tag{15.8.67}$$

one can apply the *Bayes rule*

$$E_{\theta}(Y \mid \mathcal{A}_s) = \frac{E(\Lambda_{\theta}(\tau)\, Y \mid \mathcal{A}_s)}{E(\Lambda_{\theta}(\tau) \mid \mathcal{A}_s)} \tag{15.8.68}$$

for $s \in [0, \tau]$.

15.8.10 Girsanov Transformation

The following Girsanov transformation allows us to perform a measure transformation, which transforms a drifted Wiener process into a Wiener process under a new probability measure P_{θ}. More precisely, if for $T \in (0, \infty)$ a given strictly positive Radon-Nikodym derivative process Λ_{θ} is an $(\underline{\mathcal{A}}, P)$-martingale, then the m-dimensional process $\boldsymbol{W}_{\theta} = \{\boldsymbol{W}_{\theta}(t),\ t \in [0, T]\}$, given by

$$\boldsymbol{W}_{\theta}(t) = \boldsymbol{W}_t + \int_0^t \boldsymbol{\theta}_s\, ds \tag{15.8.69}$$

for all $t \in [0, T]$, is an m-dimensional standard Wiener process on the filtered probability space $(\Omega, \mathcal{A}, \underline{\mathcal{A}}, P_{\theta})$.

Note that certain assumptions need to be satisfied before one can apply the above Girsanov transformation. The key assumption is that Λ_{θ} must be a strictly positive $(\underline{\mathcal{A}}, P)$-martingale. For instance, if the Radon-Nikodym derivative process is only a strictly positive local martingale, then this does not guarantee that P_{θ} is a probability measure, see Platen and Heath (2010).

A sufficient condition for the Radon-Nikodym derivative process Λ_{θ} to be an $(\underline{\mathcal{A}}, P)$-martingale is the *Novikov condition*, see e.g. Novikov (1972), which requires that

$$E\left(\exp\left\{\frac{1}{2}\int_0^T \boldsymbol{\theta}_s^{\top}\boldsymbol{\theta}_s\, ds\right\}\right) < \infty. \tag{15.8.70}$$

Chapter 16
Time-Homogeneous Scalar Diffusions

In this book, we pursue mostly a probabilistic approach, essentially originated in Itô (1944), who set out to generate diffusions directly from given Brownian motions. This chapter refers more to an analytic approach to diffusions, tracing its origin back to Kolmogorov (1932) and Feller (1936). The approach allows us to obtain transition densities by solving Kolmogorov equations, see Sect. 15.8.

Following Hulley (2009), we explain below how to obtain some explicit solutions for functionals of scalar diffusions, especially functionals associated with stopping times.

16.1 Basic Definitions

With the following definition we follow Hulley (2009).

Definition 16.1.1 Fix an interval $I \subseteq \Re$, with left end-point $l \geq -\infty$ and right end-point $r \leq \infty$. Denote the space of continuous I-valued paths by $\Omega := C(\Re^+, I)$ and let X be the coordinate mapping process on this space, defined by $X_t(\omega) := \omega(t)$, for all $\omega \in \Omega$ and $t \in \Re^+$. Define the filtration $\underline{\mathcal{A}}^0 = (\mathcal{A}^0_t)_{t \in \Re^+}$, by setting $\mathcal{A}^0_t := \sigma\{X_s | s \leq t\}$, for all $t \in \Re^+$, as well as the σ-algebra $\mathcal{A}^0_\infty := \sigma\{X_t \mid t \in \Re^+\}$. The shift operators $\vartheta = (\vartheta_t)_{t \in \Re^+}$ are constructed, by setting $(\vartheta_t \omega)(s) := \omega(t+s)$, for all $\omega \in \Omega$ and $t, s \in \Re^+$. Finally, let $\mathfrak{P} = \{P_x \mid x \in I\}$ be a family of probability measures on $(\Omega, \mathcal{A}^0_\infty)$, satisfying:

(i) $x \mapsto P_x(A)$ is measurable, for all $A \in \mathcal{A}^0_\infty$;
(ii) $P_x(X_0 = x) = 1$, for all $x \in I$;
(iii) $E_x(\eta \circ \vartheta_\sigma | \mathcal{A}^0_{\sigma+}) = E_{X_\sigma}(\eta) P_x$-a.s.,

for all bounded $\mathcal{A}^0_\infty$-measurable random variables η, and all $\mathcal{A}^0$-stopping times σ. The tuple $(\Omega, \mathcal{A}^0_\infty, \underline{\mathcal{A}}^0, X, \vartheta, \mathfrak{P})$ is then called a canonical diffusion on I.

The filtration $\underline{\mathcal{A}}$ used in Definition 16.1.1 is not necessarily right-continuous or complete. We remedy this by introducing the right-continuous filtration

J. Baldeaux, E. Platen, *Functionals of Multidimensional Diffusions with Applications to Finance*, Bocconi & Springer Series 5, DOI 10.1007/978-3-319-00747-2_16,

$\underline{\mathcal{A}}^+ = (\mathcal{A}_t^+)_{t\in\Re^+}$, defined by setting $\mathcal{A}_t^+ := \bigcap_{\epsilon>0} \mathcal{A}_{t+\epsilon}^0$, for all $t \in \Re^+$. Next, the family of null-sets is introduced:

$$\mathcal{A} := \{N \subseteq \Omega \mid N \subseteq A, \text{ for some } A \in \mathcal{A}_\infty^0 \text{ satisfying } P_x(A) = 0, \forall x \in I\}.$$

The filtration $\underline{\mathcal{A}} = (\mathcal{A}_t)_{t\in\Re^+}$ is constructed by setting $\mathcal{A}_t := \mathcal{A}_t^+ \vee \mathcal{N}$, for all $t \in \Re^+$. Since none of the above affects the strong Markov property of X, as expressed by Definition 16.1.1(iii), we shall henceforth regard $(\Omega, \mathcal{A}_\infty, \underline{\mathcal{A}}, X, \vartheta, \mathfrak{P})$ as the diffusion under consideration, where $\mathcal{A}_\infty := \mathcal{A}_\infty^0 \vee \mathcal{N}$. Obviously, the probability measures P_x, for all $x \in I$, are easily extended to $\mathcal{A}_\infty$, by setting $P_x(N) := 0$, for all $N \in \mathcal{N}$.

16.2 Boundary Classification

For any $z \in I$, we are interested in first-passage times of X to z, defined by

$$\tau_z := \inf\{t > 0 \mid X_t = z\}. \tag{16.2.1}$$

We shall assume that X is a *regular diffusion*, which we define below.

Definition 16.2.1 X is said to be a regular diffusion if and only if

$$P_x(\tau_z < \infty) > 0,$$

for all $x \in \text{int}(I)$ and $z \in I$.

We remark that since $P_x(\Omega) = 1$ for all $x \in I$, X is a, so-called, *honest diffusion*, meaning its behavior is completely determined by its *speed measure* m and *scale function* s. Explicit examples of speed measures and scale functions will be presented in Example 16.2.2. In particular, we classify end-points of I as *exit* or *entrance* boundaries for X. Boundaries which are both exit and entrance are called *non-singular*, while boundaries which are neither, are referred to as *natural*. If a boundary is either entrance or exit, but not both, it is referred to as *exit-not-entrance* or, respectively, as *entrance-not-exit*. We point out that natural boundaries and boundaries which are entrance-not-exit do not form part of the state space of the diffusion. In particular, if a diffusion is started from the interior of its state-space, it reaches exit-boundaries with positive probability. Diffusions can be started at entrance boundaries.

At non-singular boundaries, the behavior of the diffusion must be specified separately, as it cannot be determined from the speed measure and the scale function. Typical specifications include reflection, killing and absorption.

Example 16.2.2 For many diffusions of interest, the speed measure $m(\cdot)$ and the scale function $s(\cdot)$ are known. For example, for Brownian motion, the speed measure is given by $m(dx) = 2\,dx$, and the scale function by $s(x) = x$. For the squared

Table 16.2.1 Boundary classification for time-homogeneous scalar diffusions

	Lower Boundary l	Upper Boundary r
Exit	$\int_{(l,z)} m([y,z])s(dy) < \infty$	$\int_{(z,r)} m([z,y])s(dy) < \infty$
Entrance	$\int_{(l,z)} (s(z)-s(y))m(dy) < \infty$	$\int_{(z,r)} (s(y)-s(z))m(dy) < \infty$

Bessel process of dimension δ, the speed measure and the scale function are also known: defining the index $\nu = \frac{\delta}{2} - 1$, we obtain the speed measure

$$m(dx) = \begin{cases} \frac{x^\nu}{2|\nu|}\, dx, & \nu \neq 0 \\ \frac{1}{2}\, dx, & \nu = 0, \end{cases}$$

and scale function

$$s(x) = \begin{cases} -x^{-\nu}, & \nu > 0 \\ \log x, & \nu = 0 \\ x^{-\nu}, & \nu < 0. \end{cases}$$

For the squared Bessel process, the nature of the boundary point 0 depends on the value of ν, which can be confirmed using Table 16.2.1:

- if $\nu \geq 0$ then 0 is entrance-not-exit;
- if $-1 < \nu < 0$ then 0 is non-singular;
- if $\nu \leq -1$ then 0 is exit-not-entrance.

We remind the reader that in the literature on time-homogeneous diffusions and also in some of the preceding chapters of this book, if the boundary point at zero is non-singular, it is often specified to be reflecting.

Note that under the benchmark approach a benchmarked nonnegative portfolio is a nonnegative local martingale and hence a nonnegative supermartingale, which needs to be absorbed at zero whenever it reaches zero.

16.3 Laplace Transform Identities

We now focus on the computation of functionals associated with first passage times of X. First, we define the transition density of X with respect to its speed measure by the function $q : \Re^+ \times I \times I \to \Re^+$, where I is the state space of X, so that

$$P(X_t \in A) = \int_A q(t,x,y)m(dy),$$

for $t \in \Re^+$, $x \in I$ and $A \in \mathcal{B}(I)$. Here $\mathcal{B}(I)$ is the smallest Borel σ-algebra generated by I. The associated Green's function $G_\alpha : I \times I \to \Re^+$, for $\alpha > 0$, is given by the Laplace transform with respect to time of the transition density

$$G_\alpha(x,y) := L_\alpha\{q(t,x,y)\} = \int_0^\infty \exp\{-\alpha t\} q(t,x,y)\, dt, \qquad (16.3.2)$$

for $x, y \in I$. We have the following representation for G_α:

$$G_\alpha(x, y) = \begin{cases} w_\alpha^{-1}\psi_\alpha(x)\phi_\alpha(y) & \text{if } x \le y \\ w_\alpha^{-1}\phi_\alpha(x)\psi_\alpha(y) & \text{if } x \ge y. \end{cases} \tag{16.3.3}$$

We now make the additional assumption that the speed measure and scale function of X are absolutely continuous with respect to the Lebesgue measure. Then we can infer the existence of positive continuous functions m and s', such that

$$m(dx) = m(x)\,dx \quad \text{and} \quad s(x) = \int_c^x s'(y)\,dy,$$

for all $x, c \in I$. A further simplification occurs when s'' exists and is continuous. In this case, there are functions $a : I \to \Re^+$ and $b : I \to \Re$, with $a(x) > 0$, for all $x \in \text{int}(I)$, such that

$$m(x) = \frac{2}{a^2(x)s'(x)} \quad \text{and} \quad s'(x) = \exp\left(-\int_c^x \frac{2b(y)}{a^2(y)}\,dy\right),$$

for all $x \in I$. We point out that the processes considered in Example 16.2.2 satisfy this assumption. Under this assumption, the *Wronskian* appearing in Eq. (16.3.3) admits the representation,

$$w_\alpha = \frac{\phi_\alpha(x)\psi_\alpha'(x) - \phi'(x)\psi_\alpha(x)}{s'(x)}, \tag{16.3.4}$$

for all $x \in I$ and $\alpha > 0$.

Example 16.3.1 Continuing the Example 16.2.2, we remark that the Wronskian of Brownian motion is given by

$$w_\alpha = 2\sqrt{2\alpha},$$

whereas for the squared Bessel process, the Wronskian is given by

$$w_\alpha = \frac{1}{2|\nu|}, \quad \text{for } \nu \neq 0, \qquad w_\alpha = \frac{1}{2}, \quad \text{for } \nu = 0.$$

The functions $\psi_\alpha, \phi_\alpha : I \to \Re^+$ appearing in Eq. (16.3.3) are strictly convex, continuous, strictly monotone, positive, and finite throughout $\text{int}(I)$. Furthermore, they are the unique (up to a multiplicative constant) increasing and decreasing solutions, respectively, of the standard ordinary differential equation

$$\mathcal{G}f(x) = \alpha f(x),$$

for all $x \in \text{int}(I)$ and $\alpha > 0$, where the second-order differential operator is given by

$$\mathcal{G}f(x) := \frac{1}{2}a^2(x)f''(x) + b(x)f'(x),$$

for f in the domain of the operator.

As in Sect. 2.1, Lemma 2.1.9, we define for any $z \in I$ the density (with respect to the Lebesgue measure) of the first passage time τ_z, so that

$$P_x(\tau_z < t) = \int_0^t p_z(x,s)\,ds,$$

for all $t > 0$ and $x \in I$ representing the starting point of X. We denote by $\tilde{q}_z : [0,\infty) \times I \times I \to [0,\infty)$ the transition density (with respect to the speed measure) of X, with absorption at z. We then obtain

$$P_x(X_t \in A, \tau_z \geq t) = \int_A \tilde{q}_z(t,x,y)m(dy),$$

for all $t \in \Re^+$, $x \in I$ and $A \in \mathcal{B}$. Now we prove Lemma 2.1.9, derived in Hulley and Platen (2008).

Lemma 16.3.2 *Let $x, y, z \in I$ and suppose that $t > 0$. Then*

$$q(t,x,y) = \tilde{q}_z(t,x,y) + \int_0^t p_z(x,s)q(t-s,z,y)\,ds. \tag{16.3.5}$$

Proof From the Markov property of X, we get

$$\begin{aligned} P_x(X_t \leq y) &= P_x(X_t \leq y, \tau_z \geq t) + P_x(X_t \leq y, \tau_z < t) \\ &= P_X(X_t \leq y, \tau_z \geq t) + \int_0^t P_x(\tau_z \in ds)P_x(X_t \leq y|\tau_z = s) \\ &= P_x(X_t \leq y, \tau_z \geq t) + \int_0^t P_x(\tau_z \in ds)P_z(X_{t-s} \leq y). \end{aligned}$$

The result follows after differentiating with respect to y. □

We note that immediately from Eq. (16.3.5), we obtain

$$\tilde{G}^z_\alpha(x,y) := L_\alpha\big(\tilde{q}_z(t,x,y)\big) = L_\alpha\big(q(t,x,y)\big) - L_\alpha\big(p_z(x,t)\big)L_\alpha\big(q(t,z,y)\big). \tag{16.3.6}$$

Assume now that $t \in \Re^+$, and that $x, y, z \in I$ satisfy $x \leq z \leq y$ or $x \geq z \geq y$, from which it follows that $\tilde{q}_z(t,x,y) = 0$. Then, since the integral in Eq. (16.3.6) is a convolution, we have

$$L_\alpha\big(q(t,x,y)\big) = L_\alpha\big(p_z(x,t)\big)L_\alpha\big(q(t,z,y)\big)$$

for all $\alpha > 0$, from which we obtain that

$$E_x\big(\exp\{-\alpha\tau_z\}\big) = L_\alpha\big(p_z(x,t)\big) = \frac{G_\alpha(x,y)}{G_\alpha(z,y)} = \begin{cases} \frac{\psi_\alpha(x)}{\psi_\alpha(z)} & \text{if } x \leq z \\ \frac{\phi_\alpha(x)}{\phi_\alpha(z)} & \text{if } x \geq z. \end{cases} \tag{16.3.7}$$

Note that this is a well-known formula, see e.g. Itô and McKean (1996), p. 128. However, our derivation was purely formal, which illustrates the usefulness of Lemma 16.3.2. Returning to Eq. (16.3.6), we have

$$\begin{aligned}\tilde{G}^z_\alpha(x,y) &:= L_\alpha\big(\tilde{q}_z(t,x,y)\big) = L_\alpha\big(q(t,x,y)\big) - L_\alpha\big(p_z(x,t)\big)L_\alpha\big(q(t,z,y)\big)\\ &= \begin{cases} G_\alpha(x,y) - \frac{\psi_\alpha(x)}{\psi_\alpha(z)}G_\alpha(z,y) & \text{for } x,y \le z\\ G_\alpha(x,y) - \frac{\phi_\alpha(x)}{\phi_\alpha(z)}G_\alpha(z,y) & \text{for } x,y \ge z\end{cases}\\ &= \begin{cases} w_\alpha^{-1}\psi_\alpha(x)(\phi_\alpha(y) - \frac{\phi_\alpha(z)}{\psi_\alpha(z)}\psi_\alpha(y)) & \text{for } x \le y \le z\\ w_\alpha^{-1}\psi_\alpha(y)(\phi_\alpha(x) - \frac{\phi_\alpha(z)}{\psi_\alpha(z)}\psi_\alpha(x)) & \text{for } y \le x \le z\\ w_\alpha^{-1}(\psi_\alpha(x) - \frac{\psi_\alpha(z)}{\phi_\alpha(z)}\phi_\alpha(x))\phi_\alpha(y) & \text{for } y \ge x \ge z\\ w_\alpha^{-1}(\psi_\alpha(y) - \frac{\psi_\alpha(z)}{\phi_\alpha(z)}\phi_\alpha(y))\phi_\alpha(x) & \text{for } x \ge y \ge z,\end{cases}\end{aligned}$$

where we used Eqs. (16.3.7) and (16.3.3). Finally, we prove a main result in Hulley and Platen (2008).

Proposition 16.3.3 *Fix $\alpha > 0$ and let $t \ge 0$ and $x, z \in I$. Then*

$$L_\alpha\big(P(\tau_z \le t)\big) = \begin{cases} \frac{1}{\alpha}\frac{\psi_\alpha(x)}{\psi_\alpha(z)} & \textit{for } x \le z\\ \frac{1}{\alpha}\frac{\phi_\alpha(x)}{\phi_\alpha(z)} & \textit{for } x \ge z,\end{cases} \tag{16.3.8}$$

and

$$L_\alpha\big(E\big(\mathbf{1}_{\tau_z \le t}\exp\{-\beta\tau_z\}\big)\big) = \begin{cases} \frac{1}{\alpha}\frac{\psi_{\alpha+\beta}(x)}{\psi_{\alpha+\beta}(z)} & \textit{for } x \le z\\ \frac{1}{\alpha}\frac{\phi_{\alpha+\beta}(x)}{\phi_{\alpha+\beta}(z)} & \textit{for } x \ge z,\end{cases} \tag{16.3.9}$$

for all $\beta > 0$. Furthermore,

$$E\big((\gamma + \lambda\tau_z)^{-\rho}\big) = \begin{cases} \frac{1}{\Gamma(\rho)}L_\gamma(s^{\rho-1}\frac{\psi_{\lambda s}(x)}{\psi_{\lambda s}(z)}) & \textit{for } x \le z\\ \frac{1}{\Gamma(\rho)}L_\gamma(s^{\rho-1}\frac{\phi_{\lambda s}(x)}{\phi_{\lambda s}(z)}) & \textit{for } x \ge z,\end{cases} \tag{16.3.10}$$

and

$$L_\alpha\big(E\big(\mathbf{1}_{\tau_z \le t}(\gamma + \lambda\tau_z)^{-\rho}\big)\big) = \begin{cases} \frac{1}{\alpha\Gamma(\rho)}L_\gamma(s^{\rho-1}\frac{\psi_{\alpha+\lambda s}(x)}{\psi_{\alpha+\lambda s}(z)}) & \textit{for } x \le z\\ \frac{1}{\alpha\Gamma(\rho)}L_\gamma(s^{\rho-1}\frac{\phi_{\alpha+\lambda s}(x)}{\phi_{\alpha+\lambda s}(z)}) & \textit{for } x \ge z,\end{cases} \tag{16.3.11}$$

for all $\gamma, \lambda, \rho > 0$.

Proof To prove Eq. (16.3.8), we note that

$$L_\alpha\big(P_x(\tau_z < t)\big) = L_\alpha\bigg(\int_0^t p_z(x,s)\,ds\bigg) = \frac{1}{\alpha}L_\alpha\big(p_z(x,t)\big),$$

for all $\alpha > 0$, and one now applies Eq. (16.3.7). Regarding Eq. (16.3.9), we note that

$$\begin{aligned}L_\alpha\big(E\big(\mathbf{1}_{\tau_z \le t}\exp\{-\beta\tau_z\}\big)\big) &= L_\alpha\bigg(\int_0^t \exp\{-\beta s\}p_z(x,s)\,ds\bigg)\\ &= \int_0^\infty \exp\{-\alpha t\}\bigg(\int_0^t \exp\{-\beta s\}p_z(x,s)\,ds\bigg)\,dt\end{aligned}$$

$$
\begin{aligned}
&= \frac{1}{\alpha} \int_0^\infty \exp\bigl\{-(\alpha+\beta)s\bigr\} p_z(x,s)\, ds \\
&= \frac{1}{\alpha} L_{\alpha+\beta}\bigl(p_z(x,t)\bigr).
\end{aligned}
$$

Equation (16.3.9) now follows from Eq. (16.3.7). To prove Eqs. (16.3.10) and (16.3.11), we use the equality,

$$
\int_0^\infty s^{\rho-1} \exp\bigl\{-(\gamma+\lambda t)s\bigr\}\, ds = \frac{\Gamma(\rho)}{(\gamma+\lambda t)^\rho}, \tag{16.3.12}
$$

where $\gamma, \lambda, \rho > 0$ and Γ denotes the standard gamma function. To obtain Eq. (16.3.10), we compute

$$
\begin{aligned}
E_x\bigl((\gamma+\lambda\tau_z)^{-\rho}\bigr) &= \int_0^\infty \frac{p_z(x,t)}{(\gamma+\lambda t)^\rho}\, dt \\
&= \int_0^\infty p_z(x,t) \frac{1}{\Gamma(\rho)} \int_0^\infty s^{\rho-1} \exp\bigl\{-(\gamma+\lambda t)s\bigr\}\, ds\, dt \\
&= \frac{1}{\Gamma(\rho)} \int_0^\infty \exp\{-\gamma s\} s^{\rho-1} \int_0^\infty \exp\{-\lambda s t\} p_z(x,t)\, dt\, ds \\
&= \frac{1}{\Gamma(\rho)} L_\gamma\bigl(s^{\rho-1} L_{\lambda s}\bigl(p_z(x,t)\bigr)\bigr),
\end{aligned}
$$

and apply Eq. (16.3.7). Similarly, again employing Eq. (16.3.12), we compute

$$
\begin{aligned}
&L_\alpha\bigl(E_x\bigl(\mathbf{1}_{\tau_z \le t}(\gamma+\lambda\tau_z)^{-\rho}\bigr)\bigr) \\
&= L_\alpha\left(\int_0^t \frac{p_z(x,s)}{(\gamma+\lambda s)^\rho}\, ds\right) \\
&= \frac{1}{\alpha} L_\alpha\left(\frac{p_z(x,t)}{(\gamma+\lambda t)^\rho}\right) \\
&= \frac{1}{\alpha} \int_0^\infty \exp\{-\alpha t\} \frac{p_z(x,t)}{(\gamma+\lambda t)^\rho}\, dt \\
&= \frac{1}{\alpha} \int_0^{\int} \exp\{-\alpha t\} \frac{1}{\Gamma(\rho)} \int_0^\infty s^{\rho-1} \exp\bigl\{-(\gamma+\lambda t)s\bigr\}\, ds\, dt \\
&= \frac{1}{\alpha\Gamma(\rho)} \int_0^\infty \exp\{-\gamma s\} s^{\rho-1} \int_0^\infty \exp\bigl\{-(\alpha+\lambda s)t\bigr\} p_z(x,t)\, dt\, ds \\
&= \frac{1}{\alpha\Gamma(\rho)} L_\gamma\bigl(s^{\rho-1} L_{\alpha+\lambda s}\bigl(p_z(x,t)\bigr)\bigr),
\end{aligned}
$$

which results in (16.3.11), with the help of Eq. (16.3.7). □

Chapter 17
Detecting Strict Local Martingales

In Sect. 3.3, we presented two models for the GOP, namely the MMM and the TCEV model. For both models, we established the property that a risk-neutral measure does not exist, because the Radon-Nikodym derivative of the putative risk-neutral measure is a strict local martingale. Figure 3.3.6 seems to suggest that this is a plausible feature of our financial market, in particular, when analyzing its history over long periods of time and taking into account that investors request extra long term growth in risky securities. We established the local martingale property by making use of the explicitly available transition density of squared Bessel processes. This highlights the usefulness of squared Bessel processes in finance, they produce both tractability but also realistic models.

In this chapter, we propose another class of processes for modeling the GOP. For this class one can easily establish whether the processes allow for a risk-neutral measure or not. An argument due to Sin, see Sin (1998), allows us to determine if a particular model for the GOP admits a risk-neutral measure by studying the boundary behavior of a one-dimensional diffusion. The boundary behavior of one-dimensional diffusions is well understood, we refer the reader e.g. to our Chap. 16. As demonstrated in Chap. 16, in particular in Table 16.2.1, it simply amounts to confirming if certain integrals explode or not. Hence for the class of processes studied in this chapter, we present simple tools that allow us to answer the crucial question whether a particular local martingale is a martingale or a strict local martingale.

We remark that the question whether a local martingale is a martingale or a strict local martingale, has received much attention in the literature. In Kotani (2006), Hulley and Platen (2011), necessary and sufficient conditions have been presented for one-dimensional regular strong Markov continuous local martingales. Furthermore, in Kallsen and Muhle-Karbe (2010), Kallsen and Shiryaev (2002), and Mayerhofer et al. (2011a) an exponential semimartingale framework with focus on affine processes is considered. For further background, the reader is referred to Mijatovic and Urusov (2012). At the end of Sect. 17.1, we will present one of the main results from Mijatovic and Urusov (2012).

J. Baldeaux, E. Platen, *Functionals of Multidimensional Diffusions with Applications to Finance*, Bocconi & Springer Series 5, DOI 10.1007/978-3-319-00747-2_17,

17.1 Sin's Argument

In this section, we work on a filtered probability space $(\Omega, \mathcal{A}, \underline{\mathcal{A}}, P)$ carrying two Brownian motions W^1 and W^2. We use the following dynamics for the GOP,

$$dS_t^{\delta_*} = S_t^{\delta_*}\big((r_t + V_t)\,dt + \rho\sqrt{V_t}\,dW_t^1 + \rho^{\perp}\sqrt{V_t}\,dW_t^2\big),$$

and

$$dV_t = \kappa(\theta - V_t)\,dt + \sigma V_t^p\,dW_t^1, \tag{17.1.1}$$

where $S_0^{\delta_*} > 0$ and $V_0 > 0$. Here $r = \{r_t,\, t \geq 0\}$ is an adapted short-rate process, $\rho \in [-1, 1]$ denotes the correlation between the GOP and the variance process and $\rho^{\perp} = \sqrt{1-\rho^2}$. The parameters κ, θ, σ, and p are positive. We remark that this model is based on Andersen and Piterbarg (2007). For $p = \frac{1}{2}$, we obtain the Heston model, see also Sect. 6.3, for $b = 1$ we recover a GARCH model, and for $b = \frac{3}{2}$ we recover a 3/2-model with linear drift. Next we define the savings account $B_t = \exp\{\int_0^t r_s\,ds\}$, for $t \geq 0$, and the benchmarked savings account $\hat{B}_t = \frac{B_t}{S_t^{\delta_*}}$, $t \geq 0$, which satisfies the SDE

$$d\hat{B}_t = -\hat{B}_t\big(\rho\sqrt{V_t}\,dW_t^1 + \rho^{\perp}\sqrt{V_t}\,dW_t^2\big), \tag{17.1.2}$$

for $t \geq 0$. We recall the following properties of the process V from Andersen and Piterbarg (2007), see Proposition 2.1.

Proposition 17.1.1 *For the process V given by Eq.* (17.1.1), *the following properties hold*:

- 0 *is always an attainable boundary for* $0 < p < \frac{1}{2}$;
- 0 *is an attainable boundary for* $p = \frac{1}{2}$, *if* $2\kappa\theta < \sigma^2$;
- 0 *is an unattainable boundary for* $p > \frac{1}{2}$;
- ∞ *is an unattainable boundary for* $p > 0$.

Proof The proof is easily completed using Table 16.2.1 in Chap. 16. Recall that the speed measure is given by

$$m(dx) = m(x)\,dx$$

and the scale function by

$$s(x) = \int_c^x s'(y)\,dy,$$

where $c \in [0, \infty)$. Now,

$$m(x) = \frac{2}{a^2(x)s'(x)} \quad \text{and} \quad s'(x) = \exp\left(-\int_c^x \frac{2b(y)}{a^2(y)}\,dy\right),$$

for $x \in [0, \infty)$, where

$$a(x) = \sigma x^p \quad \text{and} \quad b(x) = \kappa(\theta - x). \qquad \square$$

For the remainder of this section, we assume that $p \geq \frac{1}{2}$ and $2\kappa\theta \geq \sigma^2$, so that V cannot reach 0. From Eq. (17.1.2), it is clear that $\hat{B}$ is a local martingale. As discussed in Sect. 3.3, if $\hat{B}$ is a martingale, a risk-neutral probability measure exists. If $\hat{B}$ is a strict local martingale, a risk-neutral measure does not exist. The following proposition identifies when $\hat{B}$ is a martingale. The proof is based on Lemma 2.3 in Andersen and Piterbarg (2007), which uses techniques from the proof of Lemma 4.2 in Sin (1998).

Lemma 17.1.2 *Let $\hat{B}$ and V be given by Eqs.* (17.1.1) *and* (17.1.2). *Denote by $\tilde{\tau}_\infty$ the explosion time for $\tilde{V}$,*

$$\tilde{\tau}_\infty = \lim_{n\to\infty} \tilde{\tau}_n, \qquad \tilde{\tau}_n = \inf\{t\colon \tilde{V}_t \geq n\}, \tag{17.1.3}$$

P-almost surely. Here the dynamics of $\tilde{V}$ under P are given by

$$d\tilde{V}_t = \left(\kappa(\theta - V_t) - \rho\sigma \tilde{V}_t^{p+\frac{1}{2}}\right)dt + \sigma \tilde{V}_t^p \, dW_t. \tag{17.1.4}$$

Then

$$E(\hat{B}_T) = \hat{B}_0 P(\tilde{\tau}_\infty > T).$$

Furthermore, when $p = \frac{1}{2}$ or $p > \frac{3}{2}$, then $\hat{B}$ is a martingale. When $\frac{1}{2} < p < \frac{3}{2}$, $\hat{B}$ is a martingale for $\rho \geq 0$ and a strict local martingale for $\rho < 0$. For $p = \frac{3}{2}$, $\hat{B}$ is a martingale for $\rho \geq -\frac{\sigma}{2}$ and a strict local martingale for $\rho < -\frac{\sigma}{2}$.

Proof We follow the technique of the proof of Lemma 2.3 in Andersen and Piterbarg (2007) and compute

$$\begin{aligned}
E(\hat{B}_T) &= \hat{B}_0 E\left(\exp\left\{-\frac{1}{2}\int_0^T V_s\,ds - \rho\int_0^T \sqrt{V_s}\,dW_s^1 - \rho^\perp \int_0^T \sqrt{V_s}\,dW_s^2\right\}\right) \\
&= \hat{B}_0 E\left(\exp\left\{-\frac{1}{2}\int_0^T V_s\,ds - \rho\int_0^T \sqrt{V_s}\,dW_s^1\right\}\right. \\
&\quad \left.\times E\left(\exp\left\{-\rho^\perp \int_0^T \sqrt{V_s}\,dW_s^2\right\} \middle| \sigma\{W_t^1, t \leq T\}\right)\right) \\
&= \hat{B}_0 E\left(\exp\left\{-\rho\int_0^T \sqrt{V_s}\,dW_s^1 - \frac{\rho^2}{2}\int_0^T V_s\,ds\right\}\right).
\end{aligned}$$

Next, introduce a sequence of stopping times

$$\tau_n := \inf\left\{t\colon \int_0^t V_s\,ds \geq n\right\}$$

and define the stochastic Doléan exponential

$$\xi_t = \exp\left\{-\rho\int_0^t \sqrt{V_s}\,dW_s^1 - \frac{\rho^2}{2}\int_0^t V_s\,ds\right\}.$$

Clearly, $\xi = \{\xi_t,\, t \geq 0\}$ is a local martingale. We now define

$$\xi_t^{(n)} = \xi_{t\wedge\tau_n},$$

which is a martingale and use it to define the auxiliary measure,

$$\tilde{P}^n(A) = E\big(\mathbf{1}_A \xi_T^{(n)}\big),$$

for $A \in \mathcal{A}_T$. We now use the argument in Lemma 4.2 in Sin (1998) and compute

$$\begin{aligned} E(\xi_T \mathbf{1}_{\tau > T}) &= E\big(\xi_T^{(n)} \mathbf{1}_{\tau_n > T}\big) \\ &= E_{\tilde{P}^n}(\mathbf{1}_{\tau_n > T}) \\ &= E_P(\mathbf{1}_{\tilde{\tau}_n > T}), \end{aligned}$$

where we recall that $\tilde{\tau}_n$ is defined in (17.1.3) and $\tilde{V}$ in (17.1.4). To justify the last equality, we note that by the Girsanov theorem, see Sect. 15.8, we obtain that the process

$$W_t^{(n)} = W_t + \rho \int_0^t \mathbf{1}_{u \le \tau_n} \sqrt{V_u}\, du$$

is a Brownian motion under $\tilde{P}^n$ and W and V satisfy

$$\begin{aligned} dW_t &= dW_t^{(n)} - \rho \mathbf{1}_{t \le \tau_n} \sqrt{V_t}\, dt, \\ dV_t &= \sigma V_t^p\, dW_t^{(n)} + \big(\kappa(\theta - V_t) - \rho \mathbf{1}_{t \le \tau_n} \sigma V_t^{p+\frac{1}{2}}\big)\, dt. \end{aligned}$$

Hence the stopped process $\tilde{V}_{t \wedge \tilde{\tau}_n}$ has the same law under P as the stopped process $V_{t \wedge \tau_n}$ under $\tilde{P}^n$. Now by Proposition 17.1.1, we have that V does not reach ∞ under P, hence

$$\begin{aligned} E(\xi_T) &= \lim_{n \to \infty} E(\xi_T \mathbf{1}_{\tau_n > T}) \\ &= \lim_{n \to \infty} E(\mathbf{1}_{\tilde{\tau}_n > T}) \\ &= P(\tilde{\tau}_\infty > T). \end{aligned}$$

The exchange of the limit and the expectation operator is justified by the monotone convergence theorem. This completes the proof of the first part of the result.

The second part follows immediately from Proposition 2.5 in Andersen and Piterbarg (2007). We replace the ρ in the statement of Proposition 2.5 in Andersen and Piterbarg (2007) by $-\rho$, as we consider the benchmarked savings account, which is essentially the inverse of the process considered in Andersen and Piterbarg (2007), and we compare Eq. (17.1.4) and Eq. (2.5) in Andersen and Piterbarg (2007). □

We have the following corollary to Lemma 17.1.2.

Corollary 17.1.3 *For important special cases, we obtain the following result:*

- *under the Heston model, which corresponds to* $p = \frac{1}{2}$, *the process* $\hat{B} = \{\hat{B}_t,\, t \ge 0\}$ *follows a martingale;*
- *under the* $\frac{3}{2}$ *model with linear drift, which corresponds to the case* $p = \frac{3}{2}$, $\hat{B} = \{\hat{B}_t,\, t \ge 0\}$ *follows a martingale for* $\rho \ge 0$ *and otherwise a strict local martingale;*

- *under a continuous limit of a GARCH model, which corresponds to* $p = 1$, $\hat{B} = \{\hat{B}_t,\ t \geq 0\}$ *follows a martingale for* $\rho \geq 0$, *otherwise a strict local martingale.*

For completeness, we present a general version of the result, which stems from Mijatovic and Urusov (2012). Consider the state space $J = (l, r)$, $-\infty \leq l < r \leq \infty$ and a J-valued diffusion $Y = \{Y_t,\ t \geq 0\}$ on $(\Omega, \mathcal{A}, \underline{\mathcal{A}}, P)$ given by the SDE

$$dY_t = \mu(Y_t)\,dt + \sigma(Y_t)\,dW_t, \tag{17.1.5}$$

where $Y_0 \in J$, W is a $\underline{\mathcal{A}}$-Brownian motion and $\mu, \sigma : J \to \Re$ are Borel functions satisfying the Engelbert-Schmidt conditions,

$$\sigma(x) \neq 0\ \forall x \in J; \qquad \frac{1}{\sigma^2}, \frac{\mu}{\sigma^2} \in L^1_{loc}(J), \tag{17.1.6}$$

where $L^1_{loc}(J)$ denotes the class of locally integrable functions, i.e. mappings from J to $\Re$ that are integrable on compact subsets of J. The SDE (17.1.5) admits a unique in law weak solution that possibly exits its state space J at the exit time ζ. Following Mijatovic and Urusov (2012), we specify that if Y can exit its state space, i.e. $P(\zeta < \infty) > 0$, then Y stays at the boundary point of J at which it exits after the time ζ, so the boundary is absorbing. We introduce the stochastic exponential

$$\xi_t = \exp\left\{\int_0^{t\wedge\zeta} b(Y_u)\,dW_u - \frac{1}{2}\int_0^{t\wedge\zeta} b^2(Y_u)\,du\right\},$$

$t \geq 0$, and set $\xi_t := 0$ for $t \geq \zeta$ on the set $\{\zeta < \infty, \int_0^\zeta b^2(Y_u)\,du = \infty\}$. Also we assume that

$$\frac{b^2}{\sigma^2} \in L^1_{loc}(J). \tag{17.1.7}$$

Remark 17.1.4 In order to connect this discussion to the results in Proposition 17.1.1 and Lemma 17.1.2, set $Y = V$ and $b(x) = -\rho\sqrt{x}$.

We now consider an auxiliary J-valued diffusion $\tilde{Y}$, where

$$d\tilde{Y}_t = \mu(\tilde{Y}_t)\,dt + b(\tilde{Y}_t)\sigma(\tilde{Y}_t)\,dt + \sigma(\tilde{Y}_t)\,dW_t.$$

Then $\tilde{Y}$ admits a unique weak solution that possibly exits its state space at $\tilde{\zeta}$. Before we present Corollary 2.2 from Mijatovic and Urusov (2012), recall from Proposition 17.1.1 that V cannot reach infinity under P, and Lemma 17.1.2 states that $\hat{B}$ is a martingale if $\tilde{V}$ cannot reach infinity under P, i.e. $\hat{B}$ is martingale if the boundary behavior of V and $\tilde{V}$ coincides. Corollary 2.2 from Mijatovic and Urusov (2012) extends this to the general case.

Corollary 17.1.5 *Assume that Y does not exit its state space and assume that μ, σ and b satisfy the assumptions* (17.1.6) *and* (17.1.7). *Then ξ is a martingale if and only if $\tilde{Y}$ does not exit its state space.*

For a more general result, see Theorem 2.1 in Mijatovic and Urusov (2012). Finally, we recall an important remark from Mijatovic and Urusov (2012).

Remark 17.1.6 The conditions in Lemma 17.1.2 (resp. Corollary 17.1.5) are necessary and sufficient conditions for $\hat{B}$ (resp. ξ) to be a martingale on the time interval $(0, \infty)$. Furthermore, they are necessary and sufficient conditions for $\hat{B}$ (resp. ξ) to be a martingale on the time interval $[0, T]$ for any fixed $T \in (0, \infty)$.

Remark 17.1.6 can intuitively be explained as follows: the process $\hat{B}$ (resp. ξ) is a time-homogeneous diffusion process, hence the process cannot loose its martingale property over time, but would have to lose it immediately.

17.2 Multidimensional Extension

This section briefly illustrates how to extend the results from the previous section to a multidimensional setting. Assume that the dynamics of the benchmarked savings account are given by

$$d\hat{B}_t = -\hat{B}_t \sum_{k=1}^{d} \sqrt{V_t^k} \left(\rho^k \, dW_t^k + \left(\rho^k\right)^{\perp} dW_t^{d+k}\right), \tag{17.2.8}$$

where $\mathbf{W} = \{\mathbf{W}_t = (W_t^1, W_t^2, \ldots, W_t^{2d}),\ t \geq 0\}$ is a vector Brownian motion and

$$dV_t^k = \kappa^k \left(\theta^k - V_t^k\right) dt + \sigma^k \left(V_t^k\right)^{p^k} dW_t^k. \tag{17.2.9}$$

Here $V_0^k, \kappa^k, \theta^k, \sigma^k, p^k$ are positive parameters, $-1 \leq \rho^k \leq 1$, and

$$\left(\rho^k\right)^{\perp} = \sqrt{1 - \left(\rho^k\right)^2}.$$

Proposition 17.2.1 *Let $\hat{B}$ and V^k, for $k = 1, \ldots, d$, be given by Eqs.* (17.2.8) *and* (17.2.9). *Denote by $\tilde{\tau}_\infty^k$ the explosion time for $\tilde{V}^k$,*

$$\tilde{\tau}_\infty^k = \lim_{n \to \infty} \tilde{\tau}_n^k, \qquad \tilde{\tau}_n^k = \inf\left\{t \colon \tilde{V}_t^k \geq n\right\},$$

where the dynamics of $\tilde{V}^k$ are given by

$$d\tilde{V}_t^k = \left(\kappa^k \left(\theta^k - \tilde{V}_t^k\right) - \rho^k \sigma^k \left(\tilde{V}_t^k\right)^{p^k + \frac{1}{2}}\right) dt + \sigma^k \left(\tilde{V}_t^k\right)^{p^k} dW_t^k. \tag{17.2.10}$$

Then

$$E(\hat{B}_T) = \hat{B}_0 \prod_{k=1}^{d} P\left(\tilde{\tau}_\infty^k > T\right).$$

Furthermore, $\hat{B}$ is a martingale if and only if for all $k \in \{1, \ldots, d\}$, one of the following conditions holds:

- $p^k = \frac{1}{2}$ *or* $p^k > \frac{3}{2}$;

- $\frac{1}{2} < p^k < \frac{3}{2}$ *and* $\rho^k \geq 0$;
- $p^k = \frac{3}{2}$ *and* $\rho^k \geq -\frac{\sigma_k}{2}$.

As in Sect. 17.1, the conditions presented in Proposition 17.2.1 are necessary and sufficient for $\hat{B}$ to be a martingale on the time interval $[0, T]$ for any fixed $T \in (0, \infty)$.

References

Abate, J., Whitt, W.: Numerical inversion of Laplace transforms of probability distributions. ORSA J. Comput. **7**(1), 36–43 (1995)

Abramowitz, M., Stegun, I.A. (eds.): Handbook of Mathematical Functions with Formulas, Graphs, and Mathematical Tables. Dover, New York (1972)

Ahdida, A., Alfonsi, A.: Exact and high order discretization schemes for Wishart processes and their affine extensions. Working paper, Université Paris-Est, Paris (2010)

Ahn, D.-H., Gao, B.: A parametric nonlinear model of term structure dynamics. Rev. Financ. Stud. **12**(4), 721–762 (1999)

Andersen, L., Piterbarg, V.: Moment explosions in stochastic volatility models. Finance Stoch. **11**(1), 29–50 (2007)

Asmussen, S., Glynn, P.W., Pitman, J.: Discretization error in simulation of one-dimensional reflecting Brownian motion. Ann. Appl. Probab. **5**(4), 875–896 (1995)

Bachelier, L.: Théorie de la spéculation. Ann. Sci. Éc. Norm. Super. (3) **17**, 21–86 (1900)

Baldeaux, J.: Exact simulation of the 3/2 model. Int. J. Theor. Appl. Finance **15**(5), 1–13 (2012a)

Baldeaux, J.: Scrambled polynomial lattice rules for infinite-dimensional integration. In: Wozniakowski, H., Plaskota, L. (eds.) Monte Carlo and Quasi-Monte Carlo Methods 2010, pp. 255–263. Springer, Berlin (2012b)

Baldeaux, J., Chan, L., Platen, E.: Derivatives on realized variance and volatility of an index under the benchmark approach. Working paper, University of Technology, Sydney (2011a)

Baldeaux, J., Chan, L., Platen, E.: Quasi-Monte Carlo methods for derivatives on realized variance of an index under the benchmark approach. In: ANZIAM Journal: Proceedings Computational Techniques and Applications Conference, pp. C727–741 (2011b)

Baldeaux, J., Ignatieva, K., Platen, E.: A tractable model for indices approximating the growth optimal portfolio. Stud. Nonlinear Dyn. Econom. (2011c, to appear). doi:10.1515/snde-2012-0054

Baldeaux, J., Gnewuch, M.: Optimal randomized multilevel algorithms for infinite-dimensional integration on function spaces with ANOVA-type decomposition. Working paper, University of Technology, Sydney (2012)

Baldeaux, J., Rutkowski, M.: Static replication of forward-start claims and realized variance swaps. Appl. Math. Finance **17**(2), 99–131 (2010)

Balduzzi, P., Das, S.R., Foresi, S., Sundaram, R.K.: A simple approach to three-factor models of interest rates. J. Fixed Income **6**(3), 43–53 (1996)

Barndorff-Nielsen, O., Stelzer, R.: Positive-definite matrix processes of finite variation. Probab. Math. Stat. **27**(1), 3–43 (2007)

Bates, D.: Jumps and stochastic volatility: exchange rate processes implicit in deutschemark options. Rev. Financ. Stud. **9**(1), 96–107 (1996)

Baumann, G.: Symmetry Analysis of Differential Equations with Mathematica. Springer, New York (1998)

J. Baldeaux, E. Platen, *Functionals of Multidimensional Diffusions with Applications to Finance*, Bocconi & Springer Series 5, DOI 10.1007/978-3-319-00747-2,

Beaglehole, D., Tenney, M.: General solutions of some interest rate contingent claim pricing equations. J. Fixed Income **1**(2), 69–83 (1991)
Becherer, D.: The numeraire portfolio for unbounded semimartingales. Finance Stoch. **5**(3), 327–341 (2001)
Benabid, A., Bensusan, H., El Karoui, N.: Wishart stochastic volatility: asymptotic smile and numerical framework. Working paper, Ecole Polytechnique (2010)
Bensoussan, A.: On the theory of option pricing. Acta Appl. Math. **2**(2), 139–158 (1984)
Beskos, A., Papaspiliopoulos, O., Roberts, G.: Retrospective exact simulation of diffusion sample paths with applications. Bernoulli **12**(6), 1077–1098 (2006)
Beskos, A., Papaspiliopoulos, O., Roberts, G.: A factorisation of diffusion measure and finite sample path constructions. Methodol. Comput. Appl. Probab. **10**(1), 85–104 (2008)
Beskos, A., Papaspiliopoulos, O., Roberts, G.: Monte-Carlo maximum likelihood estimation for discretely observed diffusion processes. Ann. Stat. **37**(1), 223–245 (2009)
Beskos, A., Roberts, G.: Exact simulation of diffusions. Ann. Appl. Probab. **15**(4), 2422–2444 (2005)
Björk, T.: Arbitrage Theory in Continuous Time. Oxford University Press, London (1998)
Black, F.: Studies in stock price volatility changes. In: Proceedings of the 1976 Business Meeting of the Business and Economic Statistics Section, pp. 177–181. Am. Statist. Assoc., Alexandria (1976)
Black, F., Scholes, M.: The pricing of options and corporate liabilities. J. Polit. Econ. **81**(3), 637–654 (1973)
Bluman, G., Kumei, S.: Symmetry and Differential Equations. Springer, New York (1989)
Borodin, A.N., Salminen, P.: Handbook of Brownian Motion—Facts and Formulae, 2nd edn. Birkhäuser, Basel (2002)
Brace, A.: Engineering BGM. Chapman & Hall/CRC, London (2008)
Broadie, M., Glasserman, P.: Pricing American-style securities using simulation. Computational financial modelling. J. Econ. Dyn. Control **21**(8–9), 1323–1352 (1997)
Broadie, M., Kaya, O.: Exact simulation of stochastic volatility and other affine jump diffusion processes. Oper. Res. **54**(2), 217–231 (2006)
Bru, M.-F.: Wishart processes. J. Theor. Probab. **4**(4), 725–751 (1991)
Buchen, P.W.: The pricing of dual-expiry exotics. Quant. Finance **4**(1), 101–108 (2004)
Bühlmann, H., Platen, E.: A discrete time benchmark approach for insurance and finance. ASTIN Bull. **33**(2), 153–172 (2003)
Bungartz, H.-J., Griebel, M.: Sparse grids. Acta Numer. **13**(1), 147–269 (2004)
Buraschi, B., Cieslak, A., Trojani, F.: Correlation risk and the term structure of interest rates. Working paper, Imperial College London (2008)
Buraschi, B., Porchia, P., Trojani, F.: Correlation risk and optimal portfolio choice. J. Finance **65**(1), 393–420 (2010)
Caister, N.C., O'Hara, J.G., Govinder, K.S.: Solving the Asian option PDE using Lie symmetry methods. Int. J. Theor. Appl. Finance **13**(8), 1265–1277 (2010)
Cantwell, B.J.: Introduction to Symmetry Analysis. Cambridge University Press, Cambridge (2002)
Carr, P., Geman, H., Madan, D., Yor, M.: Pricing options on realized variance. Finance Stoch. **9**(4), 453–475 (2005)
Carr, P., Sun, J.: A new approach for option pricing under stochastic volatility. Rev. Deriv. Res. **10**(2), 87–150 (2007)
Cesari, G., Aquilina, J., Charpillon, N., Filipović, Z., Lee, G., Manda, I.: Modelling, Pricing, and Hedging Counterparty Credit Exposure. Springer, Berlin (2009)
Chan, L., Platen, E.: Exact pricing and hedging formulas of long dated variance swaps under a $3/2$ volatility model. Working paper, University of Technology, Sydney (2011)
Chattamvelli, R.: A note on the noncentral beta distribution function. Am. Stat. **49**(2), 231–234 (1995)
Chen, N., Huang, Z.: Brownian meanders, importance sampling and unbiased simulation of diffusion extremes. Oper. Res. Lett. **40**(6), 554–563 (2012a)

Chen, N., Huang, Z.: Localization and exact simulation of Brownian motion driven stochastic differential equations. Math. Oper. Res. (2012b, to appear). doi:10.1287/moor.2013.0585

Christoffersen, P., Heston, S.L., Jacobs, K.: The shape and term structure of the index option smirk: why multifactor volatility models work so well. Manag. Sci. **55**(12), 1914–1932 (2009)

Cox, J.C.: Notes on option pricing (i): constant elasticity of variance option pricing model. J. Portf. Manag. **22**(Special Issue), 15–17 (1996)

Cox, J.C., Ingersoll, J.E., Ross, S.A.: A theory of the term structure of interest rates. Econometrica **53**(2), 385–407 (1985)

Craddock, M.: Fundamental solutions, transition densities and the integration of Lie symmetries. J. Differ. Equ. **246**(6), 2538–2560 (2009)

Craddock, M.: Lie Groups and Harmonic Analysis (2013, book manuscript)

Craddock, M., Dooley, A.H.: On the equivalence of Lie symmetries and group representations. J. Differ. Equ. **249**(3), 621–653 (2010)

Craddock, M., Heath, D., Platen, E.: Numerical inversion of Laplace transforms: a survey with applications to derivative pricing. J. Comput. Finance **4**(1), 57–81 (2000)

Craddock, M.J., Konstandatos, O., Lennox, K.A.: Some recent developments in the theory of Lie group symmetries for PDEs. In: Baswell, A.R. (eds.) Advances in Mathematics Research, pp. 1–40. Nova Science Publishers, New York (2009)

Craddock, M., Lennox, K.: Lie group symmetries as integral transforms of fundamental solutions. J. Differ. Equ. **232**(2), 652–674 (2007)

Craddock, M., Lennox, K.: The calculation of expectations for classes of diffusion processes by Lie symmetry methods. Ann. Appl. Probab. **19**(1), 127–157 (2009)

Craddock, M., Platen, E.: Symmetry group methods for fundamental solutions. J. Differ. Equ. **207**(2), 285–302 (2004)

Craddock, M., Platen, E.: On explicit probability laws for classes of scalar diffusions. Technical report, QFRC Research Paper 246, University of Technology, Sydney (2009)

Cuchiero, C., Filipović, D., Mayerhofer, E., Teichmann, J.: Affine processes on positive semidefinite matrices. Ann. Appl. Probab. **21**(2), 397–463 (2011)

Cuchiero, C., Keller-Ressel, M., Mayerhofer, E., Teichmann, J.: Affine processes on symmetric cones. Working paper, Eidgenössische Technische Hochschule Zürich (2011)

Da Fonseca, J., Grasselli, M., Ielpo, F.: Estimating the Wishart affine stochastic correlation model using the empirical characteristic function. Working paper (2008a)

Da Fonseca, J., Grasselli, M., Ielpo, F.: Hedging (co)variance risk with variance swaps. Working paper, University of Padova, Padova (2008b)

Da Fonseca, J., Grasselli, M., Tebaldi, C.: Option pricing when correlations are stochastic: an analytical framework. Rev. Deriv. Res. **10**(2), 151–180 (2007)

Da Fonseca, J., Grasselli, M., Tebaldi, C.: A multifactor volatility Heston model. Quant. Finance **8**(6), 591–604 (2008c)

Dai, Q., Singleton, K.J.: Specification analysis of affine term structure models. J. Finance **55**(5), 1943–1978 (2000)

Delbaen, F., Schachermayer, W.: The fundamental theorem of asset pricing for unbounded stochastic processes. Math. Ann. **312**(2), 215–250 (1998)

Delbaen, F., Schachermayer, W.: The Mathematics of Arbitrage. Springer, Berlin (2006)

Delbaen, F., Shirakawa, H.: A note on option pricing for the constant elasticity of variance model. Asia-Pac. Financ. Mark. **9**(2), 85–99 (2002)

Dick, J., Pillichshammer, F.: Digital Nets and Sequences. Discrepancy Theory and Quasi-Monte Carlo Integration. Cambridge University Press, Cambridge (2010)

Ding, C.G.: Algorithm AS 275: computing the non-central χ^2 distribution function. Appl. Stat. **41**(2), 478–482 (1992)

Doob, J.L.: Stochastic Processes. Wiley, New York (1953)

Du, K., Platen, E.: Benchmarked risk minimization. Working paper, University of Technology, Sydney (2012a)

Du, K., Platen, E.: Forward and futures contracts on commodities under the benchmark approach. Working paper, University of Technology, Sydney (2012b)

Duffie, D., Filipović, D., Schachermayer, W.: Affine processes and applications in finance. Ann. Appl. Probab. **3**(13), 984–1053 (2003)

Duffie, D., Glynn, P.W.: Efficient Monte Carlo simulation of security prices. Ann. Appl. Probab. **5**(4), 897–905 (1995)

Duffie, D., Kan, R.: A yield-factor model of interest rates. Math. Finance **6**(4), 379–406 (1996)

Duffie, D., Pan, J., Singleton, K.: Transform analysis and option pricing for affine jump diffusions. Econometrica **68**(6), 1343–1376 (2000)

Dyrting, S.: Evaluating the noncentral chi-square distribution for the Cox-Ingersoll-Ross process. Comput. Econ. **24**(1), 35–50 (2004)

Elliott, R.J.: Stochastic Calculus and Applications. Springer, New York (1982)

Engel, D.C., MacBeth, J.D.: Further results on the constant elasticity of variance call option pricing model. J. Financ. Quant. Anal. **17**(4), 533–554 (1982)

Ewald, C.-O., Menkens, O., Ting, S.H.M.: Asian options from an Australian perspective. Working paper, University of Sydney, Sydney (2011)

Faure, H.: Discrépance de suites associées à un système de numération (en dimension s). Acta Arith. **41**, 337–351 (1982)

Feller, W.: Zur Theorie der stochastischen Prozesse (Existenz- und Eindeutigkeitssätze). Math. Ann. **113**(1), 113–160 (1936)

Feller, W.: An Introduction to Probability Theory and Its Applications, vol. 2, 2nd edn. Wiley, New York (1971)

Fernholz, E.R., Karatzas, I.: Relative arbitrage in volatility-stabilized markets. Ann. Finance **1**(2), 149–177 (2005)

Filipović, D.: Separable term structures and the maximal degree problem. Math. Finance **12**(4), 341–349 (2002)

Filipović, D.: Term-Structure Models. Springer, Berlin (2009)

Filipović, D., Chen, L., Poor, H.V.: Quadratic term structure models for risk-free and defaultable rates. Math. Finance **14**(4), 515–536 (2004)

Filipović, D., Mayerhofer, E.: Affine diffusion processes: theory and applications. In: Albrecher, H., Runggaldier, W., Schachermayer, W. (eds.) Advanced Financial Modelling, pp. 125–165. de Gruyter, Berlin (2009)

Florens-Zmirou, D.: On estimating the diffusion coefficient from discrete observations. J. Appl. Probab. **30**(4), 790–804 (1993)

Föllmer, H., Schweizer, M.: Hedging of contingent claims under incomplete information. In: Davis, M.H.A., Elliott, R.J. (eds.) Applied Stochastic Analysis. Stochastics Monogr., vol. 5, pp. 389–414. Gordon & Breach, New York (1991)

Föllmer, H., Sondermann, D.: Hedging of non-redundant contingent claims. In: Hildebrandt, W., Mas-Colell, A. (eds.) Contributions to Mathematical Economics, pp. 205–223. North-Holland, Amsterdam (1986)

Fong, H.G., Vasiček, O.A.: Interest rate volatility as a stochastic factor. Working paper, Gifford Fong Associates (1991)

Friedman, A.: Stochastic Differential Equations and Applications, vol. I. Probability and Mathematical Statistics, vol. 28. Academic Press, New York (1975)

Galesso, G., Runggaldier, W.J.: Pricing without equivalent martingale measures under complete and incomplete observation. In: Chiarella, C., Novikov, A. (eds.) Contemporary Quantitative Finance—Essays in Honour of Eckhard Platen. Springer, Berlin (2010)

Gatheral, J.: The Volatility Surface. A Practitioner's Guide. Wiley, Chichester (2006)

Geman, H., Yor, M.: Bessel processes, Asian options and perpetuities. Math. Finance **3**(4), 349–375 (1993)

Giles, M.B.: Improved multi-level Monte Carlo convergence using Milstein scheme. In: Keller, S.H.A., Niederreiter, H. (eds.) Monte Carlo and Quasi-Monte Carlo Methods 2006, pp. 343–358. Springer, Berlin (2008a)

Giles, M.B.: Multi-level Monte Carlo path simulation. Oper. Res. **56**(3), 607–617 (2008b)

Glasserman, P.: Monte Carlo Methods in Financial Engineering. Appl. Math., vol. 53. Springer, New York (2004)

Gnewuch, M.: Infinite-dimensional integration on weighted Hilbert spaces. Math. Comput. **81**, 2175–2205 (2012a)

Gnewuch, M.: Lower error bounds for randomized multilevel and changing dimension algorithms. In: Monte Carlo and Quasi-Monte Carlo Methods 2012 (2012b, to appear)

Gnoatto, A., Grasselli, M.: The explicit Laplace transform for the Wishart process. Working paper, University of Padova, Padova (2011)

Golub, G., Van Loan, C.F.: Matrix Computations, 3rd edn. John Hopkins University Press, Baltimore (1996)

Gouriéroux, C., Monfort, A., Sufana, R.: International money and stock market contingent claims. Working paper, CREST (2007)

Gouriéroux, C., Sufana, R.: Derivative pricing with multivariate stochastic volatility: application to credit risk. Working paper, CREST (2004a)

Gouriéroux, C., Sufana, R.: Wishart quadratic term structure models. Working paper, CREST (2004b)

Grasselli, M., Tebaldi, C.: Bond price and impulse response function for the Balduzzi, Das, Foresi and Sundaram (1996) model. Econ. Notes **33**(3), 359–374 (2004a).

Grasselli, M., Tebaldi, C.: Solvable affine term structure models. Working paper, University of Padova (2004b)

Grasselli, M., Tebaldi, C.: Solvable affine term structure models. Math. Finance **18**(1), 135–153 (2008)

Gupta, A.K., Nagar, D.K.: Matrix Valued Stochastic Processes. Chapman & Hall/CRC, London (2000)

Hall, B.C.: Lie Groups, Lie Algebras, and Representations: An Elementary Introduction. Springer, New York (2003)

Hammersley, J.M.: Monte Carlo methods for solving multivariate problems. Ann. N.Y. Acad. Sci. **86**(3), 844–874 (1960)

Handley, J.C.: Variable purchase options. Rev. Deriv. Res. **4**(3), 219–230 (2000)

Handley, J.C.: An empirical test of the pricing of VPO contracts. Austral. J. Manag. **28**(1), 409–422 (2003)

Harrison, J.M., Kreps, D.M.: Martingale and arbitrage in multiperiod securities markets. J. Econ. Theory **20**(3), 381–408 (1979)

Heath, D., Platen, E.: Consistent pricing and hedging for a modified constant elasticity of variance model. Quant. Finance **2**(6), 459–467 (2002)

Heath, D., Platen, E.: Currency derivatives under a minimal market model with random scaling. Int. J. Theor. Appl. Finance **8**(8), 1157–1177 (2005)

Heath, D., Schweizer, M.: Martingales versus PDEs in finance: an equivalence result with examples. J. Appl. Probab. **37**(4), 947–957 (2000)

Heinrich, S.: Monte Carlo complexity of global solution of integral equations. J. Complex. **14**(2), 151–175 (1998)

Heinrich, S., Sindambiwe, E.: Monte Carlo complexity of parametric integration. J. Complex. **15**(3), 317–341 (1999)

Heston, S.L.: A closed-form solution for options with stochastic volatility with applications to bond and currency options. Rev. Financ. Stud. **6**(2), 327–343 (1993)

Heston, S.L.: A simple new formula for options with stochastic volatility. Working paper, Washington University of St. Louis (1997)

Hickernell, F.J., Müller-Gronbach, T.M., Niu, B., Ritter, K.: Multi-level Monte Carlo algorithms for infinite-dimensional integration on $\mathbb{R}^N$. J. Complex. **26**(3), 229–254 (2010)

Hong, H.S., Hickernell, F.: Algorithm 823: implementing scrambled digital sequences. ACM Trans. Math. Softw. **29**(2), 95–109 (2003)

Hulley, H.: Strict local martingales in continuous financial market models. PhD thesis, UTS, Sydney (2009)

Hulley, H., Platen, E.: Laplace transform identities for diffusions, with applications to rebates and barrier options. In: Stettner, L. (ed.) Advances in Mathematical Finance. Banach Center Publications, vol. 83, pp. 139–157 (2008)

Hulley, H., Platen, E.: A visual criterion for identifying Itô diffusions as martingales or strict local martingales. In: Seminar on Stochastic Analysis, Random Fields and Applications VI, vol. 63, pp. 147–157. Birkhäuser, Basel (2011)
Hulley, H., Platen, E.: Hedging for the long run. Math. Financ. Econ. **6**(2), 105–124 (2012)
Hurst, T.R., Zhou, Z.: A Fourier transform method for spread option pricing. SIAM J. Financ. Math. **1**(1), 142–157 (2010)
Ignatieva, K., Platen, E., Rendek, R.: Using dynamic copulae for modeling dependency in currency denominations of a diversified world stock index. J. Stat. Theory Pract. **5**(3), 425–452 (2011)
Ikeda, N., Watanabe, S.: Stochastic Differential Equations and Diffusion Processes, 2nd edn. North-Holland, Amsterdam (1989) (first edition 1981)
Imhof, J.P.: Density factorizations for Brownian motion, meander and the three dimensional Bessel process, and applications. J. Appl. Probab. **21**(3), 500–510 (1984)
Ingersoll, J.E.: Digital contracts: simple tools for pricing complex derivatives. J. Bus. **73**(1), 67–88 (2000)
Itkin, A., Carr, P.: Pricing swaps and options on quadratic variation under stochastic time change models—discrete observations case. Rev. Deriv. Res. **13**(2), 141–176 (2010)
Itô, K.: Stochastic integral. Proc. Imp. Acad. (Tokyo) **20**(8), 519–524 (1944)
Itô, K., McKean Jr., H.P.: Diffusion Processes and their Sample Paths. Springer, Berlin (1996)
Jäckel, P.: Monte Carlo Methods in Finance. Wiley, Chichester (2002)
Jacod, J.: Non-parametric kernel estimation of the coefficient of a diffusion. Scand. J. Stat. **27**(1), 83–96 (2000)
Jacod, J., Protter, P.: Probability Essentials. Springer, Berlin (2004)
Jacod, J., Shiryaev, A.N.: Limit Theorems for Stochastic Processes, 2nd edn. Springer, Berlin (2003)
Jeanblanc, M., Yor, M., Chesney, M.: Mathematical Methods for Financial Markets. Springer, London (2009)
Jegadesh, N., Pennachi, G.G.: The behaviour of interest rates implied by the term structure of euro-dollar futures. J. Money Credit Bank. **28**(3), 426–446 (1996)
Joe, S., Kuo, F.: Remark on algorithm 659: implementing Sobol's quasirandom sequence generator. ACM Trans. Math. Softw. **29**(1), 49–57 (2003)
Joe, S., Kuo, F.: Constructing Sobol sequences with better two-dimensional projections. SIAM J. Sci. Comput. **30**(5), 2635–2654 (2008)
Johnson, N.L., Kotz, S., Balakrishnan, N.: Continuous Univariate Distributions, 2nd edn. Wiley Series in Probability and Mathematical Statistics, vol. 1. Wiley, New York (1994)
Johnson, N.L., Kotz, S., Balakrishnan, N.: Continuous Univariate Distributions, 2nd edn. Wiley Series in Probability and Mathematical Statistics, vol. 2. Wiley, New York (1995)
Kallsen, J., Muhle-Karbe, J.: Exponentially affine martingales, affine measure changes and exponential moments of affine processes. Stoch. Process. Appl. **120**(2), 163–181 (2010)
Kallsen, J., Shiryaev, A.N.: The cumulant process and Escher's change of measure. Finance Stoch. **6**(4), 397–428 (2002)
Karatzas, I.: On the pricing of the American options. Appl. Math. Optim. **17**, 37–60 (1988)
Karatzas, I.: Optimization problems in the theory of continuous trading. SIAM J. Control Optim. **27**(6), 1221–1259 (1989)
Karatzas, I., Kardaras, C.: The numeraire portfolio in semimartingale financial models. Finance Stoch. **11**(4), 447–493 (2007)
Karatzas, I., Shreve, S.E.: Brownian Motion and Stochastic Calculus, 2nd edn. Springer, New York (1991)
Karatzas, I., Shreve, S.E.: Methods of Mathematical Finance. Appl. Math., vol. 39. Springer, New York (1998)
Kelly, J.R.: A new interpretation of information rate. Bell Syst. Tech. J. **35**, 917–926 (1956)
Kloeden, P.E., Platen, E.: Numerical Solution of Stochastic Differential Equations, Appl. Math., vol. 23. Springer, Berlin (1999). Third printing (first edition 1992)
Kolmogorov, A.: Über die analytischen Methoden in der Wahrscheinlichkeitsrechnung. Math. Ann. **104**(1), 415–458 (1932)

Korn, R., Korn, E., Kroisandt, G.: Monte Carlo Methods and Models in Finance and Insurance. Chapman & Hall/CRC, Baton Rouge (2010)

Kotani, S.: On a condition that one-dimensional diffusion processes are martingales. In: In Memoriam Paul-André Meyer: Séminaire de Probabilités XXXIX. Lecture Notes in Math., vol. 1874, pp. 149–156 (2006)

Krylov, N.V.: Controlled Diffusion Processes. Appl. Math., vol. 14. Springer, New York (1980)

Kuo, F.Y., Dunsmuir, W.T.M., Sloan, I.H., Wand, M.P., Womersley, R.: Quasi-Monte Carlo for highly structured generalised response models. Methodol. Comput. Appl. Probab. **10**(2), 239–275 (2008)

Leippold, M., Trojani, F.: Asset pricing with matrix affine jump diffusions. Working paper, Imperial College London (2008)

Lennox, K.: Lie symmetry methods for multi-dimensional linear parabolic PDEs and diffusions. PhD thesis, UTS, Sydney (2011)

Lewis, A.L.: Option Valuation Under Stochastic Volatility. Finance Press, Newport Beach (2000)

Lie, S.: Über die Integration durch bestimmate Integrale von einer Klasse linearer partieller Differentialgleichungen. Arch. Math. **6**(3), 328–368 (1881)

Loewenstein, M., Willard, G.A.: Local martingales, arbitrage, and viability: free snacks and cheap thrills. Econom. Theory **16**(1), 135–161 (2000)

Long, J.B.: The numeraire portfolio. J. Financ. Econ. **26**(1), 29–69 (1990)

MacLean, L.C., Thorp, E.O., Ziemba, W.T.: The Kelly Capital Growth Investment Criterion: Theory and Practice. World Scientific, Singapore (2011)

Mamon, R.S.: Three ways to solve for bond prices in the Vasicek model. J. Appl. Math. Decis. Sci. **8**(1), 1–14 (2004)

Margrabe, W.: The value of an option to exchange one asset for another. J. Finance **33**(1), 177–186 (1978)

Marshall, A.W., Olkin, I.: Families of multivariate distributions. J. Am. Stat. Assoc. **83**(403), 834–841 (1988)

Matoušek, J.: On the $L2$-discrepancy for anchored boxes. J. Complex. **14**(4), 527–556 (1998)

Mayerhofer, E.: Wishart Processes and Wishart Distributions: An Affine Processes Point of View. CIMPA Lecture Notes (2012, to appear)

Mayerhofer, E., Muhle-Karbe, J., Smirnov, A.: A characterization of the martingale property of exponentially affine processes. Stoch. Process. Appl. **121**(3), 568–582 (2011a)

Mayerhofer, E., Pfaffel, O., Stelzer, R.: On strong solutions for positive definite jump diffusions. Stoch. Process. Appl. **121**(9), 2072–2086 (2011b)

McKean, H.P.: Appendix: A free boundary problem for the heat equation arising from a problem in mathematical economics. Ind. Manage. Rev. **6**(2), 32–39 (1965)

McNeil, A., Frey, R., Embrechts, P.: Quantitative Risk Management. Princeton University Press, New Jersey (2005)

Merton, R.C.: Theory of rational option pricing. Bell J. Econ. Manag. Sci. **4**(1), 141–183 (1973)

Mijatovic, A., Urusov, M.: On the martingale property of certain local martingales. Probab. Theory Relat. Fields **152**(1–2), 1–30 (2012)

Moreno, M., Navas, J.F.: Australian options. Austral. J. Manag. **33**(1), 69–94 (2008)

Muirhead, R.J.: Aspects of Multivariate Statistical Theory. Wiley, New York (1982)

Musiela, M., Rutkowski, M.: Martingale Methods in Financial Modelling, 2nd edn. Appl. Math., vol. 36. Springer, Berlin (2005)

Myneni, R.: The pricing of the American option. Ann. Appl. Probab. **2**(1), 1–23 (1992)

Niederreiter, H.: Random Number Generation and Quasi-Monte-Carlo Methods. SIAM, Philadelphia (1992)

Niederreiter, H.: Constructions of (t, m, s)-nets and (t, s)-sequences. Finite Fields Appl. **11**(3), 578–600 (2005)

Niederreiter, H.: Nets, (t, s)-sequences, and codes. In: Keller, S.H.A., Niederreiter, H. (eds.) Monte Carlo and Quasi-Monte Carlo Methods 2006, pp. 83–100. Springer, Berlin (2008)

Niederreiter, H., Xing, C.P.: Global function fields with many rational places and their applications. In: Keller, S.H.A., Niederreiter, H. (eds.) Finite Fields: Theory, Applications, and Algorithms,

Waterloo, ON, 1997, pp. 87–111. Am. Math. Soc., Providence (1999)
Niu, B., Hickernell, F., Müller-Gronbach, T.M., Ritter, K.: Deterministic multi-level algorithms for infinite-dimensional integration on $\mathbb{R}^N$. J. Complex. **27**(3), 331–351 (2011)
Novikov, A.A.: On an identity for stochastic integrals. Theory Probab. Appl. **17**(4), 717–720 (1972)
Olver, P.J.: Applications of Lie Groups to Differential Equations. Graduate Texts in Mathematics. Springer, New York (1993)
Owen, A.B.: Randomly permuted (t, m, s)-nets and (t, s)-sequences. In: Niederreiter, H., Shiue, J.-S. (eds.) Monte Carlo and Quasi-Monte Carlo Methods in Scientific Computing, pp. 299–317. Springer, New York (1995)
Owen, A.B.: Monte Carlo variance of scrambled quadrature. SIAM J. Numer. Anal. **34**(5), 1884–1910 (1997)
Paskov, S., Traub, J.: Faster valuation of financial derivatives. J. Portf. Manag. **22**(1), 113–120 (1995)
Patnaik, P.B.: The non-central χ^2- and F-distributions and their applications. Biometrika **36**(1/2), 202–232 (1949)
Pfaffel, O.: Wishart processes. Technical report, Technische Universität München (2008)
Pintoux, C., Privault, N.: The Dothan pricing model revisited. Math. Finance **21**(2), 355–363 (2011)
Pirsic, G.: A software implementation of Niederreiter-Xing sequences. In: Fang, H.F.J., Niederreiter, H. (eds.) Monte Carlo and Quasi-Monte Carlo Methods 2000, pp. 434–445. Springer, Berlin (2002)
Pitman, J., Yor, M.: A decomposition of Bessel bridges. Probab. Theory Relat. Fields **59**(4), 425–457 (1982)
Platen, E.: A non-linear stochastic volatility model. Technical report, FMRR 005-97, Australian National University, Canberra, Financial Mathematics Research Reports (1997)
Platen, E.: A short term interest rate model. Finance Stoch. **3**(2), 215–225 (1999)
Platen, E.: Arbitrage in continuous complete markets. Adv. Appl. Probab. **34**(3), 540–558 (2002)
Platen, E.: A benchmark framework for risk management. In: Stochastic Processes and Applications to Mathematical Finance, Proceedings of the Ritsumeikan Intern. Symposium, pp. 305–335. World Scientific, Singapore (2004)
Platen, E.: Diversified portfolios with jumps in a benchmark framework. Asia-Pac. Financ. Mark. **11**(1), 1–22 (2005)
Platen, E., Bruti-Liberati, N.: Numerical Solution of SDEs with Jumps in Finance. Springer, Berlin (2010)
Platen, E., Heath, D.: A Benchmark Approach to Quantitative Finance. Springer, Berlin (2010)
Platen, E., Rendek, R.: Exact scenario simulation for selected multi-dimensional stochastic processes. Commun. Stoch. Anal. **3**(3), 443–465 (2009)
Platen, E., Rendek, R.: Approximating the numeraire portfolio by naive diversification. J. Asset Manag. **13**(1), 34–50 (2012)
Platen, E., Stahl, G.: A structure for general and specific market risk. Comput. Stat. **18**(3), 355–373 (2003)
Posten, H.O.: An effective algorithm for the noncentral chi-squared distribution function. Am. Stat. **43**(4), 261–263 (1989)
Posten, H.O.: An effective algorithm for the noncentral beta distribution function. Am. Stat. **47**(2), 129–131 (1993)
Press, W.H., Teukolsky, S.A., Vetterling, W.T., Flannery, B.P.: Numerical Recipes in C++. The Art of Scientific Computing, 2nd edn. Cambridge University Press, Cambridge (2002)
Protter, P.: Stochastic Integration and Differential Equations, 2nd edn. Springer, Berlin (2005)
Rao, C.R.: Linear Statistical Inference and Its Applications, 2nd edn. Wiley, New York (1973)
Revuz, D., Yor, M.: Continuous Martingales and Brownian Motion, 3rd edn. Springer, Berlin (1999)
Rogers, L.C.G., Williams, D.: Diffusions Markov Processes and Martingales: Itô Calculus, 2nd edn. Cambridge Mathematical Library, vol. 2. Cambridge University Press, Cambridge (2000)

Ross, S.A.: The arbitrage theory of capital asset pricing. J. Econ. Theory **13**(3), 341–360 (1976)
Sankaran, M.: Approximations to the non-central chi-square distribution. Biometrika **50**(1/2), 199–204 (1963)
Schroder, M.: Computing the constant elasticity of variance option pricing formula. J. Finance **44**(1), 211–219 (1989)
Schweizer, M.: Option hedging for semimartingales. Stoch. Process. Appl. **37**(2), 339–363 (1991)
Schweizer, M.: On the minimal martingale measure and the Föllmer-Schweizer decomposition. Stoch. Anal. Appl. **13**(5), 573–599 (1995)
Seber, G.A.F.: The non-central chi-squared and beta distributions. Biometrika **50**(3/4), 542–544 (1963)
Shiryaev, A.N.: Probability. Springer, Berlin (1984)
Siegel, A.F.: The noncentral chi-squared distribution with zero degrees of freedom and testing for uniformity. Biometrika **66**(2), 381–386 (1979)
Sin, C.A.: Complications with stochastic volatility models. Adv. Appl. Probab. **30**(1), 256–268 (1998)
Situ, R.: Theory of Stochastic Differential Equations with Jumps and Applications. Springer, New York (2005)
Sloan, I.H., Joe, S.: Lattice Methods for Multiple Integration. Oxford University Press, Oxford (1994)
Sobol, I.M.: On the distribution of points in a cube and the approximate evaluation of integrals. USSR Comput. Math. Math. Phys. **7**(4), 86–112 (1967)
Soulier, P.: Nonparametric estimation of the diffusion coefficient of a diffusion process. Stoch. Anal. Appl. **16**(1), 185–200 (1998)
Stelzer, R.: Multivariate continuous time stochastic volatility models driven by a Lévy process. PhD thesis, Munich University of Technology, Munich (2007)
Tang, P.C.: The power function of the analysis of variance tests with tables and illustrations of their use. Stat. Res. Mem. **2**, 126–150 (1938)
van Moerbeke, P.: On optimal stopping and free boundary value problems. Arch. Ration. Mech. Anal. **60**(2), 101–148 (1976)
Vasiček, O.A.: An equilibrium characterization of the term structure. J. Financ. Econ. **5**(2), 177–188 (1977)
Walcher, S.: Algebras and Differential Equations. Hadronic Press, Palm Harbor (1991)
Williams, D.: Path decomposition and continuity of local time for one-dimensional diffusions, I. J. Lond. Math. Soc. **3**(28), 3–28 (1974)
Yor, M.: On some exponential functionals of Brownian motion. In: On Exponential Functionals of Brownian Motion and Related Processes, pp. 23–48 (2001)
Zhu, S.-P.: An exact and explicit solution for the valuation of American put options. Quant. Finance **6**(3), 229–242 (2006)

Index

J. Baldeaux, E. Platen, *Functionals of Multidimensional Diffusions with Applications to Finance*, Bocconi & Springer Series 5, DOI 10.1007/978-3-319-00747-2,

Author Index

J. Baldeaux, E. Platen, *Functionals of Multidimensional Diffusions with Applications to Finance*, Bocconi & Springer Series 5, DOI 10.1007/978-3-319-00747-2,

D

E

F

G

H

I

J

K

L

M

N

O

P

R

S